ALGEBRAIC METHODS
IN MOLECULAR AND
NUCLEAR STRUCTURE
PHYSICS

ALGEBRAIC METHODS IN MOLECULAR AND NUCLEAR STRUCTURE PHYSICS

ALEJANDRO FRANK
Universidad National Autónoma de México
Mexico D.F., Mexico

PIETER VAN ISACKER
University of Surrey
Guildford, England
and
Grand Accélérateur National d'Ions Lourds
Caen, France

A Wiley-Interscience Publication
JOHN WILEY & SONS, INC.
New York   /   Chichester   /   Brisbane   /   Toronto   /   Singapore

This text is printed on acid-free paper.

Copyright © 1994 by John Wiley & Sons, Inc.

All rights reserved. Published simultaneously in Canada.

Reproduction or translation of any part of this work beyond that permitted by Section 107 or 108 of the 1976 United States Copyright Act without the permission of the copyright owner is unlawful. Requests for permission or further information should be addressed to the Permissions Department, John Wiley & Sons, Inc., 605 Third Avenue, New York, NY 10158-0012.

Library of Congress Cataloging in Publication Data:
Frank, A.
   Algebraic methods in molecular and nuclear structure physics / Alejandro Frank, Pieter Van Isacker.
      p.   cm.
   ISBN 0-471-52640-1 (alk. paper)
   1. Nuclear structure–Mathematics.   2. Molecular structure—Mathematics.   3. Nuclear models.   4. Molecular models.   5. Algebra.
I. I. Van Isacker, P.   II. Title.
QC793.3.S8F73 1994
539.7'4—dc20   93-41843

Printed in the United States of America

10 9 8 7 6 5 4 3 2 1

Symmetry, as wide or as narrow as you may define its meaning, is one idea by which man through the ages has tried to comprehend and create order, beauty and perfection.
H. Weyl

A mis padres con profundo agradecimiento.
A Mónica, Pablo y Dan con todo mi cariño.

Aan mijn vader en in nagedachtenis van mijn moeder.
Aan Vera, David en Thomas, mijn drie oogappels.

Contents

Preface		xi
Introduction		1
I	**Schematic Models**	**7**
1	**Identical Bosons**	**9**
1.1	The U(2) algebra	9
1.2	The s–t-boson hamiltonian	16
1.3	Symmetry limits	18
1.4	The U(1) limit	20
1.5	The SO(2) limit	23
1.6	Transformation brackets	29
1.7	Tensor calculus	30
	Summary	36
	References	37
2	**Non-identical Bosons**	**39**
2.1	The $U_1(2) \otimes U_2(2)$ algebra	39
2.2	The $U_1(2) \otimes U_2(2)$ hamiltonian	41
2.3	Symmetry limits	41
2.4	The $U_{12}(1)$ limit	50
2.5	The $SO_{12}(2)$ limit	53
2.6	Transformation brackets	55
2.7	Stretching vibrations in X–Y–X molecules	56
2.8	Stretching vibrations in polyatomic molecules	63
	Summary	71
	References	72

3 Bosons and Fermions 75
3.1 Fermion algebras 75
3.2 Boson–fermion algebras 80
3.3 Symmetry limits 82
3.4 Tensor calculus 90
Summary 95
References 96

4 Supersymmetry and F Spin 99
4.1 Superalgebras 99
4.2 The U(2/2) superalgebra 101
4.3 Supersymmetry predictions 105
4.4 F Spin and the U(4) algebra 108
Summary 113
References 114

II Molecular Models 115

5 Diatomic Molecules 117
5.1 Introduction 117
5.2 The U(4) algebra 119
5.3 Symmetry limits 125
5.4 Non-rigid molecules: The U(3) limit 132
5.5 Rigid molecules: The SO(4) limit 148
5.6 Transformation brackets 164
5.7 Examples 166
5.8 Tensor calculus 172
Summary 190
References 192

6 Triatomic Molecules 195
6.1 Introduction 195
6.2 The $U_1(4) \otimes U_2(4)$ algebra 197
6.3 Symmetry limits 199
6.4 The U(3) limits 210
6.5 The SO(4) limits 216

		6.5.1	Linear molecules 218

- 6.5.1 Linear molecules 218
- 6.5.2 Bent molecules 230
- 6.6 Examples of triatomic spectra 236
- 6.7 Polyatomic molecules 251
- Summary 259
- References 260

7 Bose–Fermi Symmetries and Molecular Electronic Spectra 263

- 7.1 Introduction 263
- 7.2 Algebra 265
- 7.3 Hamiltonian 266
- 7.4 Wave functions 269
- 7.5 Symmetry limits 275
- 7.6 Application to the hydride molecules 280
- Summary 289
- References 290

III Nuclear Models 293

8 The Interacting Boson Model 295

- 8.1 Introduction 295
- 8.2 The U(6) algebra 296
- 8.3 Symmetry limits 303
- 8.4 Vibrational nuclei: The U(5) limit 307
- 8.5 γ-unstable nuclei: The SO(6) limit 326
- 8.6 Example: The nucleus ^{196}Pt 337
- 8.7 Rotational nuclei: The SU(3) limit 342
- Summary 348
- References 349

9 The Neutron–Proton Interacting Boson Model 353

- 9.1 Introduction 353
- 9.2 The $U_\nu(6) \otimes U_\pi(6)$ algebra 354
- 9.3 F Spin 355
- 9.4 Symmetry limits 366

9.5	Symmetry limits with good F spin 371
9.6	Example: Non-symmetric states in the nucleus ^{196}Pt . . 377
	Summary 387
	References 388

10 The Interacting Boson–Fermion Model   391
   10.1 Introduction 391
   10.2 The $U^B(6) \otimes U^F(\Omega)$ algebra 393
   10.3 Symmetry limits 398
   10.4 Example 1: The gold nuclei 403
   10.5 Example 2: The platinum nuclei 417
   10.6 Supersymmetry 423
       Summary 440
       References 442

A Group Theory and the Algebraic Approach   445
   A.1 Introduction 445
   A.2 Some definitions 446
   A.3 Symmetry transformations 448
   A.4 Constants of the motion and state labeling 449
   A.5 Eigenfunctions and representations 451
   A.6 The algebraic approach 454

B The Unitary and Orthogonal Algebras   463
   B.1 Commutation relations 463
   B.2 Casimir invariants 465
   B.3 Boson bases 466
   B.4 The Gel'fand bases 471
   B.5 Outer products of $U(n)$ representations 473

C Dragt's Theorem   477

Index   481

Preface

The techniques of group theory are of considerable significance to a large proportion of current theoretical investigations in physics. This fact reflects the wide ranging applicability of symmetry considerations in both classical and quantum systems. We believe that a familiarity with these methods is not only desirable but necessary for today's researchers.

While a large number of monographs already exist dealing with this subject, it is often difficult to acquire from their study the working ability which is required to carry out actual computations. In this book we have attempted to introduce the reader to some of these methods through the study of boson–fermion systems of increasing complexity, starting with a simple model where the underlying symmetry structure is the familiar angular momentum algebra. These models have found numerous applications in molecular and nuclear structure physics and the book may be a helpful complement to the study of these subjects. However, we hope that the techniques developed here can also be useful to students and researchers whose interests lie in other domains of physics.

In the past years we have benefitted greatly from our colleagues and we would like to acknowledge their teachings and friendship. In particular, we would like to thank Marcos Moshinsky, Franco Iachello and Phil Elliott for much inspiration and for showing us what group theory can achieve. We gratefully acknowledge the role of Rick Casten, who first suggested that we should write this book and kept the pressure ever since. Many thanks are also due to Clara Alonso, Ani Aprahamian, José Arias, Roelof Bijker, Alison Bruce, Octavio Castaños, Elpidio Chacón, Jorge Dukelsky, Tony Evans, Pedro Federman, Da Hsuan Feng, Adrian

Gelberg, Bob Gilmore, Joseph Ginocchio, Peter Hess, Kris Heyde, Jan Jolie, Renato Lemus, Amiram Leviatan, Pertti Lipas, Nescio Nagarajan, Stuart Pittel, Hong Zhou Sun, Peter von Brentano, David Warner, and Kurt Bernardo Wolf for many years of scientific collaboration and much fun. Special thanks are due to S. Czitrom and E. Troya for their constant support and encouragement. This work was carried out at the Instituto de Ciencias Nucleares, UNAM, the SERC Daresbury Laboratory, the University of Surrey, the University of Seville, and the Grand Accélérateur National d'Ions Lourds. A fellowship from the Guggenheim Foundation to AF allowed us to work together for longer periods in the final stages of the writing of this book, while PVI was supported by an SERC advanced fellowship. Both fellowships are gratefully acknowledged. Finally, the skillful and patient help of Ms. Lucila González in the typing of the manuscript is also thankfully acknowledged.

ALGEBRAIC METHODS IN MOLECULAR AND NUCLEAR STRUCTURE PHYSICS

Introduction

The use of group-theoretical methods for the analysis of quantum-mechanical systems has proved a valuable tool both in terms of providing new insights into the nature of physical systems and in terms of its practical calculational capabilities. The applications range from cases where straightforward geometrical symmetries are present, such as in solid state and molecular physics, to theories where abstract "internal" or gauge symmetries are used, as in particle physics and field theory. In all these cases one explores the features of the physical system under consideration that result from an underlying symmetry and its associated group structure. In general, symmetries lead to invariances, which in turn give rise to conservation laws. In quantum mechanics they provide quantum numbers that label the wave functions and give selection rules for transition and reaction processes. In addition, even when these symmetries are broken, the interactions can be decomposed as a sum over tensor operators with well-defined properties under the symmetry transformation, and thus the mathematical machinery of group theory can be exploited in the evaluation of matrix elements.

The use of group symmetries in nuclear structure has a long history, starting from the pioneering work of Wigner and his use of U(4) in the description of spin–isospin invariance in light nuclei [1]. Many applications have been carried out since then, among which we mention here the work of Racah [2] on the classification of complex spectra and of Elliott [3] on rotational properties of nuclei and their connection with the algebra SU(3). There are several books that deal with the subject, including the ones of Moshinsky [4], Gilmore [5], Wybourne [6], Parikh [7], Elliott and Dawber [8], and Biedenharn and Louck [9,10].

More recently a surge of interest in symmetry techniques has been generated by the introduction of a new model of nuclear structure, the interacting boson model (IBM) [11], which from its inception was framed in the language of group theory and its algebraic methods. The model was originally developed to provide a simple description of the properties of even–even nuclei (i.e., nuclei with even numbers of neutrons and protons) but was later extended in numerous ways allowing, among other things, the description of odd-mass and odd–odd nuclei within the same framework.

In many of these developments the algebraic techniques of group theory have played a key role. In particular, symmetries present in the model often provide insights and set guidelines for the development of a more detailed description of nuclei. Moreover, exciting new concepts have sprung from these investigations, such as that of F-spin multiplets and nuclear supersymmetry, which predict a link among the properties of different nuclei.

The basic building blocks of the IBM are bosons of angular momentum 0 (s bosons) and 2 (d bosons), the total number of which equals the number of pairs of valence nucleons in a given even–even nucleus. The s and d bosons reflect, respectively, the monopole and quadrupole pairing character of the nuclear interaction between identical nucleons. Odd-mass and odd–odd nuclei are considered by explicitly introducing the fermionic operators of the additional nucleons occupying specific shell-model orbits, giving rise to a system of interacting bosons and fermions. All calculations carried out within the model require an appropriate mathematical treatment of such systems and the algebraic methods of group theory are ideally suited for this purpose.

As a result of the success of the algebraic approach in nuclear structure physics, similar techniques have been recently proposed for the treatment of molecular rotation–vibration spectra [12]. Symmetry groups have been used in quantum-mechanical studies of selection rules in molecular electromagnetic transitions [13]. The role of the permutation–inversion invariance of the molecular hamiltonian, which is the true symmetry present in these systems, is still a very active research area [14]. From a different point of view, the unitary group has been used for the determination of symmetry-adapted wave functions [15], useful for the description of many-electron systems in atoms and molecules and, in particular, for the development of efficient algo-

Introduction

rithms for calculating electronic matrix elements. In these applications one starts from the exact non-relativistic Schrödinger equation for the nuclei and atoms in the molecule and, in order to solve the many-body problem, carries out a number of approximations which are designed to preserve the underlying symmetry of the original hamiltonian. Even with the simplifications brought about by such considerations, the determination of wave functions remains a formidable task, requiring the diagonalization of very large matrices.

The algebraic approach to molecular rotation–vibration spectra offers an alternative route for the study of these systems inspired by the methodology of the IBM. In this case the building blocks are bosons of angular momentum 0 (s bosons) and 1 (p bosons), the latter of which reflects the dipole character of the local modes in each bond. The resulting model is referred to as the vibron model [12]. Although these methods are less accurate than *ab initio* calculations in those cases where such calculations can be performed, they nevertheless provide a much simpler framework for treating complex situations, such as those occurring in polyatomic molecules where conventional techniques are extremely difficult to apply. Moreover, the wave functions which are produced in algebraic calculations are simple enough that they can, for example, be used in combination with reaction theory to describe multistep electron scattering off molecules.

A second stage in the development of the vibron model was the introduction of electronic degrees of freedom, thus providing a unified algebraic framework for analyzing molecular spectra. Again, all calculations in the model require the analysis of a system of interacting bosons and fermions—in this case several fermions—and group-theoretical techniques are a powerful means to achieve this goal.

Besides the applications in nuclear and molecular physics the algebraic techniques are being increasingly applied in other research areas. For example, algebraic methods have also been used for the description of scattering processes [16], skyrmions [17], and hadron systems [18].

This book attempts to provide the basic technical background required in the application of these algebraic models composed of bosons and fermions, with an emphasis on group-theoretical techniques. The material is divided in three parts of increasing mathematical complexity.

Part I is designed to familiarize the reader with the concepts and terminology of group theory as well as with the basic features of boson–fermion systems. In this part a schematic two-dimensional model is analyzed in full detail, exploiting its simple SU(2) structure. Most of the group-theoretical methods necessary for the applications in Parts II and III already appear here, with the advantage that the analysis in Part I is simplified in a significant way through the use of standard angular momentum theory. Although the inclusion of these simple models has a pedagogical purpose, it has been recently shown that they are useful for the description of stretching vibrations in polyatomic molecules, as well as for the analysis of phonon spectra in crystals. In Chapter 2 some of these applications are discussed.

Part II is devoted to the study of molecular algebraic models. Since the results of U(4)-based models have as yet not been reviewed comprehensively, we present in this part an overview of what has been achieved in this research area. The necessary calculational techniques are worked out in detail, often with reference to the ideas and concepts introduced in Part I, and the results subsequently illustrated with molecular structure examples. The group-theoretical calculations, although more involved than the ones in Part I, are nevertheless simple enough that closed expressions for physically relevant matrix elements can usually be found. We include sections on diatomic, triatomic, and tetra-atomic molecules as well as on the introduction of the electronic degrees of freedom in diatomic molecules.

In Part III we turn our attention to the interacting boson model of nuclear structure. In this case a large number of specialized articles, review papers, and books [11,19,20] have been published, and it would be impossible to present a full account of the subject or even to include a complete list of references. Our aim is rather to display the analogies with the simpler U(2) and U(4) models and to guide the reader through the U(6)-based models by means of a number of selected examples which include some of the exciting new developments in nuclear structure physics. The use of these higher-dimensional algebras invariably introduce additional complications, such as the appearance of missing labels and multiplicities, as well as a growing difficulty in the evaluation of relevant matrix elements. By means of selected examples we attempt to illustrate the way these problems can be surmounted.

We hope to convince the reader of the power and beauty that the algebraic techniques offer in the interpretation of the structure of complex many-body systems.

Because the subject of this book is presented in order of increasing complexity, an appropriate way of reading it is by going through all chapters consecutively. However, an alternative way to read the book is to first concentrate on Chapters 1, 2, 5, 6, 8, and 9, which deal with systems of interacting bosons. Having done this, the reader can then turn his or her attention to the mixed systems of bosons and fermions and to the additional complications due to the latter. Also, some of the sections of the book with a more technical character might be omitted in a first reading. This is mentioned at the beginning of each such section. Because of its very nature this book contains many equations and, in view of their number, the reader might have at times some difficulty in grasping the essential message we try to convey at each stage. To alleviate this problem, the most important equations (typically three or four per section) are highlighted and placed in a box; at the end of each chapter we provide the reader with a summary of the chapter where these results are commented. Some of the more technical sections of the book can also be read by referring to the relevant part of the summary at the end of each chapter.

References

1. E.P. Wigner, "On the consequences of the symmetry of the nuclear hamiltonian on the spectroscopy of nuclei," Phys. Rev. **51** (1937) 106.
2. G. Racah, "Theory of complex spectra. I," Phys. Rev. **61** (1942) 186; **62** (1942) 438; **63** (1943) 367; **76** (1949) 1352.
3. J.P. Elliott, "Collective motion in the nuclear shell model. I. Classifications schemes for states of mixed configurations," Proc. Roy. Soc. A **245** (1958) 128; **245** (1958) 562.
4. M. Moshinsky, *Group Theory and the Many-Body Problem*, Gordon & Breach, New York, 1968.
5. R. Gilmore, *Lie Groups, Lie Algebras, and Some of Their Applications*, Wiley-Interscience, New York, 1974.

6. B.G. Wybourne, *Classical Groups for Physicists*, Wiley-Interscience, New York, 1974.
7. J.C. Parikh, *Group Symmetries in Nuclear Structure*, Plenum, New York, 1978.
8. J.P. Elliott and P.G. Dawber, *Symmetry in Physics*, Oxford University Press, New York, 1979.
9. L.C. Biedenharn and J.D. Louck, *Angular Momentum in Quantum Physics*, Addison-Wesley, Reading, MA, 1981.
10. L.C. Biedenharn and J.D. Louck, *The Racah–Wigner Algebra in Quantum Theory*, Addison-Wesley, Reading, MA, 1981.
11. F. Iachello and A. Arima, *The Interacting Boson Model*, Cambridge University Press, Cambridge, 1987.
12. F. Iachello, "Algebraic methods for molecular rotation–vibration spectra," Chem. Phys. Lett. **78** (1981) 581.
13. G. Herzberg, *Molecular Spectra and Molecular Structure. III. Electronic Spectra and Electronic Structure of Polyatomic Molecules*, van Nostrand, New York, 1966.
14. G.A. Natanson, "On invariance of localized hamiltonians under feasible elements of the nuclear permutation–inversion group," Adv. Chem. Phys. **58** (1985) 55.
15. F.A. Matzer and R. Pauncz, *The Unitary Group in Quantum Chemistry*, Elsevier, Amsterdam, 1986.
16. Y. Alhassid, F. Gürsey, and F. Iachello, "Group theory approach to scattering. II. The Euclidean connection," Ann. Phys. (NY) **167** (1986) 181.
17. M. Oka, R. Bijker, A. Bulgac, and R.D. Amado, "Algebraic approach to the two-skyrmion system," Phys. Rev. C **36** (1987) 1727.
18. A. Bohm and Y. Ne'eman, *Dynamical Groups and Spectrum Generating Algebras*, World Scientific, Singapore, 1988.
19. D. Bonatsos, *Interacting Boson Models of Nuclear Structure*, Oxford University Press, New York, 1988.
20. F. Iachello and P. Van Isacker, *The Interacting Boson–Fermion Model*, Cambridge University Press, Cambridge, 1991.

Part I

Schematic Models

Chapter 1

Identical Bosons

1.1 The U(2) algebra

The aim of this chapter is to introduce a schematic two-dimensional boson model which, in spite of its simplicity, turns out to display a rich algebraic structure. Most of the group-theoretical concepts that are required for the analysis of the molecular and nuclear models discussed in Parts II and III already appear here, with the advantage that their basic structure is related to the well-known U(2) algebra. We shall see that an essentially complete solution of the N-body system considered in this chapter is provided by a corresponding knowledge of the U(2) mathematical structure. Although the higher-dimensional algebras we consider later are more complex, the basic mathematical operations are the same and we will often refer back to the U(2) results for comparison.

We start our discussion of the U(2) algebra by considering a two-dimensional harmonic oscillator with hamiltonian

$$\boxed{\hat{H} = \tfrac{1}{2}(p_x^2 + p_y^2 + x^2 + y^2)} \quad (1.1)$$

where we have chosen units in which $\hbar$, the mass m, and the frequency ω are 1. The usual (differential) procedure to solve the time-independent Schrödinger equation is to search for a coordinate system where a separation of variables can be achieved. Thus, in terms of cartesian coordinates, we arrive at the set of equations

$$\begin{aligned}\frac{1}{2}\left(-\frac{d^2}{dx^2}+x^2\right)g(x) &= E_x g(x) \\ \frac{1}{2}\left(-\frac{d^2}{dy^2}+y^2\right)g(y) &= E_y g(y)\end{aligned} \quad (1.2)$$

which are readily recognized [1] as Hermite's equations, solved by

$$g_{n_x}(x) = A_{n_x} H_{n_x}(x) e^{-x^2/2}, \qquad g_{n_y}(y) = A_{n_y} H_{n_y}(y) e^{-y^2/2} \quad (1.3)$$

with

$$E_x = n_x + \tfrac{1}{2}, \qquad E_y = n_y + \tfrac{1}{2}, \qquad n_x, n_y = 0, 1, \ldots \quad (1.4)$$

where $A_n = (2^n \sqrt{\pi}\, n!)^{-1/2}$. Thus the eigenfunctions of (1.1) are given by

$$\boxed{\Psi_{n_x n_y}(x,y) = [2^N \pi n_x! n_y!]^{-1/2} H_{n_x}(x) H_{n_y}(y) e^{-(x^2+y^2)/2}} \quad (1.5)$$

with eigenvalues

$$E = N + 1, \qquad N = n_x + n_y = 0, 1, \ldots \quad (1.6)$$

The states (1.5) are completely specified by N and n_x (or n_y). We can also solve (1.1) by transforming to polar coordinates

$$x = r\cos\theta, \qquad y = r\sin\theta \quad (1.7)$$

in terms of which Schrödinger's equation leads to the set of differential equations

$$\begin{aligned}\frac{1}{2}\left(-\frac{1}{r}\frac{d}{dr}r\frac{d}{dr}+\frac{m^2}{r^2}+r^2\right)\Phi(r) &= E\Phi(r) \\ -\frac{d^2}{d\theta^2}\phi(\theta) &= m^2 \phi(\theta)\end{aligned} \quad (1.8)$$

We see from (1.5) that $\Phi(r)$ contains the factor $e^{-r^2/2}$ while the small-r behavior may be obtained by neglecting the terms involving r^2 and E

1.1 The U(2) algebra

in (1.8). This implies that $\Phi(r) \approx r^{|m|} f(r) e^{-r^2/2}$ where $f(r)$ satisfies Laguerre's differential equation [1]. We find the solutions

$$\Phi_{Nm}(r) = \left[\frac{2((N-m)/2)!}{((N+m)/2)!}\right]^{1/2} r^{|m|} L^{|m|}_{(N-|m|)/2}(r^2) e^{-r^2/2} \qquad (1.9)$$

again with

$$E = N+1, \qquad N = 0, 1, \ldots \qquad (1.10)$$

and

$$\phi_m(\theta) = \frac{1}{\sqrt{2\pi}} e^{im\theta}, \qquad m = \pm N, \pm(N-2), \ldots, \pm 1 \text{ or } 0 \qquad (1.11)$$

In this case the complete wave functions are thus given by

$$\boxed{\Phi_{Nm}(r,\theta) = \left[\frac{((N-m)/2)!}{\pi((N+m)/2)!}\right]^{1/2} r^{|m|} L^{|m|}_{(N-|m|)/2}(r^2) e^{-r^2/2} e^{im\theta}} \qquad (1.12)$$

with eigenvalues (1.6) or (1.10). These states are specified by the values of N and m. We can readily verify, from either (1.6) or (1.11), that each energy level, characterized by N, is $(N+1)$-fold degenerate. This degeneracy is the signature of a symmetry, the U(2) symmetry associated with the hamiltonian (1.1), as we proceed to show.

A simple way to introduce the U(2) algebra is to analyze the hamiltonian (1.1) in a different space, provided by the following creation and annihilation operators:

$$\boxed{\begin{array}{ll} s^\dagger = \sqrt{\tfrac{1}{2}}(x - ip_x), & t^\dagger = \sqrt{\tfrac{1}{2}}(y - ip_y) \\ s = \sqrt{\tfrac{1}{2}}(x + ip_x), & t = \sqrt{\tfrac{1}{2}}(y + ip_y) \end{array}} \qquad (1.13)$$

Since

$$[x, p_x] = [y, p_y] = i, \qquad [x, y] = [p_x, p_y] = 0 \qquad (1.14)$$

we find that

$$\boxed{[s, s^\dagger] = [t, t^\dagger] = 1} \qquad (1.15)$$

with all other commutators being zero. From (1.15) it is clear that the action of the annihilation operators s (t) on polynomials in $s^\dagger$ ($t^\dagger$) is analogous to the action of $\partial/\partial s^\dagger$ ($\partial/\partial t^\dagger$). The hamiltonian (1.1) can be expressed in terms of these new operators as

$$\boxed{\hat{H} = s^\dagger s + t^\dagger t + 1} \tag{1.16}$$

To study the symmetries associated with $\hat{H}$, we define [2] the set of operators (or generators)

$$\boxed{\mathcal{G}_1^1 \equiv s^\dagger s, \quad \mathcal{G}_1^2 \equiv s^\dagger t, \quad \mathcal{G}_2^1 \equiv t^\dagger s, \quad \mathcal{G}_2^2 \equiv t^\dagger t} \tag{1.17}$$

which are bilinear in the creation and annihilation operators (1.13). From (1.15) and (1.16) we find

$$\boxed{[\hat{H}, \mathcal{G}_i^j] = 0, \quad i,j = 1,2} \tag{1.18}$$

that is, the hamiltonian commutes with the operators (1.17). In addition, these operators satisfy the following commutations relations among themselves:

$$\boxed{[\mathcal{G}_i^j, \mathcal{G}_k^l] = \mathcal{G}_i^l \delta_{jk} - \mathcal{G}_k^j \delta_{il}, \quad i,j,k,l = 1,2} \tag{1.19}$$

This shows that the $\mathcal{G}_i^j$ define a Lie algebra of the general U(n) form discussed in Appendix B. In this case (1.19) defines a U(2) algebra, with generators (1.17). To see that this particular algebra is familiar, we may consider the linear combinations of (1.17)

$$\begin{aligned}
\hat{J}_x &\equiv \tfrac{1}{2}(s^\dagger s - t^\dagger t) \\
\hat{J}_y &\equiv \tfrac{1}{2}(t^\dagger s + s^\dagger t) \\
\hat{J}_z &\equiv \tfrac{i}{2}(t^\dagger s - s^\dagger t) \\
\hat{N} &\equiv s^\dagger s + t^\dagger t
\end{aligned} \tag{1.20}$$

for which we find the following commutation relations:

$$[\hat{J}_j, \hat{J}_k] = i\epsilon_{jkl}\hat{J}_l, \quad j,k,l = x,y,z \tag{1.21}$$

and

$$[\hat{N}, \hat{J}_j] = 0, \quad j = x,y,z \tag{1.22}$$

1.1 The U(2) algebra

where $\epsilon_{jkl} = +1$ if (jkl) is an even permutation of (xyz), $\epsilon_{jkl} = -1$ if (jkl) is an odd permutation of (xyz), and $\epsilon_{jkl} = 0$ otherwise. The three operators $(\hat{J}_x, \hat{J}_y, \hat{J}_z)$ satisfy the commutation relations of the SU(2) angular momentum algebra [3]. The four operators $(\hat{J}_x, \hat{J}_y, \hat{J}_z, \hat{N})$ are nothing but linear combinations of the G_i^j and thus form the same algebra as the one with which we started. Equations (1.21) and (1.22) show that this U(2) algebra, after a linear transformation of its generators, can be written in a different but equivalent way: since the operator $\hat{N}$ (which on itself trivially generates a U(1) algebra) commutes with $\hat{J}_i$, the operators $(\hat{J}_x, \hat{J}_y, \hat{J}_z, \hat{N})$ form the *direct product* $SU(2) \otimes U(1)$. We shall often encounter such direct products in the subsequent chapters of this book. In general, the direct product of two algebras $G = (\hat{g}_1, \hat{g}_2, \ldots)$ and $G' = (\hat{g}'_1, \hat{g}'_2, \ldots)$ can be formed if the generators of G commute with those of G', $[\hat{g}_i, \hat{g}'_j] = 0$, and it consists of the combined generators of G and G', $G \otimes G' = (\hat{g}_1, \hat{g}_2, \ldots, \hat{g}'_1, \hat{g}'_2, \ldots)$. A mathematically more rigorous notation would be to denote this combined collection of generators as a *direct sum* of algebras, $g \oplus g'$, and to reserve the direct-product notation, $G \otimes G'$, for the associated groups since the latter are obtained from the former through exponentiation (see Appendix A). However, as explained in Appendix A, throughout this book we use capital letters to denote Lie algebras, and we follow the customary convention to use the product sign $\otimes$ with uppercase letters (U(n), SU(n), SO(n), O(n), and Sp(n)) and the sum sign $\oplus$ with lowercase letters (u(n), su(n), so(n), o(n), and sp(n)).

We have thus shown that U(2) is equivalent to $SU(2) \otimes U(1)$. This property of equivalence after a linear transformation of the generators is called *isomorphism* and is denoted by $\simeq$. In this case U(2) is isomorphic to $SU(2) \otimes U(1)$, $U(2) \simeq SU(2) \otimes U(1)$.

According to (1.18) the harmonic-oscillator hamiltonian commutes with the generators of the U(2) algebra. Alternatively, since from (1.16) and (1.20) it follows that $\hat{H} = \hat{N} + 1$, we find from (1.22) that the hamiltonian commutes with the generators of the SU(2) algebra. Thus we may consider either U(2) or SU(2) as the symmetry algebra of the two-dimensional harmonic-oscillator hamiltonian.

The states (1.12) are partially labeled by N, a quantum number associated with U(2), and for some applications it is convenient to define an equivalent label associated with SU(2). Computing $\hat{J}^2 = \hat{J}_x^2 + \hat{J}_y^2 + \hat{J}_z^2$, we find

$$\hat{J}^2 = \tfrac{1}{4}\hat{N}(\hat{N}+2) \tag{1.23}$$

so the states in the boson realization (1.13) may be characterized by either N or $j(j+1)$, where, because of (1.23),

$$j = N/2 \tag{1.24}$$

The degeneracy of these states is then simply

$$\dim(j) \equiv 2j+1 = N+1 \tag{1.25}$$

consistent with the result previously found.

What about the wave functions in the boson representation? In a one-dimensional oscillator, corresponding to one of equations (1.2), we have the following representation of states in terms of s bosons [4]:

$$|n_s\rangle = \frac{1}{\sqrt{n_s!}}(s^\dagger)^{n_s}|0\rangle, \qquad |0\rangle = \pi^{-1/4}e^{-x^2/2} \tag{1.26}$$

The state $|0\rangle$ indeed plays the role of a vacuum, since

$$\begin{aligned} s|0\rangle &= \frac{1}{\sqrt{2}}\left(x + \frac{d}{dx}\right)\pi^{-1/4}e^{-x^2/2} \\ &= \frac{\pi^{-1/4}}{\sqrt{2}}x\left(e^{-x^2/2} - e^{-x^2/2}\right) = 0 \end{aligned} \tag{1.27}$$

with the factor $\pi^{-1/4}$ needed as a normalization. Use of the commutator property $[s, s^\dagger] = 1$ gives

$$\begin{aligned} s^\dagger s|n_s\rangle &= n_s|n_s\rangle \\ \hat{H}_s|n_s\rangle &= \tfrac{1}{2}(p_x^2 + x^2)|n_s\rangle = \left(n_s + \tfrac{1}{2}\right)|n_s\rangle \end{aligned} \tag{1.28}$$

which shows that $|n_s\rangle$ is an eigenstate of $\hat{H}_s$ with energy $E = n_s + 1/2$.

These results can be generalized to a two-dimensional harmonic oscillator, in which case the boson representation for the states becomes

$$|[N]n_t\rangle = \frac{(s^\dagger)^{N-n_t}(t^\dagger)^{n_t}}{\sqrt{(N-n_t)!n_t!}}|0\rangle \tag{1.29}$$

1.1 The U(2) algebra

where now $|0\rangle = \pi^{-1/2} e^{-(x^2+y^2)/2}$.

It is clear that the boson states (1.29) are simultaneous eigenstates of $\hat{N}$ and $t^\dagger t$ (or $s^\dagger s$). Since $t^\dagger t$ generates a U(1) subalgebra of U(2), we say that they are classified according to the

$$\begin{array}{ccc} \text{U}(2) & \supset & \text{U}(1) \equiv t^\dagger t \\ | & & | \\ [N] & & n_t \end{array} \qquad (1.30)$$

chain of algebras, where we indicate below each algebra its associated label. An alternative classification is obtained by simultaneously diagonalizing $\hat{N}$ (or $\hat{J}^2$) and $\hat{J}_z$, and the resulting states are closely related to the configuration space functions (1.12). Their explicit form is more complicated than (1.29) and we postpone their derivation until Section 1.5. Being simultaneous eigenstates of $\hat{J}^2$ and $\hat{J}_z$, however, implies that they are classified according to a chain of algebras

$$\begin{array}{ccccc} \text{U}(2) & \simeq & \text{SU}(2) & \otimes \text{U}(1) \supset & \text{SO}(2) \equiv \hat{J}_z \\ | & & | & & | \\ [N] & & j = N/2 & & \mu = m/2 \end{array} \qquad (1.31)$$

The quantum number μ associated with $\hat{J}_z$ is related to the label m (defined in (1.12)) through $\mu = m/2$, as will be shown in Section 1.3.

The harmonic-oscillator hamiltonian (1.1) acquires a very simple form in the boson realization as seen from (1.16), being (except for an additive constant) equal to the total boson number $\hat{N}$. This property is shared by harmonic-oscillator hamiltonians of any dimension. The form of the wave functions is generically given by polynomials in the creation operators, appropriately classified by a chain of algebras, in this case either (1.30) or (1.31). In the next section we consider more general hamiltonians in the two-dimensional case and their treatment in second-quantized formalism. This procedure will naturally lead us to the concept of a dynamical algebra and to the generalization of the boson calculus discussed in this introductory section.

1.2 The s–t-boson hamiltonian

The harmonic-oscillator hamiltonian (1.1) can be visualized as describing a system of non-interacting bosons of two kinds, s and t, which occupy a single energy level (in our units this energy is $\hbar\omega = 1$). The hamiltonian has the simple role of counting the total number of bosons in the system, as seen from (1.16). The *symmetry* algebra of this system is the U(2) or SU(2) algebra, as seen from (1.18) or (1.22), and this in turn explains the characteristic degeneracy associated with a given boson number. For the SU(2) algebra we know that irreducible representations are characterized by a single quantum number, j, and are $(2j+1)$-dimensional. (See Appendix A for the definition of irreducible representations and Appendix B for a discussion of U(n) algebras.)

In this section we turn our attention to more general hamiltonians for a system of N bosons, namely, hamiltonians for a system of N *interacting* bosons.

In the language of second quantization [2] a general one- and two-body hamiltonian for N particles

$$\hat{H} = \sum_{i=1}^{N} \left(\tfrac{1}{2} p_i^2 + U_i \right) + \sum_{i<j}^{N} V_{ij} \tag{1.32}$$

may be rewritten in terms of one- and two-body matrix elements by carrying out the following replacements:

$$\begin{aligned}
\sum_{i=1}^{N} \left(\tfrac{1}{2} p_i^2 + U_i \right) &\to \sum_{\alpha\alpha'} \langle \alpha' | \left(\tfrac{1}{2} p_i^2 + U_i \right) | \alpha \rangle b_{\alpha'}^\dagger b_\alpha \\
\sum_{i<j}^{N} V_{ij} &\to \tfrac{1}{2} \sum_{\alpha_1\alpha_2} \sum_{\alpha_1'\alpha_2'} \langle \alpha_1' \alpha_2' | V_{12} | \alpha_1 \alpha_2 \rangle b_{\alpha_1'}^\dagger b_{\alpha_2'}^\dagger b_{\alpha_2} b_{\alpha_1}
\end{aligned} \tag{1.33}$$

where $b_\alpha^\dagger$ and b_α are creation and annihilation operators for bosons and $|\alpha_i\rangle$ and $|\alpha_i\alpha_j\rangle$ are a complete set of one- and two-boson states in the original configuration space.

Before considering again our system of s and t bosons, it is convenient to artificially endow them with an additional property—parity—that will distinguish them. This property will be essential when considering higher-dimensional cases, where it should be carried by the

1.2 The s–t-boson hamiltonian

fundamental building blocks of the algebra. From (1.13) we see that, under hermitian conjugation, the creation and annihilation operators satisfy

$$(s^\dagger)^\dagger = s, \qquad (t^\dagger)^\dagger = t \qquad (1.34)$$

so we can impose hermiticity in the hamiltonian in a simple way. We may now forget about the original definition of these operators in terms of coordinates and simply consider (1.34) as a definition. Likewise we now *define* a "space-inversion" or parity operator $\hat{P}$ and attribute parity to the bosons through the definitions

$$\hat{P}s^\dagger\hat{P}^{-1} = s^\dagger, \qquad \hat{P}t^\dagger\hat{P}^{-1} = -t^\dagger \qquad (1.35)$$

which implies that we assign positive parity to the s bosons and negative parity to the t bosons. This assignment is arbitrary for the U(2) model, but we shall see that in higher-dimensional models boson parity arises from physical considerations.

The most general one- and two-body, hermitian, parity-invariant hamiltonian in the s–t space is thus given by

$$\begin{aligned}\hat{H} &= E_0 + \epsilon_s s^\dagger s + \epsilon_t t^\dagger t \\ &\quad + e_1 s^\dagger s^\dagger ss + e_2(s^\dagger s^\dagger tt + t^\dagger t^\dagger ss) \\ &\quad + e_3 s^\dagger t^\dagger st + e_4 t^\dagger t^\dagger tt \end{aligned} \qquad (1.36)$$

where E_0 is a constant and the parameters ϵ_s, ϵ_t, and e_i are the one- and two-body matrix elements in (1.33). We essentially recover the harmonic-oscillator hamiltonian for $\epsilon_s = \epsilon_t$ and all other interaction parameters zero. The SU(2) generators in (1.20) do not commute with the hamiltonian (1.36), which only commutes in general with the total boson number $\hat{N}$ (since all terms in (1.36) contain equal numbers of creation and annihilation operators). Thus the symmetry algebra of (1.36) is just the U(1) algebra generated by $\hat{N}$.

Since the U(2) algebra ceases to be a symmetry algebra of the hamiltonian (1.36), what is its role for this system? We note that $\hat{H}$ can be expressed as a linear combination of the U(2) generators (1.17) and products of them. Since $\hat{N}$ is a constant for the system, we may also say that $\hat{H}$ is a function of the SU(2) generators and the integer N. The U(2) (or SU(2)) algebra is referred to as the *dynamical* algebra for the system [5] since the operators associated with the dynamics, including the hamiltonian, are expressed in terms of its generators. This being so, all states of the N-boson system, in this case $N+1$ of them, are contained in the single representation $[N]$ of U(2) (or j of SU(2)). The interactions in (1.36) split the energy of the states but do not mix states in different U(2) representations.

In the next section we rewrite the hamiltonian in terms of invariant operators in the chains of algebras (1.30) and (1.31), in order to exploit the U(2) dynamical algebra associated with the system.

1.3 Symmetry limits

The solution of the s–t-boson hamiltonian (1.36) requires in general the construction of a complete set of states, the calculation of matrix elements of the hamiltonian between these states, and a subsequent (numerical) diagonalization. In this section we show how, under certain circumstances, the eigensolutions of (1.36) can be obtained by algebraic means.

In Section 1.1 we discussed the existence of two bases, characterized by the chains of algebras (1.30) and (1.31). In the U(2) $\supset$ U(1) basis the allowed values of n_t for a given N are

$$n_t = 0, 1, \ldots, N \qquad (1.37)$$

This rule can be inferred from the structure of the state $|[N]n_t\rangle$ given in (1.29) and is a special case of the more general result for U(n) $\supset$ U($n-1$) (see Appendix B). The corresponding rule in the U(2) $\supset$ SO(2) basis is well known from angular momentum theory:

$$\mu = -j, -j+1, \ldots, j \qquad (1.38)$$

Equations (1.37) and (1.38) are examples of *branching* (or *ramification*) rules which specify, for a chain of algebras $G \supset G'$, all allowed representations of G' contained in a given representation of G.

The branching rule (1.38) clarifies the relation between μ, the label associated with SO(2), and the quantum number m introduced in (1.11). Since $m = -N, -N+2, \ldots, N$ and $j = N/2$, we immediately deduce from (1.38) that $\mu = m/2$, in agreement with (1.31).

1.3 Symmetry limits

An essential ingredient in the algebraic solution of (1.36) consists of rewriting the hamiltonian in terms of operators associated with the algebras appearing in the two chains. In general, for any algebra G, a set of operators exists, referred to as *invariant operators* or *Casimir operators* of G, which commute with all the generators of G. In our example the algebras of interest are U(2), U(1), and SO(2), and the associated invariant operators are $\hat{N}$, $\hat{N}^2$ (for U(2)), $\hat{n}_t$, $\hat{n}_t^2$ (for U(1)), and $\hat{J}_z$, $\hat{J}_z^2$ (for SO(2)). In principle higher-order operators may be considered, but there is no need for them here since the hamiltonian (1.36) contains only up to two-body interactions. With the help of the commutation relations (1.15) the invariant operators can be rewritten as

$$\begin{aligned}
\hat{N} &= s^\dagger s + t^\dagger t, & \hat{N}^2 &= \hat{N} + s^\dagger s^\dagger ss + t^\dagger t^\dagger tt + 2s^\dagger t^\dagger st \\
\hat{n}_t &= t^\dagger t, & \hat{n}_t^2 &= \hat{n}_t + t^\dagger t^\dagger tt \\
\hat{J}_z &= \tfrac{i}{2}(t^\dagger s - s^\dagger t), & \hat{J}_z^2 &= \tfrac{1}{4}(\hat{N} - s^\dagger s^\dagger tt - t^\dagger t^\dagger ss + 2s^\dagger t^\dagger st)
\end{aligned} \quad (1.39)$$

Using these relations we can rewrite the hamiltonian (1.36) in the alternative form

$$\boxed{\hat{H} = E_0' + \epsilon \hat{n}_t + \alpha \hat{n}_t^2 + \beta \hat{J}_z^2} \qquad (1.40)$$

where

$$\begin{aligned}
E_0' &= E_0 + (\epsilon_s - e_1 + e_2)N + e_1 N^2 \\
\epsilon &= \epsilon_t - \epsilon_s + e_1 - e_4 - (2e_1 - 2e_2 - e_3)N \\
\alpha &= e_1 - 2e_2 - e_3 + e_4 \\
\beta &= -4e_2
\end{aligned} \qquad (1.41)$$

Since N is a constant, E_0' contributes only to the binding energy of the system.

From (1.40) we see that the general one- and two-body hamiltonian of the s–t-boson system has only three parameters, ϵ, α, and β, which contribute to the structure of the spectrum. In this form the hamiltonian is expressed in terms of invariant operators of the two chains (1.30) and (1.31).

To find the eigenvalues and eigenstates associated with (1.40), one may use the complete set of states associated with either of the chains. If one takes (1.29) as a basis, only the matrix elements of J_z^2 should be obtained, since $\hat{n}_t$ and $\hat{n}_t^2$ are already diagonal. We discuss the evaluation of these matrix elements in Section 1.4. However, we note already that analytic solutions exist for particular values of the parameters. If $\beta = 0$, the hamiltonian is diagonal in chain (1.30), while if $\epsilon = \alpha = 0$, it is diagonal in chain (1.31). These symmetry limits are known as *dynamical symmetries* and they play an important role in the development of algebraic models. In the next section we discuss the first of them, the U(1) dynamical symmetry or U(1) limit.

1.4 The U(1) limit

If in (1.40) we take $\beta = 0$ (which implies $e_2 = 0$ in (1.36)), the hamiltonian reduces to

$$\hat{H}_\mathrm{I} = E_0' + \epsilon \hat{n}_t + \alpha \hat{n}_t^2 \qquad (1.42)$$

and has the eigenvalues

$$E_\mathrm{I}(n_t) = E_0' + \epsilon n_t + \alpha n_t^2 \qquad (1.43)$$

where, due to (1.37),

$$n_t = 0, 1, \ldots, N \qquad (1.44)$$

For not too large values of $|\alpha/\epsilon|$ this eigenspectrum resembles a one-dimensional anharmonic oscillator, as shown in Figure 1.1. The corresponding wave functions are classified by the chain of algebras (see Section 1.1)

$$\begin{array}{cc} \mathrm{U}(2) \supset \mathrm{U}(1) \\ | & | \\ [N] & n_t \end{array} \qquad (1.45)$$

and thus can be denoted as $|[N]n_t\rangle$ with the following properties:

$$\hat{N}|[N]n_t\rangle = N|[N]n_t\rangle, \qquad \hat{n}_t|[N]n_t\rangle = n_t|[N]n_t\rangle \qquad (1.46)$$

The explicit form of the wave functions is given in Section 1.1,

$$|[N]n_t\rangle = \frac{(s^\dagger)^{N-n_t}(t^\dagger)^{n_t}}{\sqrt{(N-n_t)!n_t!}}|0\rangle \qquad (1.47)$$

1.4 The U(1) limit

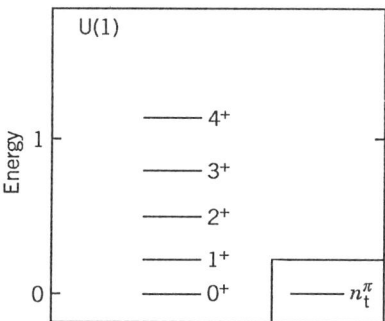

Figure 1.1: The U(1) eigenspectrum of the hamiltonian $\hat{H}_{\rm I}$ with $E'_0 = 0$, $\epsilon = 0.2$, and $\alpha = 0.02$ for $N = 4$ bosons. Levels are labeled by $n_{\rm t}$ and parity $\pi = (-)^{n_{\rm t}}$. Energies are in arbitrary units.

which carry, because of the definition (1.35), a definite parity,

$$\hat{P}|[N]n_{\rm t}\rangle = (-)^{n_{\rm t}}|[N]n_{\rm t}\rangle \tag{1.48}$$

This parity is indicated in Figure 1.1 together with the value of $n_{\rm t}$ for each level. The states (1.47) correspond to the two-dimensional harmonic-oscillator eigenstates associated with cartesian coordinates (1.5), except that for the hamiltonian (1.42) the interactions $\hat{n}_{\rm t}$ and $\hat{n}_{\rm t}^2$ split the $N+1$ levels without mixing the U(2) representations. The algebra U(2) thus changes its role from a symmetry algebra (in the harmonic-oscillator case) to a dynamical algebra for the hamiltonian (1.42).

In order to complete the discussion of this limit, it is necessary to compute the matrix elements of the U(2) (or SU(2)) generators in the basis (1.47) since all "observables" in the model are expressed in terms of them. In Parts II and III, dealing with molecular and nuclear structure models, these other observables correspond to electromagnetic moments and transitions, particle-transfer intensities, etc.

From (1.47) we find in a straightforward way

$$\begin{aligned}
s^\dagger |[N]n_t\rangle &= \sqrt{N - n_t + 1}\,|[N+1]n_t\rangle \\
t^\dagger |[N]n_t\rangle &= \sqrt{n_t + 1}\,|[N+1]n_t + 1\rangle \\
s|[N]n_t\rangle &= \sqrt{N - n_t}\,|[N-1]n_t\rangle \\
t|[N]n_t\rangle &= \sqrt{n_t}\,|[N-1]n_t - 1\rangle
\end{aligned} \qquad (1.49)$$

leading to the relations

$$\begin{aligned}
s^\dagger s|[N]n_t\rangle &= (N - n_t)|[N]n_t\rangle \\
t^\dagger t|[N]n_t\rangle &= n_t|[N]n_t\rangle \\
s^\dagger t|[N]n_t\rangle &= \sqrt{(N - n_t + 1)n_t}\,|[N]n_t - 1\rangle \\
t^\dagger s|[N]n_t\rangle &= \sqrt{(N - n_t)(n_t + 1)}\,|[N]n_t + 1\rangle
\end{aligned} \qquad (1.50)$$

The first two equations in (1.50) are simply (1.46) written in a different form while the last two equations make it clear that $t^\dagger s$ and $s^\dagger t$ play the role of raising and lowering operators for this basis.

From the definition (1.20) of the SU(2) generators and equations (1.50) we also find

$$\begin{aligned}
\hat{J}_x|[N]n_t\rangle &= \tfrac{1}{2}(N - 2n_t)|[N]n_t\rangle \\
\hat{J}_y|[N]n_t\rangle &= +\tfrac{1}{2}\sqrt{(N - n_t)(n_t + 1)}\,|[N]n_t + 1\rangle \\
&\quad +\tfrac{1}{2}\sqrt{(N - n_t + 1)n_t}\,|[N]n_t - 1\rangle \\
\hat{J}_z|[N]n_t\rangle &= +\tfrac{i}{2}\sqrt{(N - n_t)(n_t + 1)}\,|[N]n_t + 1\rangle \\
&\quad -\tfrac{i}{2}\sqrt{(N - n_t + 1)n_t}\,|[N]n_t - 1\rangle
\end{aligned} \qquad (1.51)$$

By applying $\hat{J}_z$ twice we arrive at the matrix elements of the non-diagonal operator $\hat{J}_z^2$ in this basis,

$$\begin{aligned}
\hat{J}_z^2|[N]n_t\rangle &= -\tfrac{1}{4}\sqrt{(N - n_t - 1)(N - n_t)(n_t + 1)(n_t + 2)}\,|[N]n_t + 2\rangle \\
&\quad +\tfrac{1}{4}\big(N + 2n_t(N - n_t)\big)|[N]n_t\rangle \\
&\quad -\tfrac{1}{4}\sqrt{(N - n_t + 1)(N - n_t + 2)(n_t - 1)n_t}\,|[N]n_t - 2\rangle
\end{aligned} \qquad (1.52)$$

1.5 The SO(2) limit

This result suffices to deduce the matrix elements of the model hamiltonian (1.40) and diagonalization of the matrix $\langle[N]n'_t|\hat{J}_z^2|[N]n_t\rangle$ will lead to the U(2) $\supset$ SO(2) basis $|j\mu\rangle$. In the next section we compute these states explicitly.

1.5 The SO(2) limit

For $\epsilon = \alpha = 0$ in (1.40) we arrive at the SO(2) dynamical symmetry with the hamiltonian

$$\boxed{\hat{H}_{\text{II}} = E'_0 + \beta\hat{J}_z^2} \tag{1.53}$$

with eigenvalues

$$E_{\text{II}}(\mu) = E'_0 + \beta\mu^2 \tag{1.54}$$

where, due to (1.38),

$$\mu = -j, -j+1, \ldots, j \tag{1.55}$$

and $j = N/2$. The eigenfunctions of the hamiltonian (1.53) are classified by the chain of algebras (see Section 1.1)

$$\boxed{\begin{array}{ccc} \text{U}(2) & \supset \text{SU}(2) & \supset \text{SO}(2) \\ | & | & | \\ {[N]} & j & \mu \end{array}} \tag{1.56}$$

The eigenstates are denoted as $|j\mu\rangle$ and satisfy the properties

$$\hat{J}^2|j\mu\rangle = j(j+1)|j\mu\rangle, \qquad \hat{J}_z|j\mu\rangle = \mu|j\mu\rangle \tag{1.57}$$

The eigenspectrum (1.54) is shown in Figure 1.2, where the value of μ is indicated for each level. Note that no parity is assigned to the levels. To discuss parity, it is necessary to construct the wave functions associated with this limit and study their behavior under the parity operation. A simple way to do so is to return to the definition (1.20) of the SU(2) generators,

$$\hat{J}_x = \tfrac{1}{2}(s^\dagger s - t^\dagger t), \quad \hat{J}_y = \tfrac{1}{2}(t^\dagger s + s^\dagger t), \quad \hat{J}_z = \tfrac{i}{2}(t^\dagger s - s^\dagger t) \tag{1.58}$$

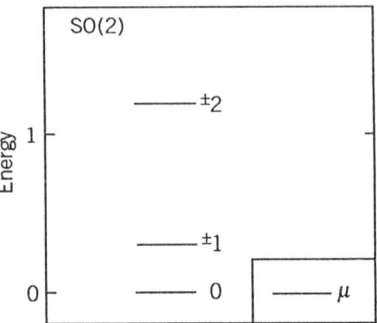

Figure 1.2: The SO(2) eigenspectrum of the hamiltonian $\hat{H}_{\text{II}}$ with $E'_0 = 0$ and $\beta = 0.3$ for $N = 4$ bosons. Levels are labeled by μ. Energies are in arbitrary units.

Note that this algebra is chosen such that $\hat{J}_z$ is diagonal in the chain $U(2) \supset SO(2)$. If we would have selected $\hat{J}_x$ instead, the corresponding SO(2) subalgebra would define a $U(2) \supset U(1)$ basis (1.45) since both $s^\dagger s$ and $t^\dagger t$ are diagonal in this basis. We denote by $\overline{SO}(2)$ this alternative subalgebra, $\overline{SO}(2) \equiv \hat{J}_x$. Denoting by σ the eigenvalue associated with $\hat{J}_x$, we find

$$\hat{J}_x |[N] n_t\rangle = \tfrac{1}{2}(\hat{N} - 2\hat{n}_t)|[N] n_t\rangle = \sigma |[N] n_t\rangle \quad (1.59)$$

from which $\sigma = N/2 - n_t$ or $n_t = j - \sigma$. This allows us to write the $U(2) \supset \overline{SO}(2)$ eigenstates as

$$|j\sigma\rangle = \frac{(s^\dagger)^{j+\sigma}(t^\dagger)^{j-\sigma}}{\sqrt{(j-\sigma)!(j+\sigma)!}}|0\rangle \quad (1.60)$$

where we distinguish them from the states (1.31) by the use of the label σ instead of μ. A possible way to find explicit expressions for these states is to carry out a transformation in the s–t space, such that $\hat{J}_x$ goes over into $\hat{J}_z$. The appropriate transformation is found by considering first a one-boson state, $N = 1$ or $j = 1/2$. According to (1.60) the s boson corresponds to $\sigma = +1/2$ and the t boson to $\sigma = -1/2$ and hence we may consider $(s^\dagger, t^\dagger)$ as the components of an

1.5 The SO(2) limit

SU(2) spinor. Restricting ourselves to rotations in the s–t space we find the following general form for this transformation:

$$\begin{pmatrix} s'^\dagger \\ t'^\dagger \end{pmatrix}_{\theta_i} = \widetilde{\mathbf{D}}^{1/2}(\theta_1, \theta_2, \theta_3) \begin{pmatrix} s^\dagger \\ t^\dagger \end{pmatrix} \tag{1.61}$$

where $\widetilde{\mathbf{M}}$ indicates the transpose of matrix $\mathbf{M}$, $\mathbf{D}^{1/2}$ is a matrix notation for the Wigner functions $D^j_{\mu\mu'}$ with $j = 1/2$, and θ_i are Euler angles [3]. The rotation that transforms $\hat{J}_x$ into $\hat{J}_z$ is a special case of (1.61) and can be obtained by choosing specific values for θ_i, as will be explained in Section 1.6. For the moment, however, we obtain this transformation by inspection of (1.58). We find that

$$s'^\dagger = \frac{e^{i\alpha}}{\sqrt{2}}(s^\dagger + it^\dagger), \qquad t'^\dagger = \frac{e^{i\alpha}}{\sqrt{2}}(s^\dagger - it^\dagger) \tag{1.62}$$

with α arbitrary, when introduced in (1.58), leads to the transformed operators

$$\hat{J}'_x = \hat{J}_z, \qquad \hat{J}'_y = \hat{J}_x, \qquad \hat{J}'_z = \hat{J}_y \tag{1.63}$$

and thus gives rise to a cyclical permutation of the angular momentum generators.

To find the $U(2) \supset SO(2)$ wave functions, we just need to rewrite the states (1.60) in terms of $s'^\dagger$ and $t'^\dagger$. Because of (1.63) these states will automatically be eigenstates of $\hat{J}_z$. Choosing the phase $\alpha = 0$ in (1.62) and returning to the label μ, we arrive at the explicit form

$$\boxed{|j\mu\rangle = \frac{(s^\dagger - it^\dagger)^{j-\mu}(s^\dagger + it^\dagger)^{j+\mu}}{2^j \sqrt{(j-\mu)!(j+\mu)!}}|0\rangle} \tag{1.64}$$

An equivalent expression may be found by writing $(s^\dagger + it^\dagger)^{j+\mu} = (s^\dagger + it^\dagger)^{2\mu}(s^\dagger + it^\dagger)^{j-\mu}$, leading to

$$\boxed{|j\mu\rangle = \frac{(s^\dagger s^\dagger + t^\dagger t^\dagger)^{j-\mu}(s^\dagger + it^\dagger)^{2\mu}}{2^j \sqrt{(j-\mu)!(j+\mu)!}}|0\rangle} \tag{1.65}$$

For $\mu < 0$ expression (1.65) seems awkward, since $(s^\dagger + it^\dagger)$ carries a negative exponent. However, this term is always cancelled by the first.

It is a simple exercise to verify directly that the states (1.64) and (1.65) are eigenstates of $\hat{N}$ and $\hat{J}_z$ with eigenvalues $2j$ and μ, respectively.

We can now return to the question of the behavior of these states under the parity transformation. According to (1.35) we find, from (1.65),

$$\hat{P}|j\mu\rangle = \frac{(s^\dagger s^\dagger + t^\dagger t^\dagger)^{j-\mu}(s^\dagger - it^\dagger)^{2\mu}}{2^j\sqrt{(j-\mu)!(j+\mu)!}}|0\rangle \qquad (1.66)$$

which shows that these states do not have good parity except for $\mu = 0$ when they carry positive parity. From (1.64), however, we see that $\hat{P}|j\mu\rangle = |j-\mu\rangle$, and since $|j\mu\rangle$ and $|j-\mu\rangle$ are degenerate (see (1.54)), we can build the linear combinations $\sqrt{\frac{1}{2}}(|j\mu\rangle \pm |j-\mu\rangle)$ or explicitly

$$|j\mu\rangle^\pm = \frac{(s^\dagger s^\dagger + t^\dagger t^\dagger)^{j-\mu}}{2^j\sqrt{2(j-\mu)!(j+\mu)!}}\{(s^\dagger + it^\dagger)^{2\mu} \pm (s^\dagger - it^\dagger)^{2\mu}\}|0\rangle, \quad \mu > 0 \qquad (1.67)$$

which satisfy

$$\hat{P}|j\mu\rangle^\pm = \pm|j\mu\rangle \qquad (1.68)$$

In Figure 1.3 we show a typical spectrum of (1.53) in this basis. Strictly speaking, the states (1.67) are no longer classified by the algebra SO(2), as they mix the representations μ and $-\mu$. They are, however,

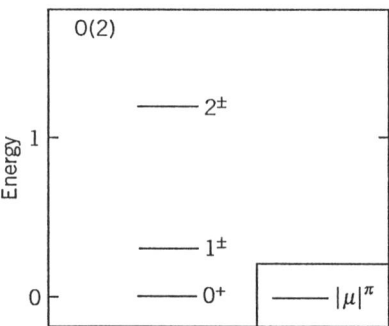

Figure 1.3: The O(2) eigenspectrum of the hamiltonian $\hat{H}_{II}$ with $E'_0(N) = 0$ and $\beta = 0.3$ for $N = 4$ bosons. Levels are labeled by $|\mu|$ and parity π. Energies are in arbitrary units.

1.5 The SO(2) limit

still classified by the two-dimensional orthogonal group O(2), which is the group of rotations and *reflections* in a plane. In the remainder of this chapter we restrict our discussion to the SO(2) states (1.64), since we can always return to the parity-preserving "physical" states through the combinations (1.67).

Before considering the evaluation of the transformation brackets between the bases (1.47) and (1.64), we wish to derive the SO(2) states in a different way which is more generally applicable for higher-dimensional examples, where the chain of algebras $U(n) \supset SO(n)$ is involved. The analysis starts from the states with $j = \mu$ in (1.64) which are called *highest-weight* states and are given by

$$|\mu\mu\rangle = \frac{(s^\dagger + it^\dagger)^{2\mu}}{2^\mu \sqrt{(2\mu)!}} |0\rangle \tag{1.69}$$

We now return to coordinate space and derive the form of the highest-weight states in the r,θ representation (1.12). From (1.31) we have that $j = N/2$ and $\mu = m/2$. Thus $j = \mu$ implies $N = m$ and the wave function (1.12) becomes

$$\Phi_{mm}(r,\theta) = \frac{1}{\sqrt{\pi m!}} r^m e^{-r^2/2} e^{im\theta} \tag{1.70}$$

and since $re^{i\theta} = (x+iy)$, it reduces to

$$\Psi_{\mu\mu}(x,y) = \frac{1}{\sqrt{\pi(2\mu)!}} (x+iy)^{2\mu} e^{-r^2/2} \tag{1.71}$$

where we have replaced m by 2μ.

Comparing (1.69) and (1.71) we note a remarkable resemblance of the expressions. By the identifications

$$\boxed{x \to \frac{s^\dagger}{\sqrt{2}}, \quad y \to \frac{t^\dagger}{\sqrt{2}}, \quad \frac{e^{-r^2/2}}{\sqrt{\pi}} \to |0\rangle} \tag{1.72}$$

we may derive the normalized highest-weight boson states (1.69) if we know their representation in terms of coordinates. A completely analogous result is found in higher-dimensional systems and is known as

Dragt's theorem [6]. The only modification necessary is to introduce the n-dimensional harmonic-oscillator vacuum states $|0\rangle = \pi^{-n/4} e^{-r^2/2}$, where r is the hyperspherical coordinate [2]. In Appendix C we give a general derivation of this theorem which is very useful for evaluating $U(n) \supset SO(n)$ boson states, with n arbitrary.

Once the highest-weight states $|\mu\mu\rangle$ have been constructed by this procedure, we can derive the general states $|j\mu\rangle$ by repeated action of $s^\dagger s^\dagger + t^\dagger t^\dagger$. We note that this operator, as well as its hermitian conjugate, commutes with $\hat{J}_z$,

$$[\hat{J}_z, s^\dagger s^\dagger + t^\dagger t^\dagger] = [\hat{J}_z, ss + tt] = 0 \tag{1.73}$$

implying that their action on states $|j\mu\rangle$ leaves μ unchanged. Furthermore, since $s^\dagger s^\dagger + t^\dagger t^\dagger$ creates two bosons, its application increases the value of j by 1. Thus repeated action of this operator on a highest-weight state gives the most general state

$$|k+\mu\,\mu\rangle \propto (s^\dagger s^\dagger + t^\dagger t^\dagger)^k (s^\dagger + it^\dagger)^{2\mu}|0\rangle \tag{1.74}$$

and in addition, action of its hermitian conjugate gives zero,

$$(ss + tt)^k (s^\dagger + it^\dagger)^{2\mu}|0\rangle = 0 \tag{1.75}$$

The normalized expression for a general state is obtained from the overlap

$$\langle 0|(s-it)^{2\mu}(ss+tt)^k(s^\dagger s^\dagger + t^\dagger t^\dagger)^k(s^\dagger + it^\dagger)^{2\mu}|0\rangle \tag{1.76}$$

which can be calculated by applying the commutation relation

$$[ss+tt, (s^\dagger s^\dagger + t^\dagger t^\dagger)^k] = (s^\dagger s^\dagger + t^\dagger t^\dagger)^{k-1} 4k(\hat{N}+k) \tag{1.77}$$

and with the help of the property (1.75) and the normalization of the highest-weight state (1.69), we arrive again at (1.64).

In order to use the $U(2) \supset SO(2)$ basis for the diagonalization of hamiltonian (1.40) and the evaluation of other "observables," we need again to compute the matrix elements of generators of the dynamical algebra $U(2)$ (or $SU(2)$). We discuss a general approach to the problem in Section 1.7. In the next section we consider an alternative way of doing so by finding the transformation from the $U(2) \supset SO(2)$ to the $U(2) \supset U(1)$ basis.

1.6 Transformation brackets

Having derived explicitly the boson realizations (1.47) and (1.64), it is of interest to find the transformation brackets between the two bases. These are defined by

$$|j\mu\rangle = \sum_{n_t=0}^{2j} \langle jn_t|j\mu\rangle |jn_t\rangle \qquad (1.78)$$

where we use the notation $|jn_t\rangle$ instead of $|[N]n_t\rangle$ to denote the U(1) states. To find their explicit expression, we use (1.64) and carry out the binomial expansions:

$$|j\mu\rangle = \frac{1}{2^j\sqrt{(j-\mu)!(j+\mu)!}} \sum_{k=0}^{j+\mu}\sum_{\ell=0}^{j-\mu} \binom{j+\mu}{k}\binom{j-\mu}{\ell}$$
$$\times (i)^{k+\ell}(-)^{\ell}(s^\dagger)^{2j-(k+\ell)}(t^\dagger)^{k+\ell}|0\rangle \qquad (1.79)$$

Defining $n_t = k + \ell$ and inverting the sums in n_t and k,

$$\sum_{k=0}^{j+\mu}\sum_{n_t=k}^{k+j-\mu} \to \sum_{n_t=0}^{2j}\sum_{k=n_t}^{n_t-j+\mu} \qquad (1.80)$$

we arrive at

$$|j\mu\rangle = \frac{1}{2^j\sqrt{(j-\mu)!(j+\mu)!}} \sum_{n_t=0}^{2j} \left\{ \sum_{k=n_t}^{n_t-j+\mu} \binom{j+\mu}{k}\binom{j-\mu}{n_t-k}(-)^k \right\}$$
$$\times (-i)^{n_t}(s^\dagger)^{2j-n_t}(t^\dagger)^{n_t}|0\rangle \qquad (1.81)$$

which by comparison with (1.47) and (1.78) gives

$$\langle jn_t|j\mu\rangle = \left[\frac{(2j-n_t)!n_t!}{2^{2j}(j-\mu)!(j+\mu)!}\right]^{1/2} (i)^{n_t} \sum_{\ell=0}^{j-\mu}(-)^\ell \binom{j+\mu}{n_t-\ell}\binom{j-\mu}{\ell} \qquad (1.82)$$

where we again introduced the index $\ell = n_t - k$ in the sum. Using (1.82) the matrix elements of U(1) operators in the SO(2) basis can be computed. For example, to diagonalize the general U(2) hamiltonian

(1.40), we need the $\hat{n}_t$ matrix elements which, according to (1.78), are given by

$$\langle j\mu'|\hat{n}_t|j\mu\rangle = \sum_{n_t}\langle jn_t|j\mu\rangle n_t \langle jn_t|j\mu'\rangle^* \qquad (1.83)$$

We have carried out explicitly all computations leading to (1.82), since similar methods will be required to derive transformation brackets in the higher-dimensional models. In the present case there is a simple group-theoretical interpretation which we now briefly discuss.

As mentioned in Section 1.5, the transformation (1.62) which connects the $\hat{J}_x$ eigenstates (1.60) with the $\hat{J}_z$ eigenstates (1.64) can be thought of as an SU(2) rotation of the form (1.61). Using the explicit form of the $\mathbf{D}^{1/2}$ matrices [3], we find

$$\widetilde{\mathbf{D}}^{1/2}(-\pi/2, -\pi/2, 0) = \frac{e^{i\pi/4}}{\sqrt{2}}\begin{pmatrix} 1 & i \\ 1 & -i \end{pmatrix} \qquad (1.84)$$

which is precisely of the form (1.62) with $\alpha = \pi/4$. The rotation carried out by (1.84) for the one-boson states $(j = 1/2)$ indicates that for arbitrary j we should write

$$|j\mu\rangle = \sum_{\sigma} D^j_{\sigma\mu}(-\pi/2, -\pi/2, 0)|j\sigma\rangle \qquad (1.85)$$

The explicit form of the D functions [3] leads back to (1.82), except for a phase factor $e^{ij\pi/2} = (i)^j$ arising from the phase $e^{i\pi/4}$ in (1.84). To compare (1.85) with (1.82), one should substitute $\sigma = j - n_t$ as done after equation (1.59). We see that, due to the simple structure of the U(2) model, the transformation brackets (1.82) may be found in terms of well-known functions. In the final section of this chapter we review some important results for the SU(2) algebra which complement the analysis carried out so far.

1.7 Tensor calculus

The discussion at the end of the last section implies that a detailed knowledge of the U(2) $\supset$ SO(2) chain of algebras, including the matrix elements of its generators, its irreducible representations, and other

1.7 Tensor calculus

properties, gives rise to a great simplification in the quantitative application of the algebraic model. We have shown that the U(1) dynamical symmetry may be thought of as an $\overline{\mathrm{SO}}(2)$ symmetry, where $\hat{J}_x$, instead of $\hat{J}_z$, is chosen to be diagonal. Equation (1.85) is a concise statement of the relationship between the two bases. Without loss of generality we therefore choose the SO(2) states (1.64) to define the tensor calculus associated with the model, following the conventions of the book by Rose [3], and summarize some of its basic results for future reference.

Instead of the cartesian components of the angular momentum generators (1.58), it is convenient to define the raising and lowering operators

$$\boxed{\hat{J}_\pm = \hat{J}_x \pm i\hat{J}_y} \qquad (1.86)$$

which, together with $\hat{J}_z$, satisfy the commutation rules

$$[\hat{J}_z, \hat{J}_\pm] = \pm\hat{J}_\pm, \qquad [\hat{J}_+, \hat{J}_-] = 2\hat{J}_z \qquad (1.87)$$

The U(2) $\supset$ SO(2) states $|j\mu\rangle$ simultaneously diagonalize $\hat{J}^2 = \hat{J}_-\hat{J}_+ + \hat{J}_z(\hat{J}_z + 1)$ and $\hat{J}_z$, with eigenvalues

$$\hat{J}^2|j\mu\rangle = j(j+1)|j\mu\rangle, \qquad \hat{J}_z|j\mu\rangle = \mu|j\mu\rangle \qquad (1.88)$$

where $j = 0, 1/2, 1, \ldots$ and $\mu = -j, -j+1, \ldots, j$. The matrix elements of $\hat{J}_\pm$ are in turn given by

$$\langle j\,\mu\pm 1|\hat{J}_\pm|j\mu\rangle = \sqrt{(j\mp\mu)(j\pm\mu+1)} \qquad (1.89)$$

with all other matrix elements being zero. Any finite rotation $\hat{R}$ acts on the basis (1.88) as $\hat{R}(\theta_1, \theta_2, \theta_3) = e^{-i\theta_1 \hat{J}_z} e^{-i\theta_2 \hat{J}_y} e^{-i\theta_3 \hat{J}_z}$, leading to

$$\hat{R}|j\mu\rangle = \sum_{\mu'}\langle j\mu'|\hat{R}|j\mu\rangle|j\mu'\rangle = \sum_{\mu'} D^j_{\mu'\mu}(\theta_1, \theta_2, \theta_3)|j\mu'\rangle \qquad (1.90)$$

where $D^j_{\mu'\mu}$ are Wigner's D functions and θ_i are the Euler angles that characterize the rotation. One may also define spherical *tensor operators* (or, in short, *tensors*) $\hat{T}^{(j)}_\mu$ that transform under rotations in the same way as the basis states,

$$\hat{R}\hat{T}^{(j)}_\mu \hat{R}^{-1} = \sum_{\mu'} D^j_{\mu'\mu}(\theta_1, \theta_2, \theta_3)\hat{T}^{(j)}_{\mu'} \qquad (1.91)$$

An alternative definition of tensor operators was given by Racah [7]. This definition arises from (1.91) in the case of infinitesimal rotations [3]. Equation (1.91) then translates into

$$[\hat{J}_\pm, \hat{T}^{(j)}_\mu] = \sqrt{(j \mp \mu)(j \pm \mu + 1)}\, \hat{T}^{(j)}_{\mu\pm 1}, \qquad [\hat{J}_z, \hat{T}^{(j)}_\mu] = \mu \hat{T}^{(j)}_\mu \qquad (1.92)$$

Comparing with relations (1.87), we see that the angular momentum operators

$$\hat{J}_0 \equiv \hat{J}_z, \qquad \hat{J}_{\pm 1} \equiv \mp\sqrt{\tfrac{1}{2}}\,\hat{J}_\pm \qquad (1.93)$$

satisfy (1.92) with $j = 1$ and thus are the spherical components of the angular momentum vector.

Denoting by $A^{(j)}_\mu$ both states and tensors, we may construct new states or tensors using *Clebsch–Gordan coefficients* [3],

$$A^{(j)}_\mu = \sum_{\mu_1 \mu_2} \langle j_1 \mu_1\, j_2 \mu_2 | j \mu \rangle A^{(j_1)}_{\mu_1} A^{(j_2)}_{\mu_2} \qquad (1.94)$$

which continue to satisfy (1.90) and (1.91), respectively. The Clebsch–Gordan coefficients are elements of a unitary transformation that connects the SU(2) coupled representation j to the uncoupled one $j_1 \otimes j_2$ and therefore satisfy orthonormality relations in rows and columns, of the form

$$\begin{aligned}
\sum_{\mu_1 \mu_2} \langle j_1 \mu_1\, j_2 \mu_2 | j_a \mu_a \rangle \langle j_1 \mu_1\, j_2 \mu_2 | j_b \mu_b \rangle &= \delta_{j_a j_b} \delta_{\mu_a \mu_b} \\
\sum_{j\mu} \langle j_1 \mu_{1a}\, j_2 \mu_{2a} | j \mu \rangle \langle j_1 \mu_{1b}\, j_2 \mu_{2b} | j \mu \rangle &= \delta_{\mu_{1a} \mu_{1b}} \delta_{\mu_{2a} \mu_{2b}}
\end{aligned} \qquad (1.95)$$

where the phase convention made for the Clebsch–Gordan coefficients is such that they are real. A closed expression for the Clebsch–Gordan coefficient exists [3].

A fundamental result for the evaluation of matrix elements of spherical tensors in the $|j\mu\rangle$ basis is the Wigner–Eckart theorem [3], which states that the dependence of the matrix elements on the SO(2) quantum numbers is entirely contained in the Clebsch–Gordan coefficient:

$$\langle j_a \mu_a | \hat{T}^{(j)}_\mu | j_b \mu_b \rangle = \langle j_b \mu_b\, j \mu | j_a \mu_a \rangle \langle j_a \| \hat{T}^{(j)} \| j_b \rangle \qquad (1.96)$$

1.7 Tensor calculus

The quantity $\langle j_a \| \hat{T}^{(j)} \| j_b \rangle$ is called the *reduced matrix element* of $\hat{T}^{(j)}$. An alternative convention [8], used in nuclear physics, is to define the reduced matrix element as follows:

$$\langle j_a \| \hat{T}^{(j)} \| j_b \rangle_{\mathrm{NP}} = (-)^{2j_a+2j_b} \sqrt{2j_a+1} \langle j_a \| \hat{T}^{(j)} \| j_b \rangle \qquad (1.97)$$

With this convention the Wigner–Eckart theorem reads

$$\langle j_a \mu_a | \hat{T}^{(j)}_\mu | j_b \mu_b \rangle = \frac{(-)^{2j_a+2j_b}}{\sqrt{2j_a+1}} \langle j_b \mu_b \, j\mu | j_a \mu_a \rangle \langle j_a \| \hat{T}^{(j)} \| j_b \rangle_{\mathrm{NP}} \qquad (1.98)$$

The result (1.96) expresses the separation of features associated with the physics of the system from those arising purely from symmetry considerations. The transformation properties (1.90) and (1.91) are reflected in the Clebsch–Gordan coefficient while the reduced matrix element distinguishes the particular operator considered. As an example, the tensor $\hat{T}^{(1)}_\mu = \hat{J}_\mu$ of (1.93), when introduced in (1.96), leads to the result

$$\langle j_a \mu_a | \hat{J}_\mu | j_b \mu_b \rangle = \langle j_b \mu_b \, 1\mu | j_a \mu_a \rangle \langle j_a \| \hat{J} \| j_b \rangle \qquad (1.99)$$

Comparing with (1.89) and inserting the expression for the Clebsch–Gordan coefficient, we find

$$\langle j_a \| \hat{J} \| j_b \rangle = \sqrt{j_a(j_a+1)}\, \delta_{j_a j_b} \qquad (1.100)$$

The Wigner–Eckart theorem (1.96) was first proved in the case considered in this section, namely, for the reduction $SU(2) \supset SO(2)$, but has been generalized to higher-dimensional algebras [5]. Illustrations of this generalization of the Wigner–Eckart theorem are given in Parts II and III.

In order to apply these results to the U(2) boson model, we start from the basic building blocks, the operators $s^\dagger$ and $t^\dagger$, as well as their hermitian conjugate partners s and t. From (1.84), (1.85), and (1.91) we see that the combinations

$$b^\dagger_{1/2\pm 1/2} = \frac{e^{i\pi/4}}{\sqrt{2}}(s^\dagger \pm it^\dagger) \qquad (1.101)$$

constitute the components of an SU(2) spinor tensor. However, its hermitian conjugate counterparts

$$b_{1/2\pm 1/2} = \frac{e^{-i\pi/4}}{\sqrt{2}}(s \mp it) \qquad (1.102)$$

do not transform according to (1.91). To see this point, we take the hermitian conjugate of (1.91) as applied to $b^\dagger_{1/2\mu}$, noting that for orthogonal transformations $\hat{R}^{-1} = \hat{R}^\dagger$,

$$\hat{R} b_{1/2\mu} \hat{R}^{-1} = \sum_{\mu'} D^{1/2}_{\mu'\mu}(\theta_1, \theta_2, \theta_3)^* b_{1/2\mu'} \quad (1.103)$$

We now use the property of the D functions [3],

$$D^{1/2}_{\mu'\mu}(\theta_1, \theta_2, \theta_3)^* = (-)^{\mu'-\mu} D^{1/2}_{-\mu'-\mu}(\theta_1, \theta_2, \theta_3) \quad (1.104)$$

and change the signs of μ and μ' in (1.103), arriving at

$$\hat{R} \left((-)^{1/2+\mu} b_{1/2-\mu} \right) \hat{R}^{-1} = \sum_{\mu'} D^j_{\mu'\mu}(\theta_1, \theta_2, \theta_3) \left((-)^{1/2+\mu'} b_{1/2-\mu'} \right) \quad (1.105)$$

where we have included a factor $(-)^{1/2}$ on both sides of (1.105) for the phases to be real. We see that it is the modified combination

$$\tilde{b}_{1/2\mu} \equiv (-)^{1/2+\mu} b_{1/2-\mu} \quad (1.106)$$

that transforms as a spinor tensor. We may now build higher-order spherical tensors using (1.94). For example,

$$\begin{aligned}
\hat{T}^{(0)}_0 &\equiv [b^\dagger \times \tilde{b}]^{(0)}_0 = \sum_\mu \langle 1/2\mu \; 1/2-\mu | 00 \rangle b^\dagger_\mu \tilde{b}_{-\mu} \\
&= \sqrt{\tfrac{1}{2}}(s^\dagger s + t^\dagger t) = \sqrt{\tfrac{1}{2}} \hat{N} \\
\hat{T}^{(1)}_\mu &\equiv [b^\dagger \times \tilde{b}]^{(1)}_\mu = \sum_{\mu_1\mu_2} \langle 1/2\mu_1 \; 1/2\mu_2 | 1\mu \rangle b^\dagger_{\mu_1} \tilde{b}_{\mu_2} \\
&= \sqrt{2} \hat{J}_\mu
\end{aligned} \quad (1.107)$$

where the fixed $1/2$ spin label is omitted from the boson operators and $b^\dagger_{1/2\mu}$ ($\tilde{b}_{1/2\mu}$) is denoted as $b^\dagger_\mu$ ($\tilde{b}_\mu$). The relations (1.107) show that the boson number $\hat{N}$ and the components $\hat{J}_\mu$ of the angular momentum operator correspond to scalar and vector SU(2) tensors, as expected. We may now compute the matrix elements of $\hat{N}$ and $\hat{J}_\mu$ on the basis of their tensor character under SU(2) and SO(2) together with the application of the Wigner–Eckart theorem (1.96) with equal SU(2) labels in

1.7 Tensor calculus

bra and ket ($j_a = j_b$) since neither $\hat{N}$ nor $\hat{J}_\mu$ can change the number of bosons. Since $\hat{T}_0^{(0)} = \sqrt{\frac{1}{2}}\hat{N}$, we find

$$\langle j\mu_a|\hat{T}_0^{(0)}|j\mu_b\rangle = \langle j\mu_b\,00|j\mu_a\rangle\langle j\,\|\,\hat{T}^{(0)}\,\|\,j\rangle = \sqrt{2}j\delta_{\mu_a\mu_b} \quad (1.108)$$

and hence
$$\langle j\,\|\,\hat{T}^{(0)}\,\|\,j\rangle = \sqrt{2}j \quad (1.109)$$

For a tensor of rank 1 we find, since $\hat{T}_0^{(1)} = \sqrt{2}\hat{J}_0$,

$$\langle j\mu_a|\hat{T}_0^{(1)}|j\mu_b\rangle = \langle j\mu_b\,10|j\mu_a\rangle\langle j\,\|\,\hat{T}^{(1)}\,\|\,j\rangle = \sqrt{2}\mu_a\delta_{\mu_a\mu_b} \quad (1.110)$$

Substituting the value of the Clebsch–Gordan coefficient [3], we derive

$$\langle j\,\|\,\hat{T}^{(1)}\,\|\,j\rangle = \sqrt{2j(j+1)} \quad (1.111)$$

and thus, from (1.96),

$$\langle j\mu_a|\hat{T}_\mu^{(1)}|j\mu_b\rangle = \langle j\mu_b\,1\mu|j\mu_a\rangle\sqrt{2j(j+1)} \quad (1.112)$$

which is equivalent to relations (1.99) and (1.100). A more interesting example concerns the evaluation of the matrix elements of $\hat{n}_t$ in the U(2) $\supset$ SO(2) basis for which an expression was given in (1.83) as a sum over transformation brackets. From (1.101) and (1.102) we have

$$\hat{n}_t = t^\dagger t = \tfrac{1}{2}(b^\dagger_{+1/2}b_{+1/2} - b^\dagger_{+1/2}b_{-1/2} - b^\dagger_{-1/2}b_{+1/2} + b^\dagger_{-1/2}b_{-1/2}) \quad (1.113)$$

which, after comparison with (1.107), gives

$$\hat{n}_t = \tfrac{1}{2}\left(\hat{T}^{(1)}_{+1} - \hat{T}^{(1)}_{-1} + \sqrt{2}\hat{T}^{(0)}_0\right) \quad (1.114)$$

This relation shows that $\hat{n}_t$ has no well-defined tensor properties under SU(2) $\supset$ SO(2) but is given as a combination of different tensors. Use of (1.108) and (1.110) leads to the matrix elements

$$\begin{aligned}\langle j\mu|\hat{n}_t|j\mu\rangle &= j \\ \langle j\,\mu\pm 1|\hat{n}_t|j\mu\rangle &= -\tfrac{1}{2}\sqrt{(j\mp\mu)(j\pm\mu+1)}\end{aligned} \quad (1.115)$$

where the values of the Clebsch–Gordan coefficients [3] have been substituted in (1.112).

In these simple examples the evaluation of the matrix elements turns out to be trivial due to the simple form of the $\hat{T}_\mu^{(j)}$ operators. We did, nevertheless, carry out the calculation in order to illustrate the general procedure followed in the case of a dynamical algebra G_1 and a chain of algebras

$$G_1 \supset G_2 \supset \cdots \supset G_n \qquad (1.116)$$

This procedure consists of the following steps (compare also with Appendix A):

1. Express the physical operators as tensor operators under the algebras in (1.116) (or linear combinations of them).

2. Use the Wigner–Eckart theorem (or its generalization) to express the matrix elements of the tensor operators in terms of algebraic coefficients (generalized coupling coefficients) and reduced matrix elements. The coefficients depend only on the tensor character of the operators while the reduced matrix elements depend on the particular operator under consideration.

3. Evaluate the reduced matrix elements by considering the simplest possible component of the tensor operator.

We shall repeatedly follow the steps outlined above in the examples discussed in this book.

In the next chapter we consider the coupling of two different U(2) boson systems, and the tensor calculus presented in this section will prove essential for the analysis of the model.

Summary

In this chapter we introduce many of the ideas that are basic to the algebraic models discussed in this book. The starting point is the two-dimensional harmonic oscillator with the hamiltonian (1.1), which can be solved in cartesian and in spherical polar coordinates with solutions given in (1.5) and (1.12), respectively. A set of s and t bosons (1.13) can be defined as linear combinations of the x and y coordinates and conjugate momenta, which satisfy the basic boson commutation relations (1.15). The harmonic-oscillator hamiltonian can be rewritten in terms of the s and t bosons as (1.16) and furthermore, operators $\mathcal{G}_i^j$ in (1.17) can be introduced which commute with the harmonic-oscillator hamiltonian (1.18) and satisfy the commutation relations (1.19). These commutators define a U(2) algebra which is thus the symmetry algebra of the two-dimensional harmonic oscillator. The harmonic-oscillator hamiltonian describes a system of non-interacting bosons and can be generalized to (1.36), which includes all two-body interactions between s and t bosons. In this case U(2) is the dynamical algebra of the system. The interacting boson hamiltonian can be written in the alternative form (1.40) which is more useful for the study of its analytical solutions. Two such solutions exist in the U(2) model. The first is the U(1) limit corresponding to the hamiltonian (1.42) with boson wave functions (1.47) classified by (1.45); the second is the SO(2) limit with the hamiltonian (1.53) and with wave functions (1.64) or (1.65) classified by (1.56). The boson wave functions of the two limits are intimately connected with two different coordinate representations of the harmonic-oscillator wave functions (i.e., in cartesian or polar coordinates). For highest-weight states the boson expression can be derived directly from the coordinate wave function with the identifications (1.72). This is known as Dragt's theorem. In the last section of this chapter the U(2) $\supset$ SO(2) tensor calculus is developed. This consists of defining the raising and lowering operators (1.87) which in turn define tensor operators through the commutation relations (1.92). States or tensors can now be coupled using Clebsch–Gordan coefficients, (1.94), which satisfy the familiar orthonormality relations (1.95). Another fundamental result of the U(2) $\supset$ SO(2) tensor calculus is the Wigner–Eckart theorem (1.96), which states that the dependence of matrix elements on the SO(2) quantum numbers is entirely contained in a Clebsch–Gordan coefficient. This result leads to considerable simplifications in the calculation of matrix elements in a U(2) $\supset$ SO(2) basis.

References

1. I.S. Gradshteyn and I.M. Ryzhik, *Table of Integrals, Series, and Products*, Academic, New York, 1965.

2. M. Moshinsky, *Group Theory and the Many-Body Problem*, Gordon & Breach, New York, 1968.
3. M.E. Rose, *Elementary Theory of Angular Momentum*, Wiley, New York, 1957.
4. M. Moshinsky, *The Harmonic Oscillator in Modern Physics: From Atoms to Quarks*, Gordon & Breach, New York, 1969.
5. B.G. Wybourne, *Classical Groups for Physicists*, Wiley-Interscience, New York, 1974.
6. A.J. Dragt, "Classification of three-particle states according to SU_3," J. Math. Phys. **6** (1965) 533.
7. G. Racah, "Theory of complex spectra. I," Phys. Rev. **61** (1942) 186.
8. A. de-Shalit and I. Talmi, *Nuclear Shell Theory*, Academic, New York, 1963.

Chapter 2

Non-identical Bosons

2.1 The $U_1(2) \otimes U_2(2)$ algebra

In Chapter 1 we analyzed the group-theoretical structure of the U(2) algebraic model whose building blocks are the s and t bosons, characterized by their null angular momentum and their positive and negative parity, respectively. This system by itself is not trivial and already displays a number of interesting properties shared by the higher-dimensional models discussed in this book.

In this chapter we consider the coupling of two independent U(2) systems of the kind studied before. The reason for carrying out such an analysis is again mainly pedagogical for the U(2) model, although in Section 2.7 we consider an application to the stretching modes of X–Y–X molecules [1]. For the U(4) and U(6) models the coupling of two (or more) algebras becomes necessary for the description of additional degrees of freedom in molecules and nuclei, respectively, and hence is of relevance in practical calculations.

We start by defining the boson creation and annihilation operators $s_\rho^\dagger$, s_ρ, $t_\rho^\dagger$, and t_ρ, with $\rho = 1, 2$, where

$$[s_\rho, s_{\rho'}^\dagger] = [t_\rho, t_{\rho'}^\dagger] = \delta_{\rho\rho'}, \qquad \rho, \rho' = 1, 2 \qquad (2.1)$$

with all other commutators being zero. This set is equivalent to two (independent) copies of the operators (1.13). Likewise we can define the set of generators

$$\mathcal{G}^{\rho 1}_{\rho 1} \equiv s^\dagger_\rho s_\rho, \quad \mathcal{G}^{\rho 2}_{\rho 1} \equiv s^\dagger_\rho t_\rho, \quad \mathcal{G}^{\rho 1}_{\rho 2} \equiv t^\dagger_\rho s_\rho, \quad \mathcal{G}^{\rho 2}_{\rho 2} \equiv t^\dagger_\rho t_\rho \qquad (2.2)$$

which satisfy the following commutation relations among themselves:

$$[\mathcal{G}^{\rho j}_{\rho i}, \mathcal{G}^{\rho' l}_{\rho' k}] = (\mathcal{G}^{\rho l}_{\rho i}\delta_{jk} - \mathcal{G}^{\rho j}_{\rho k}\delta_{il})\delta_{\rho\rho'}, \qquad i,j,k,l = 1,2 \qquad (2.3)$$

For $\rho = \rho' = 1$ or 2 we recover the commutation relations for the U(2) Lie algebra (1.19), while for $\rho \neq \rho'$ we find

$$[\mathcal{G}^{1j}_{1i}, \mathcal{G}^{2l}_{2k}] = 0, \qquad i,j,k,l = 1,2 \qquad (2.4)$$

As discussed after equation (1.22), relations (2.3) and (2.4) define the direct product of two U(2) algebras, which we denote by

$$U_1(2) \otimes U_2(2) \qquad (2.5)$$

Since (2.4) implies that the two algebras above are independent, we may also define the $SU_\rho(2)$ generators (see (1.20))

$$\hat{J}_{\rho x} \equiv \tfrac{1}{2}(s^\dagger_\rho s_\rho - t^\dagger_\rho t_\rho), \quad \hat{J}_{\rho y} \equiv \tfrac{1}{2}(t^\dagger_\rho s_\rho + s^\dagger_\rho t_\rho), \quad \hat{J}_{\rho z} \equiv \tfrac{i}{2}(t^\dagger_\rho s_\rho - s^\dagger_\rho t_\rho) \quad (2.6)$$

together with the boson number operators

$$\hat{N}_\rho \equiv s^\dagger_\rho s_\rho + t^\dagger_\rho t_\rho \qquad (2.7)$$

which satisfy commutation relations analogous to (1.21) and (1.22) for $\rho = 1$ or 2 plus the relations

$$[\hat{J}_{1i}, \hat{J}_{2j}] = [\hat{J}_{1i}, \hat{N}_2] = [\hat{N}_1, \hat{J}_{2j}] = [\hat{N}_1, \hat{N}_2] = 0, \quad i,j = x,y,z \qquad (2.8)$$

arising from (2.4). From (1.23) we also have

$$\hat{J}^2_\rho = \tfrac{1}{4}\hat{N}_\rho(\hat{N}_\rho + 2) \qquad (2.9)$$

which implies

$$j_\rho = N_\rho/2 \qquad (2.10)$$

We may thus consider, instead of (2.5), the direct-product algebra

$$SU_1(2) \otimes SU_2(2) \qquad (2.11)$$

for the description of the system. The $U_\rho(2)$ algebras in (2.5) are characterized by the boson numbers N_ρ while the $SU_\rho(2)$ are labeled by j_ρ.

2.2 The $U_1(2) \otimes U_2(2)$ hamiltonian

In this section we turn our attention to the hamiltonian of the coupled system of interacting s_1, t_1, s_2, and t_2 bosons. As for the U(2) system we assign positive parity to the s_ρ bosons and negative parity to the t_ρ bosons. As implied by the dynamical algebra (2.5) or (2.11), a further condition is the separate conservation of the number of bosons of type 1 and type 2, N_1 and N_2. In other words, no generators of the form $b_{1i}^\dagger b_{2j}$ or $b_{2i}^\dagger b_{1j}$—with $b_{\rho 1}^\dagger = s_\rho^\dagger$ and $b_{\rho 2}^\dagger = t_\rho^\dagger$ and likewise for the annihilation operators—belong to the dynamical algebra (2.5). Under this restriction the most general one- and two-body, hermitian, parity-invariant hamiltonian is given by

$$\hat{H} = \hat{H}_1 + \hat{H}_2 + \hat{V}_{12} \tag{2.12}$$

with $\hat{H}_1$ and $\hat{H}_2$ given by (1.36) (under the appropriate substitution of s and t bosons by the corresponding bosons of type 1 or 2) and the interaction $\hat{V}_{12}$ given by

$$\begin{aligned}\hat{V}_{12} =\ & v_1 s_1^\dagger s_2^\dagger s_1 s_2 + v_2 t_1^\dagger t_2^\dagger t_1 t_2 + v_3(s_1^\dagger s_2^\dagger t_1 t_2 + t_1^\dagger t_2^\dagger s_1 s_2) \\ & + v_4 s_1^\dagger t_2^\dagger s_1 t_2 + v_5 t_1^\dagger s_2^\dagger t_1 s_2 + v_6(s_1^\dagger t_2^\dagger t_1 s_2 + t_1^\dagger s_2^\dagger s_1 t_2)\end{aligned} \tag{2.13}$$

where the v_i are two-body matrix elements representing the interaction between the two kinds of bosons. It should be clear that for $v_i = 0$ (i.e., no interaction between bosons of different type) the hamiltonian (2.12) describes two independent U(2) boson systems. In this case the wave functions are simple products of the two U(2) states associated with $\hat{H}_1$ and $\hat{H}_2$. The interaction term $\hat{V}_{12}$ has in general the effect of mixing different states, as we discuss in the next section.

2.3 Symmetry limits

As for the simpler U(2) system the solution of the boson hamiltonian (2.12) requires the construction of a complete set of states in which to carry out its diagonalization. We discuss in this section the existence of dynamical symmetries for which the eigensolutions of (2.12) can be obtained algebraically and which provide suitable bases for the evaluation of matrix elements of the general one- and two-body hamiltonian. As mentioned at the end of Section 2.2, we already have at our disposal two such bases defined by the chains of algebras

$$\begin{array}{ccccc} U_1(2) \otimes U_2(2) & \supset & U_1(1) \otimes U_2(1) & \supset & U_{12}(1) \\ | & | & | & | & | \\ [N_1] & [N_2] & n_{t_1} & n_{t_2} & n_t \end{array} \tag{2.14}$$

and

$$\begin{array}{ccccc} U_1(2) \otimes U_2(2) & \supset & SO_1(2) \otimes SO_2(2) & \supset & SO_{12}(2) \\ | & | & | & | & | \\ [N_1] & [N_2] & \mu_1 & \mu_2 & \mu \end{array} \tag{2.15}$$

They correspond to the product of two $U(2) \supset U(1)$ or $U(2) \supset SO(2)$ states and thus can be written in terms of the bases introduced in Chapter 1 as

$$|[N_1]n_{t_1}, [N_2]n_{t_2}; n_t\rangle \equiv |[N_1]n_{t_1}\rangle |[N_2]n_{t_2}\rangle \tag{2.16}$$

and

$$|j_1\mu_1, j_2\mu_2; \mu\rangle \equiv |j_1\mu_1\rangle |j_2\mu_2\rangle \tag{2.17}$$

where in (2.17) we use the $SU(2) \supset SO(2)$ notation as discussed in Section 1.1 (see (1.31)). Note that, since both the U(1) and SO(2) labels are additive, the total number of t bosons $n_t = n_{t_1} + n_{t_2}$ and the total $SO_{12}(2)$ projection $\mu = \mu_1 + \mu_2$ are well-defined and hence we include these subalgebras in the chains (2.14) and (2.15), respectively. The labels n_t and μ, however, often will be omitted from the states (2.16) and (2.17). The corresponding wave functions not only diagonalize (2.12) if $V_{12} = 0$ but also are associated with particular interactions between the bosons of the two types, involving the invariant operators

$$\hat{n}_t \equiv \hat{n}_{t_1} + \hat{n}_{t_2} = t_1^\dagger t_1 + t_2^\dagger t_2 \tag{2.18}$$

and

$$\hat{J}_z \equiv \hat{J}_{1z} + \hat{J}_{2z} = \tfrac{i}{2}(t_1^\dagger s_1 - s_1^\dagger t_1) + \tfrac{i}{2}(t_2^\dagger s_2 - s_2^\dagger t_2) \tag{2.19}$$

2.3 Symmetry limits

for (2.14) and (2.15), respectively. These coupled hamiltonians may be written in a general form as

$$\hat{H}_{Ib} = \epsilon_1 \hat{n}_{t_1} + \epsilon_2 \hat{n}_{t_2} + \alpha_1 \hat{n}_{t_1}^2 + \alpha_2 \hat{n}_{t_2}^2 + \alpha \hat{n}_t^2 \qquad (2.20)$$

and

$$\hat{H}_{IIb} = \beta_1 \hat{J}_{1z}^2 + \beta_2 \hat{J}_{2z}^2 + \beta \hat{J}_z^2 \qquad (2.21)$$

where we have omitted terms that do not contribute to the excitation energy of the system or that may be effectively incorporated in the ones given in (2.20) or (2.21). The corresponding energy eigenvalues are given by

$$E_{Ib}(n_{t_1}, n_{t_2}, n_t) = \epsilon_1 n_{t_1} + \epsilon_2 n_{t_2} + \alpha_1 n_{t_1}^2 + \alpha_2 n_{t_2}^2 + \alpha n_t^2 \qquad (2.22)$$

and

$$E_{IIb}(\mu_1, \mu_2, \mu) = \beta_1 \mu_1^2 + \beta_2 \mu_2^2 + \beta \mu^2 \qquad (2.23)$$

The dynamical-symmetry hamiltonians (2.20) and (2.21) are associated with a very simple coupling of the two s–t systems, which does not change the structure of wave functions of the non-interacting system with $\hat{V}_{12} = 0$. Note that, as a consequence of the lack of well-defined parity for the $SO_1(2)$ and $SO_2(2)$ states, the eigenstates (2.15) do not carry good parity. From Section 1.5 (see relations (1.67) and (1.68)) we know that it is possible to construct states with good parity for both sets of boson operators:

$$|j_\rho \mu_\rho\rangle^\pm = \frac{1}{\sqrt{2(1+\delta_{\mu_\rho 0})}} \left(|j_\rho \mu_\rho\rangle \pm |j_\rho - \mu_\rho\rangle \right) \qquad (2.24)$$

For $\mu_\rho = 0$ only the + combination survives. From (2.15) and (2.24) it is straightforward to show that the same linear combinations of the coupled $SO_1(2) \otimes SO_2(2)$ states

$$|j_1\mu_1, j_2\mu_2\rangle^\pm = \frac{1}{\sqrt{2(1+\delta_{\mu_10}\delta_{\mu_20})}} \left(|j_1\mu_1, j_2\mu_2\rangle \pm |j_1-\mu_1, j_2-\mu_2\rangle \right) \qquad (2.25)$$

have the correct transformation properties under the inversion operator. As explained in Section 1.5, these states are no longer classified by the $SO_{12}(2)$ algebra but rather by the orthogonal algebra $O_{12}(2)$.

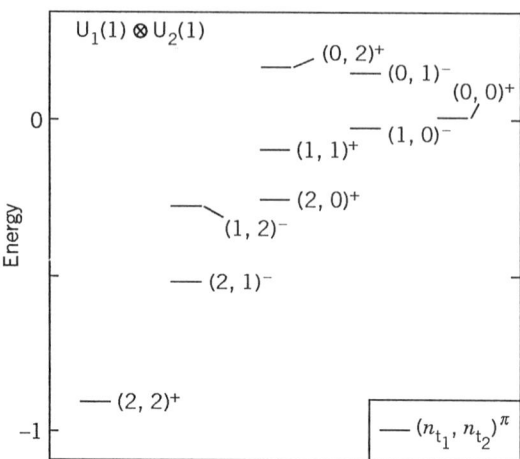

Figure 2.1: The $U_1(1) \otimes U_2(1)$ eigenspectrum of the hamiltonian $\hat{H}_{Ib}$ with $\epsilon_1 = 0.05$, $\epsilon_2 = 0.2$, $\alpha_1 = 0.01$, $\alpha_2 = 0.04$, and $\alpha = -0.1$ for $N_1 = N_2 = 2$ bosons. Levels are labeled by n_{t_1}, n_{t_2}, and parity $\pi = (-)^{n_{t_1}+n_{t_2}} = (-)^{n_t}$. Energies are in arbitrary units.

In Figures 2.1 and 2.2 we show examples of the energy spectra in the $U_1(1) \otimes U_2(1)$ and the $SO_1(2) \otimes SO_2(2)$ limits. We note that, taking into account the degeneracies in the $SO_1(2) \otimes SO_2(2)$ limit, the same number of levels is obtained in the two figures.

Are there other dynamical symmetries present in the $U_1(2) \otimes U_2(2)$ model? To answer this question, we note that, due to (2.4), the sum of the $U_1(2)$ and $U_2(2)$ generators

$$\boxed{\mathcal{G}_i^j \equiv \mathcal{G}_{1i}^{1j} + \mathcal{G}_{2i}^{2j}} \tag{2.26}$$

again satisfies the U(2) commutation relations (1.19). The four operators (2.26) thus generate a U(2) subalgebra of the dynamical algebra (2.5). This coupled $U_{12}(2)$ algebra again has $U_{12}(1)$ and $SO_{12}(2)$ subalgebras, generated by the operators (2.18) and (2.19), respectively. We thus arrive at the chains of algebras

$$U_1(2) \otimes U_2(2) \supset U_{12}(2) \supset U_{12}(1) \tag{2.27}$$

2.3 Symmetry limits

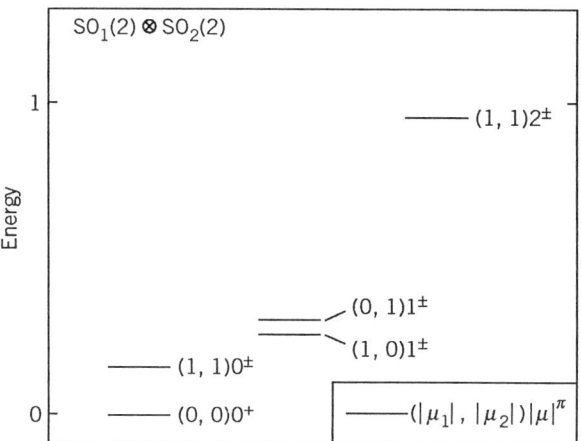

Figure 2.2: The $SO_1(2) \otimes SO_2(2)$ eigenspectrum of the hamiltonian $\hat{H}_{IIb}$ with $\beta_1 = 0.05$, $\beta_2 = 0.1$, and $\beta = 0.2$ for $N_1 = N_2 = 2$ bosons. Levels are labeled by $|\mu_1|$, $|\mu_2|$, $|\mu|$, and parity π. Energies are in arbitrary units.

and
$$U_1(2) \otimes U_2(2) \supset U_{12}(2) \supset SO_{12}(2) \qquad (2.28)$$

which, because of the $U_\rho(2) \simeq SU_\rho(2) \otimes U_\rho(1)$ isomorphism discussed in Chapter 1, may also be expressed as

$$\begin{array}{cccc} SU_1(2) \otimes SU_2(2) & \supset & SU_{12}(2) & \supset & U_{12}(1) \\ | & | & | & | \\ j_1 & j_2 & j & n_t \end{array} \qquad (2.29)$$

and

$$\begin{array}{cccc} SU_1(2) \otimes SU_2(2) & \supset & SU_{12}(2) & \supset & SO_{12}(2) \\ | & | & | & | \\ j_1 & j_2 & j & \mu \end{array} \qquad (2.30)$$

In the form (2.29) and (2.30) we can find the corresponding ramification rules by means of well-known results of angular momentum coupling [2]:

$$j = |j_1 - j_2|, |j_1 - j_2| + 1, \ldots, j_1 + j_2, \quad \mu = -j, -j+1, \ldots, j \quad (2.31)$$

To find the appropriate rule for the $SU_{12}(2) \supset U_{12}(1)$ reduction, we may use the $\overline{SO}_{12}(2)$ algebra generated by

$$\hat{J}_x \equiv \hat{J}_{1x} + \hat{J}_{2x} \tag{2.32}$$

defined in (1.59), which is isomorphic to the $U_{12}(1)$ algebra. Denoting by σ the label associated with $\hat{J}_x$, it is clear that

$$j \geq \sigma \geq -j \tag{2.33}$$

and from (2.32)

$$\sigma = \sigma_1 + \sigma_2 \tag{2.34}$$

where, from (1.59),

$$\sigma_1 = \tfrac{1}{2}(N_1 - 2n_{t_1}), \qquad \sigma_2 = \tfrac{1}{2}(N_2 - 2n_{t_2}) \tag{2.35}$$

From (2.18) and (2.33)–(2.35) we find

$$\tfrac{1}{2}(N + 2j) \geq n_t \geq \tfrac{1}{2}(N - 2j) \tag{2.36}$$

where

$$N \equiv N_1 + N_2 = 2j_1 + 2j_2 \tag{2.37}$$

is the total number of bosons.

Relations (2.31) and (2.36) are the branching rules for the chains (2.29) and (2.30). What about chains (2.27) and (2.28)? From (2.36) we realize that the $U_{12}(2)$ representations can no longer be labeled by a single number N as is the case for the s–t model of Chapter 1. We need *two* labels, which can be taken to be the N and j discussed above or, as is more common, define the integers

$$h \equiv \tfrac{1}{2}(N + 2j), \qquad h' \equiv \tfrac{1}{2}(N - 2j) \tag{2.38}$$

for which the $U_{12}(2) \supset U_{12}(1)$ reduction is

$$h \geq n_t \geq h' \tag{2.39}$$

Note that the representations of Chapter 1 are recovered if $j = N/2$ (see (1.24)). This equation is generalized to the case of two-row representations by taking the difference of the two equations in (2.38):

$$j = \tfrac{1}{2}(h - h') \tag{2.40}$$

2.3 Symmetry limits

The U(2) quantum numbers $[h, h']$ are referred to as *Gel'fand labels* and appear in the reduction

$$\begin{array}{cccc} U_1(2) \otimes U_2(2) \supset U_{12}(2) \supset U_{12}(1) \\ | & | & | & | \\ [N_1] & [N_2] & [h,h'] & n_t \end{array} \quad (2.41)$$

with, due to (2.10), (2.31), and (2.38), the following product rule applies:

$$\begin{aligned}[] [N_1] \otimes [N_2] &= [N_1 + N_2, 0] \oplus [N_1 + N_2 - 1, 1] \oplus \cdots \\ &\cdots \oplus \begin{cases} [N_1, N_2] & \text{if } N_1 \geq N_2 \\ [N_2, N_1] & \text{if } N_1 \leq N_2 \end{cases} \end{aligned} \quad (2.42)$$

Relations (2.42) and (2.40) are particular examples of the reduction rules for $U_1(n) \otimes U_2(n) \supset U_{12}(n)$ and $U(n) \supset U(n-1)$ discussed in Appendix B. For this case ($n = 2$) we have been able to establish them by direct application of angular momentum theory, through relations (2.31). We now wish to analyze the chain (2.41) from a different perspective, which is more general since the method can be readily extended to the n-dimensional case [3].

To simplify the notation in this discussion, we define the boson operators $b^\dagger_{\rho 1} = s^\dagger_\rho$ and $b^\dagger_{\rho 2} = t^\dagger_\rho$ and likewise for the annihilation operators. Now consider the $U_{12}(2)$ generators $\mathcal{G}^j_i = \sum_\rho b^\dagger_{\rho i} b_{\rho j}$ which are also defined in (2.26). As a basis for the $U_{12}(2)$ algebra we can take the set of linearly independent, homogeneous polynomials $\mathcal{Q}(b^\dagger_{\rho k})$ of degree N in the $b^\dagger_{\rho k}$, with the scalar product

$$\begin{aligned} (\mathcal{Q}_1(b^\dagger_{\rho k}), \mathcal{Q}_2(b^\dagger_{\rho k})) &\equiv \langle 0 | \left(\mathcal{Q}_1(b^\dagger_{\rho k})\right)^\dagger \mathcal{Q}_2(b^\dagger_{\rho k}) | 0 \rangle \\ &= \langle 0 | \mathcal{Q}_1(b_{\rho k}) \mathcal{Q}_2(b^\dagger_{\rho k}) | 0 \rangle \end{aligned} \quad (2.43)$$

This product can be evaluated using the commutation relations (2.1). We may characterize these polynomials by their degree ω_i in the component i of $b^\dagger_{\rho i}$ (i.e., the total number of s bosons ($i = 1$) or t bosons ($i = 2$)). The $\mathcal{Q}(b^\dagger_{\rho k})$ satisfy

$$\mathcal{G}^i_i \mathcal{Q}(b^\dagger_{\rho k}) | 0 \rangle = \omega_i \mathcal{Q}(b^\dagger_{\rho k}) | 0 \rangle \quad (2.44)$$

and each polynomial is characterized by the set of numbers $[\omega_1, \omega_2]$ which we refer to as *weights*. For two given polynomials $\mathcal{Q}(b^\dagger_{\rho k})$ and $\mathcal{P}(b^\dagger_{\rho k})$, with weights $[\omega_1, \omega_2]$ and $[v_1, v_2]$ in $U_{12}(2)$, we say that $\mathcal{Q}(b^\dagger_{\rho k})$ is of *higher weight* than $\mathcal{P}(b^\dagger_{\rho k})$ if in $[\omega_1 - v_1, \omega_2 - v_2]$ the first non-vanishing term is positive [3]. It is now easy to check that $\mathcal{G}^2_1 \mathcal{Q}(b^\dagger_{\rho k})$ has weight $[\omega_1 + 1, \omega_2 - 1]$, so it is of higher weight than $\mathcal{Q}(b^\dagger_{\rho k})$, while $\mathcal{G}^1_2 \mathcal{Q}(b^\dagger_{\rho k})$ has weight $[\omega_1 - 1, \omega_2 + 1]$, which is of lower weight than $\mathcal{Q}(b^\dagger_{\rho k})$. This means that the $U_{12}(2)$ generators can be divided in three classes:

$$\begin{array}{ll} \mathcal{G}^i_i & \text{or } \textit{weight generators} \\ \mathcal{G}^2_1 & \text{or } \textit{raising generator} \\ \mathcal{G}^1_2 & \text{or } \textit{lowering generator} \end{array} \qquad (2.45)$$

Since we have a finite basis (the degree of the homogeneous polynomials is N), there will be a *highest-weight* polynomial, $\mathcal{Q}_{\mathrm{hw}}(b^\dagger_{\rho k})$, characterized by

$$\begin{aligned} \mathcal{G}^1_1 \mathcal{Q}_{\mathrm{hw}}(b^\dagger_{\rho k}) |0\rangle &= h \mathcal{Q}_{\mathrm{hw}}(b^\dagger_{\rho k}) |0\rangle \\ \mathcal{G}^2_2 \mathcal{Q}_{\mathrm{hw}}(b^\dagger_{\rho k}) |0\rangle &= h' \mathcal{Q}_{\mathrm{hw}}(b^\dagger_{\rho k}) |0\rangle \\ \mathcal{G}^2_1 \mathcal{Q}_{\mathrm{hw}}(b^\dagger_{\rho k}) |0\rangle &= 0 \end{aligned} \qquad (2.46)$$

and we may use this highest weight $[h, h']$ to characterize the $U_{12}(2)$ representations. From the definition of the $\mathcal{G}^i_i$ we see that, for the highest-weight state, h and h' coincide with n_s and $n_t = N - n_s$, respectively. We now compute the scalar product

$$(\mathcal{G}^1_2 \mathcal{Q}_{\mathrm{hw}}(b^\dagger_{\rho k}), \mathcal{G}^1_2 \mathcal{Q}_{\mathrm{hw}}(b^\dagger_{\rho k})) = \langle 0 | \mathcal{Q}(b_{\rho k}) \mathcal{G}^2_1 \mathcal{G}^1_2 \mathcal{Q}_{\mathrm{hw}}(b^\dagger_{\rho k}) |0\rangle \qquad (2.47)$$

which equals, because of (2.46) and (1.19),

$$\begin{aligned} \langle 0 | \mathcal{Q}_{\mathrm{hw}}(b_{\rho k}) (\mathcal{G}^2_1 \mathcal{G}^1_2 &- \mathcal{G}^1_2 \mathcal{G}^2_1) \mathcal{Q}_{\mathrm{hw}}(b^\dagger_{\rho k}) |0\rangle \\ &= \langle 0 | \mathcal{Q}_{\mathrm{hw}}(b_{\rho k}) [\mathcal{G}^2_1, \mathcal{G}^1_2] \mathcal{Q}_{\mathrm{hw}}(b^\dagger_{\rho k}) |0\rangle \\ &= \langle 0 | \mathcal{Q}_{\mathrm{hw}}(b_{\rho k}) (\mathcal{G}^1_1 - \mathcal{G}^2_2) \mathcal{Q}_{\mathrm{hw}}(b^\dagger_{\rho k}) |0\rangle \\ &= (h - h')(\mathcal{Q}_{\mathrm{hw}}(b^\dagger_{\rho k}), \mathcal{Q}_{\mathrm{hw}}(b^\dagger_{\rho k})) \end{aligned} \qquad (2.48)$$

Since a scalar product is non-negative and the sum of the weights is the degree of the polynomial, we have

$$h \geq h', \qquad h + h' = N \qquad (2.49)$$

2.3 Symmetry limits

which is equivalent to (2.42). Relations (2.49) define a *Young partition* of the number N [3], which is also denoted by the *Young tableau*

$$\underbrace{\square\square\square\cdots\square}_{h} \atop \underbrace{\square\square\cdots\square}_{h'} \qquad (2.50)$$

with N squares in total [3,4]. From relations (2.38) and (2.49) we see that the U(2) model of Chapter 1 has $h = N$ and $h' = 0$ (since $j = N/2$) and corresponds to the Young tableau

$$\underbrace{\square\square\cdots\square}_{h=N} \qquad (2.51)$$

This diagram labels the *totally symmetric* representation of U(2), which indicates that, for a single kind of boson, the wave functions are fully symmetric under the interchange of any pair. For two kinds of bosons, however, as in the $U_1(2) \otimes U_2(2)$ model, the wave functions are not necessarily symmetric under the interchange of bosons of set 1 with those of set 2 [4]. For a given highest-weight polynomial $\mathcal{Q}_{\text{hw}}(b^\dagger_{\rho k})([h, h'])|0\rangle$ there will be lower-weight ones which may be obtained by successive application of the lowering generator $\mathcal{G}^1_2$. The highest-weight polynomial is analogous to the highest-weight SU(2) state $|jj\rangle$ while the lower-weight ones correspond to $|j\mu\rangle$ with $\mu < j$.

Before concluding this section, we wish to point out that it is possible to express the $U_1(2) \otimes U_2(2)$ hamiltonian (2.12) in terms of the invariant operators associated with the dynamical symmetry (2.14), (2.15), (2.29), and (2.30). We omit the explicit expressions which may be derived by straightforward, albeit tedious, calculations. To find the eigenstates and eigenvalues of (2.12), one may use the complete set of states associated with any of the chains and compute the matrix elements of all the operators in the hamiltonian. Before we discuss their evaluation, we return to the dynamical symmetries (2.27) and (2.28).

2.4 The $U_{12}(1)$ limit

In the last section the eigenspectra and wave functions of the $U_1(1) \otimes U_2(1)$ and $SO_1(2) \otimes SO_2(2)$ dynamical symmetries were analyzed. In this section we turn our attention to the $U_{12}(1)$ limit.

If in the model's hamiltonian (2.12) we take the particular choice

$$\hat{H}_{\text{Ia}} = \gamma \hat{J}^2 + \epsilon \hat{n}_t + \alpha \hat{n}_t^2 \qquad (2.52)$$

(omitting terms that depend only on $\hat{N}_1$, $\hat{N}_2$, or $\hat{N}$), we obtain the $U_{12}(1)$ dynamical-symmetry hamiltonian. Its eigenvalues are

$$E_{\text{Ia}}(j, n_t) = \gamma j(j+1) + \epsilon n_t + \alpha n_t^2 \qquad (2.53)$$

where, due to (2.10), (2.31), and (2.36),

$$j = \tfrac{1}{2}N, \tfrac{1}{2}N - 1, \ldots, \tfrac{1}{2}|N_1 - N_2| \qquad (2.54)$$

and

$$\tfrac{1}{2}(N + 2j) \geq n_t \geq \tfrac{1}{2}(N - 2j) \qquad (2.55)$$

An example of a $U_{12}(1)$ spectrum (2.53) is shown in Figure 2.3.

Comparing the $U_{12}(1)$ spectrum with Figures 2.1 and 2.2, where the $U_1(1) \otimes U_2(1)$ and $SO_1(2) \otimes SO_2(2)$ spectra are shown, we note that again the same number of states is obtained. There is, however, an additional correspondence between the $U_1(1) \otimes U_2(1)$ and $U_{12}(1)$ limits since in both cases levels can be labeled by n_t ($= n_{t_1} + n_{t_2}$ in Figure 2.1) and, in fact, in both cases the same number of levels with a given n_t is found (e.g., one $n_t^\pi = 0^+$ level, two $n_t^\pi = 1^-$ levels, etc.). This, of course, is not a coincidence but follows from the presence of $U_{12}(1)$ in the chain of algebras (2.14) and in the one that classifies the eigenstates of the hamiltonian $\hat{H}_{\text{Ia}}$,

$$\boxed{\begin{array}{cccc} U_1(2) \otimes U_2(2) & \supset U_{12}(2) & \supset U_{12}(1) \\ | & | & | & | \\ [N_1] & [N_2] & [h, h'] & n_t \end{array}} \qquad (2.56)$$

In the form (2.29) we have shown that one may resort to the isomorphic $\overline{SO}_{12}(2)$ subalgebra of $SU_{12}(1)$ to arrive at the reduction rules (2.33) and (2.36), with σ given by relations (2.34) and (2.35),

$$\sigma = \tfrac{1}{2}(N - 2n_t) \qquad (2.57)$$

2.4 The $U_{12}(1)$ limit

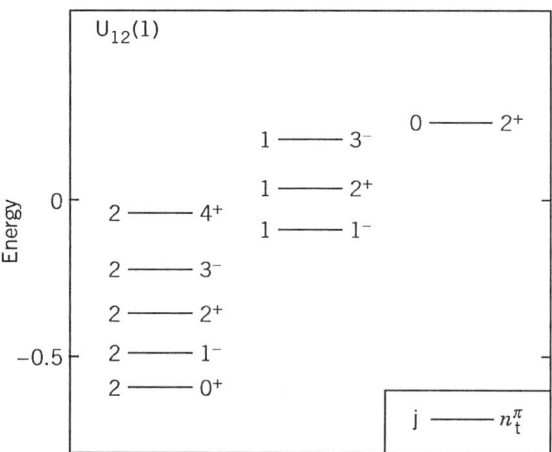

Figure 2.3: The $U_{12}(1)$ eigenspectrum of the hamiltonian $\hat{H}_{1a}$ with $\gamma = -0.1$, $\epsilon = 0.1$, and $\alpha = 0.01$ for $N_1 = N_2 = 2$ bosons. Levels are labeled by j, n_t, and parity $\pi = (-)^{n_t}$. Energies are in arbitrary units.

We may construct the appropriate wave functions by angular momentum coupling,

$$|j_1, j_2; j\sigma\rangle = \sum_{\sigma_1 \sigma_2} \langle j_1\sigma_1\, j_2\sigma_2|j\sigma\rangle |j_1\sigma_1\rangle |j_2\sigma_2\rangle \qquad (2.58)$$

where $\langle \cdot \cdot | \cdot \rangle$ is an SU(2) Clebsch–Gordan coefficient (see Section 1.7) and the states $|j_\rho\sigma_\rho\rangle$ are given in (1.60),

$$|j_\rho\sigma_\rho\rangle = \frac{(s_\rho^\dagger)^{j_\rho+\sigma_\rho}(t_\rho^\dagger)^{j_\rho-\sigma_\rho}}{\sqrt{(j_\rho-\sigma_\rho)!(j_\rho+\sigma_\rho)!}}|0\rangle_\rho \qquad (2.59)$$

Due to relations (2.37) and (2.57) the states (2.58) are eigenstates of the following invariant operators:

$$\begin{aligned}
\hat{N}|j_1, j_2; j\sigma\rangle &= 2(j_1 + j_2)|j_1, j_2; j\sigma\rangle = N|j_1, j_2; j\sigma\rangle \\
\hat{J}^2|j_1, j_2; j\sigma\rangle &= j(j+1)|j_1, j_2; j\sigma\rangle \\
\hat{n}_t|j_1, j_2; j\sigma\rangle &= \tfrac{1}{2}(N - 2\sigma)|j_1, j_2; j\sigma\rangle = n_t|j_1, j_2; j\sigma\rangle
\end{aligned} \qquad (2.60)$$

From (2.57) and (2.59) we also deduce that these states carry a definite parity given by

$$\hat{P}|j_1, j_2; j\sigma\rangle = \sum_{\sigma_1 \sigma_2} (-)^{j_1-\sigma_1+j_2-\sigma_2} \langle j_1\sigma_1\, j_2\sigma_2|j\sigma\rangle |j_1\sigma_1\rangle |j_2\sigma_2\rangle$$
$$= (-)^{N/2-\sigma}|j_1, j_2; j\sigma\rangle = (-)^{n_t}|j_1, j_2; j\sigma\rangle \quad (2.61)$$

Returning to the equivalent classification (2.56), we may use (2.40) to rewrite relations (2.60) and (2.61) in the form

$$\hat{N}|[N_1], [N_2]; [h, h']n_t\rangle = N|[N_1], [N_2]; [h, h']n_t\rangle$$
$$\hat{J}^2|[N_1], [N_2]; [h, h']n_t\rangle = \tfrac{1}{4}(h-h')(h-h'+2)|[N_1], [N_2]; [h, h']n_t\rangle$$
$$\hat{n}_t|[N_1], [N_2]; [h, h']n_t\rangle = n_t|[N_1], [N_2]; [h, h']n_t\rangle$$
$$(2.62)$$

and

$$\hat{P}|[N_1], [N_2]; [h, h']n_t\rangle = (-)^{n_t}|[N_1], [N_2]; [h, h']n_t\rangle \quad (2.63)$$

It thus follows that both $\hat{N}$ and $\hat{J}^2$ are invariant operators of the chain (2.56). It is of interest to consider a second-order invariant operator

$$\mathcal{C}_2[\mathrm{U}_{12}(2)] \equiv \tfrac{1}{2}\hat{N}^2 + 2\hat{J}^2 \quad (2.64)$$

which, due to (2.40) and (2.49), has eigenvalues

$$h(h+1) + h'(h'-1) \quad (2.65)$$

One may use the $\mathrm{U}_{12}(2)$ generators (2.26) to rewrite (2.64) in the simple form

$$\mathcal{C}_2[\mathrm{U}_{12}(2)] = \sum_{ij} \mathcal{G}_j^i \mathcal{G}_i^j \quad (2.66)$$

which can be shown to commute with all the $\mathcal{G}_l^k$ using the U(2) commutation relations (1.19). The operator $\mathcal{C}_2[\mathrm{U}_{12}(2)]$ is known as the second-order Casimir invariant of the $\mathrm{U}_{12}(2)$ algebra and the form (2.66) is valid for U(n), as discussed in Appendix B.

2.5 The $SO_{12}(2)$ limit

The $SO_{12}(2)$ limit corresponds to the particular choice for the hamiltonian (2.12)

$$\hat{H}_{IIa} = \gamma \hat{J}^2 + \beta \hat{J}_z^2 \qquad (2.67)$$

again omitting terms that contribute only to the binding energy. The eigenvalues of (2.67) are

$$E_{IIa}(j,\mu) = \gamma j(j+1) + \beta \mu^2 \qquad (2.68)$$

with ramification rules for j and μ given in (2.31). The eigenspectrum (2.68) corresponds, for the case of integer j (or even N, see (2.54)), to that of an axially symmetric, rigid rotor [2]. Before continuing the discussion, we consider the question of parity in the $SO_{12}(2)$ limit, for which we study the corresponding wave functions. The $SO_{12}(2)$ states are classified by the chain of algebras (2.28),

$$\begin{array}{|cccc|}
\hline
U_1(2) \otimes U_2(2) & \supset U_{12}(2) & \supset SO_{12}(2) \\
| & | & | & | \\
{}[N_1] & [N_2] & [h, h'] & \mu \\
\hline
\end{array} \qquad (2.69)$$

with, due to (2.40), $j = (h - h')/2$. Again, the classification (2.30) in terms of j is easier, as we may resort to angular momentum coupling of the $SO_1(2) \otimes SO_2(2)$ states (2.15),

$$\boxed{|j_1, j_2; j\mu\rangle = \sum_{\mu_1 \mu_2} \langle j_1\mu_1\, j_2\mu_2 | j\mu\rangle |j_1\mu_1\rangle |j_2\mu_2\rangle} \qquad (2.70)$$

with the states $|j_\rho\mu_\rho\rangle$ given in (1.64),

$$|j_\rho\mu_\rho\rangle = \frac{(s_\rho^\dagger - it_\rho^\dagger)^{j_\rho-\mu_\rho}(s_\rho^\dagger + it_\rho^\dagger)^{j_\rho+\mu_\rho}}{2^j \sqrt{(j_\rho - \mu_\rho)!(j_\rho + \mu_\rho)!}} |0\rangle_\rho \qquad (2.71)$$

The $SO_{12}(2)$ wave functions (2.70) are eigenstates of the operators

$$\begin{aligned}
\hat{N}|j_1, j_2; j\mu\rangle &= 2(j_1 + j_2)|j_1, j_2; j\mu\rangle = N|j_1, j_2; j\mu\rangle \\
\hat{J}^2|j_1, j_2; j\mu\rangle &= j(j+1)|j_1, j_2; j\mu\rangle \\
\hat{J}_z|j_1, j_2; j\mu\rangle &= \mu|j_1, j_2; j\mu\rangle
\end{aligned} \qquad (2.72)$$

but do not carry a definite parity since the $|j_\rho \mu_\rho\rangle$ states have no well-defined parity. Indeed, from (2.71) we see that the inversion operator $\hat{P}$ gives

$$\hat{P}|j_\rho \mu_\rho\rangle = |j_\rho - \mu_\rho\rangle \tag{2.73}$$

and applying this operator to (2.70), we find

$$\hat{P}|j_1, j_2; j\mu\rangle = \sum_{\mu_1 \mu_2} \langle j_1 \mu_1\, j_2 \mu_2 | j\mu \rangle |j_1 - \mu_1\rangle |j_2 - \mu_2\rangle \tag{2.74}$$

Since μ_1 and μ_2 take all values consistent with (2.31), we may change their sign everywhere in (2.74) and use the symmetry property of the Clebsch–Gordan coefficient [2]

$$\langle j_1 - \mu_1\, j_2 - \mu_2 | j\mu \rangle = (-)^{j_1 + j_2 - j} \langle j_1 \mu_1\, j_2 \mu_2 | j - \mu \rangle \tag{2.75}$$

to find

$$\hat{P}|j_1, j_2; j\mu\rangle = (-)^{j_1 + j_2 - j} |j_1, j_2; j - \mu\rangle \tag{2.76}$$

This shows that for $\mu = 0$ the states have well-defined parity $(-)^{j_1 + j_2 - j}$. For $\mu \neq 0$ we form combinations with good parity, namely

$$|j_1, j_2; j\mu\rangle^\pm = \sqrt{\tfrac{1}{2}} \left(|j_1, j_2; j\mu\rangle \pm (-)^{j_1 + j_2 - j} |j_1, j_2; j - \mu\rangle \right) \tag{2.77}$$

These states are degenerate in energy, as seen from (2.68), and have positive or negative parity. Again, these states are classified by $O_{12}(2)$ rather than $SO_{12}(2)$, but we shall sometimes refer to them as the $SO_{12}(2)$ wave functions. In Figure 2.4 we show the $O_{12}(2)$ spectrum (2.68) associated with the states (2.77). The $SO_1(2) \otimes SO_2(2)$ and $SO_{12}(2)$ limits have the $SO_{12}(2)$ algebra in common (see (2.15) and (2.69)) and hence its associated label μ is a good quantum number in the two limits and the same number of states with a given $|\mu|$ is found in Figures 2.2 and 2.4.

In order to discuss the evaluation of matrix elements of the general hamiltonian (2.12), we now consider the problem of finding the transformation brackets between the different basis states considered in this chapter.

2.6 Transformation brackets

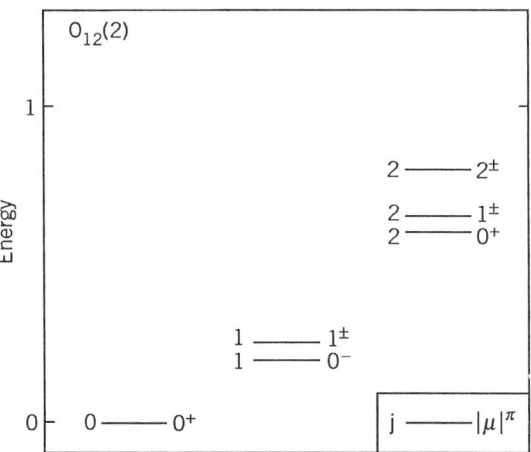

Figure 2.4: The $O_{12}(2)$ eigenspectrum of the hamiltonian $\hat{H}_{\text{IIa}}$ with $\gamma = 0.1$ and $\beta = 0.05$ for $N_1 = N_2 = 2$ bosons. Levels are labeled by j, $|\mu|$, and parity π. Energies are in arbitrary units.

2.6 Transformation brackets

The direct-product states (2.14) and (2.15) may be transformed into each other by means of the individual transformation brackets (1.82)

$$|j_1\mu_1, j_2\mu_2\rangle = \sum_{n_{t_1}=0}^{2j_1} \sum_{n_{t_2}=0}^{2j_2} \langle j_1 n_{t_1}|j_1\mu_1\rangle\langle j_2 n_{t_2}|j_2\mu_2\rangle |j_1 n_{t_1}, j_2 n_{t_2}\rangle \quad (2.78)$$

Since the $U_{12}(1)$ states (2.58) are defined in terms of the coupling of the states (2.14) (expressed in terms of $\sigma = N/2 - n_t$), the transformation brackets are simply the Clebsch–Gordan coefficients in (2.58). Likewise equation (2.70) defines the $SO_{12}(2)$ states and the corresponding transformation brackets to the uncoupled basis (2.15).

The only remaining undetermined coefficient is the one transforming between the $U_{12}(1)$ and $SO_{12}(2)$ bases. We may use the D-function form (1.84) for each of the factors in the direct-product states (2.14) and (2.15),

$$|j_\rho\mu_\rho\rangle = \sum_{\sigma_\rho} D^{j_\rho}_{\sigma_\rho\mu_\rho}(-\pi/2, -\pi/2, 0)|j_\rho\sigma_\rho\rangle \quad (2.79)$$

and use this expression in (2.70) to find

$$\begin{aligned}|j_1,j_2;j\mu\rangle &= \sum_{\mu_1\mu_2}\sum_{\sigma_1\sigma_2}\langle j_1\mu_1\ j_2\mu_2|j\mu\rangle D^{j_1}_{\sigma_1\mu_1}(-\pi/2,-\pi/2,0)\\ &\quad\times D^{j_2}_{\sigma_2\mu_2}(-\pi/2,-\pi/2,0)|j_1\sigma_1\rangle|j_2\sigma_2\rangle\end{aligned} \qquad (2.80)$$

We may now carry out the sum over μ_1 and μ_2 in (2.80) using the well-known formula for the coupling of two **D** matrices [2],

$$\sum_{\mu_1\mu_2}\langle j_1\mu_1\ j_2\mu_2|j\mu\rangle D^{j_1}_{\sigma_1\mu_1}D^{j_2}_{\sigma_2\mu_2} = \langle j_1\sigma_1\ j_2\sigma_2|j\sigma\rangle D^{j}_{\sigma\mu} \qquad (2.81)$$

where the arguments of the three **D** matrices must be the same. Comparing the result with the definition (2.58) of the $U_{12}(1)$ states, we find the final form for the transformation brackets

$$|j_1,j_2;j\mu\rangle = \sum_\sigma D^{j}_{\sigma\mu}(-\pi/2,-\pi/2,0)|j_1,j_2;j\sigma\rangle \qquad (2.82)$$

which is identical to (1.85). We could have anticipated the result (2.82) since, as explained in Section 1.6, the $SO_{12}(2)$ and $U_{12}(1)$ bases are related by a rotation in $SU(2)$—in this case characterized by the total angular momentum j and its projections. We see that, in spite of the apparent complications arising from the coupling of two s–t systems, the $SU(2)$ structure is preserved and relations such as (2.82) remain simple. The transformation brackets derived in this section are enough to compute any matrix element of interest in the model, by transforming to the basis where any given operator is diagonal.

In the last section of this chapter we present a simple model for stretching vibrations in X–Y–X molecules, using the results obtained so far.

2.7 Stretching vibrations in X–Y–X molecules

In this section we present an application of the $U_1(2)\otimes U_2(2)$ model for the description of vibrational stretching energies in triatomic molecules composed of two identical and one (in general) unlike atom, such as H_2O and CO_2 [1]. We focus on the stretching motion of the X–Y bonds and do not consider states associated with the bending motion

2.7 Stretching vibrations in X–Y–X molecules

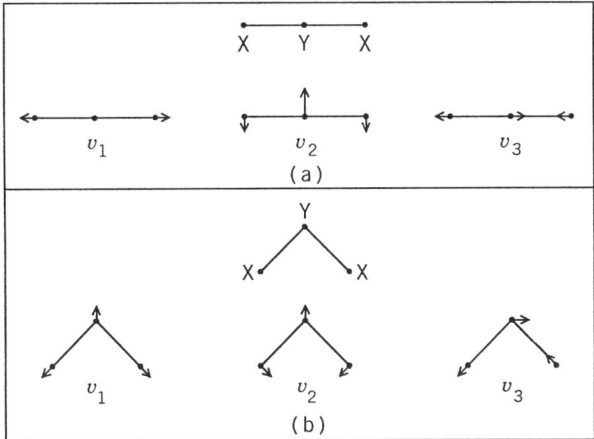

Figure 2.5: Stretching (v_1 and v_3) and bending (v_2) normal vibrations of (a) linear and (b) non-linear X–Y–X molecules. For the stretching modes both the symmetric (v_1) and the antisymmetric (v_3) oscillations are shown.

of the molecule. These two types of motion are traditionally described in terms of the normal modes depicted in Figure 2.5. The stretching modes are either symmetric or antisymmetric with respect to a plane of symmetry of the molecule (perpendicular to the X–Y–X plane and passing through the Y atom). The pure normal modes only occur for exactly harmonic motion, that is, for quadratic interaction terms in the molecular hamiltonian [5]. One should realize that in some systems it is more appropriate to describe the motion in terms of local anharmonic modes, incorporating a weak coupling of two independent oscillators, while in the general case it is a mixture of these two extreme pictures. (see [5], page 171). While a full description of rotation–vibration excitations in molecules requires the introduction of additional degrees of freedom as discussed in Part II of this book, these particular oscillations are susceptible to a description in terms of the $U_1(2) \otimes U_2(2)$ model, as we proceed to show.

The reason that this description is appropriate is due to the connection which exists between the U(2) algebra and the one-dimensional Schrödinger equation with a Morse potential [6]

$$\hat{H}_M \psi = E \psi \tag{2.83}$$

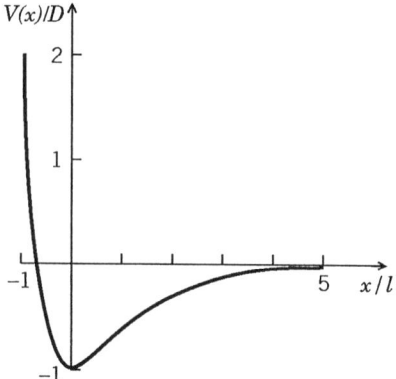

Figure 2.6: The Morse potential $V(x) = D[\exp(-2x/\ell) - 2\exp(-x/\ell)]$.

where

$$\hat{H}_{\mathrm{M}} = -\frac{\hbar^2}{2\mu}\frac{d^2}{dx^2} + D\left(e^{-2x/\ell} - 2e^{-x/\ell}\right) \quad (2.84)$$

This potential is often used as a first approximation to realistic interatomic interactions due to its characteristic shape, shown in Figure 2.6. The eigenstates of (2.83) may be put in a one-to-one correspondence with the U(2) ⊃ SO(2) states (1.9), as long as we restrict m to non-negative values. To see how this connection comes about, we return to the radial equation (1.8),

$$\frac{1}{2}\left(-\frac{1}{r}\frac{d}{dr}r\frac{d}{dr} + \frac{m^2}{r^2} + r^2\right)\Phi(r) = (N+1)\Phi(r) \quad (2.85)$$

which corresponds to a two-dimensional harmonic oscillator with eigenfunctions classified according to the U(2) ⊃ SO(2) reduction (1.31). We now carry out the transformation [7]

$$r^2 = (N+1)e^{-\rho} \quad (2.86)$$

which implies

$$\frac{d}{dr} = \frac{d\rho}{dr}\frac{d}{d\rho} = -\frac{2}{\sqrt{N+1}}e^{\rho/2}\frac{d}{d\rho} \quad (2.87)$$

2.7 Stretching vibrations in X–Y–X molecules

Substituting in (2.85) and simplifying the result leads to the following equation in ρ:

$$\left[-\frac{d^2}{d\rho^2} + \frac{(N+1)^2}{4}\left(e^{-2\rho} - 2e^{-\rho}\right)\right]\phi(\rho) = -\frac{m^2}{4}\phi(\rho) \qquad (2.88)$$

with $\phi(\rho) = \Phi(r)$. Finally, defining $x \equiv \ell\rho$ and multiplying by $\ell^2\hbar^2/2\mu$ leads to (2.83) and (2.84), provided that

$$N+1 = \left(\frac{8\mu D}{\ell^2\hbar^2}\right)^{1/2}, \qquad E = -\frac{m^2\ell^2\hbar^2}{8\mu} \qquad (2.89)$$

We have, from Chapter 1, equations (1.10) and (1.11), $N = 0, 1, \ldots$ and $m = \pm N, \pm(N-2), \ldots, \pm 1$ or 0, and thus we see that the Morse spectrum is reproduced twice and that one has to restrict m to non-negative values. Consequently, the Morse eigenstates can be classified by the $\Phi_{Nm}(r)$ states (1.9), or equivalently, the SU(2) $\supset$ SO(2) states (1.64). Note that the second of relations (2.89) implies that the Morse hamiltonian has the algebraic realization

$$\hat{H}_{\text{M}} = -\frac{\ell^2\hbar^2}{2\mu}\hat{J}_z^2 \qquad (2.90)$$

where we use the definition of $\hat{J}_z$ given in (1.20), with eigenvalue $m/2$. Comparing (2.90) with (1.53), we see that the correspondence with the SO(2) dynamical symmetry is exact. From (2.89) we also note that the U(2) eigenvalue N is related to the potential depth, as expected, since N fixes the number of bound states m. We now continue with the discussion of stretching vibrations.

Given the connection between the U(2) $\supset$ SO(2) chain and the one-dimensional Morse oscillator, we may use, for X–Y–X molecules, the two chains

$$U_1(2) \otimes U_2(2) \supset SO_1(2) \otimes SO_2(2) \supset SO_{12}(2) \qquad (2.91)$$

and

$$U_1(2) \otimes U_2(2) \supset U_{12}(2) \supset SO_{12}(2) \qquad (2.92)$$

to account for the possible interactions among two Morse oscillators which describe the X–Y bonds. While in the first coupling the nature of the individual Morse oscillators is preserved, in the second one it is not. As already mentioned, one may think of two limiting situations as one of purely local modes and one of normal modes, such as the ones shown in Figure 2.5. The first situation would arise naturally for a hypothetical molecule with the mass of the Y atom much larger than that of the X atoms while the second corresponds to purely harmonic oscillations, associated with rigid molecules where anharmonic terms may be neglected [5].

Actual molecules, in fact, correspond to an intermediate situation. States classified by chain (2.91) correspond to the local-mode basis since both Morse oscillators are well defined and interact weakly (through the diagonal term $\hat{J}_{1z}\hat{J}_{2z}$ in (2.21)). It is convenient to define the number of quanta in the local-mode basis by the relation

$$v_a = \tfrac{1}{2}(N_1 - m_1) = j_1 - \mu_1, \qquad v_c = \tfrac{1}{2}(N_2 - m_2) = j_2 - \mu_2 \qquad (2.93)$$

They take the values

$$v_a = 0, 1, \ldots, \tfrac{1}{2}N_1 \text{ or } \tfrac{1}{2}(N_1 - 1), \qquad v_c = 0, 1, \ldots, \tfrac{1}{2}N_2 \text{ or } \tfrac{1}{2}(N_2 - 1) \qquad (2.94)$$

for N_ρ even or odd. Then, recalling that we only keep non-negative values of μ, we find

$$v_a + v_c = \tfrac{1}{2}(N - 2\mu) = 0, 1, \ldots, \tfrac{1}{2}N \qquad (2.95)$$

and we may use $v_a + v_c$ in (2.95) as an alternative label to classify the $SO_{12}(2)$ states instead of μ. The local-mode basis is then classified by the chain

$$\begin{array}{ccccc} U_1(2) \otimes U_2(2) & \supset & SO_1(2) \otimes SO_2(2) & \supset & SO_{12}(2) \\ | & | & | & | & | \\ [N_1] & [N_2] & v_a & v_c & v_a + v_c \end{array} \qquad (2.96)$$

On the other hand, states (2.92) involve a strong coupling of the local-mode basis, through the non-diagonal $\hat{J}^2$ interaction in (2.67). To see whether these states correspond to normal modes, we return to the $U_{12}(2) \supset SO_{12}(2)$ states $|j_1, j_2; j\mu\rangle$ in (2.70),

$$|j_1, j_2; j\mu\rangle = \sum_{\mu_1 \mu_2} \langle j_1\mu_1\, j_2\mu_2 | j\mu\rangle |j_1\mu_1\rangle |j_2\mu_2\rangle \qquad (2.97)$$

2.7 Stretching vibrations in X–Y–X molecules

and interchange the j_1 and j_2 quantum numbers

$$|j_2, j_1; j\mu\rangle = \sum_{\mu_1 \mu_2} \langle j_2\mu_2\, j_1\mu_1 | j\mu\rangle |j_1\mu_1\rangle |j_2\mu_2\rangle$$
$$= (-)^{j_1+j_2-j}|j_1, j_2; j\mu\rangle = (-)^{N/2-j}|j_1, j_2; j\mu\rangle \quad (2.98)$$

where the symmetry properties of the Clebsch–Gordan coefficients [2] are used. This exchange is equivalent in the algebraic framework to the reflection operation on the symmetry plane passing through the Y atom. In (2.98), $N = N_1 + N_2$ is the total number of quanta. For the case of X–Y–X molecules the two bonds are identical and thus $N_1 = N_2 = N/2$, where N is an even number. If we denote the reflection operator by $\hat{\sigma}_v$, (2.98) reads

$$\hat{\sigma}_v |Nj\mu\rangle = (-)^{N/2-j}|Nj\mu\rangle \quad (2.99)$$

so we verify that the coupling (2.28) naturally involves symmetric and antisymmetric vibrational states.

We may now define the labels for the number of quanta in the normal-mode basis, as is done in (2.93) for the local-mode basis. Due to (2.99) the number of quanta in the antisymmetric mode is given by

$$v_3 = \tfrac{1}{2}(N - 2j) = 0, 1, \ldots, \tfrac{1}{2}N \quad (2.100)$$

so that we obtain symmetric or antisymmetric states under $\hat{\sigma}_v$ for v_3 even or odd, respectively. To define the number of quanta in the symmetric mode, v_1, we note that it must depend on the quantum number $\mu = \mu_1 + \mu_2$, since only N and j appear in the definition (2.100). The $SO_{12}(2)$ subalgebra is common to the two bases, (2.91) and (2.92), and thus we may fix v_1 by requiring $v_a + v_c = v_1 + v_3$, since the local-mode basis (2.96) is chosen to have $v_a + v_c = N/2 - \mu$ as $SO_{12}(2)$ label. Using (2.95) and (2.100), we find

$$v_1 = j - \mu \quad (2.101)$$

which is analogous to relations (2.93) for the coupled $SU_{12}(1)$ system. Equation (2.101) is thus consistent with the interpretation of v_1 as related to the symmetric modes which are analogous to one-dimensional vibrations. As mentioned at the beginning of this section, the number of quanta in the bending mode, traditionally denoted as v_2 or v_b [5], is taken to be zero in the present discussion.

Summarizing, we have shown that the basis states (2.91) and (2.92) may be written in the form $|Nv_av_c\rangle$ and $|Nv_1v_3\rangle$ and correspond to the local- and normal-mode bases, respectively.

The most general one- and two-body hamiltonian associated with the system is a combination of (2.21) and (2.67),

$$\hat{H} = \beta_{12}(\hat{J}_{1z}^2 + \hat{J}_{2z}^2) + \beta\hat{J}_z^2 + \gamma\hat{J}^2 \qquad (2.102)$$

where the coefficients of $\hat{J}_{1z}^2$ and $\hat{J}_{2z}^2$ are set equal due to the symmetry of the X–Y–X molecule. The hamiltonian is diagonal in the local-mode basis when $\beta = 0$ while it is diagonal in the normal-mode basis for $\gamma = 0$. In the general case we may evaluate the matrix elements of $\hat{J}^2$ in the local-mode basis and carry out the diagonalization of the hamiltonian. To compute these matrix elements, it is convenient to rewrite (2.102) in the form

$$\hat{H} = E_0(N) + \alpha\left(\hat{J}_{1z}^2 + \hat{J}_{2z}^2\right) + \alpha_{12}\hat{J}_{1z}\hat{J}_{2z} + \lambda_{12}\hat{J}_1 \cdot \hat{J}_2 \qquad (2.103)$$

where $\hat{J}_1 \cdot \hat{J}_2 = \hat{J}_{1x}\hat{J}_{2x} + \hat{J}_{1y}\hat{J}_{2y} + \hat{J}_{1z}\hat{J}_{2z}$ is a scalar product, $E_0(N) = \gamma N(N+4)/8$, $\alpha = \beta_{12} + \beta$, $\alpha_{12} = 2\beta$, and $\lambda_{12} = 2\gamma$. In this form we find that the first three terms are diagonal in the local-mode basis while $\hat{J}_1 \cdot \hat{J}_2$ gives simple matrix elements, as we proceed to show.

Defining the spherical components $\hat{J}_{\rho\mu}$ as in (1.93) and (1.86),

$$\hat{J}_{\rho 0} = \hat{J}_{\rho z}, \qquad \hat{J}_{\rho\pm 1} = \mp\sqrt{\tfrac{1}{2}}(\hat{J}_{\rho x} \pm i\hat{J}_{\rho y}), \qquad \rho = 1, 2 \qquad (2.104)$$

we find

$$\hat{J}_1 \cdot \hat{J}_2 = \sum_\mu (-)^\mu \hat{J}_{1\mu}\hat{J}_{2-\mu} \qquad (2.105)$$

which is the usual form for the scalar product of two SU(2) vectors [2]. We can now evaluate the matrix elements

$$\langle j_1\mu_1', j_2\mu_2'|\hat{J}_1 \cdot \hat{J}_2|j_1\mu_1, j_2\mu_2\rangle = \sum_\mu (-)^\mu \langle j_1\mu_1'|\hat{J}_{\mu 1}|j_1\mu_1\rangle\langle j_2\mu_2'|\hat{J}_{-\mu 2}|j_2\mu_2\rangle \qquad (2.106)$$

2.8 Stretching vibrations in polyatomic molecules

by using (1.89) and (1.93) to yield

$$\langle j_1\mu_1, j_2\mu_2|\hat{J}_1\cdot\hat{J}_2|j_1\mu_1, j_2\mu_2\rangle = \mu_1\mu_2$$
$$\langle j_1\,\mu_1\pm 1, j_2\,\mu_2\mp 1|\hat{J}_1\cdot\hat{J}_2|j_1\mu_1, j_2\mu_2\rangle$$
$$= \tfrac{1}{2}\sqrt{(j_1\mp\mu_1)(j_1\pm\mu_1+1)(j_2\pm\mu_2)(j_2\mp\mu_2+1)} \quad (2.107)$$

These matrix elements may be written in the $|Nv_av_c\rangle$ notation by using (2.93):

$$\langle Nv_av_c|\hat{J}_1\cdot\hat{J}_2|Nv_av_c\rangle = \tfrac{1}{16}(N-4v_a)(N-4v_c)$$
$$\langle N\,v_a+1\,v_c-1|\hat{J}_1\cdot\hat{J}_2|Nv_av_c\rangle$$
$$= \tfrac{1}{4}\sqrt{(N-2v_a)(N-2v_c+2)(v_a+1)v_c} \quad (2.108)$$
$$\langle N\,v_a-1\,v_c+1|\hat{J}_1\cdot\hat{J}_2|Nv_av_c\rangle$$
$$= \tfrac{1}{4}\sqrt{(N-2v_a+2)(N-2v_c)v_a(v_c+1)}$$

Note that the resulting matrix is tridiagonal. An alternative approach is to compute matrix elements of $\hat{J}_{1z}^2$ and $\hat{J}_{2z}^2$ in the normal-mode basis $|Nv_1v_3\rangle$. This can be done using the transformation brackets (2.70), which leads to a sum over Clebsch–Gordan coefficients. A simpler approach is to use the Wigner–Eckart theorem, equation (1.96), for $\hat{J}_{1z}$ and $\hat{J}_{2z}$ [1,2].

In Table 2.1 we show the results of a least-squares fit to the stretching modes of H_2O [1]. We identify the levels by the normal-mode states $|Nv_1v_2v_3\rangle$ which give maximum overlap with the resulting eigenstates, that is, the labels v_1 and v_3 quoted in the table are the maximum amplitude components in the molecular wave functions. The parameters in (2.103) are $\alpha = -80.4558$, $\alpha_{12} = -9.0554$, and $\lambda_{12} = -2.0526$ (in cm^{-1}), and $N = 88$. The root-mean-square deviation is 4.0 cm^{-1}.

The $U_1(2)\otimes U_2(2)$ model has been applied successfully to other X–Y–X molecules [1] and has recently been extended to polyatomic molecules [8,9], as we describe in the last section of this chapter.

2.8 Stretching vibrations in polyatomic molecules

Polyatomic molecules often exhibit symmetries of a different nature than the continuous ones that are used throughout this book. They

Table 2.1: Stretching vibrational energy levels of H_2O^a

v_1	v_2	v_3	Experiment	Theory	Δ
1	0	0	3 657.1	3 658.8	1.7
0	0	1	3 755.9	3 749.1	-6.8
2	0	0	7 201.5	7 205.4	3.9
1	0	1	7 249.8	7 247.0	-2.8
0	0	2	7 445.1	7 438.5	-6.6
3	0	0	10 599.7	10 604.5	4.8
2	0	1	10 613.4	10 614.7	1.3
1	0	2	10 868.9	10 865.3	-3.6
0	0	3	11 032.4	11 031.6	-0.8
2	0	2	13 828.3	13 832.8	4.5
3	0	1	13 830.9	13 834.1	3.2
4	0	0	14 221.1	14 220.7	-0.4
1	0	3	14 318.8	14 316.2	-2.6
0	0	4	14 536.9	14 546.5	9.6
3	0	2	16 898.4	16 895.2	-3.2
4	0	1	16 898.8	16 895.3	-3.5
5	0	0	17 458.2	17 459.6	1.4
2	0	3	17 495.5	17 492.6	-2.9
1	0	4	17 748.1	17 746.3	-1.8
0	0	5	17 970.9	17 971.7	0.6
4	0	2		19 795.7	
5	0	1		19 795.7	
6	0	0		20 530.0	
2	0	4		20 535.8	
1	0	5		20 897.2	
0	0	6		21 046.5	
Δ_{rms}					4.0

^{a} All energies in cm^{-1}.

2.8 Stretching vibrations in polyatomic molecules

correspond to sets of discrete transformations which leave the equilibrium configuration of the molecule invariant. Such groups are known as discrete groups, and the study of their properties is the subject of many molecular physics textbooks [5,10]. We do not attempt to review these matters here but rather, by means of a simple example, introduce some basic results which are needed for the application of the algebraic approach to polyatomic molecules.

In Figure 2.7 we show an equilateral triangle and define rotations that carry it into positions which are indistinguishable from the initial one. The operations $\hat{R}_{\pi_1}$, $\hat{R}_{\pi_2}$, and $\hat{R}_{\pi_3}$ are rotations of π around the axes in the x–y plane shown in the figure, and we also define $\hat{R}_{2\pi/3}$ and $\hat{R}_{4\pi/3}$ as rotations of $2\pi/3$ and $4\pi/3$ about the z axis. In addition, discrete groups include the identity operation, denoted by $\hat{E}$, which has no effect on the system. The set $(\hat{E}, \hat{R}_{2\pi/3}, \hat{R}_{4\pi/3}, \hat{R}_{\pi_1}, \hat{R}_{\pi_2}, \hat{R}_{\pi_3})$ satisfies all the group postulates (as defined in Appendix A) and is denoted by D_3. By successive application of these operators we arrive at the multiplication Table 2.2.

We may also consider the set of permutations of three objects 1, 2, and 3 and use the notation

$$\hat{P} = \begin{pmatrix} 1 & 2 & 3 \\ p_1 & p_2 & p_3 \end{pmatrix} \quad (2.109)$$

to denote a permutation in which the object labeled i is replaced by the one labeled p_i. This is the group S_3, or the symmetric group of three

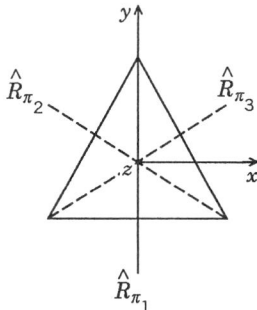

Figure 2.7: The rotations $\hat{R}_{\pi_i}$ of an equilateral triangle.

Table 2.2: The D_3 multiplication table

	$\hat{E}$	$\hat{R}_{2\pi/3}$	$\hat{R}_{4\pi/3}$	$\hat{R}_{\pi_1}$	$\hat{R}_{\pi_2}$	$\hat{R}_{\pi_3}$
$\hat{E}$	$\hat{E}$	$\hat{R}_{2\pi/3}$	$\hat{R}_{4\pi/3}$	$\hat{R}_{\pi_1}$	$\hat{R}_{\pi_2}$	$\hat{R}_{\pi_3}$
$\hat{R}_{2\pi/3}$	$\hat{R}_{2\pi/3}$	$\hat{R}_{4\pi/3}$	$\hat{E}$	$\hat{R}_{\pi_2}$	$\hat{R}_{\pi_3}$	$\hat{R}_{\pi_1}$
$\hat{R}_{4\pi/3}$	$\hat{R}_{4\pi/3}$	$\hat{E}$	$\hat{R}_{2\pi/3}$	$\hat{R}_{\pi_3}$	$\hat{R}_{\pi_1}$	$\hat{R}_{\pi_2}$
$\hat{R}_{\pi_1}$	$\hat{R}_{\pi_1}$	$\hat{R}_{\pi_3}$	$\hat{R}_{\pi_2}$	$\hat{E}$	$\hat{R}_{4\pi/3}$	$\hat{R}_{2\pi/3}$
$\hat{R}_{\pi_2}$	$\hat{R}_{\pi_2}$	$\hat{R}_{\pi_1}$	$\hat{R}_{\pi_3}$	$\hat{R}_{2\pi/3}$	$\hat{E}$	$\hat{R}_{4\pi/3}$
$\hat{R}_{\pi_3}$	$\hat{R}_{\pi_3}$	$\hat{R}_{\pi_2}$	$\hat{R}_{\pi_1}$	$\hat{R}_{4\pi/3}$	$\hat{R}_{2\pi/3}$	$\hat{E}$

objects. The operations

$$\hat{E} = \begin{pmatrix} 1 & 2 & 3 \\ 1 & 2 & 3 \end{pmatrix}, \quad \hat{P}_1 = \begin{pmatrix} 1 & 2 & 3 \\ 2 & 3 & 1 \end{pmatrix}, \quad \hat{P}_2 = \begin{pmatrix} 1 & 2 & 3 \\ 3 & 1 & 2 \end{pmatrix}$$
$$\hat{P}_3 = \begin{pmatrix} 1 & 2 & 3 \\ 2 & 1 & 3 \end{pmatrix}, \quad \hat{P}_4 = \begin{pmatrix} 1 & 2 & 3 \\ 1 & 3 & 2 \end{pmatrix}, \quad \hat{P}_5 = \begin{pmatrix} 1 & 2 & 3 \\ 3 & 2 & 1 \end{pmatrix}$$
(2.110)

satisfy the multiplication Table 2.3.

The groups D_3 and S_3 are isomorphic since we can establish a one-to-one correspondence between the elements: $\hat{R}_i \leftrightarrow \hat{P}_i$, such that if $\hat{R}_i\hat{R}_j = \hat{R}_k$, then $\hat{P}_i\hat{P}_j = \hat{P}_k$. This isomorphism is established by the correspondence $\hat{R}_{2\pi/3} \leftrightarrow \hat{P}_1$, $\hat{R}_{4\pi/3} \leftrightarrow \hat{P}_2$, $\hat{R}_{\pi_1} \leftrightarrow \hat{P}_3$, $\hat{R}_{\pi_2} \leftrightarrow \hat{P}_4$, and $\hat{R}_{\pi_3} \leftrightarrow \hat{P}_5$, as can be readily verified from Tables 2.2 and 2.3. Every discrete group is in fact isomorphic to a subgroup of the symmetric group S_n, a result which is known as Cayley's theorem [10].

Table 2.3: The S_3 multiplication table

	$\hat{E}$	$\hat{P}_1$	$\hat{P}_2$	$\hat{P}_3$	$\hat{P}_4$	$\hat{P}_5$
$\hat{E}$	$\hat{E}$	$\hat{P}_1$	$\hat{P}_2$	$\hat{P}_3$	$\hat{P}_4$	$\hat{P}_5$
$\hat{P}_1$	$\hat{P}_1$	$\hat{P}_2$	$\hat{E}$	$\hat{P}_4$	$\hat{P}_5$	$\hat{P}_3$
$\hat{P}_2$	$\hat{P}_2$	$\hat{E}$	$\hat{P}_1$	$\hat{P}_5$	$\hat{P}_3$	$\hat{P}_4$
$\hat{P}_3$	$\hat{P}_3$	$\hat{P}_5$	$\hat{P}_4$	$\hat{E}$	$\hat{P}_2$	$\hat{P}_1$
$\hat{P}_4$	$\hat{P}_4$	$\hat{P}_3$	$\hat{P}_5$	$\hat{P}_1$	$\hat{E}$	$\hat{P}_2$
$\hat{P}_5$	$\hat{P}_5$	$\hat{P}_4$	$\hat{P}_3$	$\hat{P}_2$	$\hat{P}_1$	$\hat{E}$

2.8 Stretching vibrations in polyatomic molecules

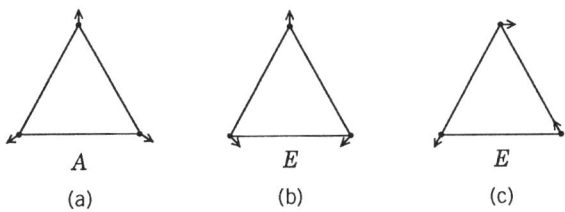

(a) A   (b) E   (c) E

Figure 2.8: Normal vibrations of a triangular X_3 molecule.

Why are these groups relevant in the description of molecular vibrations? If we assume that the atoms in a molecule undergo harmonic vibrations around their equilibrium positions, it is well known that collective oscillations (normal modes) occur [10,11], as depicted in Figure 2.5 for the case of X–Y–X molecules. In Figure 2.8 we show the normal modes for the triangular configuration of Figure 2.7.

While all operations of D_3 (or S_3) leave invariant the symmetric mode (a) in Figure 2.8, the modes (b) and (c) are combined by these operators, that is, the action of the $\hat{R}_i$ on (b) leads to a linear combination of (b) and (c). Since the hamiltonian for the system should be invariant under the symmetry group, this means that the modes (b) and (c) correspond to the same energy. The normal modes in Figure 2.8 are then a basis for the irreducible representations of the group (see Appendix A). In this case there are only the symmetric one, which is denoted by A, the doubly degenerate one, denoted by E, and a third, non-degenerate representation, denoted by B, which is non-symmetric. The latter mode only appears for higher excitations. In general, a harmonically interacting molecule whose equilibrium position has a discrete symmetry undergoes normal oscillations which are associated with the representations of the corresponding discrete group and which display degeneracies equal to their dimensionality.

After this brief analysis of the symmetry of molecular normal modes, we now return to the U(2) model and its application to stretching modes in polyatomic molecules.

For a polyatomic molecule with m chemical bonds, each bond is described by a $U_\rho(2)$ algebra related to a Morse potential as in (2.90).

As for the X–Y–X molecules of the previous section, for every pair of bonds ρ and σ we consider interactions diagonal in the chains (2.91) and (2.92) (with the indices 1 and 2 substituted by ρ and σ), which clearly leads to a generalization of the hamiltonian (2.103) of the form [8,9]

$$\hat{H} = \sum_{\rho=1}^{m} \alpha_\rho \hat{J}_{\rho z}^2 + \sum_{\rho<\sigma}^{m} \alpha_{\rho\sigma} \hat{J}_{\rho z} \hat{J}_{\sigma z} + \sum_{\rho<\sigma}^{m} \lambda_{\rho\sigma} \hat{J}_\rho \cdot \hat{J}_\sigma \qquad (2.111)$$

The local-mode basis is simply given by the product states

$$|[N_1]n_1, [N_2]n_2, \ldots, [N_m]n_m\rangle = |[N_1]n_1\rangle |[N_2]n_2\rangle \cdots |[N_m]n_m\rangle \quad (2.112)$$

were $|[N_\rho]n_\rho$ are the $U_\rho(2) \supset SO_\rho(2)$ eigenfunctions. The matrix elements of all operators in (2.111) in this basis can be copied from the results of the last section, again replacing the indices $(1,2)$ by (ρ,σ).

The parameters in (2.111) are in principle arbitrary, but they should be chosen in such a way that the hamiltonian is invariant under the operations of the (discrete) symmetry group of the molecule under consideration. For example, for the hypothetical triangular molecule of Figures 2.7 and 2.8 it should be clear that $\alpha_1 = \alpha_2 = \alpha_3$, $\alpha_{12} = \alpha_{13} = \alpha_{23}$, and $\lambda_{12} = \lambda_{13} = \lambda_{23}$, since all interactions between the bonds are symmetric. The resulting hamiltonian will then certainly commute with the operations of D_3 (or the permutations in S_3). Furthermore, the diagonalization of (2.111) automatically provides the discrete quantum labels and degeneracies associated with the normal modes [8,9]. For a simple system such as X_3 the conditions on the hamiltonian parameters are obvious and can be deduced by inspection. For more complicated configurations, however, it is necessary to systematize the procedure. We now indicate how this is done [9].

Given a molecular symmetry associated to a discrete group G, we start by establishing an isomorphism between G and a subgroup of the symmetric group S_m:

$$\hat{R}_i \leftrightarrow \begin{pmatrix} 1 & 2 & \ldots & m \\ p_1 & p_2 & \ldots & p_m \end{pmatrix} \equiv (p_1, p_2, \ldots, p_m) \qquad (2.113)$$

as explained above. We then consider an arbitrary bond, which we may denote by $\rho = 1$, and a next neighbor, denoted by $\rho = 2$, and apply

2.8 Stretching vibrations in polyatomic molecules

all permutations in the group to the three interactions $\hat{J}_{1z}^2$, $\hat{J}_{1z} \cdot \hat{J}_{2z}$, and $\hat{J}_1 \cdot \hat{J}_2$, arriving at three sets of terms associated with the same parameters, which give a first contribution to the hamiltonian:

$$\hat{H}_1 = \alpha_1 \sum_{(p_1,p_2,\ldots,p_m) \in G} (p_1, p_2, \ldots, p_m) \hat{J}_{1z}^2$$
$$+ \alpha_{12} \sum_{(p_1,p_2,\ldots,p_m) \in G} (p_1, p_2, \ldots, p_m) \hat{J}_{1z} \cdot \hat{J}_{2z}$$
$$+ \lambda_{12} \sum_{(p_1,p_2,\ldots,p_m) \in G} (p_1, p_2, \ldots, p_m) \hat{J}_1 \cdot \hat{J}_2 \quad (2.114)$$

By inspection of (2.114) we identify any bond ρ' and/or pair of bonds (ρ', σ') not already contained in this expression and apply again the permutations $(p_1, p_2, \ldots, p_m) \in G$, arriving at additional sets of interactions, $\hat{H}_2$, with common parameters. The procedure is repeated until all possible interactions are included. The molecular hamiltonian $\hat{H} = \hat{H}_1 + \hat{H}_2 + \cdots$ will then have the full symmetry of G.

As an example, consider the octahedral molecule XY_6 [8], shown in Figure 2.9, which also defines the numbering of the bonds from 1 to 6. The symmetry group of these molecules is the octahedral group which is denoted by O_h and is the group of transformations leaving a cube invariant [10]. This group has 48 elements which can be put into a one-to-one correspondence with a subgroup of S_6, the symmetric group of six objects [9]. We do not write down the isomorphism here, however, since for this simple example it is not necessary. From Figure 2.9 we

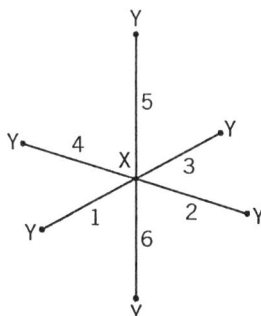

Figure 2.9: The bond coordinates of octahedral XY_6 molecules.

see that all bonds are equivalent and that there are just two kinds of interactions, the "first-neighbor" ones, such as 1–2, 1–4, 1–5, and 1–6, and the "opposite-neighbor" ones 1–3, 2–4, and 5–6. The appropriate hamiltonian is thus

$$\hat{H} = \alpha_1 \sum_{\rho=1}^{6} \hat{J}_{\rho z}^2 + \alpha_{12} \sum_{\rho<\sigma}^{6}{}' \hat{J}_{\rho z}\hat{J}_{\sigma z} + \lambda_{12} \sum_{\rho<\sigma}^{6}{}' \hat{J}_{\rho} \cdot \hat{J}_{\sigma}$$
$$+\alpha_{13}(\hat{J}_{1z}\hat{J}_{3z} + \hat{J}_{2z}\hat{J}_{4z} + \hat{J}_{5z}\hat{J}_{6z})$$
$$+\lambda_{13}(\hat{J}_1 \cdot \hat{J}_3 + \hat{J}_2 \cdot \hat{J}_4 + \hat{J}_5 \cdot \hat{J}_6) \qquad (2.115)$$

where the prime in the sums indicate that the opposite neighbors are excluded from the sum. The octahedral group has 10 different irreducible representations of dimensions 1, 2, and 3, which are denoted by A, E, and F, respectively [5,10]. Only five of them appear in connection with stretching vibrations in XY_6 molecules, denoted by A_{1g}, E_g, F_{1u}, F_{2g}, and F_{2u}. The symbols g and u denote the behavior of the wave functions under reflection on any plane containing four Y atoms: g implies no change of sign while u corresponds to states which change sign under this symmetry operation.

As for the X–Y–X molecules of the previous section, the model hamiltonian (2.115) contains interactions which emphasize the normal $\hat{J}_\rho \cdot \hat{J}_\sigma$ or local $\hat{J}_{\rho z}^2$ and $\hat{J}_{\rho z} \cdot \hat{J}_{\sigma z}$ character of the vibrations. In Table 2.4 we compare the results of fits [8] to the vibrational levels of the octahedral molecules SF_6, WF_6, and UF_6. The parameters α_{12} and α_{13} are taken to be equal in the three calculations. The three fundamental modes A_{1g}, E_g, and F_{1u} (which occur for one-phonon excitations) are denoted by v_1, v_2, and v_3, respectively. The values of the parameters and the number of bosons can be determined by a least-squares fit. A good agreement is obtained and, as explained above, automatically gives the correct degeneracy and symmetry species.

We end this chapter with the following observation. The main purpose of the first part of this book is to introduce the ideas of algebraic models without recourse to sophisticated group-theoretical techniques but relying entirely on standard angular momentum theory. The contents of Part I thus have a predominantly pedagogical value. More recent developments have shown, however, that the U(2)-based models are in fact useful in the description of vibrational spectra of large

Table 2.4: Stretching vibrational energy levels of SF_6, WF_6, and UF_6[a]

v_1	v_2	v_3		Expt	Theory	Expt	Theory	Expt	Theory
0	1	0	E_g	645.4	643.4	678.2	678.0	534.1	533.5
1	0	0	A_{1g}	774.5	774.1	772.1	772.1	667.1	666.4
0	0	1	F_{1u}	948.1	948.2	712.4	712.6	625.5	625.7
0	2	0	A_{1g}		1 288.2	1 354.0	1 354.1	1 066.5	1 065.9
			E_g		1 289.5	1 354.0[b]	1 354.8	1 066.3	1 066.3
1	1	0	E_g		1 416.7		1 448.3	1 197.0	1 198.5
2	0	0	A_{1g}		1 546.8		1 543.0		1 331.9
0	1	1	F_{1u}	1 588.1	1 588.3	1 387.1	1 387.2	1 156.9	1 157.1
			F_{2u}		1 593.7		1 390.2		1 159.0
1	0	1	F_{1u}	1 719.6	1 719.7	1 482.8	1 482.8	1 290.9	1 290.7
0	0	2	E_g	1 889.1	1 890.9	1 422.4	1 422.3		1 249.4
			A_{1g}	1 889.1	1 890.9	1 422.4[b]	1 422.4		1 249.4
			F_{2g}	1 896.5	1 896.5	1 422.4[b]	1 424.8		1 251.2

[a] All energies in cm^{-1}.
[b] Estimated values.

molecules, examples of which have been discussed at the end of this chapter, with others available in the literature [12,13]. Applications to other physical systems, such as a linear chain of atoms composing a crystal [14], have also been carried out. Given their simplicity, these models are ideal for the development of new concepts and techniques, such as in the introduction of point symmetries into the algebraic framework, discussed in Section 2.8, which may then be generalized to the more realistic higher-dimensional models.

Summary

This chapter deals with the coupling of two U(2) algebras to describe a system consisting of two different sets of interacting s and t bosons. The corresponding algebraic structure is the direct-product algebra $U_1(2) \otimes U_2(2)$ which is defined by relations (2.1) to (2.4). A large part of the chapter is devoted to the study of the subalgebra chains of $U_1(2) \otimes U_2(2)$ and their associated symmetry limits which can be of two types. The limits of the first type, specified in (2.14) and (2.15), are uncoupled; the corresponding basis states (2.16) and (2.17) are simple products of the states constructed in Chapter 1. The limits of the second type are given in (2.56) and (2.69); the basis states in this case are linear combinations of the uncoupled products with Clebsch–Gordan coefficients as expansion coefficients, (2.58) and (2.70). The characteristic feature of the latter two limits is the appearance of the coupled $U_{12}(2)$ algebra generated by the sum (2.26) of the $U_1(2)$ and $U_2(2)$ generators. Since one is dealing with a system of non-identical bosons, the basis states are not necessarily symmetric under $U_{12}(2)$ nor are they characterized by a single label. By introducing the weight, raising, and lowering generators (2.45), it is shown that two (Gel'fand) labels are needed which are the weights of a highest-weight polynomial defined in (2.46). The $U_{12}(2)$ symmetry character of a state can be conveniently summarized with a Young tableau (2.50). A system of identical bosons is characterized by a one-row Young tableau (2.51) which corresponds to a symmetric representation. The last two sections of this chapter deal with applications to stretching vibrations in triatomic and polyatomic molecules. The connection with the U(2) model can be established by showing that the eigenstates of the Morse hamiltonian (2.84) can be put in a one-to-one correspondence with the $U(2) \supset SO(2)$ eigenstates. In the case of triatomic molecules (i.e., the coupling of two Morse oscillators) the coupled and uncoupled bases, introduced earlier, can be shown to correspond to the normal and local (stretching) vibrations of the molecule. By considering the coupling of n Morse oscillators, this scheme can be extended to describe stretching vibrations of polyatomic molecules, and in particular, a description can be obtained which includes the discrete-symmetry character of the molecule.

References

1. O.S. van Roosmalen, I. Benjamin, and R.D. Levine, "A unified algebraic model description for interacting vibrational modes in ABA molecules," J. Chem. Phys. **81** (1984) 5986.
2. M.E. Rose, *Elementary Theory of Angular Momentum*, Wiley, New York, 1957.

References

3. M. Moshinsky, *Group Theory and the Many-Body Problem*, Gordon & Breach, New York, 1968.
4. B.G. Wybourne, *Classical Groups for Physicists*, Wiley-Interscience, New York, 1974.
5. G. Herzberg, *Molecular Spectra and Molecular Structure. II. Infrared and Raman Spectra of Polyatomic Molecules*, van Nostrand, New York, 1945.
6. P.M. Morse, "Diatomic molecules according to the wave mechanics. II. Vibrational levels," Phys. Rev. **34** (1929) 57.
7. Y. Alhassid, F. Gursey, and F. Iachello, "Group theory approach to scattering," Ann. Phys. (NY) **148** (1983) 346.
8. F. Iachello and S. Oss, "Model of n coupled anharmonic oscillators and applications to octahedral molecules," Phys. Rev. Lett. **66** (1991) 2976.
9. A. Frank and R. Lemus, "Comment on 'Model of n coupled anharmonic oscillators and applications to octahedral molecules'," Phys. Rev. Lett. **68** (1992) 413.
10. M. Hamermesh, *Group Theory and Its Application to Physical Problems*, Addison-Wesley, Reading, MA, 1962.
11. J.P. Elliott and P.G. Dawber, *Symmetry in Physics. I. Principles and Simple Applications*, Oxford University Press, New York, 1979.
12. F. Iachello and S. Oss, "Stretching vibrations of benzene in the algebraic model," Chem. Phys. Lett. **187** (1991) 500.
13. J.M. Arias, A. Frank, R. Lemus, and F. Perez-Bernal, "Algebraic description of stretching and bending modes in non-linear molecules," Phys. Rev. A (1994) to be published.
14. R. Lemus and A. Frank, "Algebraic approach for vibrational excitations in an anharmonic linear chain," in *Symmetries in Science VII*, edited by B. Gruber and T. Otsuka, Plenum, New York, 1994, p. 357.

Chapter 3

Bosons and Fermions

3.1 Fermion algebras

Up until this chapter we have considered algebraic hamiltonians and other physical operators built in terms of boson creation and annihilation operators. On the other hand, many algebraic procedures have been developed to study many-fermion systems—such as atoms, molecules, and nuclei—where the methods of second quantization and unitary algebras have significantly simplified the analysis.

The algebraic techniques presented in this book involve the use of both boson and fermion algebras as well as a combination of both, which are known as boson–fermion algebras. The latter are required for the analysis of electronic excitations in molecules [1,2] and the spectroscopic properties of odd-mass and odd–odd nuclei [3,4]) In this chapter we exploit the simplicity of the schematic U(2) model to illustrate these methods and to introduce the basic concepts involved.

We start by defining spin $1/2$ particles, to which we associate fermion creation and annihilation operators $a^\dagger_{1/2\sigma}$ and $a_{1/2\sigma}$, satisfying the *anti-commutation relations*

$$\{a_\sigma, a^\dagger_\tau\} \equiv a_\sigma a^\dagger_\tau + a^\dagger_\tau a_\sigma = \delta_{\sigma\tau}, \qquad \{a^\dagger_\sigma, a^\dagger_\tau\} = \{a_\sigma, a_\tau\} = 0 \qquad (3.1)$$

where the fixed $1/2$ spin label is omitted and only its projection $\sigma, \tau = \pm 1/2$ is kept. The next step is to define the fermion generators

$$\mathcal{G}^\tau_\sigma \equiv a^\dagger_\sigma a_\tau, \qquad \sigma, \tau = \pm 1/2 \qquad (3.2)$$

and evaluate their commutation relations. We find, for example,

$$\begin{aligned}[\mathcal{G}_{+1/2}^{-1/2},\mathcal{G}_{-1/2}^{-1/2}] &= [a^\dagger_{+1/2}a_{-1/2}, a^\dagger_{-1/2}a_{-1/2}] \\ &= a^\dagger_{+1/2}\underbrace{a_{-1/2}a^\dagger_{-1/2}}a_{-1/2} - a^\dagger_{-1/2}\underbrace{a_{-1/2}a^\dagger_{+1/2}}a_{-1/2}\end{aligned} \tag{3.3}$$

We now apply the anticommutators (3.1) to the pairs indicated by the curly brackets to find

$$\begin{aligned}[\mathcal{G}_{+1/2}^{-1/2},\mathcal{G}_{-1/2}^{-1/2}] &= a^\dagger_{+1/2}(1 - a^\dagger_{-1/2}a_{-1/2})a_{-1/2} + \underbrace{a^\dagger_{-1/2}a^\dagger_{+1/2}}a_{-1/2}a_{-1/2} \\ &= a^\dagger_{+1/2}(1 - a^\dagger_{-1/2}a_{-1/2})a_{-1/2} - a^\dagger_{+1/2}a^\dagger_{-1/2}a_{-1/2}a_{-1/2} \\ &= a^\dagger_{+1/2}a_{-1/2}\end{aligned} \tag{3.4}$$

Note that the higher-order terms cancel since $a^\dagger_\sigma a^\dagger_\tau = -a^\dagger_\tau a^\dagger_\sigma$. In dealing with fermion systems one has to be particularly careful with sign changes and cancellations. Following the same procedure for all generators, we arrive at

$$\boxed{[\mathcal{G}^\tau_\sigma, \mathcal{G}^\gamma_\beta] = \mathcal{G}^\gamma_\sigma\delta_{\tau\beta} - \mathcal{G}^\tau_\beta\delta_{\sigma\gamma}, \qquad \sigma,\tau,\beta,\gamma = \pm 1/2} \tag{3.5}$$

which is identical to (1.19). The $\mathcal{G}^\tau_\sigma$ thus generate a U(2) algebra, which we denote by $U^F(2)$, while in this chapter we attach a superscript B to the corresponding boson algebra (1.19). Using (3.5) it is straightforward to show that the linear combinations of the generators (3.2),

$$\begin{aligned}\hat{S}_x &\equiv \tfrac{1}{2}(a^\dagger_{+1/2}a_{+1/2} - a^\dagger_{-1/2}a_{-1/2}) \\ \hat{S}_y &\equiv \tfrac{1}{2}(a^\dagger_{+1/2}a_{-1/2} - a^\dagger_{-1/2}a_{+1/2}) \\ \hat{S}_z &\equiv \tfrac{i}{2}(a^\dagger_{-1/2}a_{+1/2} - a^\dagger_{+1/2}a_{-1/2}) \\ \hat{M} &\equiv a^\dagger_{-1/2}a_{-1/2} + a^\dagger_{+1/2}a_{+1/2}\end{aligned} \tag{3.6}$$

satisfy the commutation relations

$$[\hat{S}_j, \hat{S}_k] = i\epsilon_{jkl}\hat{S}_l, \qquad j,k,l = x,y,z \tag{3.7}$$

and

$$[\hat{M}, \hat{S}_j] = 0, \qquad j = x,y,z \tag{3.8}$$

3.1 Fermion algebras

which define an $SU^F(2) \otimes U^F(1)$ algebra, isomorphic to $U^F(2)$. Note that the formal identification

$$s^\dagger s \mapsto a^\dagger_{+1/2} a_{+1/2}, \qquad s^\dagger t \mapsto a^\dagger_{+1/2} a_{-1/2} \\ t^\dagger s \mapsto a^\dagger_{-1/2} a_{+1/2}, \qquad t^\dagger t \mapsto a^\dagger_{-1/2} a_{-1/2} \tag{3.9}$$

transforms the boson operators (1.20) into (3.6). In analogy with the definition (1.106) of covariant operators it is again convenient to introduce

$$\tilde{a}_\sigma \equiv (-)^{1/2+\sigma} a_{-\sigma} \tag{3.10}$$

which transform under $SU^F(2)$ in the same way as the $a^\dagger_\sigma$. In terms of these modified operators, the spherical components of the spin (i.e., fermion angular momentum) operator may be written as

$$\hat{S}_\mu = \sqrt{\tfrac{1}{2}} [a^\dagger \times \tilde{a}]^{(1)}_\mu \tag{3.11}$$

while the fermion number operator is

$$\hat{M} = \sqrt{2} [a^\dagger \times \tilde{a}]^{(0)}_0 \tag{3.12}$$

which are again formally identical to the relations (1.107). The total spin operator is given by

$$\hat{S}^2 = \sum_\mu (-)^\mu \hat{S}_\mu \hat{S}_{-\mu} = \hat{S}_x^2 + \hat{S}_y^2 + \hat{S}_z^2 \tag{3.13}$$

which by direct substitution of (3.6) leads to

$$\hat{S}^2 = \tfrac{3}{4} \hat{M}(2 - \hat{M}) \tag{3.14}$$

This relation differs from the corresponding one in the boson case, (1.23), and reflects the fact that permissible fermion states in a $j = 1/2$ orbit are limited to $M = 0, 1, 2$. For $M = 1$ we obtain from (3.14) an eigenvalue of $3/4$ for $\hat{S}^2$, corresponding to $s = 1/2$, while for $M = 2$ we obtain $s = 0$, consistent with the fact that the $s = 1$ state is not allowed by the Pauli principle. In other words, the *representations* allowed for the fermion algebra $U^F(2)$ are different from the ones obtained with $U^B(2)$. In Chapter 1 it is shown that for a single type of boson the relevant representations $[N,0]$ are fully symmetric while in Chapter 2 the general representations $[h,h']$ are introduced, which appear when the U(2) algebra is realized in terms of two types of bosons. In the fermion case the allowed representations may be found using relations

$$h = \tfrac{1}{2}(M+2s), \qquad h' = \tfrac{1}{2}(M-2s) \qquad (3.15)$$

obtained from (2.38) with M (the fermion number) and s (the fermion spin) instead of N and j. Obviously, for the trivial case of zero fermions we have $[h,h'] = [0,0]$. For $M = 1$, $s = 1/2$ we find $[h,h'] = [1,0]$ and for $M = 2$, $s = 0$ we obtain $[h,h'] = [1,1]$. This result states the Pauli principle in terms of the representations of unitary algebras: For a many-fermion system only fully antisymmetric representations are allowed. It can be generalized to the representations of higher-order U(n) fermion algebras, as we shall see in the next sections. In terms of the Young tableaux defined in (2.50), the allowed representations are

$$\left.\begin{array}{c}\square\\\square\\\vdots\\\square\end{array}\right\}M \qquad (3.16)$$

where M cannot be greater than the n in U(n) (see Appendix B). In the following we also denote these antisymmetric representations as $[1^M]$, where it is understood that $[1^0] = [0,0,\ldots]$, $[1^1] = [1,0,\ldots]$, $[1^2] = [1,1,0,\ldots]$, and so on. For the $U^F(2)$ model, only $[0,0]$, $[1,0]$, and $[1,1]$ are admissible.

We now turn our attention to the construction of the fermion hamiltonian. Since our aim is to later combine the fermion algebras with the boson algebras of Chapter 1, we do not attach a physical meaning to the spin and allow fermion hamiltonians that are not necessarily scalar under $SU^F(2)$. In this case the most general one- and two-body, M-conserving, hermitian hamiltonian in the $a^\dagger_{-1/2}, a^\dagger_{+1/2}$ space takes the form

$$\boxed{\hat{H} = E_0 + \epsilon_+ a^\dagger_{+1/2} a_{+1/2} + \epsilon_- a^\dagger_{-1/2} a_{-1/2} + e\, a^\dagger_{+1/2} a^\dagger_{-1/2} a_{+1/2} a_{-1/2}}$$
$$(3.17)$$

3.1 Fermion algebras

All other two-body interactions vanish due to antisymmetry (e.g., $a^\dagger_\sigma a^\dagger_\sigma \mapsto 0$) and the structure of the $U^F(2)$ hamiltonian is thus very simple.

Next, we consider the dynamical symmetries present in the system. Since the $U^F(2)$ algebra (3.6) is isomorphic to the $U^B(2)$ algebra of Chapter 1, we may copy the algebra chains from that chapter. The first defines the $U^F(1)$ limit,

$$\begin{array}{ccc} U^F(2) & \supset & U^F(1) \equiv \hat{n}_- \\ | & & | \\ [1^M] & & n_- \end{array} \tag{3.18}$$

where we introduce the number operators $\hat{n}_\pm \equiv a^\dagger_{\pm 1/2} a_{\pm 1/2}$ with associated eigenvalues $n_\pm$. As shown in Chapter 1, the same basis states are obtained with the algebra chain

$$\begin{array}{ccc} U^F(2) & \supset & \overline{SO}^F(2) \equiv \hat{S}_x \\ | & & | \\ [1^M] & & (M - 2n_-)/2 \end{array} \tag{3.19}$$

The second chain of algebras gives rise to the $SO^F(2)$ limit,

$$\begin{array}{cccc} U^F(2) & \supset & SU^F(2) & \supset & SO^F(2) \equiv \hat{S}_z \\ | & & | & & | \\ [1^M] & & s & & \mu \end{array} \tag{3.20}$$

The states in each case are very simple to construct, since $M = 1$ or $M = 2$. For the $U^F(1)$ chain they are

$$|[1,0]n_-\rangle = (a^\dagger_{+1/2})^{1-n_-}(a^\dagger_{-1/2})^{n_-}|0\rangle, \qquad n_- = 0, 1 \tag{3.21}$$

and

$$|[1,1]1\rangle = a^\dagger_{+1/2} a^\dagger_{-1/2}|0\rangle \tag{3.22}$$

For the $SO^F(2)$ chain we may proceed again in analogy with the analysis carried out in Chapter 1, rotating the $\overline{SO}^F(2)$ (or $U^F(1)$) states. We only write down the result:

$$\begin{aligned} |[1,0]1/2\mu\rangle &= \sqrt{\tfrac{1}{2}}(a^\dagger_{+1/2} - ia^\dagger_{-1/2})^{1/2-\mu} \\ &\quad \times (a^\dagger_{+1/2} + ia^\dagger_{-1/2})^{1/2+\mu}|0\rangle, \quad \mu = \pm 1/2 \end{aligned} \tag{3.23}$$

and
$$|[1,1]00\rangle = a^\dagger_{+1/2} a^\dagger_{-1/2}|0\rangle \qquad (3.24)$$

Note that the states (3.22) and (3.24) are identical. This is always the case for a unique state: it always is simultaneously classified by all possible chains of algebras.

In terms of the invariant operators in the chains (3.18) and (3.20), the hamiltonian (3.17) can be expressed as

$$\hat{H} = E'_0 + \epsilon \hat{n}_- + \beta \hat{S}_z^2 \qquad (3.25)$$

where

$$\begin{aligned} E'_0 &= E_0 + \left(\epsilon_+ - \tfrac{1}{2}e\right) M \\ \epsilon &= \epsilon_- - \epsilon_+ \\ \beta &= 2e \end{aligned} \qquad (3.26)$$

Comparing (3.25) with the corresponding boson hamiltonian (1.40), we note that the former does not contain an $\hat{n}_-^2$ term which vanishes because of antisymmetry. Also, the $\hat{S}_z^2$ operator does not contribute to the energy spectrum as it gives identical matrix elements for the two states with $M = 1$ and no contribution for the state with $M = 2$. The $U^F(1)$ symmetry arises for $\beta = 0$ in (3.25) while the $SO^F(2)$ symmetry corresponds to $\epsilon = 0$. Because of the isomorphism that exists between $U^F(2)$ and $U^B(2)$, the definition of tensor operators, the evaluation of matrix elements, and all other algebraic operations in the fermion space are analogous to the ones described in Chapter 1.

Having introduced fermion algebras in the context of U(2), we proceed in the next section to study the combination of boson and fermion algebras.

3.2 Boson–fermion algebras

Algebraic techniques in both molecular and nuclear structure physics were originally applied to boson systems to describe rotation–vibration spectra. In the later stages of the development of algebraic models it became necessary to incorporate fermion degrees of freedom. In odd-mass and odd–odd nuclei (i.e., nuclei with odd numbers of neutrons

3.2 Boson–fermion algebras

and/or protons) the unpaired fermion(s) must be treated explicitly, and a similar approach is needed for the electronic degrees of freedom in molecules. The treatment of these more complex systems or situations requires, in an algebraic framework, the introduction of boson–fermion algebras. The purpose of this section is to investigate boson–fermion algebras in the context of the schematic U(2) model.

In Section 3.1 it is shown that $U^B(2)$ and $U^F(2)$ are isomorphic algebras. For the purpose of the present discussion it is useful to introduce a slightly different notation for the generators of these algebras. Instead of characterizing the generators of $U^B(2)$ as $\mathcal{G}_i^j$ with indices $i,j = 1,2$, we denote them here by $\mathcal{G}_{B\sigma}^{B\tau}$ with $\sigma, \tau = \pm 1/2$, in analogy with the notation (3.2) introduced for the fermion generators. In addition, boson and fermion generators are distinguished by extra indices B and F. To avoid confusion, we give the explicit definitions in the new notation:

$$\mathcal{G}_{B+1/2}^{B+1/2} \equiv s^\dagger s, \qquad \mathcal{G}_{B+1/2}^{B-1/2} \equiv s^\dagger t, \qquad \mathcal{G}_{B-1/2}^{B+1/2} \equiv t^\dagger s, \qquad \mathcal{G}_{B-1/2}^{B-1/2} \equiv t^\dagger t \tag{3.27}$$

and

$$\mathcal{G}_{F\sigma}^{F\tau} \equiv a_\sigma^\dagger a_\tau, \qquad \sigma, \tau = \pm 1/2 \tag{3.28}$$

Since boson and fermion operators commute,

$$[\mathcal{G}_{B\sigma}^{B\tau}, \mathcal{G}_{F\beta}^{F\gamma}] = 0, \qquad \sigma, \tau, \beta, \gamma = \pm 1/2 \tag{3.29}$$

the set of operators $(\mathcal{G}_{B\sigma}^{B\tau}, \mathcal{G}_{F\beta}^{F\gamma})$ define the direct-product algebra

$$U^B(2) \otimes U^F(2) \tag{3.30}$$

which is the dynamical algebra for the combined boson–fermion system. Note again the analogy with the two-boson system discussed in Chapter 2. Alternatively, we may use the isomorphism $U(2) \simeq SU(2) \otimes U(1)$ and consider

$$SU^B(2) \otimes SU^F(2) \tag{3.31}$$

as the dynamical algebra. The $SU^B(2)$ representation is fixed by the number of bosons according to the relation (see (1.24))

$$j = N/2 \tag{3.32}$$

while the $SU^F(2)$ representation is determined by the fermion number, as explained in the previous section.

The most general one- and two-body hamiltonian for the coupled system may be written as

$$\hat{H} = \hat{H}^B + \hat{H}^F + \hat{V}^{BF} \qquad (3.33)$$

where $\hat{H}^B$ is the $U^B(2)$ hamiltonian (1.36) and $\hat{H}^F$ is the $U^F(2)$ hamiltonian (3.17). The interaction term $\hat{V}^{BF}$, which should separately conserve M and N, is given by

$$\begin{aligned}
\hat{V}^{BF} =\ & v_1 s^\dagger a^\dagger_{+1/2} s a_{+1/2} + v_2 t^\dagger a^\dagger_{-1/2} t a_{-1/2} \\
& + v_3 (s^\dagger a^\dagger_{+1/2} t a_{-1/2} + t^\dagger a^\dagger_{-1/2} s a_{+1/2}) \\
& + v_4 s^\dagger a^\dagger_{-1/2} s a_{-1/2} + v_5 t^\dagger a^\dagger_{+1/2} t a_{+1/2} \\
& + v_6 (s^\dagger a^\dagger_{-1/2} t a_{+1/2} + t^\dagger a^\dagger_{+1/2} s a_{-1/2}) \\
& + v_7 (s^\dagger a^\dagger_{+1/2} s a_{-1/2} + s^\dagger a^\dagger_{-1/2} s a_{+1/2}) \\
& + v_8 (t^\dagger a^\dagger_{+1/2} t a_{-1/2} + t^\dagger a^\dagger_{-1/2} t a_{+1/2}) \\
& + v_9 (s^\dagger a^\dagger_{+1/2} t a_{+1/2} + t^\dagger a^\dagger_{+1/2} s a_{+1/2}) \\
& + v_{10} (s^\dagger a^\dagger_{-1/2} t a_{-1/2} + t^\dagger a^\dagger_{-1/2} s a_{-1/2}) \qquad (3.34)
\end{aligned}$$

Since we have not defined parity in the fermion space, the condition of parity invariance in the boson space is relaxed in the expansion (3.34), and as a consequence more interaction terms are possible than in the corresponding expression (2.13) for a two-boson system. Since in this approach parity is irrelevant, we will, in the the discussion of the symmetry limits in the next section, not worry about the $O^B(2)$ boson states (1.67) with good parity but consider the simpler $SO^B(2)$ states (1.64) instead.

3.3 Symmetry limits

As for the other systems discussed in the previous chapters, the solution of the boson–fermion hamiltonian (3.33) requires the construction of complete sets of states in which to diagonalize it. In analogy with

3.3 Symmetry limits

Section 2.3 we proceed to study for the boson–fermion case the dynamical symmetries present in the system.

Two of the boson–fermion dynamical symmetries are of a trivial nature, namely, those arising from the direct product of $U(2) \supset U(1)$ or $U(2) \supset SO(2)$ algebras:

$$\begin{array}{ccccc} U^B(2) \otimes U^F(2) & \supset & U^B(1) \otimes U^F(1) & \supset & U^{BF}(1) \\ | & & | \quad\quad | & & | \\ [N] & [1^M] & n_t \quad\quad n_- & & n_t + n_- \end{array} \tag{3.35}$$

and

$$\begin{array}{ccccc} U^B(2) \otimes U^F(2) & \supset & SO^B(2) \otimes SO^F(2) & \supset & SO^{BF}(2) \\ | & & | \quad\quad | & & | \\ [N] & [1^M] & \mu_b \quad\quad \mu_f & & \mu_b + \mu_f \end{array} \tag{3.36}$$

where we use the notation μ_b and μ_f to distinguish the SO(2) quantum numbers for bosons and fermions.

The corresponding states are simple products of the dynamical-symmetry states introduced in Chapter 1 and Section 3.1,

$$|[N]n_t, [1^M]n_-; n \equiv n_t + n_-\rangle \equiv |[N]n_t\rangle|[1^M]n_-\rangle \tag{3.37}$$

and

$$|[N]\mu_b, [1^M]\mu_f; \mu \equiv \mu_b + \mu_f\rangle \equiv |[N]\mu_b\rangle|[1^M]\mu_f\rangle \tag{3.38}$$

which diagonalize the hamiltonian (3.33) not only for $\hat{V}^{BF} = 0$ but also for particular interactions involving the operators

$$\hat{n} \equiv \hat{n}_t + \hat{n}_- \tag{3.39}$$

and

$$\hat{K}_z \equiv \hat{J}_z + \hat{S}_z \tag{3.40}$$

respectively. The dynamical-symmetry hamiltonians take the form

$$\hat{H}_{Ib} = \epsilon_b \hat{n}_t + \epsilon_f \hat{n}_- + \alpha_b \hat{n}_t^2 + \alpha \hat{n}^2 \tag{3.41}$$

and

$$\hat{H}_{IIb} = \beta_b \hat{J}_z^2 + \beta_f \hat{S}_z^2 + \beta \hat{K}_z^2 \tag{3.42}$$

where we have omitted terms that do not contribute to the binding energy. The corresponding energy eigenvalues are given by

$$E_{Ib}(n_t, n_-, n) = \epsilon_b n_t + \epsilon_f n_- + \alpha_b n_t^2 + \alpha n^2 \qquad (3.43)$$

and

$$E_{IIb}(\mu_b, \mu_f, \mu) = \beta_b \mu_b^2 + \beta_f \mu_f^2 + \beta \mu^2 \qquad (3.44)$$

The relevant branching rules are given in Chapter 1 and Section 3.1 and allow one to construct the associated energy spectra, examples of which are shown in Figures 3.1 and 3.2 for $M = 1$.

Due to the commutation of boson and fermion generators, (3.29), it is straightforward to show that the sum of the $U^B(2)$ and $U^F(2)$ generators

$$\boxed{\mathcal{G}_\sigma^\tau \equiv \mathcal{G}_{B\sigma}^{B\tau} + \mathcal{G}_{F\sigma}^{F\tau}, \qquad \sigma, \tau = \pm 1/2} \qquad (3.45)$$

also satisfies the U(2) commutation relations (1.19). The four operators (3.45) generate a U(2) subalgebra of the boson–fermion algebra (3.30) which contains as subalgebras U(1) and SO(2) generated by the operators $\hat{n}$ and $\hat{K}_z$ defined in (3.39) and (3.40). We thus find two additional

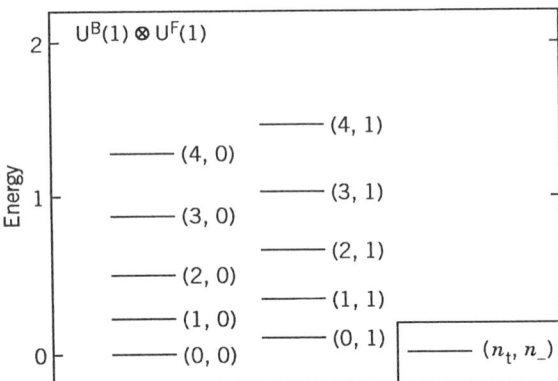

Figure 3.1: The $U^B(1) \otimes U^F(1)$ eigenspectrum of the hamiltonian $\hat{H}_{Ib}$ with $\epsilon_b = 0.2$, $\epsilon_f = 0.1$, $\alpha_b = 0.02$, and $\alpha = 0.01$ for $N = 4$ bosons and $M = 1$ fermion. Levels are labeled by n_t and n_-. Energies are in arbitrary units.

3.3 Symmetry limits

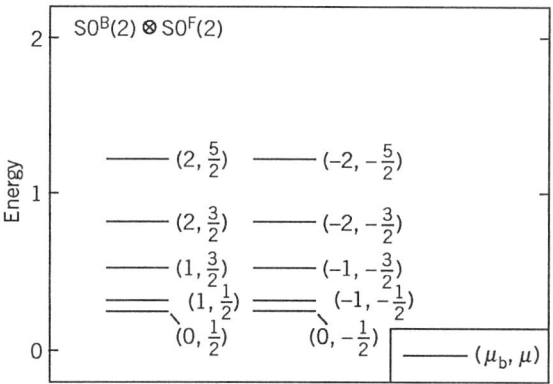

Figure 3.2: The $\mathrm{SO^B}(2) \otimes \mathrm{SO^F}(2)$ eigenspectrum of the hamiltonian $\hat{H}_{\mathrm{II}b}$ with $\beta_b = 0.3$, $\beta_f = 0$, and $\beta = 0.1$ for $N = 4$ bosons and $M = 1$ fermion. Levels are labeled by μ_b and μ. Energies are in arbitrary units.

dynamical symmetries

$$\boxed{\begin{array}{cccc} \mathrm{U^B}(2) \otimes \mathrm{U^F}(2) & \supset & \mathrm{U^{BF}}(2) & \supset \mathrm{U^{BF}}(1) \\ | & | & | & | \\ [N] & [1^M] & [h,h'] & n \end{array}} \qquad (3.46)$$

and

$$\boxed{\begin{array}{cccc} \mathrm{U^B}(2) \otimes \mathrm{U^F}(2) & \supset & \mathrm{U^{BF}}(2) & \supset \mathrm{SO^{BF}}(2) \\ | & | & | & | \\ [N] & [1^M] & [h,h'] & \mu \end{array}} \qquad (3.47)$$

Using the isomorphism with $\mathrm{U}(2) \simeq \mathrm{SU}(2) \otimes \mathrm{U}(1)$, we may also express these chains as

$$\begin{array}{cccc} \mathrm{SU^B}(2) \otimes \mathrm{SU^F}(2) & \supset & \mathrm{SU^{BF}}(2) & \supset \mathrm{U^{BF}}(1) \\ | & | & | & | \\ j & s & k & n \end{array} \qquad (3.48)$$

and

$$\begin{array}{cccc} \mathrm{SU^B}(2) \otimes \mathrm{SU^F}(2) & \supset & \mathrm{SU^{BF}}(2) & \supset \mathrm{SO^{BF}}(2) \\ | & | & | & | \\ j & s & k & \mu \end{array} \qquad (3.49)$$

To find the various branching rules in (3.46)–(3.49), we begin with (3.49) in which case they follow from angular momentum coupling rules:

$$k = |j - s|, |j - s| + 1, \ldots, j + s, \quad \mu = -k, -k+1, \ldots, k \quad (3.50)$$

The $SU^{BF}(2) \supset U^{BF}(1)$ branching rule in (3.48) may be obtained, following the procedure described in Section 2.3, by means of the $\overline{SO}^{BF}(2)$ algebra generated by

$$\hat{K}_x \equiv \hat{J}_x + \hat{S}_x \quad (3.51)$$

We find

$$\tfrac{1}{2}(N + M + 2k) \geq n \geq \tfrac{1}{2}(N + M - 2k) \quad (3.52)$$

Since N is related to j ($N = 2j$) and M is determined by s ($M = 1$ for $s = 1/2$ and $M = 2$ for $s = 0$), (3.52) gives us the desired branching rule. This rule also shows that *two* labels are needed to characterize a $U^{BF}(2)$ representation and, as in Section 2.3, we define

$$h \equiv \tfrac{1}{2}(N + M + 2k), \qquad h' \equiv \tfrac{1}{2}(N + M - 2k) \quad (3.53)$$

such that the $U^{BF}(2) \supset U^{BF}(1)$ branching rule can be written as

$$h \geq n \geq h' \quad (3.54)$$

The branching rules in (3.46) and (3.47) can now be derived from those in (3.47) and (3.48) since all labels in the former can be related to those in the latter. This allows, for example, to derive the $U^B(2) \otimes U^F(2)$ multiplication rule $[N] \otimes [1^M]$. For $M = 1$ we have $s = 1/2$, hence $k = j \pm 1/2$, which, using (3.53), implies

$$[N] \otimes [1] = [N+1, 0] \oplus [N, 1] \quad (3.55)$$

which is a particular case of the multiplication (2.42). For $M = 2$ we have $s = 0$ and $k = j$ and deduce

$$[N] \otimes [1,1] = [N+1, 1] \quad (3.56)$$

or, in terms of Young tableaux,

$$\underbrace{\square\square\cdots\square}_{N} \otimes \begin{array}{c}\square\\\square\end{array} = \begin{array}{c}\overbrace{\square\square\square\cdots\square}^{N+1}\\\square\end{array} \quad (3.57)$$

3.3 Symmetry limits

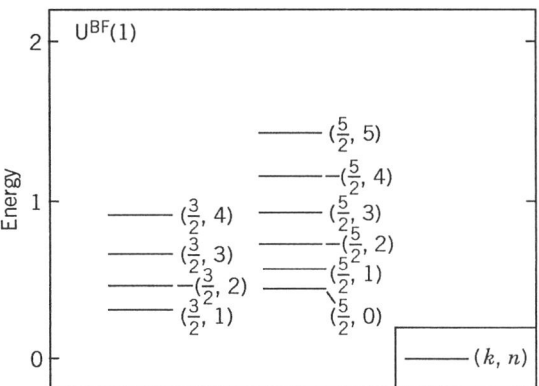

Figure 3.3: The $U^{BF}(1)$ eigenspectrum of the hamiltonian $\hat{H}_{Ia}$ with $\gamma = 0.5$, $\epsilon = 0.1$, and $\alpha = 0.02$ for $N = 4$ bosons and $M = 1$ fermion. Levels are labeled by k and n. Energies are in arbitrary units.

This kind of product does not occur in Chapter 2, where only symmetric-with-symmetric product representations are considered. For a discussion of products of arbitrary $U(n)$ representations we refer the reader to Appendix B.

The $U^B(2) \otimes U^F(2)$ hamiltonian (3.33) may be expressed in terms of the invariant operators associated with the chains of algebras (3.46)–(3.49) and then diagonalized in any of these bases. The $U^{BF}(1)$ hamiltonian is

$$\hat{H}_{Ia} = \gamma \hat{K}^2 + \epsilon \hat{n} + \alpha \hat{n}^2 \tag{3.58}$$

with eigenvalues

$$E_{Ia}(k,n) = \gamma k(k+1) + \epsilon n + \alpha n^2 \tag{3.59}$$

where the values of the quantum numbers k and n are given by (3.50) and (3.52). An example of a $U^{BF}(1)$ spectrum is shown in Figure 3.3.

The $U^{BF}(1)$ wave functions may be constructed again resorting to the $\overline{SO}^{BF}(2)$ subalgebra of $U^{BF}(2)$, with quantum number σ, associated with the operator $\hat{K}_x$ in (3.51). From its definition and the definitions of $\hat{J}_x$ and $\hat{S}_x$ in (1.20) and (3.6) we find

$$\sigma = \tfrac{1}{2}(N + M - 2n) \tag{3.60}$$

a result used previously to arrive at the reduction (3.53). We may now construct the $U^{BF}(1)$ wave functions by angular momentum coupling,

$$|j,s;k\sigma\rangle = \sum_{\sigma_b \sigma_f f} \langle j\sigma_b\, s\sigma_f | k\sigma \rangle |j\sigma_b\rangle |s\sigma_f\rangle \qquad (3.61)$$

where

$$|j\sigma_b\rangle = \frac{(s^\dagger)^{j+\sigma_b}(t^\dagger)^{j-\sigma_b}}{\sqrt{(j-\sigma_b)!(j+\sigma_b)!}}|0\rangle \qquad (3.62)$$

and $|s\sigma_f\rangle$ denotes the fermion states (3.21) and (3.22), with $\sigma_f = (M - 2n_-)/2$. In the $U^{BF}(1)$ limit both $\hat{N}+\hat{M}$ and $\hat{K}^2$ are invariant operators. Alternatively, we may use the second-order Casimir invariant of $U^{BF}(2)$ which has the form

$$\mathcal{C}_2[U^{BF}(2)] \equiv \sum_{\sigma\tau} \mathcal{G}_\tau^\sigma \mathcal{G}_\sigma^\tau \qquad (3.63)$$

with the $\mathcal{G}_\sigma^\tau$ defined in (3.45). In terms of $\hat{K}^2$ and $\hat{N}+\hat{M}$ it takes the form

$$\mathcal{C}_2[U^{BF}(2)] = \tfrac{1}{2}(\hat{N}+\hat{M})^2 + 2\hat{K}^2 \qquad (3.64)$$

From the definition of the Gel'fand labels (3.53) we find the relations

$$N+M = h+h', \qquad k = \tfrac{1}{2}(h-h') \qquad (3.65)$$

which implies the following eigenvalue of $\mathcal{C}_2[U^{BF}(2)]$:

$$\tfrac{1}{2}(N+M)^2 + 2K^2 = h(h+1) + h'(h'-1) \qquad (3.66)$$

This eigenvalue is of the general U(2) form (2.65), as expected. The second-order Casimir invariant (3.64) can shift the relative positions of symmetric and non-symmetric representations (e.g., the $U^{BF}(2)$ representations $[N+1,0]$ and $[N,1]$ that occur for $M=1$). A similar result turns out to be very useful in the case of the U(6) boson–fermion algebras used in the nuclear interacting boson model.

The $SO^{BF}(2)$ limit corresponds to the choice of hamiltonian

$$\hat{H}_{IIa} = \gamma \hat{K}^2 + \beta \hat{K}_z^2 \qquad (3.67)$$

with eigenvalues

$$E_{IIa}(k,\mu) = \gamma k(k+1) + \beta \mu^2 \qquad (3.68)$$

3.3 Symmetry limits

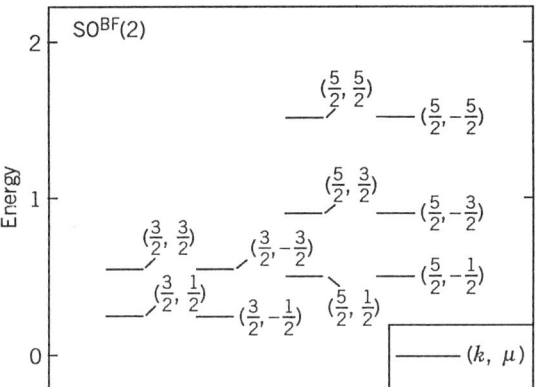

Figure 3.4: The $SO^{BF}(2)$ eigenspectrum of the hamiltonian $\hat{H}_{IIa}$ with $\gamma = 0.05$ and $\beta = 0.2$ for $N = 4$ bosons and $M = 1$ fermion. Levels are labeled by k and μ. Energies are in arbitrary units.

and branching rules for k and μ given in (3.50). An example of the energy spectrum is shown in Figure 3.4.

The $SO^{BF}(2)$ wave functions are obtained by angular momentum coupling,

$$|j,s;k\mu\rangle = \sum_{\mu_b \mu_f} \langle j\mu_b \, s\mu_f | k\mu\rangle |j\mu_b\rangle |s\mu_f\rangle \qquad (3.69)$$

with $|j\mu_b\rangle$ and $|s\mu_f\rangle$ given by (1.64), (3.23), and (3.24).

Relations (3.61) and (3.69) establish the transformation brackets between the $U^B(1) \otimes U^F(1)$ and $U^{BF}(1)$ states, and between the $SO^B(2) \otimes SO^F(2)$ and $SO^{BF}(2)$ states, respectively. Using the unitary character of the Clebsch–Gordan coefficients, we may invert these relationships and find, for example, in the case of (3.69)

$$|j\mu_b\rangle |s\mu_f\rangle = \sum_{k\mu} \langle j\mu_b \, s\mu_f | k\mu\rangle |j,s;k\mu\rangle \qquad (3.70)$$

Finally, the transformation brackets between $U^{BF}(1)$ and $SO^{BF}(2)$ are a particular case of those found in Sections 1.6 and 2.6,

$$|j,s;k\mu\rangle = \sum_\sigma D^k_{\sigma\mu}(-\pi/2, -\pi/2, 0)|j,s;k\sigma\rangle \qquad (3.71)$$

Using these results, one may compute all matrix elements of interest in one or another of the limiting bases. In the next section we analyze an alternative way to evaluate matrix elements, based on a boson–fermion tensor calculus.

3.4 Tensor calculus

The concept of tensor operators and the associated tensor calculus introduced in Section 1.7 can be generalized to other chains of algebras and, in particular, to the boson–fermion algebras discussed in this chapter. Since the SU(2) version of this calculus was reviewed in Chapter 1 and because it can be applied with minor modifications to the $SU^{BF}(2)$ model, it is desirable to carry out the discussion in this section directly in terms of the U(2) algebras. In this way the results, although equivalent to the SU(2) calculus introduced previously, are more easily seen to be in close correspondence with the results in higher-dimensional U(n) algebras that are treated in later chapters. We use the algebra chain (3.46)

$$\begin{array}{cccc} U^B(2) \otimes U^F(2) & \supset U^{BF}(2) & \supset U^{BF}(1) \\ | & | & | & | \\ [N] & [1^M] & [h,h'] & n \end{array} \quad (3.72)$$

to illustrate the technique, although we could have equally worked in any of the other three limits discussed in the previous section.

In analogy with the SU(2) case we start by defining the $U^{BF}(2) \supset U^{BF}(1)$ tensor operators $\hat{T}_n^{[h,h']}$, obeying the transformation property

$$\hat{U}\hat{T}_n^{[h,h']}\hat{U}^{-1} = \sum_{n'} D_{n'n}^{[h,h']} \hat{T}_{n'}^{[h,h']} \quad (3.73)$$

where $\hat{U}$ denotes a unitary transformation and the $D_{n'n}^{[h,h']}$ correspond to the representations of the group U(2),

$$D_{n'n}^{[h,h']} = \langle [h,h']n'|\hat{U}|[h,h']n \rangle \quad (3.74)$$

Likewise, the $U^{BF}(1)$ states transform in the same way as (3.73), in analogy with the SU(2) result (1.90). These matrices can in fact be identified with the Wigner **D** matrices in the SU(2) $\supset \overline{SO}(2)$ chain.

3.4 Tensor calculus

In higher dimensions there is no such simple relationship with SU(2) and the matrices are only known explicitly in some cases. The $\hat{U}$ operators can always be written in terms of exponentials of the group generators. Thus, for an infinitesimal transformation, we may write (see Appendix B)

$$\hat{U} = 1 + \epsilon^i \hat{X}_i + \cdots \qquad (3.75)$$

where ϵ^i are infinitesimal parameters and the $\hat{X}_i$ are generators of the algebra, in our case given by the four operators (3.45). Inserting this expansion into (3.73) and keeping terms to first order in ϵ^i, we find

$$[\hat{X}_i, \hat{T}_n^{[h,h']}] = \sum_{n'} \langle [h,h']n' | \hat{X}_i | [h,h']n \rangle \hat{T}_{n'}^{[h,h']} \qquad (3.76)$$

While the tensor operators transform under the generators as in (3.76), it should be clear that under infinitesimal operations the wave functions transform in a similar manner:

$$\hat{X}_i | [h,h']n \rangle = \sum_{n'} \langle [h,h']n' | \hat{X}_i | [h,h']n \rangle | [h,h']n' \rangle \qquad (3.77)$$

Relations (3.76) and (3.77) are very useful to identify the $U^{BF}(2) \supset U^{BF}(1)$ tensor character of boson and fermion operators. As a simple example, from (3.21) we identify the fermion tensor operators

$$\hat{F}_0^{[1,0]} = a_{+1/2}^\dagger, \qquad \hat{F}_1^{[1,0]} = a_{-1/2}^\dagger, \qquad (3.78)$$

while, from the states (1.47) and the relations (2.49), we find

$$\hat{B}_0^{[1,0]} = s^\dagger, \qquad \hat{B}_1^{[1,0]} = t^\dagger, \qquad (3.79)$$

where boson and fermion tensor operators are distinguished by the notation $\hat{B}$ and $\hat{F}$, respectively. Note, however, that boson and fermion creation operators have identical properties under $U^{BF}(2)$. Powers of boson creation operators have tensor properties directly derivable from the eigenstates (1.47), which transform as the $[N,0]$ representation of $U^{BF}(2)$. In the case of powers of fermion creation operators, only the tensor operator

$$\hat{F}_1^{[1,1]} = a_{+1/2}^\dagger a_{-1/2}^\dagger \qquad (3.80)$$

exists. For combined powers of boson and fermion creation operators we note from the $U^B(2) \otimes U^F(2)$ coupling rules (3.55) and (3.56) that both symmetric $[N+1,0]$ and non-symmetric tensors $[N,1]$ exist for $M=1$ while only antisymmetric tensors $[N+1,1]$ exist for $M=2$.

The antisymmetric tensor (3.80) is unique, that is, there is a single state associated with this representation, corresponding to the full $s = 1/2$ fermion shell. In this sense it transforms in the same way as the vacuum, and this enables us to derive the transformation properties of the fermion annihilation operators. These may be thought of as acting on the "full-shell" state $|\tilde{0}\rangle \equiv a^\dagger_{+1/2} a^\dagger_{-1/2}|0\rangle$, so

$$a_{+1/2}|\tilde{0}\rangle = a^\dagger_{-1/2}|0\rangle, \qquad a_{-1/2}|\tilde{0}\rangle = a^\dagger_{+1/2}|0\rangle \qquad (3.81)$$

and hence they also transform as the $[1,0]$ representation of $U^{BF}(2)$. However, we see from (3.81) that the $U^{BF}(1)$ label is not the same as the one of the corresponding creation operator. The modified states (3.10), on the other hand, do transform as creation operators of the same projection:

$$\tilde{a}_{+1/2}|\tilde{0}\rangle = a^\dagger_{+1/2}|0\rangle, \qquad \tilde{a}_{-1/2}|\tilde{0}\rangle = a^\dagger_{-1/2}|0\rangle \qquad (3.82)$$

This is the $U(2)$ version of the previously mentioned $SU(2)$ tensor character of the modified operators.

The above argument uses the fact that $|0\rangle$ and $|\tilde{0}\rangle$ transform identically, that is, $[1,1]$ and $[0,0]$ are equivalent, where the "identity representation" is associated to the (fermion) vacuum. This is a familiar result from atomic and nuclear physics, where closed shells are identified with the vacuum. The argument can be generalized to $U(n)$, where the a_σ annihilation operator transforms as the $[1^{n-1}] = [1,1,\ldots,1,0]$ representation of $U(n)$.

The $U^{BF}(2)$ generators, being products of creation and annihilation operators, must correspond to the product

$$[1,0] \otimes [1,0] = [2,0] \oplus [1,1] \qquad (3.83)$$

so they must belong to a mixture of $[2,0]$ and $[1,1]$ representations. In terms of the $SU(2)$ label k, the $[1,1]$ representation corresponds to $k=0$ while $[2,0]$ corresponds to $k=1$. This means that we may

3.4 Tensor calculus

readily identify the combination $\hat{M} = a^\dagger_{+1/2} a_{+1/2} + a^\dagger_{-1/2} a_{-1/2}$ (which is an SU(2) scalar) with $\hat{F}_1^{[1,1]}$,

$$\hat{T}_1^{[1,1]} = a^\dagger_{+1/2} a_{+1/2} + a^\dagger_{-1/2} a_{-1/2} = \hat{M} \tag{3.84}$$

The $[2,0]$ tensors with definite U(1) projection are $\hat{T}_n^{[2,0]}$, with $n = 0, 1, 2$, and should be linear combinations of the fermion angular momentum operators $\hat{S}_\mu$, since they correspond to the $k = 1$ SU(2) tensor components. To find their precise form, we need to know the U(2) $\supset$ U(1) coupling coefficients. These are defined in analogy with the SU(2) Clebsch–Gordan coefficients as elements of a unitary transformation which takes the direct-product representation $|[h_1, h'_1]n_1\rangle |[h_2, h'_2]n_2\rangle$ into a direct sum of U(2) $\supset$ U(1) representations $|[h, h']n\rangle$:

$$|[h_1, h'_1]n_1\rangle|[h_2, h'_2]n_2\rangle = \sum_{hh'n} \left\langle \begin{array}{cc} [h_1, h'_1] & [h_2, h'_2] \\ n_1 & n_2 \end{array} \middle| \begin{array}{c} [h, h'] \\ n \end{array} \right\rangle |[h, h']n\rangle \tag{3.85}$$

where the symbol in angle brackets is a coupling coefficient [5], in this case pertaining to U(2) $\supset$ U(1). For notational convenience we denote this coefficient as $\langle [h_1, h'_1]n_1 \otimes [h_2, h'_2]n_2 | [h, h']n \rangle$ when it appears in the text (as opposed to equations). Because of the unitary character of the transformation, the coupling coefficients satisfy the usual orthonormality relations in rows and columns:

$$\sum_{n_1 n_2} \left\langle \begin{array}{cc} [h_1, h'_1] & [h_2, h'_2] \\ n_1 & n_2 \end{array} \middle| \begin{array}{c} [h_a, h'_a] \\ n_a \end{array} \right\rangle$$
$$\times \left\langle \begin{array}{cc} [h_1, h'_1] & [h_2, h'_2] \\ n_1 & n_2 \end{array} \middle| \begin{array}{c} [h_b, h'_b] \\ n_b \end{array} \right\rangle = \delta_{h_a h_b} \delta_{h'_a h'_b} \delta_{n_a n_b}$$
$$\sum_{hh'n} \left\langle \begin{array}{cc} [h_1, h'_1] & [h_2, h'_2] \\ n_{1a} & n_{2a} \end{array} \middle| \begin{array}{c} [h, h'] \\ n \end{array} \right\rangle$$
$$\times \left\langle \begin{array}{cc} [h_1, h'_1] & [h_2, h'_2] \\ n_{1b} & n_{2b} \end{array} \middle| \begin{array}{c} [h, h'] \\ n \end{array} \right\rangle = \delta_{n_{1a} n_{1b}} \delta_{n_{2a} n_{2b}} \tag{3.86}$$

which are analogous to relations (1.95). We may thus invert (3.85) to find

$$|[h, h']n\rangle = \sum_{n_1 n_2} \left\langle \begin{array}{cc} [h_1, h'_1] & [h_2, h'_2] \\ n_1 & n_2 \end{array} \middle| \begin{array}{c} [h, h'] \\ n \end{array} \right\rangle |[h_1, h'_1]n_1\rangle|[h_2, h'_2]n_2\rangle \tag{3.87}$$

We have taken the trouble to define the U(2) ⊃ U(1) coupling coefficients since they will appear in similar form in the higher-dimensional models. In this case, as should be clear from the previous sections, we may express them directly in terms of SU(2) ⊃ $\overline{\text{SO}}$(2) coefficients. Using (3.60) and (3.65), we obtain the relation

$$\left\langle \begin{matrix} [h_1, h_1'] & [h_2, h_2'] \\ n_1 & n_2 \end{matrix} \middle| \begin{matrix} [h, h'] \\ n \end{matrix} \right\rangle = \langle j_1 \sigma_1 \, j_2 \sigma_2 | j\sigma \rangle \tag{3.88}$$

where

$$\begin{aligned} j_1 &= \tfrac{1}{2}(h_1 - h_1'), & \sigma_1 &= \tfrac{1}{2}(h_1 + h_1' - 2n_1) \\ j_2 &= \tfrac{1}{2}(h_2 - h_2'), & \sigma_2 &= \tfrac{1}{2}(h_2 + h_2' - 2n_2) \\ j &= \tfrac{1}{2}(h - h'), & \sigma &= \tfrac{1}{2}(h + h' - 2n) \end{aligned} \tag{3.89}$$

Using the known values of the Clebsch–Gordan coefficients, we may now construct the $\hat{T}_n^{[2,0]}$ fermion tensors,

$$\hat{T}_n^{[2,0]} = \sum_{n_1 n_2} \left\langle \begin{matrix} [1,0] & [1,0] \\ n_1 & n_2 \end{matrix} \middle| \begin{matrix} [2,0] \\ n \end{matrix} \right\rangle a^\dagger_{1/2 - n_1} \tilde{a}_{1/2 - n_2} \tag{3.90}$$

and compute them explicitely. We find

$$\begin{aligned} \hat{T}_0^{[2,0]} &= -a^\dagger_{+1/2} a_{-1/2} = -\hat{S}_y - i\hat{S}_z \\ \hat{T}_1^{[2,0]} &= \sqrt{\tfrac{1}{2}}(a^\dagger_{+1/2} a_{+1/2} - a^\dagger_{-1/2} a_{-1/2}) = \sqrt{2}\hat{S}_x \\ \hat{T}_2^{[2,0]} &= a^\dagger_{-1/2} a_{+1/2} = \hat{S}_y - i\hat{S}_z \end{aligned} \tag{3.91}$$

with $\hat{S}_x$, $\hat{S}_y$, and $\hat{S}_z$ defined in (3.6). In similar fashion we can construct higher-order boson–fermion tensors.

We may now compute matrix elements in a straigthforward fashion. The Wigner–Eckart theorem for the U(2) ⊃ U(1) chain becomes

$$\langle [h_a, h_a'] n_a | \hat{T}_n^{[h,h']} | [h_b, h_b'] n_b \rangle$$
$$= \left\langle \begin{matrix} [h_b, h_b'] & [h, h'] \\ n_b & n \end{matrix} \middle| \begin{matrix} [h_a, h_a'] \\ n_a \end{matrix} \right\rangle \langle [h_a, h_a'] \| \hat{T}^{[h,h']} \| [h_b, h_b'] \rangle \tag{3.92}$$

Summary

and thus the evaluation of matrix elements involves the knowledge of the coupling coefficients and reduced matrix elements. The latter can usually be computed directly by taking particular cases. For example, the matrix elements of $\hat{T}_1^{[2,0]}$ are

$$\langle [h_a, h'_a] n_a | \hat{T}_1^{[2,0]} | [h_b, h'_b] n_b \rangle = \left\langle \begin{array}{cc} [h_b, h'_b] & [2,0] \\ n_b & 1 \end{array} \middle| \begin{array}{c} [h_a, h'_a] \\ n_a \end{array} \right\rangle \langle [h_a, h'_a] \| \hat{T}^{[2,0]} \| [h_b, h'_b] \rangle \tag{3.93}$$

Since $\hat{T}^{[2,0]}_1 = \sqrt{2}\hat{S}_x = (\hat{M} - 2\hat{n})/\sqrt{2}$, the matrix element is diagonal and this leads to the expression

$$\langle [h_a, h'_a] \| \hat{T}^{[2,0]} \| [h_b, h'_b] \rangle = \frac{(h_a + h'_a - 2n)\delta_{h_a h_b} \delta_{h'_a h'_b}}{\sqrt{2} \left\langle \begin{array}{cc} [h_b, h'_b] & [2,0] \\ n_b & 1 \end{array} \middle| \begin{array}{c} [h_a, h'_a] \\ n_a \end{array} \right\rangle} \tag{3.94}$$

Introducing the explicit value of the coupling coefficient using (3.88) and (3.89), we find [6]

$$\langle [h_a, h'_a] \| \hat{T}^{[2,0]} \| [h_b, h'_b] \rangle = \left[\frac{(h-h')(h-h'+2)}{2} \right]^{1/2} \delta_{h_a h_b} \delta_{h'_a h'_b} \tag{3.95}$$

All other matrix elements in the model may be computed using the $U^{BF}(2) \supset U^{BF}(1)$ Wigner–Eckart theorem.

Applications of the formalism presented in this chapter have not been proposed so far. We note, however, that boson–fermion algebras are used extensively in molecular and nuclear algebraic models, as discussed in Chapters 7 and 10.

Summary

The main purpose of this chapter is to introduce fermion operators and to describe some of their algebraic properties. The most basic of these is the fact that the fermion creation and annihilation operators anticommute, (3.1), rather than commute as the corresponding boson operators do. Nevertheless, it is shown that the bilinear combinations

$\mathcal{G}^\tau_\sigma$ in (3.2) satisfy the same commutation relations (3.5) as those derived for bosons in Chapter 1 and that they define the same algebraic structure. An important difference between boson and fermion systems is the symmetry of the many-body states: for identical bosons the wave function is symmetric whereas for identical fermions it is antisymmetric. As a result, a fermion state corresponds to a one-column Young tableau (3.16). In many other respects, however, fermions can be dealt with in much the same way as bosons. An analogous hamiltonian (3.17) can be proposed which includes all possible non-vanishing two-body interactions, and U(1) and SO(2) limits (3.18) and (3.20) can be defined, corresponding to analytical solutions of the fermion many-body system. Sections 3.2 and 3.3 then investigate the logical generalization of the previous material, namely, the algebraic structure of mixed systems composed of bosons and fermions. Many of the results derived in Chapter 2 can be used in this context, and in particular, the occurrence of symmetries is entirely analogous. Two uncoupled limits (3.35) and (3.36) exist, which have basis states (3.37) and (3.38) that correspond to simple products of boson and fermion basis states. The summed boson-plus-fermion generators (3.45) define two coupled limits (3.46) and (3.47) whose basis states (3.61) and (3.69) are linear combinations of the uncoupled products with Clebsch–Gordan coefficients as expansion coefficients. The last section of the chapter develops the tensor calculus necessary to deal with this boson–fermion system.

References

1. A. Frank, F. Iachello, and R. Lemus, "Algebraic methods for molecular electronic spectra," Chem. Phys. Lett. **131** (1986) 380.
2. R. Lemus and A. Frank, "An algebraic model for molecular electronic excitations in diatomic molecules," Ann. Phys. (NY) **206** (1991) 122.
3. F. Iachello and O. Scholten, "Interacting boson–fermion model of collective states in odd-A nuclei," Phys. Rev. Lett. **43** (1979) 679.
4. F. Iachello and P. Van Isacker, *The Interacting Boson–Fermion Model*, Cambridge University Press, Cambridge, 1991.

References

5. B.G. Wybourne, *Classical Groups for Physicists*, Wiley-Interscience, New York, 1974.
6. M.E. Rose, *Elementary Theory of Angular Momentum*, Wiley, New York, 1957.

Chapter 4
Supersymmetry and F Spin

4.1 Superalgebras

The mathematical structures studied in the previous chapters are based on U(2) algebras realized in terms of either bosons or fermions. Although in Chapter 3 we considered the direct product of a boson and a fermion algebra $U^B(2) \otimes U^F(2)$, where the boson and fermion generators are taken together in a single set, no cross terms of the type $s^\dagger a_{\pm 1/2}$ or $a^\dagger_{\pm 1/2} t$ are included in the set. By construction, the hamiltonian (3.33) separately conserves the boson and fermion numbers. The question arises as to whether one may define a generalized dynamical algebra where these kinds of terms are included and, if so, to study the consequences of this generalization. From the standpoint of fundamental processes, where bosons correspond to forces (i.e., photons, gluons, etc.) and fermions to matter (i.e., electrons, nucleons, quarks, etc.), it may seem strange at first sight to consider symmetries which mix such intrinsically different particles. However, in recent years such symmetries—known as supersymmetries—have given rise to schemes which hold great promise in quantum field theory in regards to the unification of the fundamental interactions [1–3]. In a different context, the consideration of such "higher" symmetries in nuclear structure physics has provided a remarkable unification of the spectroscopic properties of neighboring even–even and even–odd nuclei [4], as we shall see in Part III of this book. For this reason we consider in this chapter the effects on the $U^B(2) \otimes U^F(2)$ model arising from embedding its dynamical algebra into a superalgebra.

To start our discussion of superalgebras, it is convenient to consider a simpler system consisting of a single boson and a single fermion, denoted by $b^\dagger$ and $a^\dagger$, respectively. In this case the bilinear products $\mathcal{G}_B^B = b^\dagger b$ and $\mathcal{G}_F^F = a^\dagger a$ each generate a U(1) algebra. Taken together, these generators conform the

$$U^B(1) \otimes U^F(1) \tag{4.1}$$

dynamical algebra, in analogy with the boson–fermion algebra of the previous chapter, (3.30). Let us now consider the introduction of the mixed terms $b^\dagger a$ and $a^\dagger b$. Computing the commutator of these operators, we find

$$[a^\dagger b, b^\dagger a] = a^\dagger b b^\dagger a - b^\dagger a a^\dagger b = a^\dagger a - b^\dagger b + 2b^\dagger b a^\dagger a \tag{4.2}$$

which does not close into the original set $(a^\dagger a, b^\dagger b, a^\dagger b, b^\dagger a)$. This means that the inclusion of the cross terms does not lead to a Lie algebra. We note, however, that the bilinear operators $b^\dagger a$ and $a^\dagger b$ do not behave like bosons, but rather as fermion operators, in contrast to $a^\dagger a$ and $b^\dagger b$, both of which have bosonic character (in the sense that, e.g., $a_i^\dagger a_j$ *commutes* with $a_k^\dagger a_l$). This suggests the separation of the generators in two *sectors*, the bosonic sector $(a^\dagger a, b^\dagger b)$ and the fermionic sector $(a^\dagger b, b^\dagger a)$. Computing the *anti*commutators of the latter, we find

$$\{a^\dagger b, a^\dagger b\} = 0, \quad \{b^\dagger a, b^\dagger a\} = 0, \quad \{a^\dagger b, b^\dagger a\} = a^\dagger a + b^\dagger b \tag{4.3}$$

which indeed close into the same set. The commutators between the bosonic and fermionic sectors give

$$\begin{aligned}{}[a^\dagger b, a^\dagger a] &= -a^\dagger b, & [b^\dagger a, a^\dagger a] &= b^\dagger a \\ [a^\dagger b, b^\dagger b] &= a^\dagger b, & [b^\dagger a, b^\dagger b] &= -b^\dagger a \end{aligned} \tag{4.4}$$

The operations defined in (4.3) and (4.4), together with the (in this case) trivial $U^B(1) \otimes U^F(1)$ commutators

$$[a^\dagger a, a^\dagger a] = [b^\dagger b, b^\dagger b] = [a^\dagger a, b^\dagger b] = 0 \tag{4.5}$$

define the *superalgebra* U(1/1). To maintain the closure property for the enlarged set of generators belonging to the boson and fermion sectors, we are forced to include both commutators and anticommutators in the definition of a superalgebra. In general, superalgebras then involve boson-sector generators $\hat{B}_i$ and fermion-sector generators $\hat{F}_j$, satisfying the generalized relations

$$[\hat{B}_i, \hat{B}_j] = \sum_k c_{ij}^k \hat{B}_k, \quad [\hat{B}_i, \hat{F}_j] = \sum_k d_{ij}^k \hat{F}_k, \quad \{\hat{F}_i, \hat{F}_j\} = \sum_k e_{ij}^k \hat{B}_k \tag{4.6}$$

where c_{ij}^k, d_{ij}^k, and e_{ij}^k are complex constants defining the structure of the superalgebra, hence their denomination as *structure constants* of the superalgebra [5]. We shall only be concerned in this book with superalgebras of the form U(n/m), where n and m denote the dimensions of the boson and fermion subalgebras $U^B(n)$ and $U^F(n)$. The next step is to consider the *embedding* of the dynamical boson–fermion algebra $U^B(2) \otimes U^F(2)$ of the previous chapter into the superalgebra U(2/2).

4.2 The U(2/2) superalgebra

In addition to the set of generators $\mathcal{G}_{B\sigma}^{B\tau}$ and $\mathcal{G}_{F\sigma}^{F\tau}$ defining the $U^B(2) \otimes U^F(2)$ dynamical algebra and comprising the boson sector of the superalgebra, we consider the fermion-sector operators

$$\begin{array}{ll} \mathcal{G}_{B+1/2}^{F+1/2} \equiv s^\dagger a_{+1/2}, & \mathcal{G}_{B+1/2}^{F-1/2} \equiv s^\dagger a_{-1/2} \\ \mathcal{G}_{B-1/2}^{F+1/2} \equiv t^\dagger a_{+1/2}, & \mathcal{G}_{B-1/2}^{F-1/2} \equiv t^\dagger a_{-1/2} \end{array} \tag{4.7}$$

and their hermitian conjugates

$$\begin{array}{ll} \mathcal{G}_{F+1/2}^{B+1/2}(2) \equiv a_{+1/2}^\dagger s, & \mathcal{G}_{F+1/2}^{B-1/2}(2) \equiv a_{+1/2}^\dagger t \\ \mathcal{G}_{F-1/2}^{B+1/2}(2) \equiv a_{-1/2}^\dagger s, & \mathcal{G}_{F-1/2}^{B-1/2}(2) \equiv a_{-1/2}^\dagger t \end{array} \tag{4.8}$$

where we follow a notation consistent with (3.27) and (3.28). In this case the first set of relations in (4.6) corresponds to the $U^B(2) \otimes U^F(2)$

commutators discussed in Chapter 3 while the second set translates into

$$\begin{array}{ll} [\mathcal{G}^{B\tau}_{B\sigma}, \mathcal{G}^{B\gamma}_{F\beta}] = -\mathcal{G}^{B\tau}_{F\beta}\delta_{\sigma\gamma}, & [\mathcal{G}^{F\tau}_{F\sigma}, \mathcal{G}^{B\gamma}_{F\beta}] = \mathcal{G}^{B\gamma}_{F\sigma}\delta_{\tau\beta} \\ [\mathcal{G}^{B\tau}_{B\sigma}, \mathcal{G}^{F\gamma}_{B\beta}] = \mathcal{G}^{F\gamma}_{B\sigma}\delta_{\tau\beta}, & [\mathcal{G}^{F\tau}_{F\sigma}, \mathcal{G}^{F\gamma}_{B\beta}] = -\mathcal{G}^{F\tau}_{B\beta}\delta_{\sigma\gamma} \end{array} \qquad (4.9)$$

and the third set corresponds to

$$\begin{array}{l} \{\mathcal{G}^{B\tau}_{F\sigma}, \mathcal{G}^{B\gamma}_{F\beta}\} = \{\mathcal{G}^{F\tau}_{B\sigma}, \mathcal{G}^{F\gamma}_{B\beta}\} = 0 \\ \{\mathcal{G}^{B\tau}_{F\sigma}, \mathcal{G}^{F\gamma}_{B\beta}\} = \mathcal{G}^{B\tau}_{B\beta}\delta_{\sigma\gamma} + \mathcal{G}^{F\gamma}_{F\sigma}\delta_{\tau\beta} \end{array} \qquad (4.10)$$

where in all equations σ, τ, β, and γ take the values $\pm 1/2$. The same *supercommutation* relations are valid for $U(n/m)$, but with the boson and fermion indices taking n and m different values, respectively.

The $U(2/2)$ generators may be conveniently written in the matrix form

$$\mathbf{G} \equiv \left[\begin{array}{ccc} \mathcal{G}^{B\sigma}_{B\tau} & \vdots & \mathcal{G}^{F\gamma'}_{B\beta} \\ \cdots & \cdot & \cdots \\ \mathcal{G}^{B\gamma}_{F\beta'} & \vdots & \mathcal{G}^{F\sigma'}_{F\tau'} \end{array} \right] \qquad (4.11)$$

where the boson sector corresponds to the diagonal blocks and the fermion sector to the antidiagonal ones. The 4×4 supermatrix $\mathbf{G}$ may be thought of as acting on a four-component state of the form

$$\left[\begin{array}{c} \psi^B_{+1/2} \\ \psi^B_{-1/2} \\ \psi^F_{+1/2} \\ \psi^F_{-1/2} \end{array} \right] \qquad (4.12)$$

The boson sector of $\mathbf{G}$ only mixes the $s^\dagger$ and $t^\dagger$ wave functions and the $a^\dagger_{+1/2}$ and $a^\dagger_{-1/2}$ wave functions among themselves, that is, it corresponds to the $U^B(2) \otimes U^F(2)$ subalgebra of $U(2/2)$. The fermion sector, however, mixes all four components.

The general theory of representations for the $U(n/m)$ superalgebras is somewhat involved, although it is similar to the corresponding theory for $U(n+m)$ algebras. We do not discuss this subject here, as it is not necessary for the applications considered in this book. We refer the

4.2 The U(2/2) superalgebra

interested reader to a review article by Bars [5]. The representations of U(2/2) are in general labeled by four integer numbers, in analogy with U(4). Depending on the choice of subalgebras of U(2/2), one must then derive the branching rules for the corresponding labels. These subalgebras may be either other superalgebras or normal Lie algebras. We are interested only in the latter case, since our purpose is to study the consequences of embedding the $U^B(2) \otimes U^F(2)$ dynamical algebra. Thus the algebra chain of interest is

$$U(2/2) \supset U^B(2) \otimes U^F(2) \qquad (4.13)$$

where we have the restriction that only the fully symmetric representations of $U^B(2)$ and the fully antisymmetric representations of $U^F(2)$ are physically allowed. Taking the trace of the superalgebra matrix **G** in (4.11), we find

$$\hat{\mathcal{N}} \equiv \text{trace}(\mathbf{G}) = \hat{N} + \hat{M} \qquad (4.14)$$

and one may easily verify that $\hat{\mathcal{N}}$ commutes with all generators of the superalgebra. For example,

$$\begin{aligned}[\hat{\mathcal{N}}, s^\dagger a_{+1/2}] &= [s^\dagger s, s^\dagger]a_{+1/2} + s^\dagger[a^\dagger_{+1/2}a_{+1/2}, a_{+1/2}] \\ &= s^\dagger a_{+1/2} - s^\dagger a_{+1/2} = 0 \end{aligned} \qquad (4.15)$$

This shows that $\hat{\mathcal{N}}$ is a linear invariant of U(2/2). In addition, one may verify that the quadratic operator

$$\mathcal{C}_2[U(2/2)] \equiv \sum_{\sigma\tau} \left(\mathcal{G}^{B\sigma}_{B\tau}\mathcal{G}^{B\tau}_{B\sigma} - \mathcal{G}^{F\sigma}_{F\tau}\mathcal{G}^{F\tau}_{F\sigma} + \mathcal{G}^{B\sigma}_{F\tau}\mathcal{G}^{F\tau}_{B\sigma} - \mathcal{G}^{F\sigma}_{B\tau}\mathcal{G}^{B\tau}_{F\sigma} \right) \qquad (4.16)$$

is also an invariant of the superalgebra U(2/2). Using the symmetry in the bosons and the antisymmetry in the fermions, this Casimir invariant can be rewritten as

$$\begin{aligned}\mathcal{C}_2[U(2/2)] &= \sum_{\sigma\tau}(b^\dagger_\tau b_\sigma b^\dagger_\sigma b_\tau - a^\dagger_\tau a_\sigma a^\dagger_\sigma a_\tau + a^\dagger_\tau b_\sigma b^\dagger_\sigma a_\tau - b^\dagger_\tau a_\sigma a^\dagger_\sigma b_\tau) \\ &= \sum_{\sigma\tau} \left(b^\dagger_\tau b_\sigma(b_\tau b^\dagger_\sigma - \delta_{\sigma\tau}) - a^\dagger_\tau a_\sigma(-a_\tau a^\dagger_\sigma + \delta_{\sigma\tau}) \right. \\ &\quad \left. + a^\dagger_\tau a_\tau(b^\dagger_\sigma b_\sigma + 2) - b^\dagger_\tau b_\tau(-a^\dagger_\sigma a_\sigma + 2) \right) \\ &= (\hat{N} + \hat{M})(\hat{N} + \hat{M} - 1) = \hat{\mathcal{N}}(\hat{\mathcal{N}} - 1) \end{aligned} \qquad (4.17)$$

that is, it reduces to a quadratic function of the first-order Casimir invariant. It thus follows that only one label is needed to characterize those U(2/2) representations that only contain symmetric $U^B(2)$ and antisymmetric $U^F(2)$ representations. For this label we may use the total-particle number $\mathcal{N}$, the eigenvalue of $\hat{\mathcal{N}}$. The corresponding representations are denoted as $[\mathcal{N}\}$ and are called the *(fully) supersymmetric* representations of U(2/2) [6]. Using (4.14) we arrive at the result that the supersymmetric representation $[\mathcal{N}\}$ of U(2/2) contains the $U^B(2) \otimes U^F(2)$ representations

$$\overbrace{\square\square\cdots\square}^{\mathcal{N}} \oplus \left(\overbrace{\square\square\cdots\square}^{\mathcal{N}-1} \otimes \square \right) \oplus \left(\overbrace{\square\square\cdots\square}^{\mathcal{N}-2} \otimes {\square \atop \square} \right) \quad (4.18)$$

that is, it contains representations with zero, one, and two fermions. This result can be considered as a particular $U(2/2) \supset U^B(2) \otimes U^F(2)$ branching rule.

What are the consequences of assuming that the boson–fermion system of Chapter 3 possesses this higher symmetry? An immediate consequence of the supersymmetry is that the boson–fermion systems (N, M) with $(\mathcal{N}, 0)$, $(\mathcal{N} - 1, 1)$, and $(\mathcal{N} - 2, 2)$ are members of the same (super)multiplet $[\mathcal{N}\}$. Whereas in Chapters 1 and 3 the pure boson system $(M = 0)$ and the boson–fermion systems with $M = 1$ and $M = 2$ are taken as separate and unrelated, supersymmetry imposes a link between the "spectroscopic" properties of these systems. In particular, the *same* hamiltonian (3.33) should simultaneously describe their energy spectra. Also, one expects that other operators, which in higher-dimensional models correspond to electromagnetic transition and particle-transfer operators, are the same for all members of the supermultiplet. We may think of the U(2/2) supersymmetry as providing a dynamical (super)algebra, where different physical entities are treated as a single system. For example, in the nuclear physics case of Part III, the spectroscopic properties of neighboring nuclei are simultaneously described by supersymmetry. The different systems linked by supersymmetry are referred to as *superpartners*.

4.3 Supersymmetry predictions

We illustrate the ideas of the previous section by considering the particular case of the $U^{BF}(2) \supset U^{BF}(1)$ dynamical symmetry (3.46), although it should be stressed that supersymmetry is not related to any particular chain of algebras and that any other dynamical symmetry or combination of symmetries can be chosen. We consider the chain

$$
\begin{array}{ccccc}
U(2/2) \supset & U^B(2) \otimes U^F(2) \supset & U^{BF}(2) \supset & U^{BF}(1) \\
| & | & | & | & | \\
\{\mathcal{N}\} & [N] & [1^M] & [h, h'] & n
\end{array}
\quad (4.19)
$$

where the relevant branching rules can be deduced from (4.18), (3.53), and (3.54). The corresponding hamiltonian is a linear combination of invariant operators of the chain (4.19),

$$\hat{H}_{Ia} = E_0 + \gamma \mathcal{C}_2[U^{BF}(2)] + \epsilon \hat{n} + \alpha \hat{n}^2 \quad (4.20)$$

where we put into E_0 all contributions that do not affect the spectrum but only the binding energy of the systems under consideration. From the explicit expression for $\mathcal{C}_2[U^{BF}(2)]$ and its associated eigenvalue, (3.64) and (3.66), we find the eigenspectrum

$$E_{Ia}(h, h', n) = E_0 + \gamma[h(h+1) + h'(h'-1)] + \epsilon n + \alpha n^2 \quad (4.21)$$

In Figure 4.1 we show an example of the spectra for the three superpartners

$$
\begin{array}{llll}
N = 4, & M = 0, & [h, h'] = [4, 0], & 0 \leq n \leq 4 \\
N = 3, & M = 1, & [h, h'] = [4, 0], & 0 \leq n \leq 4 \\
& & [3, 1], & 1 \leq n \leq 3 \\
N = 2, & M = 2, & [h, h'] = [3, 1], & 1 \leq n \leq 3
\end{array}
\quad (4.22)
$$

Note that not only a simultaneous description of the spectra is achieved, but one finds in addition that the corresponding wave functions are linked by means of the operators in the fermion sector of the superalgebra. These operators correspond to "particle-transfer" operators, that is, to operators that produce "reactions" in which the systems transform into each other.

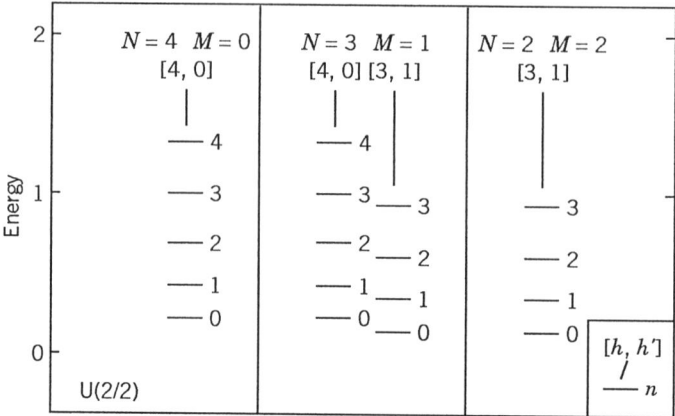

Figure 4.1: The eigenspectra of the U(2/2) superpartners calculated with the hamiltonian $\hat{H}_{Ia}$ with $E_0 = 0$, $\gamma = 0.1$, $\epsilon = 0.2$, and $\alpha = 0.02$ for $\mathcal{N} = 4$ particles. Levels are labeled by N, M, $[h, h']$, and n. Energies are in arbitrary units.

In close analogy with the evaluation of fermion tensor operators carried out in Section 3.4, we may now construct the boson–fermion tensors $\hat{T}_n^{[h,h']}$,

$$\begin{aligned}
\hat{T}_1^{[1,1]} &= a_{+1/2}^\dagger s + a_{-1/2}^\dagger t \\
\hat{T}_0^{[2,0]} &= -a_{+1/2}^\dagger t \\
\hat{T}_1^{[2,0]} &= \sqrt{\tfrac{1}{2}}(a_{+1/2}^\dagger s - a_{-1/2}^\dagger t) \\
\hat{T}_2^{[2,0]} &= a_{-1/2}^\dagger s
\end{aligned} \qquad (4.23)$$

and their corresponding hermitian conjugates. From their form it is seen that these operators indeed connect states associated with different superpartners. Furthermore, because of their well-defined $U^{BF}(2) \supset U^{BF}(1)$ character, we may explicitly calculate their action on these wave functions. Thus, for example, using the Wigner–Eckart theorem for the $U^{BF}(2) \supset U^{BF}(1)$ chain in conjunction with the result (3.88) and (3.89), we find for the transfer from an N-boson system to a system with $N-1$ bosons and one fermion

$$\langle [N-1,1]n-1|\hat{T}_0^{[2,0]}|[N,0]n\rangle$$

4.3 Supersymmetry predictions

$$\begin{aligned}
&= \left\langle \begin{array}{cc|c} [N,0] & [2,0] & [N-1,1] \\ n & 0 & n-1 \end{array} \right\rangle \langle [N-1,1] \| \hat{T}^{[2,0]} \| [N,0] \rangle \\
&= \left[\frac{n(n-1)}{N(N+1)} \right]^{1/2} \langle [N-1,1] \| \hat{T}^{[2,0]} \| [N,0] \rangle
\end{aligned} \qquad (4.24)$$

On the other hand, the state $|[N-1,1]n-1\rangle$ may be obtained directly by the coupling

$$|[N-1,1]n-1\rangle = \sum_{n_b n_f} \left\langle \begin{array}{cc|c} [N-1,0] & [1,0] & [N-1,1] \\ n_b & n_f & n-1 \end{array} \right\rangle \\ \times |[N-1,0]n_b\rangle |[1,0]n_f\rangle \qquad (4.25)$$

which, after inserting the expressions for the coupling coefficients again from (3.88) and (3.89), leads to

$$\begin{aligned}
|[N-1,1]n-1\rangle &= - \left[\frac{n-1}{N} \right]^{1/2} |[N-1,0]n-1\rangle |[1,0]0\rangle \\
&+ \left[\frac{N-n+1}{N} \right]^{1/2} |[N-1,0]n-2\rangle |[1,0]1\rangle
\end{aligned} \qquad (4.26)$$

We now compute the action of the operator $\hat{T}_0^{[2,0]} = -a_{+1/2}^\dagger t$ on the $|[N,0]n\rangle$ states (1.47) to find

$$\begin{aligned}
-a_{+1/2}^\dagger t \frac{(s^\dagger)^{N-n}(t^\dagger)^n}{\sqrt{(N-n)!n!}} |0\rangle &= -\sqrt{n} \frac{(s^\dagger)^{N-n}(t^\dagger)^{n-1}}{\sqrt{(N-n)!(n-1)!}} a_{+1/2}^\dagger |0\rangle \\
&= -\sqrt{n} |[N-1,0]n-1\rangle |[1,0]0\rangle \qquad (4.27)
\end{aligned}$$

Combining the previous two equations, we find

$$\langle [N-1,1]n-1| \hat{T}_0^{[2,0]} |[N,0]n\rangle = \left[\frac{n(n-1)}{N} \right]^{1/2} \qquad (4.28)$$

By comparing (4.24) and (4.28), we arrive at the reduced matrix element of the $\hat{T}^{[2,0]}$ tensors,

$$\langle [N-1,1] \| \hat{T}^{[2,0]} \| [N,0] \rangle = \sqrt{N+1} \qquad (4.29)$$

which may be used again in the Wigner–Eckart theorem to evaluate the matrix elements of the other tensor components $\hat{T}_1^{[2,0]}$ and $\hat{T}_2^{[2,0]}$. The supersymmetry prediction is thus that the probability amplitude for the "reaction" from the boson state $|[N,0]n\rangle$ to the boson–fermion state $|[N-1,1]n-1\rangle$ is proportional to $\sqrt{n(n-1)/N}$.

The calculations carried out above are typical of the kind of computations arising in the molecular and nuclear algebraic models, as we shall see in Parts II and III.

We now turn our attention to another embedding, namely that of the algebra $U_1(2) \otimes U_2(2)$ introduced in Chapter 2 into a larger dynamical algebra.

4.4 F Spin and the U(4) algebra

In the previous sections we studied the embedding of the $U^B(2) \otimes U^F(2)$ model into a U(2/2) superalgebra, which plays the role of a larger dynamical algebra describing the properties of different boson–fermion systems. We now return to the $U_1(2) \otimes U_2(2)$ boson system of Chapter 2 and study the effects of its embedding into a larger dynamical algebra. In many respects this is analogous to the introduction of U(2/2) for the boson–fermion systems, although in this case the larger structure turns out to be the U(4) Lie algebra, as we show below.

We start by noting that the type-1 and type-2 bosons of Chapter 2, in analogy with the isospin formalism for neutrons and protons, may be considered as identical bosons with different "charge" states. We introduce a σ label and designate the $U_1(2) \otimes U_2(2)$ bosons by $b^\dagger_{1/2\sigma,i}$, where $i = 1, 2$ indicates whether $b^\dagger$ creates an s or t boson, respectively, and $\sigma = -1/2$ and $\sigma = +1/2$ are the two projections of the spin quantum number, which by convention correspond to type-1 and type-2 bosons, respectively. A single boson thus carries an intrinsic-spin label of $1/2$, which we refer to as F spin. To construct annihilation operators with the same transformation properties under F-spin rotations, we define

$$\tilde{b}_{1/2\sigma,i} \equiv (-)^{1/2+\sigma} b_{1/2-\sigma,i} \tag{4.30}$$

in analogy with the definition (1.106). Note, however, that in this case the σ label distinguishes type-1 from type-2 bosons and not s from t

4.4 F Spin and the U(4) algebra

bosons, as was the case in Section 1.7. We may now couple bilinear products of creation and annihilation operators to a total F spin,

$$[b^\dagger_{1/2,i} \times \tilde{b}_{1/2,j}]^{(\kappa)}_\zeta = \sum_{\sigma\sigma'} \langle 1/2\sigma\, 1/2\sigma' | \kappa\zeta \rangle b^\dagger_{1/2\sigma,i} \tilde{b}_{1/2\sigma',j} \qquad (4.31)$$

where F can take the values 0 and 1. For example, for $i = j = 1$ (two s bosons) we find

$$\begin{aligned}
{[b^\dagger_{1/2,1} \times \tilde{b}_{1/2,1}]^{(0)}_0} &= \sqrt{\tfrac{1}{2}}(s^\dagger_1 s_1 + s^\dagger_2 s_2) \\
{[b^\dagger_{1/2,1} \times \tilde{b}_{1/2,1}]^{(1)}_{-1}} &= s^\dagger_1 s_2 \\
{[b^\dagger_{1/2,1} \times \tilde{b}_{1/2,1}]^{(1)}_0} &= \sqrt{\tfrac{1}{2}}(s^\dagger_2 s_2 - s^\dagger_1 s_1) \\
{[b^\dagger_{1/2,1} \times \tilde{b}_{1/2,1}]^{(1)}_{+1}} &= -s^\dagger_2 s_1
\end{aligned} \qquad (4.32)$$

and similar results for other values of i and j in (4.31). We note that in the F-spin vector components (i.e., for $\kappa = 1$) we find products of the form $s^\dagger_1 s_2$ and $s^\dagger_2 s_1$, which transform between type-1 and type-2 bosons. These operators do not belong to the set of eight generators (2.2) of $U_1(2) \otimes U_2(2)$. One may show, however, that all operators (4.31) can be written in terms of the 16 generators

$$\boxed{\begin{array}{lll}
(\mathcal{G}^{1j}_{1i}, i,j = 1,2) & \text{or} \quad (s^\dagger_1 s_1, s^\dagger_1 t_1, t^\dagger_1 s_1, t^\dagger_1 t_1) & 4 \\
(\mathcal{G}^{2j}_{1i}, i,j = 1,2) & \text{or} \quad (s^\dagger_1 s_2, s^\dagger_1 t_2, t^\dagger_1 s_2, t^\dagger_1 t_2) & 4 \\
(\mathcal{G}^{1j}_{2i}, i,j = 1,2) & \text{or} \quad (s^\dagger_2 s_1, s^\dagger_2 t_1, t^\dagger_2 s_1, t^\dagger_2 t_1) & 4 \\
(\mathcal{G}^{2j}_{2i}, i,j = 1,2) & \text{or} \quad (s^\dagger_2 s_2, s^\dagger_2 t_2, t^\dagger_2 s_2, t^\dagger_2 t_2) & 4
\end{array}} \qquad (4.33)$$

of which (2.2) is a subset. These are operators that carry out unitary transformations among four boson states and hence generate a U(4) algebra, as may be verified by direct computation of the commutators. We thus arrive at the chain of algebras

$$\mathrm{U}(4) \supset \mathrm{U}_1(2) \otimes \mathrm{U}_2(2) \qquad (4.34)$$

The first-order linear invariant of the U(4) algebra is given by

$$\hat{N} = \hat{N}_1 + \hat{N}_2 \qquad (4.35)$$

where $\hat{N}_1 = s_1^\dagger s_1 + t_1^\dagger t_1$ and $\hat{N}_2 = s_2^\dagger s_2 + t_2^\dagger t_2$. One may verify directly that $\hat{N}$ commutes with the 16 generators (4.33), implying that $\hat{N}$ is the first-order Casimir invariant of U(4). The second-order Casimir invariant of U(4),

$$\mathcal{C}_2[\mathrm{U}(4)] = \sum_{\rho\rho'}\sum_{ij} \mathcal{G}^{\rho i}_{\rho' j} \mathcal{G}^{\rho' j}_{\rho i} \qquad (4.36)$$

can be shown to reduce, in this particular realization of U(4), to

$$\mathcal{C}_2[\mathrm{U}(4)] = \hat{N}(\hat{N} + 3) \qquad (4.37)$$

and is thus a quadratic function of the linear Casimir invariant. We conclude that U(4) is characterized by a single label, for which we may use the eigenvalue of $\hat{N}$. Thus N labels the (fully symmetric) representations of U(4). From (4.35) we see that the boson numbers N_1 and N_2 are not separately conserved by general U(4) transformations, but only their sum is, and from this we conclude that the representation $[N]$ of U(4) contains the

$$\boxed{([0] \otimes [N]) \oplus ([1] \otimes [N-1]) \oplus \cdots \oplus ([N] \otimes [0])} \qquad (4.38)$$

representations of $\mathrm{U}_1(2) \otimes \mathrm{U}_2(2)$. In analogy with the previous discussion on U(2/2) this means that the U(4) algebra can be used to describe the $N+1$ different $\mathrm{U}_1(2) \otimes \mathrm{U}_2(2)$ systems contained in the decomposition (4.38). We omit here showing any examples of the embedding of $\mathrm{U}_1(2) \otimes \mathrm{U}_2(2)$ into U(4) but refer the reader to Chapter 9, where an analogous generalization arises for the U(6) nuclear model.

We now return to F spin. From (4.32) we note that the F-scalar operator (i.e., with $\kappa = 0$) is proportional to the sum of type-1 and type-2 generators $s_\rho^\dagger s_\rho$, a property which is true for all F scalars. It is simple to prove that the set of four operators (4.31) with $F = 0$ close under commutation, giving a U(2) subalgebra of U(4), which we denote by $\mathrm{U}_\mathrm{st}(2)$. The subscript st indicates that only the s or t identity is retained in the generators (through the indices i and j) while all information on the F-spin character (type 1 or type 2) is lost by summation over the σ labels. We may also define a subalgebra of U(4) by *contracting* over these indices, that is, by summing over the i labels in (4.31):

$$\sum_i [b^\dagger_{1/2,i} \times \tilde{b}_{1/2,i}]^{(\kappa)}_\zeta \qquad (4.39)$$

4.4 F Spin and the U(4) algebra

which give, explicitly,

$$
\begin{array}{ll}
\kappa=0, \zeta=0 & : \quad \sqrt{\frac{1}{2}}(s_1^\dagger s_1 + s_2^\dagger s_2 + t_1^\dagger t_1 + t_2^\dagger t_2) = \sqrt{\frac{1}{2}} \hat{N} \\
\kappa=1, \zeta=-1 & : \quad s_1^\dagger s_2 + t_1^\dagger t_2 \equiv \hat{F}_- \\
\kappa=1, \zeta=0 & : \quad \sqrt{\frac{1}{2}}(s_2^\dagger s_2 - s_1^\dagger s_1 + t_2^\dagger t_2 - t_1^\dagger t_1) \equiv \sqrt{2} \hat{F}_z \\
\kappa=1, \zeta=+1 & : \quad -s_2^\dagger s_1 - t_2^\dagger t_1 = -(\hat{F}_-)^\dagger \equiv -\hat{F}_+
\end{array}
\tag{4.40}
$$

Note that $\hat{F}_z = (\hat{N}_2 - \hat{N}_1)/2$. The F scalar found in this way is thus proportional to $\hat{N}$. Being a scalar, it is also contained in $U_{st}(2)$. As mentioned before, $\hat{N}$ commutes with all the generators of $U(4)$ and in particular with the operators (4.40). The latter commute with all F-scalar operators and satisfy the following commutation rules among themselves:

$$[\hat{F}_z, \hat{F}_\pm] = \pm \hat{F}_\pm, \qquad [\hat{F}_+, \hat{F}_-] = 2\hat{F}_z \tag{4.41}$$

which are identical to the SU(2) commutators (1.87). The four operators (4.40) generate a U(2) algebra which we denote as $U_F(2)$. The italic subscript in this notation refers to F spin and should not be confused with the roman superscript F which is used to denote fermion algebras. We thus arrive at the alternative chain of algebras

$$U(4) \supset U_{st}(2) \otimes (U_F(2) \supset SU_F(2)) \tag{4.42}$$

arising from the separation of the U(4) degrees of freedom into its s–t and F-spin parts. Note that $U_{st}(2)$ is identical to the coupled U(2) algebra defined in Chapter 2 by the generators (2.26). Thus its representations are labeled by the $[h, h']$ quantum numbers (2.38), which can take the values given in (2.42):

$$[h, h'] = \oplus \sum_f [N-f, f] \tag{4.43}$$

where f runs from zero to $\min(N_1, N_2)$. What are the corresponding $U_F(2)$ representations? It can be shown that these representations must be identical to the $U_{st}(2)$ ones and hence we arrive at the decomposition

$$
\boxed{\begin{array}{ccc}
U(4) & \supset & U_{st}(2) \otimes U_F(2) \\
| & & | \qquad\qquad | \\
[N] & & [h, h'] \qquad [h, h']
\end{array}}
\tag{4.44}
$$

where $[h, h']$ are given in (4.43). This result can be proved by evaluating the linear and quadratic invariants of $U_{st}(2)$ and $U_F(2)$. The linear one is simply given by $\hat{N}$ for all the algebras in (4.44) while it is not difficult to show that for the second-order one (see the definition (2.66))

$$\mathcal{C}_2[U_{st}(2)] = \mathcal{C}_2[U_F(2)] \tag{4.45}$$

Denoting by $[h, h']$ and $[k, k']$ the $U_{st}(2)$ and $U_F(2)$ representations, the above results together with equations (2.49) and (2.65) imply

$$h + h' = k + k' \tag{4.46}$$

and

$$h(h+1) + h'(h'-1) = k(k+1) + k'(k'-1) \tag{4.47}$$

whose unique solution is $[k, k'] = [h, h']$, as stated above.

Returning to the $SU_F(2)$ subalgebra, we note that F plays the same role in $U_F(2)$ as the angular momentum j does in the $U(2)$ coupled algebra of Chapter 2. Thus, in analogy with (2.40), we find

$$F = \tfrac{1}{2}(h - h') = \tfrac{1}{2}(N - 2f) \tag{4.48}$$

where in the last equality the notation of (4.43) is used. Thus F may take the range of values

$$F = \tfrac{1}{2}N, \tfrac{1}{2}N - 1, \ldots, \tfrac{1}{2}N - \min(N_1, N_2) = \tfrac{1}{2}|N_1 - N_2| \tag{4.49}$$

Since F takes its largest value $F = N/2$ for $[h, h'] = [N, 0]$, this particular representation corresponds to the fully symmetric representation of $U_{st}(2)$ (and $U_F(2)$), and the associated states are analogous to the ones of the $U(2)$ model of Chapter 1. Other values of F spin correspond to non-symmetric representations, with states that have no counterpart in the $U(2)$ model. We conclude that the F-spin classification (4.42) is associated with states where the symmetry between type-1 and type-2 bosons is well defined. This kind of labeling turns out to be significant when dealing with the nuclear $U(6)$ model with neutrons and protons [7].

This concludes our discussion of the schematic $U(2)$ models. We have studied in this first part most of the concepts and techniques

Summary

which are needed for the higher-dimensional applications. Although the U(2) models are of a schematic nature, they can be applied to physical systems which are described in terms of coupled Morse oscillators and we hope to have convinced the reader that realistic examples of such systems exist (Sections 2.7 and 2.8). Other ideas that have been discussed in this part, especially those presented in Chapters 3 and 4, have not yet found applications in the context of a schematic U(2) model. The main purpose of the latter chapters has been to give a pedagogical introduction to algebraic techniques dealing with systems of bosons and fermions, which have found widespread application in molecular and nuclear models, discussed in Parts II and III.

Summary

If, in addition to the generators considered in Chapter 3, bilinear combinations are formed from boson creation and fermion annihilation operators (and *vice versa*), the resulting set of operators does not close under commutation. A closed algebraic structure can be defined, however, through the separation of the set into two sectors and the generalization to a mixture of commutation and anticommutation—referred to as supercommutation—relations. The resulting mathematical construction has been named a superalgebra. Its generalized (anti)commutation relations are defined in (4.6). For the bosons and fermions introduced in Chapter 3 this superalgebra reduces to $U(2/2)$, and contains a boson sector, consisting of the $U^B(2) \otimes U^F(2)$ generators, and a fermion sector defined in (4.7) and (4.8). The combined set of generators satisfies the supercommutation relations (4.9) and (4.10). Among the different representations of $U(2/2)$ the supersymmetric one is the most important: it imposes symmetry on the bosons and antisymmetry on the fermions. The main purpose for introducing the $U(2/2)$ superalgebra is that a single one of its supersymmetric representations contains states that do not have fixed numbers of bosons and fermions separately. This property is formally expressed by (4.18). As a consequence, $U(2/2)$ can be used as a dynamical algebra to simultaneously describe the boson–fermion systems specified by (4.18). Just as $U^B(2) \otimes U^F(2)$ can be extended to the superalgebra $U(2/2)$, it is possible to enlarge the coupled algebra $U_1(2) \otimes U_2(2)$ by adding "off-diagonal" generators which transform type-1 into type-2 bosons and *vice versa*. The U(4) algebra defined in (4.33) results. A single symmetric representation of U(4) contains states that do not have fixed numbers of type-1 and type-2 bosons; only the total number of bosons is conserved, as expressed by (4.38). Thus U(4) can be used for the simultaneous description of several coupled boson sytems, each of them separately described by $U_1(2) \otimes U_2(2)$. Within U(4) is contained a U(2) algebra with generators (4.40). To this U(2) algebra corresponds a quantum number called the F spin. This quantum number arises by associating with each boson an F spin of $F = 1/2$ and a projection $\sigma = -1/2$ or $\sigma = +1/2$ for type-1 and type-2 bosons, respectively. The F spin provides a convenient classification of the symmetry character under the exchange of type-1 and type-2 bosons and furthermore can be related to the Gel'fand labels introduced in Chapter 2 through the classification (4.44).

References

1. J. Wess and B. Zumino, "Supergauge transformations in four dimensions," Nucl. Phys. B **70** (1974) 39.
2. P. Fayet and S. Ferrara, "Supersymmetry," Phys. Rep. **32** (1977) 249.
3. P. van Nieuwenhuizen, "Supergravity," Phys. Rep. **68** (1981) 189.
4. F. Iachello, "Dynamical supersymmetries in nuclei," Phys. Rev. Lett. **44** (1980) 772.
5. I. Bars, "Supergroups and their representations," in *Introduction to Supersymmetry in Particle and Nuclear Physics*, edited by O. Castaños, A. Frank, and L. Urrutia, Plenum, New York, 1983, p. 107.
6. A.B. Balantekin and I. Bars, "Dimension and character formulas for Lie supergroups," J. Math. Phys. **22** (1981) 1149.
7. T. Otsuka, A. Arima, F. Iachello, and I. Talmi, "Shell model description of interacting bosons," Phys. Lett. B **76** (1978) 139.

Part II

Molecular Models

Chapter 5
Diatomic Molecules

5.1 Introduction

In Chapter 1 we showed how the algebra U(2) is related to the harmonic oscillator in two dimensions and how it can be realized in terms of s and t bosons. This equivalence between the *coordinate* realization of a Lie algebra on the one hand and its *operator* realization on the other hand persists in higher dimensions, and it is largely a matter of taste which of the two should be preferred. For the applications discussed in Parts II and III of this book (an algebraic approach to the phenomenology of molecular and nuclear rotation–vibration spectra) the operator realization is, in general, more convenient. For this reason we introduce here the U(4) algebra by way of operators and indicate later its connection with coordinates.

The U(4) algebra can be realized in terms of four creation and four annihilation operators $b_i^\dagger$ and b_i ($i = 1, 2, 3, 4$). Restrictions on the form of these operators are obtained from physical considerations. First, we wish to describe molecules that are not placed in an external electric or magnetic field and which have, as a consequence, states with definite values of the angular momentum L. In addition, the eigenstates must be characterized by a definite parity π. These restrictions require that the U(4) model hamiltonian be invariant under rotations and reflections, and to construct such a hamiltonian, we must introduce operators that have definite transformation properties under these op-

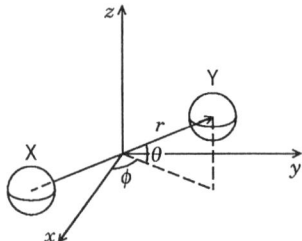

Figure 5.1: Schematic illustration of the geometric structure of a diatomic molecule X–Y and the definition of the spherical coordinates r, θ, ϕ to characterize the distance vector **r** between X and Y.

erations, that is, introduce spherical tensor operators (as defined in (1.91)) with a definite parity (as defined in (1.35)). Even with these restrictions several realizations of U(4) remain possible (e.g., four scalar operators, two spinor operators). Our further choice is determined by the importance of the dipole degree of freedom in the description of rotations and vibrations of diatomic molecules. Many of the essential features of such molecules can be characterized in terms of the distance vector **r** between the atoms (see Figure 5.1) which behaves as a rank-1 tensor under rotations and has negative parity. It is therefore natural to introduce operators with similar transformation properties, which leads us to propose a realization of U(4) in terms of a vector (rank-1) operator with components $p_m^\dagger$ and a scalar (rank-0) operator $s^\dagger$. At times, we use the generic notation $b_{lm}^\dagger$ referring to $p_m^\dagger$ if $l = 1$ and to $s^\dagger$ if $l = m = 0$. Since these operators have integer (intrinsic) angular momentum l, they create bosonic states. While the p boson is used to describe the dipole degree of freedom, the physical role of the s boson is less clear; its introduction is justified mainly on phenomenological grounds as a way to generate interactions between bosonic states with different numbers of p bosons.

The p and s bosons are the basic building blocks of the U(4) molecular vibron model as proposed by Iachello and Levine [1,2]. In the following sections we review the algebraic properties of the vibron model, which can be done in close analogy with the discussion of Chapter 1.

5.2 The U(4) algebra

The creation and annihilation operators introduced in Section 5.1 satisfy the usual boson commutation relations,

$$[b_i, b_j^\dagger] = \delta_{ij}, \qquad [b_i, b_j] = [b_i^\dagger, b_j^\dagger] = 0, \qquad i,j = 1,2,3,4 \qquad (5.1)$$

where the following notation is introduced:

$$\begin{aligned} b_1^\dagger &\equiv p_{-1}^\dagger, & b_2^\dagger &\equiv p_0^\dagger, & b_3^\dagger &\equiv p_{+1}^\dagger, & b_4^\dagger &\equiv s^\dagger \\ b_1 &\equiv p_{-1}, & b_2 &\equiv p_0, & b_3 &\equiv p_{+1}, & b_4 &\equiv s \end{aligned} \qquad (5.2)$$

In analogy with the discussion in Section 1.1 we introduce the set of bilinear operators

$$\mathcal{G}_i^j \equiv b_i^\dagger b_j, \qquad i,j = 1,2,3,4 \qquad (5.3)$$

satisfying the commutation relations

$$[\mathcal{G}_i^j, \mathcal{G}_k^l] = \mathcal{G}_i^l \delta_{jk} - \mathcal{G}_k^j \delta_{il}, \qquad i,j,k,l = 1,2,3,4 \qquad (5.4)$$

The operators $\mathcal{G}_i^j$ are the generators of a U(4) algebra (see Appendix B).

The rotation and reflection properties of $p^\dagger$ and $s^\dagger$ can be summarized by giving their transformation under a rotation $\hat{R}(\theta_1, \theta_2, \theta_3)$ characterized by the three Euler angles θ_i,

$$\hat{R}^{-1} p_m^\dagger \hat{R} = \sum_{m'} D_{m'm}^1(\theta_1, \theta_2, \theta_3) p_{m'}^\dagger, \qquad \hat{R}^{-1} s^\dagger \hat{R} = s^\dagger \qquad (5.5)$$

as well as their transformation under reflection $\hat{P}$,

$$\hat{P}^{-1} p_m^\dagger \hat{P} = -p_m^\dagger, \qquad \hat{P}^{-1} s^\dagger \hat{P} = s^\dagger \qquad (5.6)$$

These transformation relations pertain to the creation operators $p_m^\dagger$ and $s^\dagger$. As discussed in Section 1.7, the annihilation operators do *not*, in general, transform as spherical tensor operators, but the modified combinations

$$\tilde{p}_m \equiv (-)^{1+m} p_{-m}, \qquad \tilde{s} \equiv s \qquad (5.7)$$

do satisfy the transformation property (5.5). We see from (5.7) that, for the scalar s boson, the modified operator $\tilde{s}$ does not differ from the annihilation operator s, but for consistency we shall use the notation with the tilde.

Since rotational invariance is of central importance in the vibron model, it is useful to write the U(4) generators in a different form, such that they too transform as spherical tensors under rotations. This can be achieved, using standard Clebsch–Gordan coefficients, by coupling the creation operators $b_{lm}^\dagger$ and the modified annihilation operators $\tilde{b}_{lm}$ to bilinear operators with definite angular momentum λ and projection μ on the z axis. We define

$$\mathcal{B}_\mu^{(\lambda)}(l,l') \equiv [b_l^\dagger \times \tilde{b}_{l'}]_\mu^{(\lambda)} = \sum_{mm'} \langle lm\, l'm' | \lambda\mu \rangle b_{lm}^\dagger \tilde{b}_{l'm'} \qquad (5.8)$$

The operators $\mathcal{B}_\mu^{(\lambda)}(l,l')$ are linear combinations of $\mathcal{G}_i^j$ and hence generate the same U(4) algebra as before. In the coupled-tensor form most of the generators of U(4) have a simple physical interpretation. For instance, we find

$$\mathcal{B}_0^{(0)}(1,1) = [p^\dagger \times \tilde{p}]_0^{(0)} = \sum_m \langle 1m\, 1-m | 00 \rangle p_m^\dagger \tilde{p}_{-m} = \sqrt{\tfrac{1}{3}} \sum_m p_m^\dagger p_m \qquad (5.9)$$

Since the p-boson number operator $\hat{n}_{\mathrm{p}}$ is given by

$$\hat{n}_{\mathrm{p}} = \sum_m p_m^\dagger p_m \qquad (5.10)$$

we have established the following correspondence:

$$\hat{n}_{\mathrm{p}} = \sqrt{3}\, [p^\dagger \times \tilde{p}]_0^{(0)} \qquad (5.11)$$

The physical interpretation of other generators (5.8) will be discussed in the course of this section.

We now turn to the problem of finding the subalgebras of U(4). For this purpose it is useful to write the generators $\mathcal{B}_\mu^{(\lambda)}(l,l')$ in explicit

5.2 The U(4) algebra

form:

$$\begin{aligned}
\mathcal{B}_0^{(0)}(1,1) &= [p^\dagger \times \tilde{p}]_0^{(0)} & 1 \\
\mathcal{B}_\mu^{(1)}(1,1) &= [p^\dagger \times \tilde{p}]_\mu^{(1)} & 3 \\
\mathcal{B}_\mu^{(2)}(1,1) &= [p^\dagger \times \tilde{p}]_\mu^{(2)} & 5 \\
\mathcal{B}_\mu^{(1)}(1,0) &= [p^\dagger \times \tilde{s}]_\mu^{(1)} & 3 \\
\mathcal{B}_\mu^{(1)}(0,1) &= [s^\dagger \times \tilde{p}]_\mu^{(1)} & 3 \\
\mathcal{B}_0^{(0)}(0,0) &= [s^\dagger \times \tilde{s}]_0^{(0)} & 1
\end{aligned} \quad (5.12)$$

On the right-hand side the number of components of each tensor is given, adding up to 16, the number of generators of U(4). Many possible subalgebra chains of U(4) can be constructed. In the context of the vibron model, however, we are only interested in those subalgebras that contain the SO(3) angular momentum algebra. This restriction is needed in order to construct bases that have L, the angular momentum, as a good quantum number. The components $\hat{L}_\mu$ of the angular momentum operator are among the generators in (5.12) and can be readily identified from general principles: they must form a tensor operator of rank 1 and have positive parity. Only the generators $[p^\dagger \times \tilde{p}]_\mu^{(1)}$ in (5.12) satisfy both conditions, and from this we infer that $\hat{L}_\mu$ is proportional to $[p^\dagger \times \tilde{p}]_\mu^{(1)}$. In order to find the proportionality coefficient, the SU(2) commutation relations satisfied by $\hat{L}_\mu$,

$$[\hat{L}_0, \hat{L}_{\pm 1}] = \pm \hat{L}_{\pm 1}, \qquad [\hat{L}_{-1}, \hat{L}_{+1}] = \hat{L}_0 \quad (5.13)$$

should be compared to those valid for the operators $[p^\dagger \times \tilde{p}]_\mu^{(1)}$. For example, we find

$$\begin{aligned}
{[[p^\dagger \times \tilde{p}]_{-1}^{(1)}, [p^\dagger \times \tilde{p}]_{+1}^{(1)}]} &= -\tfrac{1}{2}[p^\dagger_{-1}\tilde{p}_0 + p^\dagger_0\tilde{p}_{-1}, p^\dagger_0\tilde{p}_{+1} + p^\dagger_{+1}\tilde{p}_0] \\
&= -\tfrac{1}{2}[p^\dagger_{-1}p_0, p^\dagger_0 p_{-1}] - \tfrac{1}{2}[p^\dagger_0 p_{+1}, p^\dagger_{+1} p_0] \\
&= -\tfrac{1}{2}(p^\dagger_{-1}\tilde{p}_{+1} - p^\dagger_{+1}\tilde{p}_{-1}) \\
&= \sqrt{\tfrac{1}{2}}[p^\dagger \times \tilde{p}]_0^{(1)}
\end{aligned} \quad (5.14)$$

where we have used the explicit values [3] for the Clebsch–Gordan coefficients appearing in the expansion (5.8). Comparison with the last commutator in (5.13) shows that

$$\hat{L}_\mu = \sqrt{2}\mathcal{B}_\mu^{(1)}(1,1) = \sqrt{2}[p^\dagger \times \tilde{p}]_\mu^{(1)} \quad (5.15)$$

This result is confirmed by considering other commutators in (5.13). Note that the commutation relations (5.13) satisfied by the operators $\hat{L}_\mu$ are identical to those satisfied by the $\hat{J}_\mu$ introduced in (1.93), and hence both sets of operators generate the same Lie algebra. The fact that we refer to one set as SO(3) and to the other as SU(2) has to do with the angular momenta which are considered in the two cases. In the U(4) model only integer angular momentum values occur, whereas in the application of Section 1.7 both integer and half-integer angular momenta are possible. The notation SO(3) and SU(2) is often used to distinguish between these two situations.

The commutation relations between the generators $\mathcal{B}_\mu^{(\lambda)}(l,l')$ can be derived in general. Since it will be useful throughout this chapter, we quote here the general result:

$$[\mathcal{B}_\mu^{(\lambda)}(l,l'), \mathcal{B}_{\mu'}^{(\lambda')}(l'',l''')]$$
$$= \sum_{\lambda''\mu''} \sqrt{(2\lambda+1)(2\lambda'+1)} \langle \lambda\mu\,\lambda'\mu'|\lambda''\mu''\rangle$$
$$\times \left[(-)^{\lambda''+l+l'''} \left\{ \begin{array}{ccc} \lambda & \lambda' & \lambda'' \\ l''' & l & l' \end{array} \right\} \delta_{l'l''} \mathcal{B}_{\mu''}^{(\lambda'')}(l,l''') \right.$$
$$\left. -(-)^{\lambda+\lambda'+l'+l''} \left\{ \begin{array}{ccc} \lambda & \lambda' & \lambda'' \\ l'' & l' & l \end{array} \right\} \delta_{ll'''} \mathcal{B}_{\mu''}^{(\lambda'')}(l'',l') \right] \quad (5.16)$$

In this expression $\langle \lambda\mu\,\lambda'\mu'|\lambda''\mu''\rangle$ is a Clebsch–Gordan coefficient, introduced in Section 1.7, and the symbol in curly brackets is a $6j$ coefficient [3].

Returning to the question of finding all rotationally invariant subalgebra chains of U(4), we have shown so far that they are found by solving for all algebras G that satisfy

$$\mathrm{U}(4) \supset G \supset \mathrm{SO}(3) \quad (5.17)$$

where U(4) contains the generators (5.12) and SO(3) consists of the operators $\hat{L}_\mu$. This is a purely algebraic problem that can be solved with the help of the commutation relations (5.16).

One subalgebra G of U(4) is obtained after deleting from (5.12) all

5.2 The U(4) algebra

generators that contain an $s^\dagger$ or $\tilde{s}$ operator. The remaining generators

$$\begin{aligned}\hat{n}_p &\equiv \sqrt{3}[p^\dagger \times \tilde{p}]^{(0)}_0 & 1 \\ \hat{L}_\mu &\equiv \sqrt{2}[p^\dagger \times \tilde{p}]^{(1)}_\mu & 3 \\ \hat{Q}_\mu &\equiv [p^\dagger \times \tilde{p}]^{(2)}_\mu & 5\end{aligned} \quad (5.18)$$

form the U(3) algebra since, in the uncoupled notation (5.3), they satisfy the commutation relations (5.4) but with the indices i, j, k, l running over three instead of four possible values. Hence the following chain of algebras is established:

$$\begin{array}{cccc} U(4) \supset & U(3) \supset & SO(3) \supset & SO(2) \\ | & | & | & | \\ [N] & n_p & L & M_L \end{array} \quad (5.19)$$

Since we are dealing with a system of bosons, the U(4) representation must be totally symmetric; it is denoted by $[N]$ in (5.19) with N the total number of p and s bosons. Also indicated are the angular momentum L and its projection M_L on the z axis. The quantum number n_p is associated with U(3) and is discussed in detail in Section 5.4.

There exists a second algebra G that satisfies (5.17). It consists of the generators

$$\begin{aligned}\hat{L}_\mu &\equiv \sqrt{2}[p^\dagger \times \tilde{p}]^{(1)}_\mu & 3 \\ \hat{D}_\mu &\equiv i[p^\dagger \times \tilde{s} + s^\dagger \times \tilde{p}]^{(1)}_\mu & 3\end{aligned} \quad (5.20)$$

as it can be shown with the help of (5.16) that the set of operators (5.20) closes under commutation. The number of generators of SO(n) being $n(n-1)/2$ (see Appendix B), we may conclude that the operators (5.20) generate the algebra SO(4). Thus a second chain of algebras is established:

$$\begin{array}{cccc} U(4) \supset & SO(4) \supset & SO(3) \supset & SO(2) \\ | & | & | & | \\ [N] & \omega & L & M_L \end{array} \quad (5.21)$$

The same quantum numbers N, L, and M_L appear as in (5.19); the label ω is associated with SO(4) and is discussed in Section 5.5.

Note that a factor i is included in the definition of the dipole operator $\hat{D}_\mu$ in (5.20). This is a matter of convention since any other c

number (in place of i) would also result in a closed Lie algebra. The factor i is chosen because it makes $\hat{D}_0$ hermitian, $\hat{D}_0^\dagger = \hat{D}_0$, and in general, we will define any generator $\hat{T}_\mu^{(\lambda)}$ such that its $\mu = 0$ component is hermitian. In many cases an electromagnetic moment of a state can be written as the matrix element of $d\hat{T}_0^{(\lambda)}$ and our convention will guarantee that the proportionality coefficient d is real.

The existence of SO(4) as a subalgebra of U(4) is not a coincidence only occurring for p and s bosons. Rather, any Lie algebra U(n) can be shown to contain an SO(n) subalgebra. In terms of the uncoupled generators $b_{lm}^\dagger \tilde{b}_{l'm'}$ we may introduce the *antisymmetric* combinations

$$b_{lm}^\dagger \tilde{b}_{l'm'} - b_{l'm'}^\dagger \tilde{b}_{lm}, \qquad l, l' = 1, 0 \qquad (5.22)$$

which can be shown to close under commutation. Clearly, this can be done for any value of n, and such antisymmetric operators define a subalgebra of U(n) with $n(n-1)/2$ independent generators. For $n = 4$ with p and s bosons the operators (5.22) reduce in coupled-tensor notation to

$$\begin{aligned} \hat{L}_\mu &\equiv \sqrt{2}[p^\dagger \times \tilde{p}]_\mu^{(1)} & 3 \\ \hat{D}'_\mu &\equiv [p^\dagger \times \tilde{s} - s^\dagger \times \tilde{p}]_\mu^{(1)} & 3 \end{aligned} \qquad (5.23)$$

The argument leading to (5.23) shows that SO(4) occurs as a subalgebra of U(4) by taking antisymmetric combinations of the generators $\mathcal{G}_i^j$ and that this is a special case of the general reduction U(n) $\supset$ SO(n). Note, however, that the last three generators in (5.23) differ from those in (5.20), the two choices giving rise to an algebra which closes under commutation. Both are possible realizations of SO(4) in terms of p and s bosons (in fact, the *only* possible realizations if we require the associated Casimir invariant to be hermitian), and to distinguish one from the other, we follow the convention to denote by SO(4) the algebra containing $i[p^\dagger \times \tilde{s} + s^\dagger \times \tilde{p}]_\mu^{(1)}$ while we reserve the notation $\overline{\text{SO}}(4)$ for the one containing $[p^\dagger \times \tilde{s} - s^\dagger \times \tilde{p}]_\mu^{(1)}$. The SO(4) and $\overline{\text{SO}}(4)$ algebras are related through the transformation

$$p_m^\dagger \to -ip_m^\dagger, \qquad \tilde{p}_m \to i\tilde{p}_m \qquad (5.24)$$

This is a *canonical* transformation in the sense that it leaves invariant the commutation relations (5.1), and as a consequence, the properties

5.3 Symmetry limits

of the two algebras are very similar. Throughout this and the following chapters we are mainly concerned with SO(4) and only occasionally give a separate discussion of $\overline{\mathrm{SO}}(4)$. Our preference of SO(4) is due to its simpler treatment in coordinate space, which is presented in Section 5.5. We note, however, that Iachello and collaborators [1,2] follow a different convention and discuss the properties of the algebra (5.23) instead. (In comparing the two conventions, note also that they use a different convention for the $\tilde{p}_\mu$ operators.) The two algebras are isomorphic, and a consistent treatment leads to the same results for all matrix elements in the model.

We have demonstrated in this section the existence of the U(3) and SO(4) subalgebras of U(4), a situation reminiscent of the one encountered in the schematic U(2) model where the subalgebras U(1) and SO(2) occur. There is, however, one important difference between the two cases. The subalgebras of U(4) are required here to contain SO(3), the algebra of angular momentum in three dimensions, whereas the equivalent condition is absent from the U(2) model. The condition is essential in the context of the vibron model to construct rotationally invariant hamiltonians and to obtain eigenstates with good angular momentum L. In the next section we show how rotational invariance in the vibron model restricts the choice of the hamiltonian and how it influences its symmetry character.

5.3 Symmetry limits

The most general hamiltonian of the vibron model containing one- and two-body terms is of the form

$$\hat{H} = E_0 + \sum_{ij} \epsilon_{ij} b_i^\dagger b_j + \sum_{ijkl} e_{ijkl} b_i^\dagger b_j^\dagger b_k b_l \qquad (5.25)$$

where E_0 is a constant and ϵ_{ij} and e_{ijkl} are parameters that define the single-boson energies and the interactions between the bosons. According to the boson commutation relations (5.1), the two-body parameters satisfy the relations $e_{ijkl} = e_{ijlk} = e_{jikl} = e_{jilk}$, but otherwise they are independent. Imposing hermiticity on the hamiltonian introduces the additional relations $\epsilon_{ij} = \epsilon_{ji}$ and $e_{ijkl} = e_{klij}$ and reduces the number of independent parameters from 117 to 66. A further substantial reduction of this number is obtained by requiring the hamiltonian to be invariant under rotations and reflections. The hamiltonian then reduces to

$$
\begin{aligned}
\hat{H} = {} & E_0 + \epsilon_{\mathrm{p}} \hat{n}_{\mathrm{p}} + \epsilon_{\mathrm{s}} \hat{n}_{\mathrm{s}} \\
& + e_1 [[p^\dagger \times p^\dagger]^{(0)} \times [\tilde{p} \times \tilde{p}]^{(0)}]_0^{(0)} + e_2 [[p^\dagger \times p^\dagger]^{(2)} \times [\tilde{p} \times \tilde{p}]^{(2)}]_0^{(0)} \\
& + e_3 [[p^\dagger \times p^\dagger]^{(0)} \times [\tilde{s} \times \tilde{s}]^{(0)} + [s^\dagger \times s^\dagger]^{(0)} \times [\tilde{p} \times \tilde{p}]^{(0)}]_0^{(0)} \\
& + e_4 [[p^\dagger \times s^\dagger]^{(1)} \times [\tilde{p} \times \tilde{s}]^{(1)}]_0^{(0)} + e_5 [[s^\dagger \times s^\dagger]^{(0)} \times [\tilde{s} \times \tilde{s}]^{(0)}]_0^{(0)}
\end{aligned}
$$

(5.26)

where $\hat{n}_{\mathrm{p}}$ and $\hat{n}_{\mathrm{s}}$ are the number operators for p and s bosons. A final simplification is obtained by noting that the total number of bosons is conserved by $\hat{H}$ and that, as a consequence, the *operator* $\hat{N} \equiv \hat{n}_{\mathrm{p}} + \hat{n}_{\mathrm{s}}$ can be replaced by the *number* N which in the representation $[N]$ reduces to a constant that can be absorbed in E_0. This procedure further reduces the number of independent parameters. For example, the s-boson number operator $\hat{n}_{\mathrm{s}}$ undergoes the following replacement:

$$\hat{n}_{\mathrm{s}} = \hat{N} - \hat{n}_{\mathrm{p}} \mapsto N - \hat{n}_{\mathrm{p}} \qquad (5.27)$$

Similar manipulations with the two-body terms in (5.26) transform $\hat{H}$ into

$$
\begin{aligned}
\hat{H} = {} & E_0' + \epsilon_{\mathrm{p}}' \hat{n}_{\mathrm{p}} \\
& + e_1' [[p^\dagger \times p^\dagger]^{(0)} \times [\tilde{p} \times \tilde{p}]^{(0)}]_0^{(0)} + e_2' [[p^\dagger \times p^\dagger]^{(2)} \times [\tilde{p} \times \tilde{p}]^{(2)}]_0^{(0)} \\
& + e_3' [[p^\dagger \times p^\dagger]^{(0)} \times [\tilde{s} \times \tilde{s}]^{(0)} + [s^\dagger \times s^\dagger]^{(0)} \times [\tilde{p} \times \tilde{p}]^{(0)}]_0^{(0)}
\end{aligned}
\qquad (5.28)
$$

with the primed and unprimed parameter sets related through

$$
\begin{aligned}
E_0' &= E_0 + (\epsilon_{\mathrm{s}} - e_5) N + e_5 N^2 \\
\epsilon_{\mathrm{p}}' &= \epsilon_{\mathrm{p}} - \epsilon_{\mathrm{s}} - \sqrt{\tfrac{1}{3}} e_4 + 2 e_5 + \left(\sqrt{\tfrac{1}{3}} e_4 - 2 e_5 \right) N \\
e_1' &= e_1 - \sqrt{\tfrac{1}{3}} e_4 + e_5 \\
e_2' &= e_2 - \sqrt{\tfrac{5}{3}} e_4 + \sqrt{5} e_5 \\
e_3' &= e_3
\end{aligned}
\qquad (5.29)
$$

5.3 Symmetry limits

Table 5.1: Number of independent parameters in the vibron hamiltonian

	Zero body	One body	Two body
General	1	16	100
$\hat{H}^\dagger = \hat{H}$	1	10	55
And $\hat{R}^{-1}\hat{H}\hat{R} = \hat{H}$	1	2	5
And $\hat{P}^{-1}\hat{H}\hat{P} = \hat{H}$	1	2	5
And $\hat{N} \to N$	1	1	3

The number of independent parameters in the vibron hamiltonian under various restrictions such as hermiticity, rotational invariance, etc., is shown in Table 5.1. It illustrates the considerable reduction in complexity achieved on the basis of simple symmetry arguments.

The hamiltonian (5.26) can be written equivalently as an expansion in invariant operators of U(4) and its subalgebras. This form is more useful for the applications in Sections 5.4 and 5.5 and therefore we discuss it here in detail. The first task is to derive the different invariant operators of the algebras in (5.19) and (5.21), a procedure which we illustrate with some examples.

As a first example we consider the linear Casimir invariant of U(n) which can be defined as (see (2.66))

$$\mathcal{C}_1[\mathrm{U}(n)] \equiv \sum_{i=1}^{n} \mathcal{G}_i^i \tag{5.30}$$

This operator commutes with all generators of U(n), as is shown easily with the help of the commutation relations (5.4):

$$[\sum_{i=1}^{n} \mathcal{G}_i^i, \mathcal{G}_k^r] = \sum_{i=1}^{n} \mathcal{G}_k^i \delta_{ir} - \sum_{i=1}^{n} \mathcal{G}_i^r \delta_{ik} = \mathcal{G}_k^r - \mathcal{G}_k^r = 0 \tag{5.31}$$

For $n = 3$ the linear Casimir invariant $\mathcal{C}_1[\mathrm{U}(3)]$ can be written in terms of p bosons,

$$\mathcal{C}_1[\mathrm{U}(3)] = \sum_{i=1}^{3} \mathcal{G}_i^i = \sum_m p_m^\dagger p_m = \hat{n}_\mathrm{p} \tag{5.32}$$

which shows that the linear Casimir invariant of U(3) is nothing but the p-boson number operator. Similarly, the linear Casimir invariant of

U(4) becomes the total boson number operator:

$$C_1[U(4)] = \sum_{i=1}^{4} \mathcal{G}_i^i = \sum_m p_m^\dagger p_m + s^\dagger s = \hat{n}_p + \hat{n}_s = \hat{N} \qquad (5.33)$$

A quadratic invariant of $U(n)$ is trivially constructed by taking the square of the linear invariant. This operator again commutes with all generators of $U(n)$ and is, as such, an invariant of the algebra. Since it is constructed from a lower-order Casimir invariant, we refer to it as an invariant rather than a *Casimir* invariant. The quadratic Casimir invariant of $U(n)$ is defined as

$$\mathcal{C}_2[U(n)] \equiv \sum_{ij=1}^{n} \mathcal{G}_i^j \mathcal{G}_j^i \qquad (5.34)$$

Again, it can be shown that this operator commutes with all generators of $U(n)$:

$$\begin{aligned}
\left[\sum_{ij=1}^{n} \mathcal{G}_i^j \mathcal{G}_j^i, \mathcal{G}_k^r\right] &= \sum_{ij=1}^{n} \mathcal{G}_i^j [\mathcal{G}_j^i, \mathcal{G}_k^r] + \sum_{ij=1}^{n} [\mathcal{G}_i^j, \mathcal{G}_k^r]\mathcal{G}_j^i \\
&= \sum_{ij=1}^{n} \mathcal{G}_i^j (\mathcal{G}_j^r \delta_{ik} - \mathcal{G}_k^i \delta_{jr}) + \sum_{ij=1}^{n} (\mathcal{G}_i^r \delta_{jk} - \mathcal{G}_k^j \delta_{ir})\mathcal{G}_j^i \\
&= \sum_{j=1}^{n} \mathcal{G}_k^j \mathcal{G}_j^r - \sum_{i=1}^{n} \mathcal{G}_i^r \mathcal{G}_k^i + \sum_{i=1}^{n} \mathcal{G}_i^r \mathcal{G}_k^i - \sum_{j=1}^{n} \mathcal{G}_k^j \mathcal{G}_j^r \\
&= 0 \qquad (5.35)
\end{aligned}$$

In terms of the generators $b_{lm}^\dagger \tilde{b}_{l'm'}$ the quadratic Casimir invariant can also be written as

$$\sum_{ll'} \sum_{mm'} b_{lm}^\dagger b_{l'm'} b_{l'm'}^\dagger b_{lm} = \sum_{ll'} \sum_{mm'} (-)^{l+m+l'+m'} b_{lm}^\dagger \tilde{b}_{l'-m'} b_{l'm'}^\dagger \tilde{b}_{l-m} \qquad (5.36)$$

where l and l' run over the angular momentum values of the bosons that form the $U(n)$ algebra. For $n = 3$ the quadratic Casimir invariant assumes the form

$$\mathcal{C}_2[U(3)] = \sum_{mm'} (-)^{m+m'} p_m^\dagger \tilde{p}_{-m'} p_{m'}^\dagger \tilde{p}_{-m}$$

5.3 Symmetry limits

$$\begin{aligned}
&= \sum_{mm'}(-)^{m+m'}\left(\sum_{\lambda\mu}\langle 1m\ 1-m'|\lambda\mu\rangle[p^\dagger\times\tilde{p}]^{(\lambda)}_\mu\right) \\
&\qquad\times\left(\sum_{\lambda'\mu'}\langle 1m'\ 1-m|\lambda'\mu'\rangle[p^\dagger\times\tilde{p}]^{(\lambda')}_{\mu'}\right) \\
&= \sum_{\lambda\mu}\sum_{\lambda'\mu'}(-)^\mu[p^\dagger\times\tilde{p}]^{(\lambda)}_\mu[p^\dagger\times\tilde{p}]^{(\lambda')}_{\mu'} \\
&\qquad\times\sum_{mm'}\langle 1m\ 1-m'|\lambda\mu\rangle\langle 1m'\ 1-m|\lambda'\mu'\rangle \\
&= \sum_{\lambda\mu}(-)^\mu[p^\dagger\times\tilde{p}]^{(\lambda)}_\mu[p^\dagger\times\tilde{p}]^{(\lambda)}_{-\mu} \quad (5.37)
\end{aligned}$$

where the inverse of relation (5.8) and the symmetry and orthonormality properties of the Clebsch–Gordan coefficients are used. Thus, according to the definition of a scalar product, we obtain

$$\mathcal{C}_2[\mathrm{U}(3)] = \sum_\lambda [p^\dagger\times\tilde{p}]^{(\lambda)}\cdot[p^\dagger\times\tilde{p}]^{(\lambda)} = \tfrac{1}{3}\hat{n}_\mathrm{p}^2 + \tfrac{1}{2}\hat{L}^2 + \hat{Q}^2 \quad (5.38)$$

In a similar way an expression is derived for the quadratic Casimir invariant of U(4):

$$\begin{aligned}
\mathcal{C}_2[\mathrm{U}(4)] &= \sum_{ll'}(-)^{l+l'}\sum_\lambda \mathcal{B}^{(\lambda)}(l,l')\cdot\mathcal{B}^{(\lambda)}(l',l) \\
&= \tfrac{1}{3}\hat{n}_\mathrm{p}^2 + \tfrac{1}{2}(\hat{L}^2 + \hat{D}^2 + \hat{D}'^2) + \hat{Q}^2 + \hat{n}_\mathrm{s}^2 \quad (5.39)
\end{aligned}$$

Clearly, this procedure for constructing Casimir invariants of U(n) can be extended to higher orders, $\sum_{ijk}\mathcal{G}^j_i\mathcal{G}^k_j\mathcal{G}^i_k$ being the cubic Casimir invariant, $\sum_{ijkl}\mathcal{G}^j_i\mathcal{G}^k_j\mathcal{G}^l_k\mathcal{G}^i_l$ the quartic Casimir invariant, etc. These sums can be shown to commute with the generators of U(n). The complete set of Casimir invariants of any Lie algebra is given by Racah [4] and reproduced in Chapter 15 of the book by Wybourne [5]. The Casimir invariants appearing in this set are, in general, independent, but they might become dependent if a specific representation of the Lie algebras is considered. A simple example of this dependence is given for U(3) at the end of this section.

Next, we discuss the construction of Casimir invariants of orthogonal algebras $\overline{\mathrm{SO}}(n)$ generated by the antisymmetric operators $b^\dagger_{lm}\tilde{b}_{l'm'} -$

$b^\dagger_{l'm'}\tilde{b}_{lm}$. No linear Casimir invariant of $\overline{SO}(n)$ exists since the contraction of a single antisymmetric operator vanishes identically:

$$\sum_{lm}(-)^{l+m}(b^\dagger_{lm}\tilde{b}_{lm} - b^\dagger_{lm}\tilde{b}_{lm}) = 0 \qquad (5.40)$$

Note that this result is true for both sets of generators (5.20) and (5.23). The quadratic Casimir invariant is, in analogy with (5.36), defined as

$$\mathcal{C}_2[\overline{SO}(n)] \equiv \tfrac{1}{2}\sum_{ll'}\sum_{mm'}(-)^{l+m+l'+m'}(b^\dagger_{lm}\tilde{b}_{l'm'} - b^\dagger_{l'm'}\tilde{b}_{lm})$$
$$\times (b^\dagger_{l'-m'}\tilde{b}_{l-m} - b^\dagger_{l-m}\tilde{b}_{l'-m'}) \qquad (5.41)$$

Specific cases are obtained after inserting the appropriate values for the summation indices l and l'. For example, for $n = 4$ we have $l, l' = 0, 1$, and after a derivation analogous to that in (5.37) we find

$$\mathcal{C}_2[\overline{SO}(4)] = \hat{L}^2 + \hat{D}'^2 \qquad (5.42)$$

The canonical transformation (5.24) leads to the expression for the SO(4) Casimir invariant:

$$\mathcal{C}_2[SO(4)] = \hat{L}^2 + \hat{D}^2 \qquad (5.43)$$

The quadratic Casimir invariant of SO(3) is obtained for $l = l' = 1$ and reduces to

$$\mathcal{C}_2[SO(3)] = \hat{L}^2 \qquad (5.44)$$

as expected.

In Table 5.2 we summarize all linear and quadratic invariants of U(4) and of its subalgebras containing the angular momentum algebra SO(3). These operators can be used to produce an alternative expansion of the vibron hamiltonian. Since Table 5.2 gives two linear and eight quadratic invariants as opposed to two-plus-five terms in (5.26), it is clear that the set of operators given in the table is not independent. Using the explicit expressions for the various Casimir invariants, one may, for instance, derive the operator relation

$$\mathcal{C}_2[U(4)] = \hat{N}^2 - 2\hat{N}\hat{n}_p + \hat{n}_p^2 + \mathcal{C}_2[U(3)]$$
$$+ \tfrac{1}{2}(\mathcal{C}_2[SO(4)] + \mathcal{C}_2[\overline{SO}(4)]) - \mathcal{C}_2[SO(3)] \qquad (5.45)$$

5.3 Symmetry limits

Table 5.2: Linear and quadratic invariants of U(4) and of its subalgebras containing SO(3)

Algebra	Linear	Quadratic
U(4)	$\mathcal{C}_1[\mathrm{U}(4)] = \hat{N}$	$\hat{N}^2$
		$\mathcal{C}_2[\mathrm{U}(4)] = \frac{1}{3}\hat{n}_\mathrm{p}^2 + \frac{1}{2}(\hat{L}^2 + \hat{D}^2 + \hat{D}'^2) + \hat{Q}^2 + \hat{n}_\mathrm{s}^2$
U(3)	$\mathcal{C}_1[\mathrm{U}(3)] = \hat{n}_\mathrm{p}$	$\hat{n}_\mathrm{p}^2$
		$\hat{N}\hat{n}_\mathrm{p}$
		$\mathcal{C}_2[\mathrm{U}(3)] = \frac{1}{3}\hat{n}_\mathrm{p}^2 + \frac{1}{2}\hat{L}^2 + \hat{Q}^2$
SO(4)		$\mathcal{C}_2[\mathrm{SO}(4)] = \hat{L}^2 + \hat{D}^2$
$\overline{\mathrm{SO}}(4)$		$\mathcal{C}_2[\overline{\mathrm{SO}}(4)] = \hat{L}^2 + \hat{D}'^2$
SO(3)		$\mathcal{C}_2[\mathrm{SO}(3)] = \hat{L}^2$

and hence eliminate one of the invariants from the expansion. There exists, however, a more subtle kind of dependence which we illustrate here with an example. We first note the absence from (5.26) of a term $[[p^\dagger \times p^\dagger]^{(1)} \times [\tilde{p} \times \tilde{p}]^{(1)}]_0^{(0)}$. It is natural to exclude this two-body interaction from the vibron hamiltonian since, acting on a many-body state containing identical p bosons, it will always give zero. Expressed in a different way: two identical p bosons cannot couple to angular momentum $L = 1$. By interchanging the second $p^\dagger$ and the first $\tilde{p}$, we may write this term differently,

$$\mathcal{P}_1^\dagger \cdot \tilde{\mathcal{P}}_1 \equiv [p^\dagger \times p^\dagger]^{(1)} \cdot [\tilde{p} \times \tilde{p}]^{(1)} = -\hat{n}_\mathrm{p} - \tfrac{1}{3}\hat{n}_\mathrm{p}^2 + \tfrac{1}{4}\hat{L}^2 + \tfrac{1}{2}\hat{Q}^2 \qquad (5.46)$$

and hence derive the following operator identity:

$$\mathcal{C}_2[\mathrm{U}(3)] - 2\mathcal{P}_1^\dagger \cdot \tilde{\mathcal{P}}_1 = \hat{n}_\mathrm{p}^2 + 2\hat{n}_\mathrm{p} \qquad (5.47)$$

Since we know that in the *symmetric* space of identical p bosons the $\mathcal{P}_1^\dagger \cdot \tilde{\mathcal{P}}_1$ term reduces to zero, we conclude that in this space the quadratic Casimir invariant of U(3) can be effectively replaced by a combination of $\hat{n}_\mathrm{p}$ and $\hat{n}_\mathrm{p}^2$:

$$\mathcal{C}_2[\mathrm{U}(3)] \mapsto \hat{n}_\mathrm{p}(\hat{n}_\mathrm{p} + 2) \qquad (5.48)$$

We emphasize that this relation, in contrast to (5.45) or (5.47), is *not* an operator identity and hence the symbol $\mapsto$ is used rather than an equality sign. Thus the replacement (5.48) is only valid in the symmetric representation of U(3) while, in general, the quadratic Casimir invariant of U(3) cannot be written as a function of the linear invariant. In the same way it can be shown that, in the symmetric representation of U(4), the following replacement is valid:

$$\mathcal{C}_2[\mathrm{U}(4)] \mapsto \hat{N}(\hat{N}+3) \tag{5.49}$$

If we consistently replace $\hat{N}$ by N and make use of (5.45), (5.48), and (5.49), we find that the most general vibron hamiltonian expanded in terms of the invariants of Table 5.2 reduces to

$$\boxed{\hat{H} = E_0'' + \epsilon \hat{n}_\mathrm{p} + \alpha \hat{n}_\mathrm{p}^2 + \beta(\hat{L}^2 + \hat{D}^2) + \gamma \hat{L}^2} \tag{5.50}$$

where E_0'' is an N-dependent constant and the parameters ϵ, α, β, and γ are linear combinations of those appearing in (5.28).

In the next two sections we study the dynamical symmetries of the vibron hamiltonian which occur if some of the parameters in (5.50) are zero. We conclude this section by giving a brief summary of the algebraic character of the vibron hamiltonian. Since $\hat{H}$ is expressed in terms of the generators of U(4), this algebra plays the role of the dynamical algebra of the vibron model, similar to the role of U(2) in the schematic model of Chapter 1. The new element introduced in this chapter is the simultaneous existence of a symmetry algebra of the hamiltonian, namely the algebra SO(3) of rotations in three dimensions, giving rise to significant simplifications in the structure of the vibron model.

5.4 Non-rigid molecules: The U(3) limit

The U(3) limit is obtained by taking $\beta = 0$ in the hamiltonian (5.50), which then reduces to

$$\boxed{\hat{H}_\mathrm{I} = E_0'' + \epsilon \hat{n}_\mathrm{p} + \alpha \hat{n}_\mathrm{p}^2 + \gamma \hat{L}^2} \tag{5.51}$$

5.4 Non-rigid molecules: The U(3) limit

This hamiltonian contains invariants of the algebras U(3) and SO(3) only, and hence its eigenstates are classified according to

$$\begin{array}{ccc} U(4) \supset & U(3) \supset & SO(3) \\ | & | & | \\ [N] & n_{\text{p}} & L \end{array} \qquad (5.52)$$

The reduction from the total angular momentum to its projection on the z axis is not included in (5.52). By virtue of rotational invariance this reduction is common to all labeling schemes for molecular and nuclear spectra discussed in Parts II and III, and from now on it will not be indicated explicitly but rather implicitly assumed. The hamiltonian (5.51) has the eigenvalues

$$E_{\text{I}}(n_{\text{p}}, L) = E_0'' + \epsilon n_{\text{p}} + \alpha n_{\text{p}}^2 + \gamma L(L+1) \qquad (5.53)$$

The eigenspectrum is further determined by $U(4) \supset U(3)$ and $U(3) \supset SO(3)$ branching rules. The first gives the values of n_{p} contained in the symmetric representation $[N]$:

$$n_{\text{p}} = 0, 1, \ldots, N \qquad (5.54)$$

The second branching rule determines the allowed values of angular momentum L for a given number of p bosons n_{p}:

$$L = n_{\text{p}}, n_{\text{p}} - 2, \ldots, 1 \text{ or } 0 \qquad (5.55)$$

Equation (5.54) is a particular example of the general rule for $U(n) \supset U(n-1)$ discussed in Appendix B. The $U(4) \supset U(3)$ branching rule states that the number of p bosons can have any non-negative integer value smaller than or equal to the total number of bosons and is thus self-evident. The $U(3) \supset SO(3)$ branching rule limits the allowed values of angular momentum: states with even parity (or with an even number of p bosons) can only have even L while odd-parity states (with an odd number of p bosons) have odd L. This rule is less self-evident but will become clear when we discuss the construction of the wave functions in the U(3) limit. Note also that (5.54) and (5.55) are straightforward generalizations of the corresponding rules in Chapter 1.

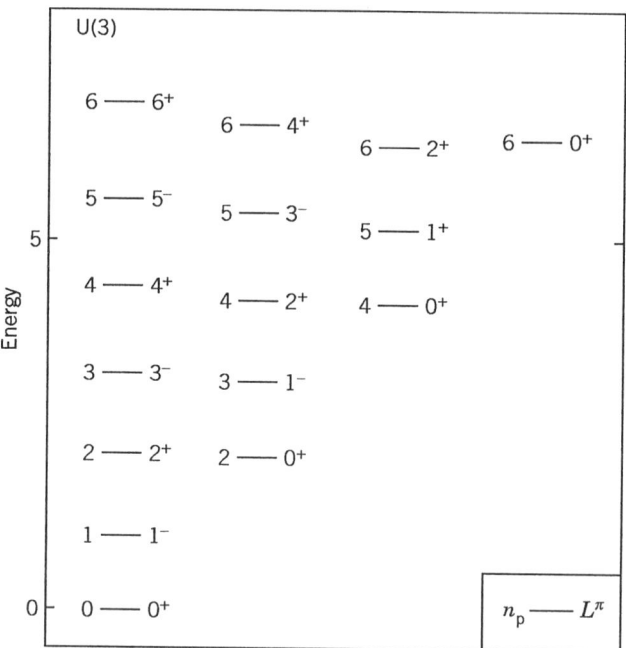

Figure 5.2: The U(3) eigenspectrum of the hamiltonian $\hat{H}_\mathrm{I}$ with $E_0'' = 0$, $\epsilon = 1$, and $\alpha = \gamma = 0.01$ for $N = 6$ bosons. Levels are labeled by n_p, angular momentum L, and parity $\pi = (-)^{n_\mathrm{p}} = (-)^L$. Energies are in arbitrary units.

The branching rules (5.54) and (5.55) together with the eigenvalue expression (5.53) completely determine the spectrum of the hamiltonian $\hat{H}_\mathrm{I}$. An example is shown in Figure 5.2. For small values of α/ϵ and γ/ϵ (as is the case in Figure 5.2) the energy spectrum resembles that of a three-dimensional isotropic harmonic oscillator typical of non-rigid molecules [6]. In this context it is convenient to define the rigidity parameter γ_rig [7],

$$\gamma_\mathrm{rig} = 2E_\mathrm{r}/E_\mathrm{v} \tag{5.56}$$

which is twice the ratio of the energy of the first rotational to that of the first vibrational level. For a rigid rotational molecule we expect γ_rig to be small. Using the energy expression (5.53), we find

$$\gamma_\mathrm{rig} = \frac{\epsilon + \alpha + 2\gamma}{\epsilon + 2\alpha} \approx 1, \qquad \alpha, \gamma \ll \epsilon \tag{5.57}$$

5.4 Non-rigid molecules: The U(3) limit

which confirms the non-rigid character of the U(3) limit.

States in the U(3) limit are thus characterized by the quantum numbers N, n_p, L, and M_L and are denoted as

$$|[N]n_p L M_L\rangle \tag{5.58}$$

We now turn to the problem of finding the explicit realization of these states in terms of polynomials in the $p_m^\dagger$ and $s^\dagger$ operators. We begin by summarizing the general idea of the procedure [8] and then proceed by giving a detailed derivation. In a first step the operator equations defining the basis (5.58) are converted into a set of coupled differential equations in the variables that are associated with the operators $p_m^\dagger$, $s^\dagger$, and their hermitian conjugates. After a coordinate transformation these differential equations can be made separable and solved in terms of standard mathematical functions. We thus obtain the U(3) states in terms of coordinates. The next step consists in the application of Dragt's theorem, which relates a particular class of coordinate solutions (i.e., those of *highest weight*) to corresponding states in the boson space, in the same way as is done in the U(2) model at the end of Section 1.5. We find in this way a boson realization of those U(3) states which have highest weight. In a final step this realization is generalized to an arbitrary U(3) state through the application of a boson operator that connects highest-weight states to all others.

We begin our derivation by writing the operator equations defining the U(3) basis:

$$\begin{aligned}
\hat{N}|[N]n_p L M_L\rangle &= N|[N]n_p L M_L\rangle \\
\hat{n}_p|[N]n_p L M_L\rangle &= n_p|[N]n_p L M_L\rangle \\
\hat{L}^2|[N]n_p L M_L\rangle &= L(L+1)|[N]n_p L M_L\rangle \\
\hat{L}_0|[N]n_p L M_L\rangle &= M_L|[N]n_p L M_L\rangle
\end{aligned} \tag{5.59}$$

The first two of these equations can be combined to give

$$\hat{n}_s|[N]n_p L M_L\rangle = (N - n_p)|[N]n_p L M_L\rangle \tag{5.60}$$

and hence the original equations are separated into two sets: those written in terms of operators depending on the p boson only (the last three in (5.59)) and one involving only s-boson operators, (5.60). Consequently, the corresponding differential equations are separable in the coordinates associated with the p bosons on the one hand and s bosons on the other and can be treated consecutively. We first concentrate on the solution of the p-boson equations and at the end multiply with the solution of the s-boson equation already known from Chapter 1.

The last three equations in (5.59) are converted into a set of differential equations by carrying out the substitutions

$$p_m^\dagger = \frac{1}{\sqrt{2}}\left(x_m - (-)^m \frac{\partial}{\partial x_m}\right), \qquad \tilde{p}_m = -\frac{1}{\sqrt{2}}\left(x_m + (-)^m \frac{\partial}{\partial x_m}\right) \tag{5.61}$$

with the coordinates (x_{-1}, x_0, x_{+1}) related to cartesian coordinates (x, y, z) in the following way:

$$x_{\pm 1} = \mp\sqrt{\tfrac{1}{2}}(x \pm iy), \qquad x_0 = z \tag{5.62}$$

One may verify that the operators (5.61) satisfy the correct commutation relations (5.1). The resulting differential equations in (x, y, z) are made separable by transforming to spherical coordinates (r, θ, ϕ) defined through the relations

$$x = r\sin\theta\cos\phi, \qquad y = r\sin\theta\sin\phi, \qquad z = r\cos\theta \tag{5.63}$$

These successive replacements and transformations are applied to the operators $\hat{n}_p$, $\hat{L}^2$, and $\hat{L}_0$ in (5.59). For example, the operator $\hat{L}_0$ becomes

$$\hat{L}_0 = p_{+1}^\dagger \tilde{p}_{-1} - p_{-1}^\dagger \tilde{p}_{+1} = -i\left(x\frac{\partial}{\partial y} - y\frac{\partial}{\partial x}\right) = -i\frac{\partial}{\partial \phi} \tag{5.64}$$

which, incidentally, is a direct justification of our earlier choice (5.15) as the angular momentum operator. Similarly, we find

$$\hat{L}^2 = \hat{L}_-\hat{L}_+ + \hat{L}_z + \hat{L}_z^2 = -\frac{1}{\sin\theta}\frac{\partial}{\partial\theta}\left(\sin\theta\frac{\partial}{\partial\theta}\right) + \frac{1}{\sin^2\theta}\hat{L}_0^2 \tag{5.65}$$

where use is made of

$$\hat{L}_\pm = \mp\sqrt{2}\hat{L}_{\pm 1} = \mp 2[p^\dagger \times \tilde{p}]_{\pm 1}^{(1)} = e^{\pm i\phi}\left(\pm\frac{\partial}{\partial\theta} + \frac{i}{\tan\theta}\frac{\partial}{\partial\phi}\right) \tag{5.66}$$

5.4 Non-rigid molecules: The U(3) limit

Finally, the p-boson number operator $\hat{n}_p$ becomes

$$\begin{aligned}\hat{n}_p &= p_{+1}^\dagger \tilde{p}_{-1} + p_{-1}^\dagger \tilde{p}_{+1} - p_0^\dagger \tilde{p}_0 \\ &= \frac{1}{2}\left(x^2 + y^2 + z^2 - \frac{\partial^2}{\partial x^2} - \frac{\partial^2}{\partial y^2} - \frac{\partial^2}{\partial z^2} - 3\right) \\ &= \frac{1}{2}\left[-\frac{1}{r^2}\frac{\partial}{\partial r}\left(r^2 \frac{\partial}{\partial r}\right) + r^2 + \frac{1}{r^2}\hat{L}^2 - 3\right]\end{aligned} \quad (5.67)$$

where the cartesian form of $\hat{n}_p$ shows that its eigensolutions are those of the isotropic three-dimensional harmonic oscillator. The solutions of (5.59) are separable in the spherical coordinates,

$$\Psi_{n_p L M_L}(r, \theta, \phi) = f_L^{n_p}(r) Y_{M_L}^L(\theta, \phi) \quad (5.68)$$

where the $Y_{M_L}^L(\theta, \phi)$ are the well-known spherical harmonics [9] which are the eigensolutions of the $\hat{L}^2$ and $\hat{L}_0$ operators in (5.65) and (5.64) with eigenvalues $L(L+1)$ and M_L, respectively. For future reference we give here their explicit expression:

$$Y_{M_L}^L(\theta, \phi) = \left[\frac{(2L+1)(L-M_L)!}{4\pi(L+M_L)!}\right]^{1/2} P_L^{M_L}(\cos\theta)e^{iM_L\phi} \quad (5.69)$$

where $P_n^m(x)$ is the associated Legendre function (see equations 8.810 and 8.910 of [10]),

$$\begin{aligned}P_n^m(x) &= (-)^m(1-x^2)^{m/2}\frac{d^m}{dx^m}P_n(x) \\ &= \frac{1}{2^n n!}(-)^m(1-x^2)^{m/2}\frac{d^{n+m}}{dx^{n+m}}(x^2-1)^n\end{aligned} \quad (5.70)$$

The radial part of the wave function satisfies, according to (5.67),

$$\frac{1}{2}\left[-\frac{1}{r^2}\frac{d}{dr}\left(r^2\frac{d}{dr}\right) + r^2 + \frac{L(L+1)}{r^2} - 3\right]f_L^{n_p}(r) = n_p f_L^{n_p}(r) \quad (5.71)$$

and its solutions are found by making the substitution $f_L^{n_p}(r) = r^L \exp(-r^2/2)h(r^2)$ where the function $h(x)$ satisfies a differential equation of the Laguerre type [10]:

$$\left[x\frac{d^2}{dx^2} + \left(L + \tfrac{3}{2} - x\right)\frac{d}{dx} + \frac{n_p - L}{2}\right]h(x) = 0 \quad (5.72)$$

The coordinate representation of the states defined by (5.59) is thus

$$\boxed{\Psi_{n_p L M_L}(r,\theta,\phi) = A_{n_p L}\, r^L e^{-r^2/2} L^{L+1/2}_{(n_p-L)/2}(r^2) Y^L_{M_L}(\theta,\phi)}$$ (5.73)

with coefficients $A_{n_p L}$ obtained from normalization:

$$A_{n_p L} = \left[\frac{2^{L+2}(n_p - L)!!}{\sqrt{\pi}(n_p + L + 1)!!}\right]^{1/2}$$ (5.74)

Having derived the form of the most general U(3) state, we now specify to the highest-weight state with $n_p = L$:

$$\Psi_{LLM_L}(r,\theta,\phi) = A_{LL}\, r^L e^{-r^2/2} Y^L_{M_L}(\theta,\phi)$$ (5.75)

To find the coordinate representation of this state we apply Dragt's theorem (see (1.72) and Appendix C) and make the replacements

$$x_m \to \frac{p^\dagger_m}{\sqrt{2}}, \qquad \frac{e^{-r^2/2}}{\pi^{3/4}} \to |0\rangle$$ (5.76)

which in spherical coordinates become

$$r = \sqrt{-x_{-1}x_{+1} - x_{+1}x_{-1} + x_0^2} \to \left(\frac{p^\dagger \cdot p^\dagger}{2}\right)^{1/2}$$

$$\cos\theta = \frac{z}{r} \to \frac{p^\dagger_0}{\sqrt{p^\dagger \cdot p^\dagger}}$$ (5.77)

$$e^{\pm i\phi} = \mp\frac{\sqrt{2}x_{\pm 1}}{\sqrt{x^2+y^2}} \to \mp\frac{\sqrt{2}p^\dagger_{\pm 1}}{\sqrt{p^\dagger \cdot p^\dagger - p^\dagger_0 p^\dagger_0}}$$

Carrying out these replacements in the highest-weight state, we find

$$\Psi_{LLM_L}(r,\theta,\phi) \to \left[\frac{4\pi}{(2L+1)!!}\right]^{1/2} \mathcal{Y}^L_{M_L}(\mathbf{p}^\dagger)|0\rangle$$ (5.78)

where $\mathcal{Y}^L_{M_L}(\mathbf{r})$ are the solid spherical harmonics

$$\mathcal{Y}^L_{M_L}(\mathbf{r}) = r^L Y^L_{M_L}(\theta,\phi)$$ (5.79)

5.4 Non-rigid molecules: The U(3) limit

To find the operator representation of the most general state, we can start from (5.78) and act on it with an operator that increases the number of p bosons without changing the angular momentum (i.e., a scalar in SO(3)). The simplest of such operators is $p^\dagger \cdot p^\dagger$, and hence we conclude that the most general U(3) state is of the form

$$|n_p L M_L\rangle \propto (p^\dagger \cdot p^\dagger)^{(n_p-L)/2} \mathcal{Y}^L_{M_L}(\mathbf{p}^\dagger)|0\rangle \qquad (5.80)$$

where the proportionality coefficient is found from normalization. Note that an SO(3)-scalar creation operator is at least quadratic in $p^\dagger$, and hence it increases the number of p bosons by at least 2. More generally, an SO(3)-scalar operator increasing the number of identical p bosons by k can only exist for even k. For example, for $k = 3$ the operator is of the form $[[p^\dagger \times p^\dagger]^{(1)} \times p^\dagger]^{(0)}_0$ and vanishes identically acting on a symmetric state. We conclude that the difference $n_p - L$ must be even (consistent with the branching rule (5.55)) and that the expression (5.80) is well defined. In Chapter 8 a similar argument is used in a more general context to derive the U(n) to SO(n) branching rule.

Finally, the normalization of (5.80) is obtained from the highest-weight character of the state $\mathcal{Y}^L_{M_L}(\mathbf{p}^\dagger)|0\rangle$, which implies

$$(\tilde{p} \cdot \tilde{p}) \mathcal{Y}^L_{M_L}(\mathbf{p}^\dagger)|0\rangle = 0 \qquad (5.81)$$

and the application of the commutator relation

$$[\tilde{p} \cdot \tilde{p}, (p^\dagger \cdot p^\dagger)^k] = 4k(p^\dagger \cdot p^\dagger)^{k-1}\left(\hat{n}_p + k + \tfrac{1}{2}\right) \qquad (5.82)$$

This commutator identity can be proven by induction since explicit evaluation shows it to be valid for $k = 1$,

$$[\tilde{p} \cdot \tilde{p}, p^\dagger \cdot p^\dagger] = 4\left(\hat{n}_p + \tfrac{3}{2}\right) \qquad (5.83)$$

and using the induction hypothesis, we find

$$\begin{aligned}
[\tilde{p} \cdot \tilde{p}, (p^\dagger \cdot p^\dagger)^k] &= (p^\dagger \cdot p^\dagger)^{k-1}[\tilde{p} \cdot \tilde{p}, p^\dagger \cdot p^\dagger] + [\tilde{p} \cdot \tilde{p}, (p^\dagger \cdot p^\dagger)^{k-1}](p^\dagger \cdot p^\dagger) \\
&= 4(p^\dagger \cdot p^\dagger)^{k-1}\left(\hat{n}_p + \tfrac{3}{2}\right) \\
&\quad + 4(k-1)(p^\dagger \cdot p^\dagger)^{k-2}\left(\hat{n}_p + k - \tfrac{1}{2}\right)(p^\dagger \cdot p^\dagger) \\
&= 4(p^\dagger \cdot p^\dagger)^{k-1}\left[\left(\hat{n}_p + \tfrac{3}{2}\right) + (k-1)\left(\hat{n}_p + k + \tfrac{3}{2}\right)\right]
\end{aligned}$$
$$(5.84)$$

leading to the result (5.82). From successive applications of the relations (5.81) and (5.82) we deduce the normalization of the state (5.80) and obtain the operator representation of the U(3) states:

$$|n_{\mathrm{p}}LM_L\rangle = B_{n_{\mathrm{p}}L}(p^\dagger \cdot p^\dagger)^{(n_{\mathrm{p}}-L)/2}\mathcal{Y}^L_{M_L}(\mathbf{p}^\dagger)|0\rangle \tag{5.85}$$

with the normalization coefficients

$$B_{n_{\mathrm{p}}L} = (-)^{(n_{\mathrm{p}}-L)/2}\left[\frac{4\pi}{(n_{\mathrm{p}}-L)!!(n_{\mathrm{p}}+L+1)!!}\right]^{1/2} \tag{5.86}$$

The phase factor $(-)^{(n_{\mathrm{p}}-L)/2}$ is taken to conform with the conventions followed by Moshinsky [8].

The complete U(3) state, including the s-boson dependence, is obtained by multiplying (5.85) with the one-dimensional oscillator result:

$$\boxed{|[N]n_{\mathrm{p}}LM_L\rangle = B_{n_{\mathrm{p}}L}(p^\dagger \cdot p^\dagger)^{(n_{\mathrm{p}}-L)/2}\mathcal{Y}^L_{M_L}(\mathbf{p}^\dagger)B'_{Nn_{\mathrm{p}}}(s^\dagger)^{N-n_{\mathrm{p}}}|0\rangle} \tag{5.87}$$

where

$$B'_{Nn_{\mathrm{p}}} = \left[\frac{1}{(N-n_{\mathrm{p}})!}\right]^{1/2} \tag{5.88}$$

In order to compare the observed transition properties of diatomic molecules to the predictions of the vibron model, it is necessary to compute matrix elements of U(4) generators in a U(3) basis. Given the explicit form (5.87) of the U(3) states $|[N]n_{\mathrm{p}}LM_L\rangle$, this is now reduced to a problem of elementary algebraic manipulations. In a first step we deduce the matrix elements of the $p^\dagger_m$ and $s^\dagger$ operators. For example, we find

$$\begin{aligned}\langle[N+1]n_{\mathrm{p}}LM_L|s^\dagger|[N]n_{\mathrm{p}}LM_L\rangle \\ = \langle 0|B'_{N+1\,n_{\mathrm{p}}}(s)^{N-n_{\mathrm{p}}+1}s^\dagger B'_{Nn_{\mathrm{p}}}(s^\dagger)^{N-n_{\mathrm{p}}}|0\rangle \\ = \sqrt{N-n_{\mathrm{p}}+1}\end{aligned} \tag{5.89}$$

Results are summarized in a more compact form by giving the *reduced* matrix elements (as defined in (1.96)) instead of the matrix elements

5.4 Non-rigid molecules: The U(3) limit

themselves. From (5.89) and similar expressions for the matrix elements of $p_m^\dagger$ we find

$$\langle [N+1]n_\mathrm{p}+1\ L-1 \| p^\dagger \| [N]n_\mathrm{p}L\rangle = \left[\frac{(n_\mathrm{p}-L+2)L}{2L-1}\right]^{1/2}$$
$$\langle [N+1]n_\mathrm{p}+1\ L+1 \| p^\dagger \| [N]n_\mathrm{p}L\rangle = \left[\frac{(n_\mathrm{p}+L+3)(L+1)}{2L+3}\right]^{1/2}$$
$$\langle [N+1]n_\mathrm{p}L \| s^\dagger \| [N]n_\mathrm{p}L\rangle = \sqrt{N-n_\mathrm{p}+1}$$
(5.90)

The reduced matrix elements of $\tilde{p}$ and $\tilde{s}$ can be derived from (5.90) and the relation

$$\langle \alpha'L' \| \tilde{b}_l \| \alpha L\rangle = (-)^{L-L'+l} \left[\frac{2L+1}{2L'+1}\right]^{1/2} \langle \alpha L \| b_l^\dagger \| \alpha'L'\rangle \qquad (5.91)$$

where α and α' represent all quantum numbers (other than the angular momentum) necessary to characterize the states. The relation (5.91) is easily derived from the definition (1.96) of the reduced matrix element.

These results can be used to derive the reduced matrix elements of the U(4) generators summarized in Table 5.3 and lead to predictions concerning the intensities of electromagnetic transitions between molecular rotation–vibration levels. As an example we show in Figure 5.3 all non-zero matrix elements of the operator $\hat{D}$ between U(3) states. The squares of these matrix elements are related to the intensities of electric dipole transitions, as is shown in Section 5.7.

The results in Table 5.3 can be used in turn to compute matrix elements of operators quadratic in the generators such as, for instance, quadratic invariants of U(4) and subalgebras. All operators listed in Table 5.2, except for the Casimir invariants of SO(4) and $\overline{\mathrm{SO}}(4)$, are diagonal in the U(3) basis with matrix elements that are trivially deduced from those of the generators. The derivation of the matrix elements of $\mathcal{C}_2[\mathrm{SO}(4)]$, on the other hand, requires a lengthy though straightforward calculation. The diagonal one equals

$$\langle [N]n_\mathrm{p}LM_L|\hat{L}^2+\hat{D}^2|[N]n_\mathrm{p}LM_L\rangle = N(2n_\mathrm{p}+3) - 2n_\mathrm{p}(n_\mathrm{p}+1) + L(L+1) \qquad (5.92)$$

and the off-diagonal matrix element is given by

$$\langle [N]n_\mathrm{p}+2\ LM_L|\hat{L}^2+\hat{D}^2|[N]n_\mathrm{p}LM_L\rangle$$

Table 5.3: Matrix elements of U(4) generators in a U(3) basis[a]

$$\langle [N]n_p L \| \hat{n}_p \| [N]n_p L \rangle = n_p$$

$$\langle [N]n_p L \| \hat{n}_s \| [N]n_p L \rangle = N - n_p$$

$$\langle [N]n_p L \| \hat{L} \| [N]n_p L \rangle = \sqrt{L(L+1)}$$

$$\langle [N]n_p-1\ L-1 \| \hat{D}' \| [N]n_p L \rangle = -\left[\frac{(N-n_p+1)(n_p+L+1)L}{2L-1}\right]^{1/2}$$

$$\langle [N]n_p-1\ L+1 \| \hat{D}' \| [N]n_p L \rangle = -\left[\frac{(N-n_p+1)(n_p-L)(L+1)}{2L+3}\right]^{1/2}$$

$$\langle [N]n_p+1\ L-1 \| \hat{D}' \| [N]n_p L \rangle = +\left[\frac{(N-n_p)(n_p-L+2)L}{2L-1}\right]^{1/2}$$

$$\langle [N]n_p+1\ L+1 \| \hat{D}' \| [N]n_p L \rangle = +\left[\frac{(N-n_p)(n_p+L+3)(L+1)}{2L+3}\right]^{1/2}$$

$$\langle [N]n_p L \| \hat{Q} \| [N]n_p L \rangle = (2n_p + 3)\left[\frac{L(L+1)}{6(2L-1)(2L+3)}\right]^{1/2}$$

$$\langle [N]n_p\ L+2 \| \hat{Q} \| [N]n_p L \rangle = \left[\frac{(n_p-L)(n_p+L+3)(L+1)(L+2)}{(2L+3)(2L+5)}\right]^{1/2}$$

[a] Matrix elements of $\hat{D}_\mu$ are identical to those of $\hat{D}'_\mu$ except for the overall signs + or −, which should be replaced by i.

$$= \sqrt{(N-n_p-1)(N-n_p)(n_p-L+2)(n_p+L+3)} \quad (5.93)$$

The same expression as (5.92) results for the diagonal matrix element of $\mathcal{C}_2[\overline{SO}(4)] = \hat{L}^2 + \hat{D}'^2$ while the off-diagonal one differs from (5.93) by a sign.

The procedure to derive the previous results rests entirely on the boson realization (5.87) of the U(3) states $|[N]n_p LM_L\rangle$. The derivation is similar in spirit to the one in the U(2) model discussed in Chapter 1, and we therefore indicated only the essential steps in the calculation as well as a number of intermediate results useful for later purposes (e.g., the matrix elements of the dipole operators $\hat{D}_\mu$ or $\hat{D}'_\mu$). If one is only interested in the end result (5.92) and (5.93), a more concise and elegant derivation exists as we now proceed to show.

One first should recognize the *pairing* character of the quadratic invariants of SO(4) and $\overline{SO}(4)$. Defining the pairing operators $\mathcal{P}^\dagger_\pm$ and

5.4 Non-rigid molecules: The U(3) limit

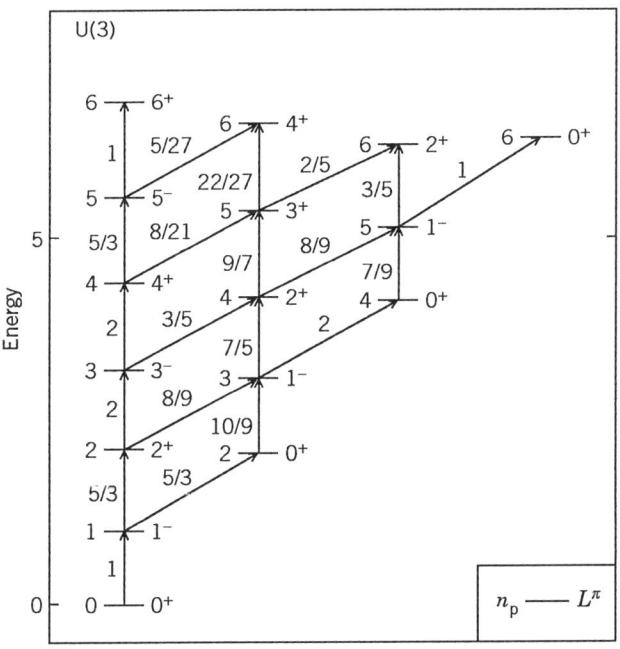

Figure 5.3: Allowed transitions in the U(3) limit induced by the dipole operator $\hat{D}_\mu$ (or $\hat{D}'_\mu$). The numbers alongside the arrows are the squares of the reduced matrix elements in Table 5.3 normalized to the $0^+_1 \to 1^-_1$ transition. Parameters, labels, and energy units are as in Figure 5.2.

$\tilde{\mathcal{P}}_\pm$ as

$$\mathcal{P}^\dagger_\pm \equiv \tfrac{1}{2}(p^\dagger \cdot p^\dagger \pm s^\dagger s^\dagger), \qquad \tilde{\mathcal{P}}_\pm \equiv \tfrac{1}{2}(\tilde{p} \cdot \tilde{p} \pm \tilde{s}\tilde{s}) \qquad (5.94)$$

and using the replacement (see discussion at the end of Section 5.3)

$$p^\dagger \cdot p^\dagger \tilde{p} \cdot \tilde{p} = -\hat{n}_\mathrm{p} + \tfrac{1}{3}\hat{n}_\mathrm{p}^2 - \tfrac{1}{2}\hat{L}^2 + \hat{Q}^2 \mapsto \hat{n}_\mathrm{p} + \hat{n}_\mathrm{p}^2 - \hat{L}^2 \qquad (5.95)$$

we rewrite the pairing operators $4\mathcal{P}^\dagger_\pm \tilde{\mathcal{P}}_\pm$ as follows:

$$\begin{aligned} 4\mathcal{P}^\dagger_+ \tilde{\mathcal{P}}_+ &= p^\dagger \cdot p^\dagger \tilde{p} \cdot \tilde{p} + (p^\dagger \cdot p^\dagger \tilde{s}\tilde{s} + s^\dagger s^\dagger \tilde{p} \cdot \tilde{p}) + s^\dagger s^\dagger \tilde{s}\tilde{s} \\ &\mapsto \hat{n}_\mathrm{p} + \hat{n}_\mathrm{p}^2 - \hat{L}^2 - (\hat{D}^2 + p^\dagger \cdot \tilde{p}\tilde{s}s^\dagger + s^\dagger \tilde{s}\tilde{p} \cdot p^\dagger) - \hat{n}_\mathrm{s} + \hat{n}_\mathrm{s}^2 \\ &= \hat{N}(\hat{N} + 2) - (\hat{L}^2 + \hat{D}^2) \end{aligned} \qquad (5.96)$$

and

$$4\mathcal{P}^\dagger_- \tilde{\mathcal{P}}_- \mapsto \hat{N}(\hat{N} + 2) - (\hat{L}^2 + \hat{D}'^2) \qquad (5.97)$$

This shows that in a symmetric space of p and s bosons the matrix elements of the pairing operator $4\mathcal{P}_+^\dagger \tilde{\mathcal{P}}_+$ ($4\mathcal{P}_-^\dagger \tilde{\mathcal{P}}_-$) are related to those of $\mathcal{C}_2[\mathrm{SO}(4)]$ ($\mathcal{C}_2[\overline{\mathrm{SO}}(4)]$). The off-diagonal ones are the same up to a sign while the diagonal matrix elements differ by an N-dependent constant.

Our original problem is thus reduced to one of finding the matrix elements of $\mathcal{P}_\pm^\dagger$ or of $p^\dagger \cdot p^\dagger$ since the s-boson part in the pairing operator is trivially dealt with. In order to do so, we define the operators

$$\hat{T}_+ \equiv \tfrac{1}{2}(p^\dagger \cdot p^\dagger), \quad \hat{T}_- \equiv (\hat{T}_+)^\dagger = \tfrac{1}{2}(\tilde{p} \cdot \tilde{p}), \quad \hat{T}_z \equiv -\tfrac{1}{4}(p^\dagger \cdot \tilde{p} + \tilde{p} \cdot p^\dagger) \quad (5.98)$$

which can be shown to form a closed algebra with commutation relations

$$[\hat{T}_z, \hat{T}_\pm] = \pm \hat{T}_\pm, \qquad [\hat{T}_+, \hat{T}_-] = -2\hat{T}_z \quad (5.99)$$

But for the sign in the last commutator of (5.99) these are identical to the SU(2) relations (1.87). The algebraic structure associated with (5.98) is related—but by no means identical—to SU(2); it has been named SU(1,1) and belongs to the class of so-called *non-compact* algebras [5,11]. Due to the similarities between the usual angular momentum or spin algebra SU(2) and the algebra SU(1,1), the latter is sometimes referred to as the *quasi-spin* algebra [12]. We emphasize, however, that this equivalence between SU(2) and SU(1,1) only holds for particular properties whereas in other aspects the two algebras are drastically different.

The quadratic Casimir invariant of SU(1,1) is

$$\hat{T}^2 = -\tfrac{1}{2}(\hat{T}_+\hat{T}_- + \hat{T}_-\hat{T}_+) + \hat{T}_z^2 \quad (5.100)$$

since it commutes with all generators (5.98). This operator $\hat{T}^2$, together with $\hat{T}_z$, defines a set of eigenstates $|tm_t\rangle$, analogous to the SU(2) $\supset$ SO(2) classification in angular momentum. Standard expressions are available for the matrix elements of raising and lowering operators in an SU(1,1) basis, but since they are not as well known than corresponding ones in SU(2) (see (1.89)), we give here a brief derivation. With the help of the commutation relations (5.99) we find

$$\begin{aligned}\langle tm_t| - 2\hat{T}_z |tm_t\rangle &= -2m_t = \langle tm_t|\hat{T}_+\hat{T}_- - \hat{T}_-\hat{T}_+|tm_t\rangle \\ &= |\langle tm_t|\hat{T}_+|t\,m_t-1\rangle|^2 - |\langle tm_t|\hat{T}_-|t\,m_t+1\rangle|^2\end{aligned}$$
$$(5.101)$$

5.4 Non-rigid molecules: The U(3) limit

leading to the following recurrence relation for the matrix elements of $\hat{T}_-$ (since $\hat{T}_-$ and $\hat{T}_+$ are hermitian conjugate operators):

$$|\langle t\, m_t-1|\hat{T}_-|t m_t\rangle|^2 - |\langle t m_t|\hat{T}_-|t\, m_t+1\rangle|^2 = -2m_t \quad (5.102)$$

This difference equation has the general solution

$$|\langle t\, m_t-1|\hat{T}_-|t m_t\rangle|^2 = C + m_t(m_t - 1) \quad (5.103)$$

where the constant C is found by requiring the matrix element to vanish for $m_t = t$ (its minimum value), giving $C = -t(t-1)$. In this way we obtain

$$\langle t\, m_t \pm 1|\hat{T}_\pm|t m_t\rangle = \sqrt{(t \pm m_t)(-t \pm m_t + 1)} \quad (5.104)$$

The recurrence relation (5.102) determines the modulus of the matrix elements of $\hat{T}_\pm$ which are thus known up to a phase factor only. In (5.104) this phase is chosen to be $+1$. The entire derivation of (5.104)— as well as the end result itself—closely resembles the discussion and solution of the corresponding problem in SU(2) (see, e.g., Section 2.3 of [3]). Differences originate from a sign change in the recurrence relation (5.102) which, in SU(2), has the term $-2m_t$ replaced by $2m_t$. This difference in sign influences the boundary conditions of the recurrence relation and ultimately determines why SU(2) has a finite number of allowed m_t values whereas in SU(1,1) this number is infinite.

Because the operators $\hat{T}^2$ and $\hat{T}_z$ are realized by way of $p^\dagger$ and $\tilde{p}$ operators, it must also be possible to find an interpretation of the labels t and m_t in terms of p bosons or, more specifically, to associate them with the quantum numbers n_p and L of the U(3) basis. This can be achieved by rewriting the two operators in terms of U(3) generators:

$$\hat{T}_z = \tfrac{1}{2}\left(\hat{n}_\text{p} + \tfrac{3}{2}\right) \quad (5.105)$$

and

$$\begin{aligned}\hat{T}^2 &= -\hat{T}_+\hat{T}_- - \hat{T}_z + \hat{T}_z^2 \\ &= -\tfrac{1}{4}p^\dagger \cdot p^\dagger \tilde{p} \cdot \tilde{p} - \tfrac{1}{4}(2\hat{n}_\text{p} + 3) + \tfrac{1}{16}(2\hat{n}_\text{p} + 3)^2 \\ &\mapsto \tfrac{1}{4}\left(\hat{L}^2 - \tfrac{3}{4}\right)\end{aligned} \quad (5.106)$$

The connection is found by comparing these expressions with the eigenvalues of $\hat{T}_z$ and $\hat{T}^2$ in the $|tm_t\rangle$ basis,

$$\langle tm_t|\hat{T}_z|tm_t\rangle = m_t \qquad (5.107)$$

and

$$\begin{aligned}
\langle tm_t|\hat{T}^2|tm_t\rangle \\
&= -\tfrac{1}{2}\left(\langle tm_t|\hat{T}_+|t\,m_t-1\rangle^2 + \langle tm_t|\hat{T}_-|t\,m_t+1\rangle^2\right) + m_t^2 \\
&= -\tfrac{1}{2}[-t(t-1)+m_t(m_t-1)-t(t-1)+m_t(m_t+1)]+m_t^2 \\
&= t(t-1) \qquad (5.108)
\end{aligned}$$

The first of these comparisons implies

$$m_t = \tfrac{1}{2}\left(n_p + \tfrac{3}{2}\right) \qquad (5.109)$$

while the second implies the following relation between t and L:

$$t = \tfrac{1}{2}\left(L + \tfrac{3}{2}\right) \qquad (5.110)$$

This, together with the U(3) $\supset$ SO(3) branching rule $L = n_p, n_p - 2, \ldots, 1$ or 0, determines the allowed values of m_t for a given t:

$$m_t = t, t+1, \ldots \qquad (5.111)$$

In contrast to the corresponding SU(2) rule m_t is not bounded from above but can take an infinite number of discrete values. The interpretation of this result is clear: the raising operator $\hat{T}_+$ (which increases m_t by 1) creates a pair of p bosons and hence its action can be continued indefinitely. In contrast, the lowering operator $\hat{T}_-$, destroying two p bosons, must necessarily yield zero for $n_p \leq L$ since $n_p = L$ is the minimum number of p bosons required to form a state with angular momentum L. This condition is equivalent with a lower bound for m_t at $m_t = t$. The existence of infinite-dimensional (unitary) representations is characteristic for all non-compact algebras and also clarifies the nomenclature chosen for them. Such infinities make non-compact algebras a great deal more difficult to handle mathematically as compared with their compact siblings, and apart from some isolated discussions at an elementary level, we do not consider them in this book.

5.4 Non-rigid molecules: The U(3) limit

At this point it is perhaps useful to recall our original objective, which was to derive the matrix elements of $\mathcal{P}_\pm^\dagger$ or $\tilde{\mathcal{P}}_\pm$ in a U(3) basis. We have shown that these are essentially identical to the matrix elements of $\hat{T}_+$ or $\hat{T}_-$ in the SU(1,1) basis $|tm_t\rangle$ and that the latter can be obtained using simple algebraic manipulations involving the SU(1,1) commutation relations (5.99). The only remaining task is to rewrite the result (5.104) in (n_p, L) notation using the correspondence (5.109) and (5.110), and this leads to

$$\langle n_p+2\,LM_L|p^\dagger \cdot p^\dagger|n_p LM_L\rangle = \sqrt{(n_p - L + 2)(n_p + L + 3)}$$
$$\langle n_p-2\,LM_L|\tilde{p} \cdot \tilde{p}|n_p LM_L\rangle = \sqrt{(n_p - L)(n_p + L + 1)}$$
(5.112)

The off-diagonal matrix element of the pairing operators $4\mathcal{P}_\pm^\dagger \tilde{\mathcal{P}}_\pm$ now follows immediately,

$$\langle [N]n_p+2\,LM_L|4\mathcal{P}_\pm^\dagger \tilde{\mathcal{P}}_\pm|[N]n_p LM_L\rangle$$
$$= \pm\langle [N]n_p+2\,LM_L|p^\dagger \cdot p^\dagger \tilde{s}\tilde{s}|[N]n_p LM_L\rangle$$
$$= \pm\sqrt{(N - n_p - 1)(N - n_p)}\langle n_p+2\,LM_L|p^\dagger \cdot p^\dagger|n_p LM_L\rangle$$
$$= \pm\sqrt{(N - n_p - 1)(N - n_p)(n_p - L + 2)(n_p + L + 3)}$$
(5.113)

and coincides—up to a sign—with our previous result (5.93) while for the diagonal matrix element we obtain

$$\langle [N]n_p LM_L|4\mathcal{P}_\pm^\dagger \tilde{\mathcal{P}}_\pm|[N]n_p LM_L\rangle$$
$$= (N - n_p - 1)(N - n_p) + (n_p - L)(n_p + L + 1) \quad (5.114)$$

which, taking into account (5.96) or (5.97), agrees with (5.92).

In principle we have now fully determined the algebraic structure of the vibron model. The most general hamiltonian (assuming it to be at most quadratic in the generators of U(4)) can be diagonalized in the U(3) basis $|[N]n_p LM_L\rangle$ and its eigenvalues obtained numerically; transition matrix elements between the resulting eigenstates (known as an expansion in terms of U(3) states) can be computed on the basis of the results derived in this section. Specifically, using this approach we might study in this way the properties of the SO(4) limit of the

5.5 Rigid molecules: The SO(4) limit

If in (5.50) we put $\epsilon = \alpha = 0$, the hamiltonian reduces to

$$\hat{H}_{\mathrm{II}} = E_0'' + \beta(\hat{L}^2 + \hat{D}^2) + \gamma \hat{L}^2 \qquad (5.115)$$

and the SO(4) limit of the vibron model is obtained. This hamiltonian contains invariants of the algebras SO(4) and SO(3) only, and its eigenstates can thus be classified according to

$$\begin{array}{ccc} U(4) \supset SO(4) \supset SO(3) \\ | & | & | \\ [N] & \omega & L \end{array} \qquad (5.116)$$

where, as in (5.52), the SO(2) subalgebra and its related label M_L is implicitly assumed. Unless otherwise stated we analyze throughout this section the algebra SO(4) rather than $\overline{\mathrm{SO}}(4)$; the properties of the two algebras are essentially identical and eventual differences (such as their coordinate realization) will be pointed out occasionally.

Before proceeding with the discussion of the energy spectrum of $\hat{H}_{\mathrm{II}}$ and of the branching rules in (5.116), we point out a remarkable property of the algebra SO(4) which is very useful in this section and throughout the entire Part II of this book. As shown in Section 5.2, SO(4) is composed of the generators $\hat{L}_\mu$ and $\hat{D}_\mu$ satisfying the commutation relations

$$[\hat{L}_\mu, \hat{L}_{\mu'}] = a_{\mu\mu'} \hat{L}_{\mu+\mu'}, \quad [\hat{L}_\mu, \hat{D}_{\mu'}] = a_{\mu\mu'} \hat{D}_{\mu+\mu'}, \quad [\hat{D}_\mu, \hat{D}_{\mu'}] = a_{\mu\mu'} \hat{L}_{\mu+\mu'} \qquad (5.117)$$

where $a_{\mu\mu'} = -\sqrt{2} \langle 1\mu \, 1\mu' | 1 \, \mu+\mu' \rangle$. Defining the operators $\hat{R}_\mu$ and $\hat{R}'_\mu$ as the sum and the difference of these generators,

$$\hat{R}_\mu = \tfrac{1}{2}(\hat{L}_\mu + \hat{D}_\mu), \qquad \hat{R}'_\mu = \tfrac{1}{2}(\hat{L}_\mu - \hat{D}_\mu) \qquad (5.118)$$

5.5 Rigid molecules: The SO(4) limit

or inversely,
$$\hat{R}_\mu + \hat{R}'_\mu = \hat{L}_\mu, \qquad \hat{R}_\mu - \hat{R}'_\mu = \hat{D}_\mu, \tag{5.119}$$
we find that they satisfy
$$[\hat{R}_\mu, \hat{R}_{\mu'}] = a_{\mu\mu'}\hat{R}_{\mu+\mu'}, \quad [\hat{R}'_\mu, \hat{R}'_{\mu'}] = a_{\mu\mu'}\hat{R}'_{\mu+\mu'}, \quad [\hat{R}_\mu, \hat{R}'_{\mu'}] = 0 \tag{5.120}$$
where use is made of the antisymmetry property of the structure constants, $a_{\mu\mu'} = -a_{\mu'\mu}$. The last relation shows that $\hat{R}_\mu$ and $\hat{R}'_\mu$ generate a direct product while the first two commutators in (5.120) indicate that it is a product of two SU(2) algebras. In carrying out the simple basis transformation (5.118), we have thus established the isomorphism

$$\boxed{\begin{array}{ccc} \mathrm{SO}(4) & \simeq & \mathrm{SU}_r(2) \otimes \mathrm{SU}_{r'}(2) \\ | & & | \qquad | \\ (\omega, \omega') & & r \qquad r' \end{array}} \tag{5.121}$$

where, to distinguish one SU(2) algebra from the other, they are ascribed subscripts r and r' taking integer or half-integer values.

On the basis of its isomorphism with $\mathrm{SU}_r(2) \otimes \mathrm{SU}_{r'}(2)$, many of the properties of SO(4) can be deduced from elementary arguments. For example, we immediately conclude that since each of the SU(2) algebras is characterized by a single index, a general representation of SO(4) is labeled by two indices. This indeed is confirmed by the general representation theory for SO(n) (see Appendix B); the labels ω and ω' conventionally used for SO(4) are not identical to the SU(2) labels r and r' but are related to them in a simple way:

$$\omega = r + r', \qquad \omega' = r - r' \tag{5.122}$$

or inversely,
$$r = \tfrac{1}{2}(\omega + \omega'), \qquad r' = \tfrac{1}{2}(\omega - \omega') \tag{5.123}$$

If a general representation of SO(4) needs two labels, why then do we use only one index in the classification (5.116)? The answer to this question is provided by considering the scalar product $\hat{L} \cdot \hat{D}$, which can be rewritten in two different ways:

$$\begin{aligned}\hat{L} \cdot \hat{D} &= i\sqrt{6}[[p^\dagger \times p^\dagger]^{(1)} \times [\tilde{p} \times \tilde{s}]^{(1)} - [p^\dagger \times s^\dagger]^{(1)} \times [\tilde{p} \times \tilde{p}]^{(1)}]_0^{(0)} \\ &= \hat{R}^2 - \hat{R}'^2 \end{aligned} \tag{5.124}$$

The first form of $\hat{L}\cdot\hat{D}$ shows that the expectation value of this operator necessarily vanishes in the symmetric space of p and s bosons. It follows that the expectation value of $\hat{R}^2 - \hat{R}'^2$ must also vanish, and since it is given by $r(r+1) - r'(r'+1)$, this in turn implies $r = r'$ or, equivalently, $\omega' = 0$. We conclude that a symmetric U(4) representation can only contain symmetric (i.e., single-labeled) SO(4) representations with $\omega = 2r$ taking integer values.

An equally simple argument gives us the eigenvalue of the quadratic Casimir invariant of SO(4). Since $\mathcal{C}_2[\text{SO}(4)]$ can be rewritten as

$$\hat{L}^2 + \hat{D}^2 = 2\hat{R}^2 + 2\hat{R}'^2 \tag{5.125}$$

we immediately find its eigenvalues as

$$2r(r+1) + 2r'(r'+1) = \omega(\omega+2) + \omega'^2 \tag{5.126}$$

This again agrees with the general expression valid for SO(n) as given in Appendix B.

Using these results we are able to construct the eigenspectrum of the hamiltonian (5.115). In the symmetric space of p and s bosons $\hat{H}_{\text{II}}$ has the eigenvalues

$$E_{\text{II}}(\omega, L) = E_0'' + \beta\omega(\omega+2) + \gamma L(L+1) \tag{5.127}$$

The allowed values of ω and L are obtained from $\text{U}(4) \supset \text{SO}(4)$ and $\text{SO}(4) \supset \text{SO}(3)$ branching rules. The first determines the values of ω contained in the symmetric representation $[N]$:

$$\omega = N, N-2, \ldots, 1 \text{ or } 0 \tag{5.128}$$

The second branching rule gives the values of angular momentum L contained in the SO(4) representation ω:

$$L = 0, 1, \ldots, \omega \tag{5.129}$$

These rules can be derived in several ways. They follow from the general $\text{U}(n) \supset \text{SO}(n)$ and $\text{SO}(n) \supset \text{SO}(n-1)$ branching rules as given in Chapter 8 and Appendix B. Alternatively, they are obtained from the explicit form of the wave functions in the SO(4) limit, as will be

5.5 Rigid molecules: The SO(4) limit

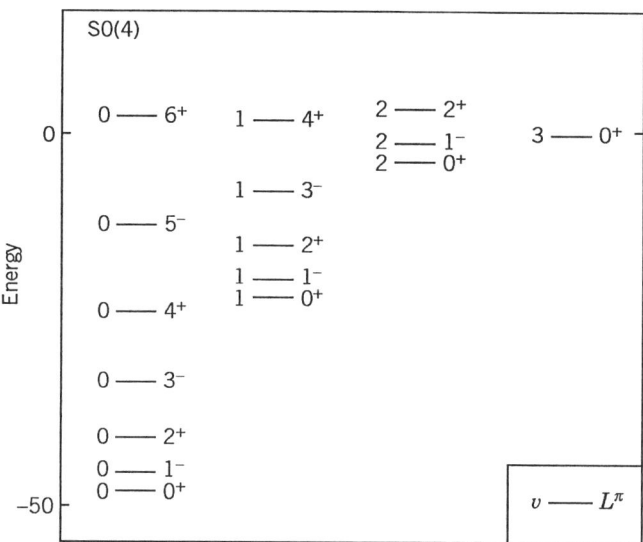

Figure 5.4: The SO(4) eigenspectrum of the hamiltonian $\hat{H}_{\text{II}}$ with $E_0'' = 0$, $\beta = -1$, and $\gamma = 1.2$ for $N = 6$ bosons. Levels are labeled by $v = (N-\omega)/2$, angular momentum L, and parity $\pi = (-)^L$. Energies are in arbitrary units.

shown later in this section. They can also be found from the isomorphism between SO(4) and $\text{SU}_r(2) \otimes \text{SU}_{r'}(2)$. For example, to derive the SO(4) ⊃ SO(3) rule, we observe that since $\hat{L}_\mu = \hat{R}_\mu + \hat{R}'_\mu$, the angular momentum vector **L** results from the coupling of **R** and **R'**, two vectors equal in magnitude ($r = r'$) but arbitrary otherwise. The allowed values of L are thus $L = 0, 1, \ldots, 2r$, in agreement with (5.129).

A schematic example of the eigenspectrum of $\hat{H}_{\text{II}}$ is shown in Figure 5.4. To contrast it with the spectrum in the U(3) limit, shown in Figure 5.2, we have taken the same number of bosons ($N = 6$). More realistic spectra, however, require a larger number of bosons, as is shown in Section 5.7. Figure 5.4 suggests that the SO(4) limit of the vibron model might be of use in the description of rigid diatomic molecules [6]. This is confirmed by calculating from the energy expression (5.127) the rigidity parameter defined in (5.56):

$$\gamma_{\text{rig}} = -\frac{\gamma}{N\beta} \ll 1 \qquad (5.130)$$

The last inequality is valid since the number of bosons N is usually large while β and γ are of the same order.

In Figure 5.4 the label v is introduced which is related to ω in the following way:

$$v = \tfrac{1}{2}(N - \omega), \qquad v = 0, 1, \ldots, \tfrac{1}{2}(N-1) \text{ or } \tfrac{1}{2}N \qquad (5.131)$$

In terms of this alternative quantum number v the eigenvalue expression can be rewritten as

$$E_{\mathrm{II}}(v, L) = E_0''' - 4(N+2)\beta \left(v + \tfrac{1}{2}\right) + 4\beta \left(v + \tfrac{1}{2}\right)^2 + \gamma L(L+1) \qquad (5.132)$$

with

$$E_0''' = E_0'' + \beta(N+1)(N+3) \qquad (5.133)$$

In the form (5.132) the energy eigenvalue becomes a particular case of the general expression

$$E(v, L) = \sum_{ij} Y_{ij} \left(v + \tfrac{1}{2}\right)^i \left(L(L+1)\right)^j \qquad (5.134)$$

known as the *Dunham expansion* [13] and used in simple analyses of molecular rotation–vibration spectra. From this analogy it thus appears that the SO(4) limit of the vibron model contains the essential ingredients for a correct interpretation of such spectra. The fact that the hamiltonian (5.115) corresponds to only part of the Dunham expansion is easily remedied by adding higher-order invariant operators that retain SO(4) as a dynamical symmetry. In this way a complete equivalence between the Dunham expansion (5.134) and the SO(4) limit of the vibron model is obtained. The advantage of the algebraic approach is that the hamiltonian $\hat{H}_{\mathrm{II}}$—in contrast to expansions such as (5.134)—also determines a molecular wave function which can be used to compute additional observables.

To find the explicit form of the SO(4) states

$$|[N]\omega L M_L\rangle \qquad (5.135)$$

5.5 Rigid molecules: The SO(4) limit

we follow the approach of [14] and start from the defining equations

$$\begin{aligned}
\hat{N}|[N]\omega L M_L\rangle &= N|[N]\omega L M_L\rangle \\
\left(\hat{L}^2 + \hat{D}^2\right)|[N]\omega L M_L\rangle &= \omega(\omega + 2)|[N]\omega L M_L\rangle \\
\hat{L}^2|[N]\omega L M_L\rangle &= L(L+1)|[N]\omega L M_L\rangle \\
\hat{L}_0|[N]\omega L M_L\rangle &= M_L|[N]\omega L M_L\rangle
\end{aligned} \quad (5.136)$$

Unlike the U(3) equations (5.59), these cannot be separated into two sets and treated consecutively but the p–s-boson dependence must be dealt with simultaneously. We convert (5.136) into a set of differential equations through the substitutions

$$s^\dagger = \frac{1}{\sqrt{2}}\left(\bar{x} - \frac{\partial}{\partial \bar{x}}\right), \qquad \tilde{s} = \frac{1}{\sqrt{2}}\left(\bar{x} + \frac{\partial}{\partial \bar{x}}\right) \quad (5.137)$$

in addition to the ones for the p boson, (5.61), and arrive at a set of separable differential equations via a transformation from $(x, y, z, \bar{x})$ to the *hyperspherical* coordinates $(\rho, \varphi, \theta, \phi)$:

$$\begin{aligned}
x &= \rho \sin\varphi \sin\theta \cos\phi, \\
y &= \rho \sin\varphi \sin\theta \sin\phi, \\
z &= \rho \sin\varphi \cos\theta, \\
\bar{x} &= \rho \cos\varphi
\end{aligned} \quad (5.138)$$

where ρ is the hyperspherical radius. The boundary values for the variables ρ, φ, θ, and ϕ are defined to cover the entire four-dimensional space in x, y, z, and $\bar{x}$:

$$0 \leq \rho < \infty, \qquad 0 \leq \varphi < \pi, \qquad 0 \leq \theta < \pi, \qquad 0 \leq \phi < 2\pi \quad (5.139)$$

with the four-dimensional infinitesimal volume element transformed into

$$dx\, dy\, dz\, d\bar{x} = \rho^3 \sin^2\varphi \sin\theta\, d\rho\, d\varphi\, d\theta\, d\phi \quad (5.140)$$

The replacements (5.61), (5.137), and (5.138) are applied to the operators in the defining equations (5.136). Transformation of $\hat{L}_0$ and $\hat{L}^2$

leads to the same differential operators in ϕ and θ as the ones of Section 5.4 (see (5.64) and (5.65)). The differential form of the quadratic Casimir invariant of SO(4) is obtained from

$$\hat{D}^2 = \hat{D}_0^2 - 2\hat{D}_{-1}\hat{D}_{+1} + \hat{L}_0 \tag{5.141}$$

together with

$$\begin{aligned}
\hat{D}_0 &= -i\cos\theta \frac{\partial}{\partial\varphi} + i\frac{\sin\theta}{\tan\varphi}\frac{\partial}{\partial\theta} \\
\hat{D}_{\pm 1} &= \pm\frac{e^{\pm i\phi}}{\sqrt{2}}\left(i\sin\theta\frac{\partial}{\partial\varphi} + i\frac{\cos\theta}{\tan\varphi}\frac{\partial}{\partial\theta} \mp \frac{1}{\tan\varphi\sin\theta}\frac{\partial}{\partial\phi}\right)
\end{aligned} \tag{5.142}$$

leading to

$$\hat{L}^2 + \hat{D}^2 = -\frac{\partial^2}{\partial\varphi^2} - \frac{2}{\tan\varphi}\frac{\partial}{\partial\varphi} + \frac{1}{\sin^2\varphi}\hat{L}^2 \tag{5.143}$$

Likewise, we find, for the total boson number operator,

$$\begin{aligned}
\hat{N} &= p_{+1}^\dagger\tilde{p}_{-1} + p_{-1}^\dagger\tilde{p}_{+1} - p_0^\dagger\tilde{p}_0 + s^\dagger\tilde{s} \\
&= \frac{1}{2}\left[-\frac{1}{\rho^3}\frac{\partial}{\partial\rho}\left(\rho^3\frac{\partial}{\partial\rho}\right) + \rho^2 + \frac{1}{\rho^2}\left(\hat{L}^2 + \hat{D}^2\right) - 4\right]
\end{aligned} \tag{5.144}$$

The differential equations are thus separable in hyperspherical coordinates with solutions of the form

$$\Psi_{N\omega LM_L}(\rho,\varphi,\theta,\phi) = f_\omega^N(\rho)g_L^\omega(\varphi)Y_{M_L}^L(\theta,\phi) \tag{5.145}$$

where the "angular" dependence on θ and ϕ is the same as in the U(3) case and the functions $g_L^\omega(\varphi)$ and $f_\omega^N(\rho)$ satisfy the separate equations

$$\left(-\frac{d^2}{d\varphi^2} - \frac{2}{\tan\varphi}\frac{d}{d\varphi} + \frac{L(L+1)}{\sin^2\varphi}\right)g_L^\omega(\varphi) = \omega(\omega+2)g_L^\omega(\varphi)$$

$$\frac{1}{2}\left[-\frac{1}{\rho^3}\frac{d}{d\rho}\left(\rho^3\frac{d}{d\rho}\right) + \rho^2 + \frac{\omega(\omega+2)}{\rho^2} - 4\right]f_\omega^N(\rho) = Nf_\omega^N(\rho)$$

$$\tag{5.146}$$

After a substitution $f_\omega^N(\rho) = \rho^\omega \exp(-\rho^2/2)h(\rho^2)$ the second is reduced to a differential equation of the Laguerre type (with $x = \rho^2$),

$$\left[x\frac{d^2}{dx^2} + (\omega+2-x)\frac{d}{dx} + \frac{N-\omega}{2}\right]h(x) = 0 \tag{5.147}$$

5.5 Rigid molecules: The SO(4) limit

while the first equation is brought into standard form with the transformation $g_L^\omega(\varphi) = (\sin\varphi)^L p(\cos\varphi)$ where $p(y)$ satisfies

$$\left[(1-y^2)\frac{d^2}{dy^2} - (2L+3)y\frac{d}{dy} + (\omega-L)(\omega+L+2)\right]p(y) = 0 \quad (5.148)$$

where $y = \cos\varphi$. This is the differential equation defining the Gegenbauer polynomial $C_{\omega-L}^{L+1}(x)$ [10]. The complete SO(4) wave function in hyperspherical coordinates is thus of the form

$$\boxed{\begin{aligned}\Psi_{N\omega LM_L}(\rho,\varphi,\theta,\phi) &= A_{N\omega L}\,\rho^\omega e^{-\rho^2/2} L_{(N-\omega)/2}^{\omega+1}(\rho^2)(\sin\varphi)^L \\ &\quad \times C_{\omega-L}^{L+1}(\cos\varphi) Y_{M_L}^L(\theta,\phi)\end{aligned}} \quad (5.149)$$

The coefficients $A_{N\omega L}$ are obtained from the normalization condition

$$\int_0^\infty \rho^3\,d\rho \int_0^\pi \sin^2\varphi\,d\varphi \int_0^\pi \sin\theta\,d\theta \int_0^{2\pi} d\phi\, |\Psi_{N\omega LM_L}(\rho,\varphi,\theta,\phi)|^2 = 1 \quad (5.150)$$

The spherical harmonics $Y_{M_L}^L(\theta,\phi)$ are normalized in θ and ϕ, and furthermore using the tabulated integrals [10]

$$\int_0^\pi (\sin\varphi)^{2L+2}\left(C_{\omega-L}^{L+1}(\cos\varphi)\right)^2 d\varphi = \frac{\pi(\omega+L+1)!}{2^{2L+1}(\omega+1)(\omega-L)!L!L!}$$

$$\int_0^\infty \rho^{2\omega+3} e^{-\rho^2}\left(L_{(N-\omega)/2}^{\omega+1}(\rho^2)\right)^2 d\rho = \frac{(N+\omega+2)!!}{2^{\omega+2}(N-\omega)!!} \quad (5.151)$$

we find the result

$$A_{N\omega L} = 2^{L+1} L! \left[\frac{2^{\omega+1}(\omega+1)(N-\omega)!!(\omega-L)!}{\pi(N+\omega+2)!!(\omega+L+1)!}\right]^{1/2} \quad (5.152)$$

The next step, as in the U(3) derivation of Section 5.4, is to apply Dragt's theorem to the highest-weight state

$$\Psi_{\omega\omega LM_L}(\rho,\varphi,\theta,\phi) = A_{\omega\omega L}\,\rho^\omega e^{-\rho^2/2}(\sin\varphi)^L C_{\omega-L}^{L+1}(\cos\varphi) Y_{M_L}^L(\theta,\phi) \quad (5.153)$$

which amounts to carrying out the replacements

$$x_m \to \frac{p_m^\dagger}{\sqrt{2}}, \quad \bar{x} \to \frac{s^\dagger}{\sqrt{2}}, \quad \frac{e^{-\rho^2/2}}{\pi} \to |0\rangle \quad (5.154)$$

or in hyperspherical coordinates,

$$\rho \to \left(\frac{p^\dagger \cdot p^\dagger + s^\dagger s^\dagger}{2}\right)^{1/2}$$

$$\cos\varphi \to \frac{s^\dagger}{\sqrt{p^\dagger \cdot p^\dagger + s^\dagger s^\dagger}} \qquad (5.155)$$

$$\sin\varphi \to \left(\frac{p^\dagger \cdot p^\dagger}{p^\dagger \cdot p^\dagger + s^\dagger s^\dagger}\right)^{1/2}$$

and expressions for $\cos\theta$ and $e^{\pm i\phi}$ given in (5.77). These replacements transform the state $\Psi_{\omega\omega LM_L}(\rho,\varphi,\theta,\phi)$ of highest weight into

$$\frac{\pi A_{\omega\omega L}}{2^{\omega/2}}(p^\dagger \cdot p^\dagger + s^\dagger s^\dagger)^{(\omega-L)/2} C_{\omega-L}^{L+1}(t^\dagger)\mathcal{Y}_{M_L}^L(\mathbf{p}^\dagger)|0\rangle \qquad (5.156)$$

where the notation

$$t^\dagger = \frac{s^\dagger}{\sqrt{p^\dagger \cdot p^\dagger + s^\dagger s^\dagger}} \qquad (5.157)$$

is used.

The final step is to generalize the operator representation (5.156), valid for highest weight, to an arbitrary SO(4) state. Again, the derivation is very similar to the corresponding one in Section 5.4 (see also Chapter 8): the state (5.156) is used as a starting point and acted upon with an operator increasing the total number of bosons of the state without changing the SO(4) label ω. The simplest SO(4)-scalar operator is of the form $p^\dagger \cdot p^\dagger + s^\dagger s^\dagger$, and a general SO(4) state is thus of the form

$$|[N]\omega LM_L\rangle \propto (p^\dagger \cdot p^\dagger + s^\dagger s^\dagger)^{(N-L)/2} C_{\omega-L}^{L+1}(t^\dagger)\mathcal{Y}_{M_L}^L(\mathbf{p}^\dagger)|0\rangle \qquad (5.158)$$

To find the normalized expression for this state, we use

$$(\tilde{p}\cdot\tilde{p} + \tilde{s}\tilde{s})(p^\dagger \cdot p^\dagger + s^\dagger s^\dagger)^{(\omega-L)/2} C_{\omega-L}^{L+1}(t^\dagger)\mathcal{Y}_{M_L}^L(\mathbf{p}^\dagger)|0\rangle = 0 \qquad (5.159)$$

together with the commutation relation

$$[\tilde{p}\cdot\tilde{p} + \tilde{s}\tilde{s}, (p^\dagger \cdot p^\dagger + s^\dagger s^\dagger)^k] = 4k(p^\dagger \cdot p^\dagger + s^\dagger s^\dagger)^{k-1}\left(\hat{N} + k + 1\right) \qquad (5.160)$$

5.5 Rigid molecules: The SO(4) limit

to arrive at the final result

$$|[N]\omega L M_L\rangle = B_{N\omega L}(p^\dagger \cdot p^\dagger + s^\dagger s^\dagger)^{(N-L)/2} C_{\omega-L}^{L+1}(t^\dagger) \mathcal{Y}_{M_L}^L(\mathbf{p}^\dagger)|0\rangle \tag{5.161}$$

with

$$B_{N\omega L} = (-)^{(N-\omega)/2} 2^L L! \left[\frac{8\pi(\omega+1)(\omega-L)!}{(N-\omega)!!(N+\omega+2)!!(\omega+L+1)!} \right]^{1/2} \tag{5.162}$$

Once the explicit form (5.161) of the SO(4) states is obtained, matrix elements of the $p^\dagger$ and $s^\dagger$ operators in this basis are derived by making use of known recurrence relations in the Gegenbauer polynomials and solid spherical harmonics. To illustrate the procedure, we begin by computing the matrix elements of $s^\dagger$ and consider its action on $|[N]\omega L M_L\rangle$:

$$s^\dagger |[N]\omega L M_L\rangle = B_{N\omega L}(p^\dagger \cdot p^\dagger + s^\dagger s^\dagger)^{(N-L+1)/2} t^\dagger C_{\omega-L}^{L+1}(t^\dagger) \mathcal{Y}_{M_L}^L(\mathbf{p}^\dagger)|0\rangle \tag{5.163}$$

where use is made of (5.157). With the help of the recurrence relation [10]

$$t C_{n+1}^\lambda(t) = \frac{2\lambda+n}{2(\lambda+n+1)} C_n^\lambda(t) + \frac{n+2}{2(\lambda+n+1)} C_{n+2}^\lambda(t) \tag{5.164}$$

this can be converted into

$$\begin{aligned} s^\dagger |[N]\omega L M_L\rangle &= \frac{B_{N\omega L}}{B_{N+1\,\omega-1\,L}} \frac{\omega+L+1}{2(\omega+1)} |[N+1]\omega-1\,L M_L\rangle \\ &+ \frac{B_{N\omega L}}{B_{N+1\,\omega+1\,L}} \frac{\omega-L+1}{2(\omega+1)} |[N+1]\omega+1\,L M_L\rangle \end{aligned} \tag{5.165}$$

and after inserting the $B_{N\omega L}$ coefficients, we arrive at the following reduced matrix elements:

$$\begin{aligned} \langle [N+1]\omega-1\,L \| s^\dagger \| [N]\omega L\rangle &= -\left[\frac{(N-\omega+2)(\omega-L)(\omega+L+1)}{4\omega(\omega+1)} \right]^{1/2} \\ \langle [N+1]\omega+1\,L \| s^\dagger \| [N]\omega L\rangle &= \left[\frac{(N+\omega+4)(\omega-L+1)(\omega+L+2)}{4(\omega+1)(\omega+2)} \right]^{1/2} \end{aligned} \tag{5.166}$$

The derivation in the case of the p boson is more complicated since it involves recurrence relations in both the Gegenbauer polynomials and the solid spherical harmonics. We first note that, from (5.156) for $N = \omega = L = 1$, it follows that

$$\mathcal{Y}_m^1(\mathbf{p}^\dagger) = \left(\frac{3}{4\pi}\right)^{1/2} p_m^\dagger \tag{5.167}$$

Thus $p_m^\dagger$ is proportional to a solid spherical harmonic with $L = 1$, and this property can be used to determine its action on the state (5.161). The general multiplication rule for spherical harmonics [9],

$$Y_m^l(\theta,\phi) Y_{M_L}^L(\theta,\phi) = \sum_{L'M_L'} \langle lm\, LM_L | L'M_L' \rangle \langle l0\, L0 | L'0 \rangle$$
$$\times \left[\frac{(2L+1)(2l+1)}{4\pi(2L'+1)}\right]^{1/2} Y_{M_L'}^{L'}(\theta,\phi) \tag{5.168}$$

reduces for $l = 1$ to

$$Y_m^1(\theta,\phi) Y_{M_L}^L(\theta,\phi)$$
$$= -\langle 1m\, LM_L | L-1\, m+M_L \rangle \left[\frac{3L}{4\pi(2L-1)}\right]^{1/2} Y_{m+M_L}^{L-1}(\theta,\phi)$$
$$+ \langle 1m\, LM_L | L+1\, m+M_L \rangle \left[\frac{3(L+1)}{4\pi(2L+3)}\right]^{1/2} Y_{m+M_L}^{L+1}(\theta,\phi) \tag{5.169}$$

and after multiplication by $(p^\dagger \cdot p^\dagger)^{(L+1)/2}$ and use of (5.167), we find the corresponding relation for the solid spherical harmonics:

$$p_m^\dagger \mathcal{Y}_{M_L}^L(\mathbf{p}^\dagger)$$
$$= -\langle 1m\, LM_L | L-1\, m+M_L \rangle \left[\frac{L}{2L-1}\right]^{1/2} (p^\dagger \cdot p^\dagger) \mathcal{Y}_{m+M_L}^{L-1}(\mathbf{p}^\dagger)$$
$$+ \langle 1m\, LM_L | L+1\, m+M_L \rangle \left[\frac{L+1}{2L+3}\right]^{1/2} \mathcal{Y}_{m+M_L}^{L+1}(\mathbf{p}^\dagger) \tag{5.170}$$

5.5 Rigid molecules: The SO(4) limit

The action of the operator $p^\dagger \cdot p^\dagger$ on a state $|[N]\omega L M_L\rangle$ is found by noting that

$$p^\dagger \cdot p^\dagger = (p^\dagger \cdot p^\dagger + s^\dagger s^\dagger)(1 - t^\dagger t^\dagger) \tag{5.171}$$

which, used in conjunction with two recurrence relations in the Gegenbauer polynomials [10],

$$\begin{aligned}
C_n^\lambda(t) &= \frac{\lambda}{\lambda + n}\left(C_n^{\lambda+1}(t) - C_{n-2}^{\lambda+1}(t)\right) \\
(1-t^2)C_n^\lambda(t) &= \frac{(2\lambda + n - 2)(2\lambda + n - 1)}{4(\lambda - 1)(\lambda + n)}C_n^{\lambda-1}(t) \\
&\quad - \frac{(n+1)(n+2)}{4(\lambda-1)(\lambda+n)}C_{n+2}^{\lambda-1}(t)
\end{aligned} \tag{5.172}$$

leads to the following expansion:

$$\begin{aligned}
&p_m^\dagger|[N]\omega L M_L\rangle \\
&= -\langle 1m\, LM_L | L-1\, m+M_L\rangle \left[\frac{L}{2L-1}\right]^{1/2} \\
&\quad \times \Bigg[\frac{B_{N\omega L}}{B_{N+1\,\omega-1\,L-1}}\frac{(\omega+L)(\omega+L+1)}{4L(\omega+1)}|[N+1]\omega-1\, L-1\, m+M_L\rangle \\
&\quad - \frac{B_{N\omega L}}{B_{N+1\,\omega+1\,L-1}}\frac{(\omega-L+1)(\omega-L+2)}{4L(\omega+1)} \\
&\quad \times |[N+1]\omega+1\, L-1\, m+M_L\rangle\Bigg] \\
&\quad - \langle 1m\, LM_L | L+1\, m+M_L\rangle \left[\frac{L+1}{2L+3}\right]^{1/2} \\
&\quad \times \Bigg[\frac{B_{N\omega L}}{B_{N+1\,\omega-1\,L+1}}\frac{L+1}{\omega+1}|[N+1]\omega-1\, L+1\, m+M_L\rangle \\
&\quad - \frac{B_{N\omega L}}{B_{N+1\,\omega+1\,L+1}}\frac{L+1}{\omega+1}|[N+1]\omega+1\, L+1\, m+M_L\rangle\Bigg]
\end{aligned} \tag{5.173}$$

With the help of the expression for the coefficient $B_{N\omega L}$ and the definition (1.96), the following reduced matrix elements result:

$$\langle [N+1]\omega-1\, L-1 \| p^\dagger \| [N]\omega L \rangle$$
$$= \left[\frac{(N-\omega+2)(\omega+L)(\omega+L+1)L}{4\omega(\omega+1)(2L-1)} \right]^{1/2}$$
$$\langle [N+1]\omega-1\, L+1 \| p^\dagger \| [N]\omega L \rangle$$
$$= \left[\frac{(N-\omega+2)(\omega-L-1)(\omega-L)(L+1)}{4\omega(\omega+1)(2L+3)} \right]^{1/2}$$
$$\langle [N+1]\omega+1\, L-1 \| p^\dagger \| [N]\omega L \rangle \qquad (5.174)$$
$$= \left[\frac{(N+\omega+4)(\omega-L+1)(\omega-L+2)L}{4(\omega+1)(\omega+2)(2L-1)} \right]^{1/2}$$
$$\langle [N+1]\omega+1\, L+1 \| p^\dagger \| [N]\omega L \rangle$$
$$= \left[\frac{(N+\omega+4)(\omega+L+2)(\omega+L+3)(L+1)}{4(\omega+1)(\omega+2)(2L+3)} \right]^{1/2}$$

The reduced matrix elements of the $\tilde{p}$ and $\tilde{s}$ operators can be obtained from (5.174) and (5.166) together with the relation (5.91).

Matrix elements of all generators of U(4) can now be computed by a straightforward expansion over intermediate states. This expansion assumes a simpler form when written in terms of reduced matrix elements,

$$\langle [N]\omega'L' \| \mathcal{B}^{(\lambda)}(l,l') \| [N]\omega L \rangle$$
$$= (-)^{L+L'+\lambda} \sqrt{(2\lambda+1)(2L+1)}$$
$$\times \sum_{\omega''L''} \langle [N]\omega'L' \| b_l^\dagger \| [N-1]\omega''L'' \rangle$$
$$\times \langle [N]\omega L \| b_{l'}^\dagger \| [N-1]\omega''L'' \rangle \begin{Bmatrix} l & l' & \lambda \\ L & L' & L'' \end{Bmatrix} \quad (5.175)$$

where $\mathcal{B}^{(\lambda)}_\mu(l,l')$ ($l,l' = 0,1$) is a U(4) generator in the notation introduced in (5.8). With the help of this expression, together with (5.166) and (5.174), the calculation of the matrix elements of U(4) generators is reduced to a mechanical process, the results of which are summarized in Table 5.4. In the table we use the short-hand notation

5.5 Rigid molecules: The SO(4) limit

Table 5.4: Matrix elements of U(4) generators in an SO(4) basis[a]

$$\langle [N]\omega L \| \hat{n}_{\mathrm{p}} \| [N]\omega L \rangle = \frac{N-1}{2} + \frac{(N+2)L(L+1)}{2\omega(\omega+2)}$$

$$\langle [N]\omega+2L \| \hat{n}_{\mathrm{p}} \| [N]\omega L \rangle = \left[\frac{(N-\omega)(N+\omega+4)(\omega-L+1)_2(\omega+L+2)_2}{16(\omega+1)_3(\omega+2)} \right]^{1/2}$$

$$\langle [N]\omega L \| \hat{L} \| [N]\omega L \rangle = \sqrt{L(L+1)}$$

$$\langle [N]\omega L+1 \| \hat{D} \| [N]\omega L \rangle = i \left[\frac{(\omega-L)(\omega+L+2)(L+1)}{2L+3} \right]^{1/2}$$

$$\langle [N]\omega L+1 \| \hat{D}' \| [N]\omega L \rangle = \frac{(N+2)(L+1)}{\omega(\omega+2)} \left[\frac{(\omega-L)(\omega+L+2)(L+1)}{2L+3} \right]^{1/2}$$

$$\langle [N]\omega+2L-1 \| \hat{D}' \| [N]\omega L \rangle$$
$$= -\left[\frac{(N-\omega)(N+\omega+4)(\omega-L+1)_3(\omega+L+2)L}{4(\omega+1)_3(\omega+2)(2L-1)} \right]^{1/2}$$

$$\langle [N]\omega+2L+1 \| \hat{D}' \| [N]\omega L \rangle$$
$$= -\left[\frac{(N-\omega)(N+\omega+4)(\omega-L+1)(\omega+L+2)_3(L+1)}{4(\omega+1)_3(\omega+2)(2L+3)} \right]^{1/2}$$

$$\langle [N]\omega L \| \hat{Q} \| [N]\omega L \rangle = (N+2)\left(1+\frac{L(L+1)}{\omega(\omega+2)}\right) \left[\frac{L(L+1)}{6(2L-1)(2L+3)} \right]^{1/2}$$

$$\langle [N]\omega L+2 \| \hat{Q} \| [N]\omega L \rangle = (N+2)\left[\frac{(\omega-L-1)_2(\omega+L+2)_2(L+1)(L+2)}{4\omega^2(\omega+2)^2(2L+3)(2L+5)} \right]^{1/2}$$

$$\langle [N]\omega+2L-2 \| \hat{Q} \| [N]\omega L \rangle = \left[\frac{(N-\omega)(N+\omega+4)(\omega-L+1)_4 L(L-1)}{16(\omega+1)_3(\omega+2)(2L-3)(2L-1)} \right]^{1/2}$$

$$\langle [N]\omega+2L \| \hat{Q} \| [N]\omega L \rangle$$
$$= \left[\frac{(N-\omega)(N+\omega+4)(\omega-L+1)_2(\omega+L+2)_2 L(L+1)}{24(\omega+1)_3(\omega+2)(2L-1)(2L+3)} \right]^{1/2}$$

$$\langle [N]\omega+2L+2 \| \hat{Q} \| [N]\omega L \rangle$$
$$= \left[\frac{(N-\omega)(N+\omega+4)(\omega+L+2)_4(L+1)(L+2)}{16(\omega+1)_3(\omega+2)(2L+3)(2L+5)} \right]^{1/2}$$

[a] $(\cdots)_s$ is the Pochhammer symbol defined in the text.

$$(a)_s \equiv a(a+1)\cdots(a+s-1) \qquad (5.176)$$

which is known as a Pochhammer symbol [10]. Other matrix elements can be derived from the ones given Table 5.4 using

$$\langle \alpha' L' \| \hat{T}^{(\lambda)} \| \alpha L \rangle = (-)^{L-L'} \left[\frac{2L+1}{2L'+1}\right]^{1/2} \langle \alpha L \| \hat{T}^{(\lambda)} \| \alpha' L' \rangle^* \qquad (5.177)$$

This relation follows from the definition (1.96) of the reduced matrix element and is valid for any tensor operator $\hat{T}^{(\lambda)}_\mu$ if λ is integer and $\hat{T}^{(\lambda)}_0$ is hermitian. It thus holds for all U(4) generators as they are defined at the end of Section 5.2 and appear in Table 5.4.

In principle, we can continue this computational scheme and derive matrix elements of more complicated operators that are quadratic, cubic, etc., in the generators. It is clear, however, that soon this becomes a tedious procedure involving a great deal of lengthy algebra (even the derivation of the results of Table 5.4 requires a non-negligible amount of algebra) and that we would profit from a better and quicker approach. Such an alternative method is introduced in Section 5.8 and is based on a generalization to U(4) of the familiar SU(2) tensor calculus discussed in Chapter 1. We will find that many of the results presented in this section (e.g., expressions for matrix elements of U(4) generators) can be derived in a more concise fashion using U(4) tensor calculus.

From the results given in Table 5.4 we can, in principle, derive the intensities of electromagnetic transitions as they are predicted in the SO(4) limit. As an example we show in Figure 5.5 all non-zero matrix elements of the operator $\hat{D}_\mu$ between SO(4) states. The figure illustrates that electric dipole transitions induced by the operator $\hat{D}_\mu$ only occur between levels belonging to the same rotational band (i.e., with the same v or ω) or, in other words, that there exists an SO(4) selection rule $\Delta v = 0$ for such transitions. This selection rule is, to some extent, observed in rigid molecules in the sense that transitions within a band are weaker than those between bands and, in fact, decrease exponentially with Δv. Nevertheless, a description of electric dipole transitions solely based on the properties of the operator $\hat{D}_\mu$ in the SO(4) limit would be a poor one and is in need of refinements, some of which are discussed in Section 5.7. We note that $\hat{D}'_\mu$, *not* being a generator of SO(4), has off-diagonal matrix elements with $\Delta v = \pm 1$

5.5 Rigid molecules: The SO(4) limit

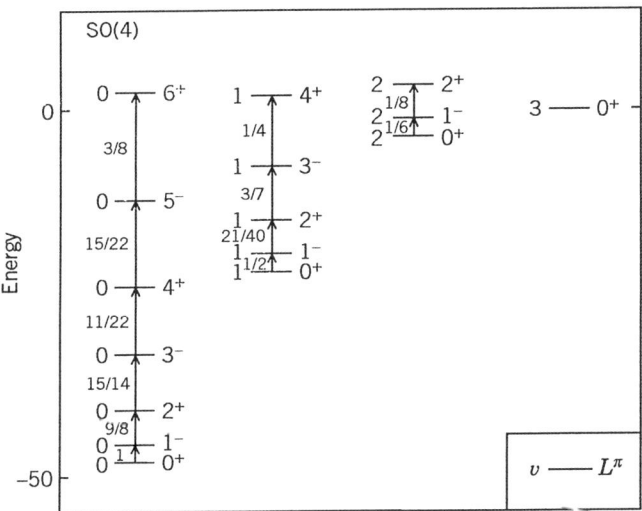

Figure 5.5: Allowed transitions in the SO(4) limit induced by the dipole operator $\hat{D}_\mu$. The numbers alongside the arrows are the squares of the reduced matrix elements in Table 5.4 normalized to the $0_1^+ \to 1_1^-$ transition. Parameters, labels, and energy units are as in Figure 5.4.

and might be useful for the description of molecular rotation–vibration properties. In general, the description of infrared intensities requires the use of a more realistic transition operator, as discussed in [19,20].

In this and the previous section we have given a complete and detailed overview of the algebraic properties of the two limits of the vibron model, U(3) and SO(4). The one remaining task is establishing a link between the two, that is, finding the transformation brackets. In view of the many explicit results we have already derived, we will find this a readily solvable problem, and therefore the discussion in the next section shall be brief.

5.6 Transformation brackets

The transformation brackets between the U(3) and SO(4) limits of the vibron model are defined through the expansion

$$|[N]\omega L M_L\rangle = \sum_{n_\mathrm{p}} \langle [N]n_\mathrm{p} L | [N]\omega L \rangle |[N]n_\mathrm{p} L M_L\rangle \qquad (5.178)$$

and are found by comparing the expressions (5.87) and (5.161). We start from the SO(4) wave functions (5.161) and rewrite them using the following expansion for the Gegenbauer polynomial [10]:

$$C_n^\lambda(t) = \frac{1}{\Gamma(\lambda)} \sum_{k=0}^{\lfloor n/2 \rfloor} (-)^k \frac{\Gamma(\lambda + n - k)}{k!(n-2k)!}(2t)^{n-2k} \qquad (5.179)$$

where $\lfloor x \rfloor$ denotes the integer part of x. The resulting expression is rewritten by carrying out the binomial expansion in $p^\dagger \cdot p^\dagger + s^\dagger s^\dagger$ leading to

$$\begin{aligned}
&|[N]\omega L M_L\rangle \\
&= B_{N\omega L} \sum_{k=0}^{(\omega-L)/2} \sum_{\ell=0}^{(N-\omega+2k)/2} (-)^k 2^{\omega-L-2k} \\
&\quad \times \frac{(\omega-k)!((N-\omega)/2+k)!}{L!k!(\omega-L-2k)!\ell!((N-\omega)/2+k-\ell)!} \\
&\quad \times (p^\dagger \cdot p^\dagger)^{(N-\omega+2k-2\ell)/2}(s^\dagger)^{\omega-L-2k+2\ell}\mathcal{Y}_{M_L}^L(\mathbf{p}^\dagger)|0\rangle
\end{aligned} \qquad (5.180)$$

This result is brought into the appropriate form by a change of variables from (k,ℓ) to (n_p,k) with $n_\mathrm{p} \equiv N - \omega + L + 2k - 2\ell$ and an inversion of summations,

$$\sum_{k=0}^{(\omega-L)/2} \sum_{\ell=0}^{(N-\omega+2k)/2} \to \sum_{n_\mathrm{p}=L}^{N} \sum_{k=\max(0,(-N+n_\mathrm{p}+\omega-L)/2)}^{(\omega-L)/2} \qquad (5.181)$$

with n_p increasing in steps of 2, $n_\mathrm{p} = L, L+2, \ldots, N$. The boundary values of the summation variables in (5.181) are illustrated in Figure 5.6. With these successive transformations we arrive at an expansion of

5.6 Transformation brackets

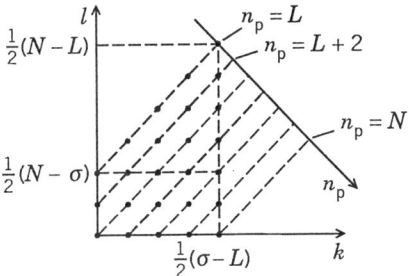

Figure 5.6: Illustration of the change of variables and the summation inversion in (5.87). Each term in the sum is represented by a dot.

SO(4) states in terms of U(3) states similar to (5.178). A term-by-term comparison of the two expansions then leads to the following expression for the transformation brackets:

$$\langle [N]n_\mathrm{p}L|[N]\omega L\rangle = \frac{B_{N\omega L}}{B_{n_\mathrm{p}L}B'_{Nn_\mathrm{p}}}$$
$$\times \sum_k (-)^k 2^{\omega-L-2k} \binom{\omega-k}{L+k}\binom{L+k}{k}\binom{(N-\omega)/2+k}{(n_\mathrm{p}-L)/2}$$
(5.182)

where the summation range of k is determined by the zeros of the binomial coefficients (i.e., is as in (5.181)). Note that the entire derivation of this result is very similar to the corresponding one in the schematic U(2) model presented in Section 1.6.

This completes our overview of the algebraic properties of U(4) that are of relevance to the vibron model. The next section is devoted to the question of the extent to which the rotation–vibration excitations of diatomic molecules can be parametrized in terms of p and s bosons and, more specifically, to what degree of approximation the U(3) and SO(4) limits of the vibron model can be applied to their description. Subsequently, in the last section of this chapter, we reanalyze the group-theoretical structure of the vibron model, not with the purpose of deriving any new results but rather to cast the formalism presented so far into a form that is more readily generalized to the more complex situations considered in Chapters 6 and 7.

5.7 Examples

In presenting the applications of the vibron model to molecular rotation–vibration spectra we follow here the discussion given by van Roosmalen [15].

A comparison of the data with exact symmetry calculations is rather restricted in scope and mainly confined to the SO(4) limit. We already noted in Section 5.5 that the SO(4) energy expression (5.127) is a particular case of the Dunham expansion (5.134), the exact correspondence between the parameters being

$$Y_{00} = E_0''', \quad Y_{10} = -4(N+2)\beta, \quad Y_{20} = 4\beta, \quad Y_{01} = \gamma \qquad (5.183)$$

with all other parameters Y_{ij} zero. Although certainly not all molecular rotation–vibration spectra can be parametrized with just the four Dunham parameters (5.183), for some it gives a fair approximation. Another complication is that the vibron model, in the simple form discussed so far, is not appropriate to describe homonuclear molecules as it does not incorporate the quantum labels associated with the interchange of nuclei or electron inversion. We nevertheless present here a comparison of the global characteristics of the H_2 molecule in its electronic ground-state configuration $^1\Sigma_g^+$ [6] with those of the SO(4) symmetry in order to gauge its general features.

The SO(4) parametrization—apart from being restricted to four Dunham parameters—displays another constraint which can be illustrated with $H_2(^1\Sigma_g^+)$. We first note that the number of bosons N determines the number of vibrational levels that occur in the molecule (see (5.131)). On the other hand, from (5.183) we predict $Y_{10}/Y_{20} = -(N+2)$, independent of other parameters appearing in the SO(4) hamiltonian. We thus find that in the SO(4) limit a relation exists between the number of vibrational levels and the ratio Y_{20}/Y_{10}. In $H_2(^1\Sigma_g^+)$ there are 14 vibrational states [16], and hence the number of bosons N must be 28 or 29, giving a ratio Y_{20}/Y_{10} of -30 or -31. This is in reasonable agreement with the value of -37.28 obtained to fit the experimental spectrum [17]. It thus appears that the two features characteristic of the SO(4) limit of the vibron model (i.e., up to quadratic terms in the Dunham expansion and the relation between N and

5.7 Examples

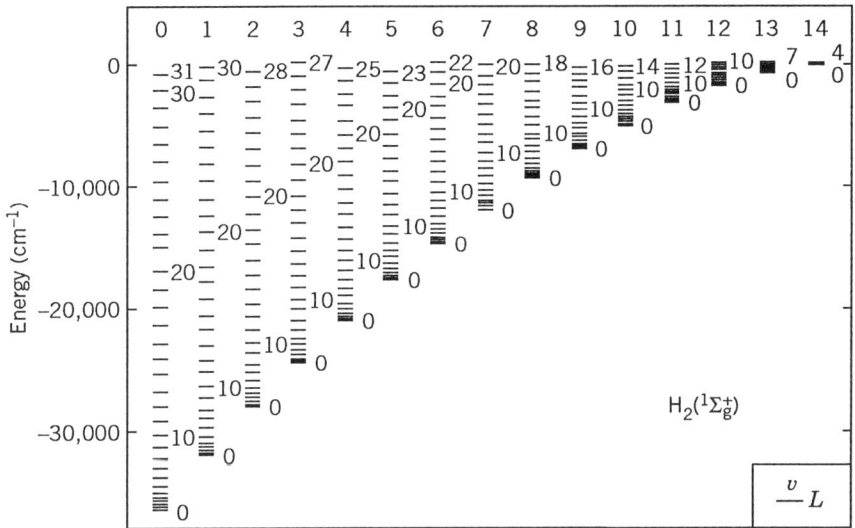

Figure 5.7: The SO(4) eigenspectrum of the hamiltonian $\hat{H}_{\mathrm{II}}$ for $N = 29$ bosons. The spectrum is obtained from (5.127) with the parameters (in cm^{-1}) $E_0'' = 0$, $\beta = -40$, and $\gamma = 41.5$. Levels are labeled by $v = (N-\omega)/2$ and the angular momentum L.

Y_{20}/Y_{10}) are, to a good approximation, realized in H$_2$($^1\Sigma_g^+$). This is further illustrated in Figures 5.7 and 5.8, where the result of a best fit using the SO(4) energy expression (5.127) is compared to the complete bound-state rotation–vibration spectrum of this molecule in its $^1\Sigma_g^+$ state as obtained in [17]. Comparison of the figures also illustrates the limitations of the SO(4) hamiltonian (5.115). It predicts a constant inertia parameter for the different vibrational bands in contrast to the observed one which varies with v [16]. Another characteristic feature of the SO(4) limit is that it predicts band terminations at specific L values which change with v. In H$_2$($^1\Sigma_g^+$) these terminations approximately agree with the observed ones, but for many other diatomic molecules this is not the case.

Besides investigating the properties of the SO(4) limit of the vibron model as regards energy spectra, we must also study its predictions

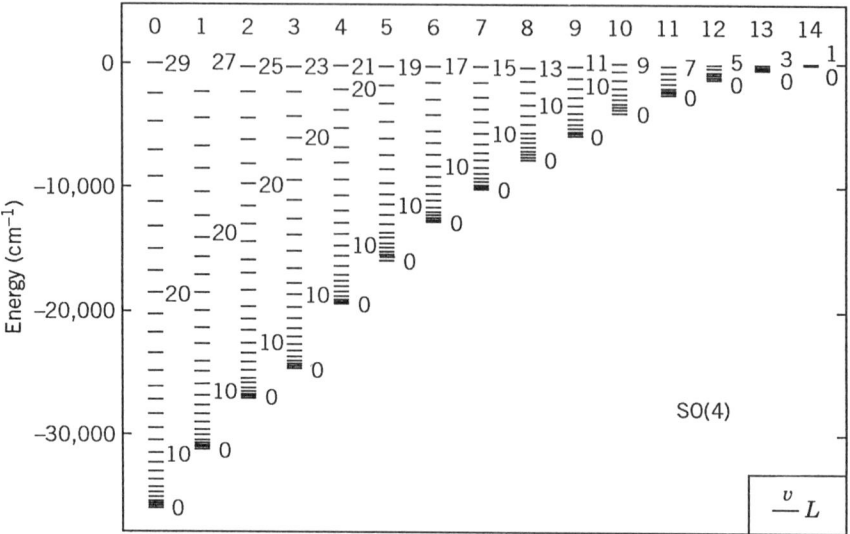

Figure 5.8: The bound-state rotation–vibration spectrum of $H_2(^1\Sigma_g^+)$. Levels are labeled by v and the angular momentum L.

concerning electromagnetic transitions—specifically electric dipole or E1 transitions—which generally constitute a more sensitive test of molecular models. The intensity of an electromagnetic transition between levels with vL and $v'L'$ is mainly determined by the thermal distribution of molecules over the rotational and vibrational states. Not taking into account this effect, the intensity is directly proportional to the absolute square of the corresponding electric or magnetic operator summed over all possible magnetic substates [6],

$$I(vL \to v'L') \propto \sum_{M_L M'_L m} |\langle v'L'M'_L | \hat{T}^{(\lambda)}_\mu | vLM_L \rangle|^2$$
$$= (2L'+1)|\langle v'L' \| \hat{T}^{(\lambda)} \| vL \rangle|^2 \qquad (5.184)$$

where the proportionality factor involves the number of molecules in the initial state. In the case of E1 radiation several types of transitions occur: (i) $\Delta v = 0$ far-infrared transitions and (ii) $\Delta v \neq 0$ near-infrared transitions, which are further subdivided in two classes, those belonging

5.7 Examples

to the R branch, which have $L' = L - 1$, and the P-branch transitions, with $L' = L + 1$. As a further convention, the intensity (5.184) for near-infrared transitions is usually written as [18]

$$I(\text{E1}; vL \to v'L') \propto \frac{1}{2L+1} |m| F_v^{v'}(m) | R_v^{v'} |^2 \qquad (5.185)$$

where $m = L$ for the R branch and $m = -L'$ for P-branch transitions. The factorization (5.185) is useful because it displays explicitly $R_v^{v'}$, the matrix element for the electric dipole moment of a non-rotating anharmonic oscillator, and $F_v^{v'}(m)$, a correction factor taking into account the rotation–vibration interaction, which often is expanded in powers of m,

$$F_v^{v'}(m) = 1 + C(v, v')m + D(v, v')m^2 + \cdots \qquad (5.186)$$

In all cases it is found that $I(\text{E1}; vL \to v'L')/|m|$ is only weakly dependent on the angular momenta L and L'.

The form of the E1 transition operator is restricted by requiring it to carry angular momentum $L = 1$ and parity $\pi = -$. Although more complicated combinations should be considered in general [19,20], we confine ourselves here to operators that are linear in the generators of U(4), in which case two choices are possible: the SO(4) generators $\hat{D}_\mu$ and the $\overline{\text{SO}}(4)$ generators $\hat{D}'_\mu$ defined in (5.20) and (5.23), respectively. Their matrix elements in an SO(4) basis are calculated in Section 5.5 and the resulting expressions given in Table 5.4, and from them the predicted intensities can be derived. The results are shown in Figure 5.9, where $I(\text{E1}; vL \to v'L')/|m|$ is plotted as a function of the initial angular momentum L for all transitions induced by $\hat{D}_\mu$ or $\hat{D}'_\mu$. The dipole operator $\hat{D}_\mu$ gives a strong $\Delta v = 0$ transition, the intensity of which decreases smoothly with increasing angular momentum, but no other transitions are allowed. The $\overline{\text{SO}}(4)$ generator $\hat{D}'_\mu$ has much smaller matrix elements both for $\Delta v = 0$ and $\Delta v = \pm 1$. Thus it appears that an appropriate combination of $\hat{D}_\mu$ and $\hat{D}'_\mu$ might lead to $\Delta v = 0, \pm 1$ transitions with non-zero intensities varying smoothly with L. It is clear, however, that this combination strictly forbids all $\Delta v > 1$ transitions. Experimentally one finds that although intensities decrease exponentially as a function of Δv, they are nevertheless non-zero, and hence

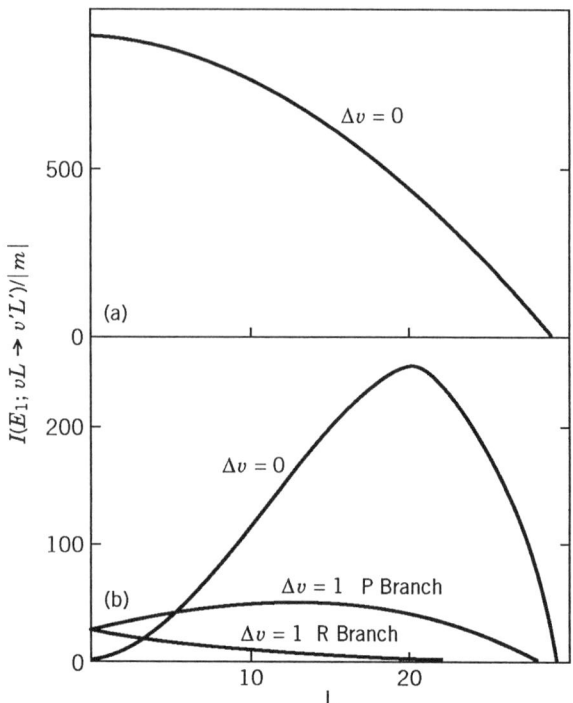

Figure 5.9: The quantity $I(\text{E1}; vL \to v' = 0\ L')/|m|$ ($|m| = \max(L, L')$) plotted as a function of the initial angular momentum L for (a) $\hat{T}_\mu^{(\text{E1})} = \hat{D}_\mu$ and (b) $\hat{T}_\mu^{(\text{E1})} = \hat{D}'_\mu$. The number of bosons $N = 29$.

the selection rules $\Delta v = 0, \pm 1$ of the dipole operators $\hat{D}_\mu$ and $\hat{D}'_\mu$ are not satisfactory. As already pointed out, the E1 transition operator needed for carrying out detailed fits of infrared transitions has a more complicated form [19,20], but the evaluation of matrix elements can be carried out using the same kind of methods discussed in this chapter.

From the previous discussion it is clear that the symmetry description in the SO(4) limit of the vibron model can only give a first-order account of spectroscopic properties of molecules and is in need of improvement. This can be achieved in two different ways. In the first, higher-order SO(4)-invariant operators are added to the hamiltonian

5.7 Examples

(5.115), which then becomes an expansion in powers of the SO(3) and SO(4) Casimir invariants $\mathcal{C}_2[\text{SO}(3)]$ and $\mathcal{C}_2[\text{SO}(4)]$. The resulting energy expression is very similar to the Dunham expansion (5.134), and realistic fits to molecular rotation–vibration spectra are obtained, albeit at the expense of introducing more parameters. The inclusion of additional invariants does not, however, alter the SO(4) wave functions, and as a consequence the limitations of the SO(4) limit as regards the description of transition properties remain unchanged unless higher-order terms are also added to the dipole operator [19]. A second, more promising method is to add to the hamiltonian (5.115) SO(4)-breaking terms which, for example, can be of the form $\hat{n}_\text{p}$. In this approach one therefore considers the hamiltonian

$$\hat{H} = E_0'' + \epsilon \hat{n}_\text{p} + \beta(\hat{L}^2 + \hat{D}^2) + \gamma \hat{L}^2 \qquad (5.187)$$

which can either diagonalized numerically or, if ϵ is small, be treated in perturbation theory since all matrix elements of this operator in an SO(4) basis are known. The latter approach has been investigated by van Roosmalen [15] and the interested reader is referred to Chapter 3 of his thesis.

The calculations for diatomic molecules reported in this section are rather limited in scope. The main reason for this is that unless many higher-order corrections are considered, the results of the U(4) vibron model are of similar quality as those obtained by solving the Schrödinger equation associated with an interatomic potential. As a consequence not many detailed U(4) calculations have been carried out for diatomic molecules, and it may thus seem that the formalism developed in the preceding sections is rather heavy in comparison with the few applications presented in this section. We emphasize, however, that this formalism forms the basis of the applications of the U(4) model to the more complicated molecules discussed in Chapters 6 and 7, for which the *ab initio* calculations are difficult—if not impossible—to carry out. Thus the main purpose of the present chapter is preparing the ground for the applications of the next two chapters, rather than a study of the properties of diatomic molecules themselves.

5.8 Tensor calculus

The purpose of this section is similar to that of Section 1.7 of Chapter 1 and consists of analyzing the algebraic properties of the U(4) vibron model from a different perspective using tensor calculus. Considerable simplifications (especially in subsequent chapters) result from this approach, which can be developed in close analogy with the U(2) case. Because of this analogy, it is easy to understand the general features of the generalized tensor calculus presented in this section, but a complete derivation of its properties, which we endeavor to present here, requires a rather lengthy analysis. The reader might not wish, in a first reading, to go through all the detailed derivations presented in this section but may wish to be confined to the essential results which are highlighted in the usual way and summarized at the end of this chapter.

A tensor calculus can only be developed with respect to a given group-theoretical basis which, in general, is specified by a dynamical algebra G_1 and its subalgebras G_i, $i > 1$,

$$G_1 \supset G_2 \supset \cdots \supset G_n \qquad (5.188)$$

In the U(4) model two possible choices exist: the U(3) basis and the SO(4) basis, discussed in Sections 5.4 and 5.5. As many molecular spectra can approximately be interpreted in terms of SO(4)—rather than SU(3)—quantum numbers, the former limit is of considerably more practical interest, and therefore we confine ourselves here to developing a tensor calculus with respect to the SO(4) basis. We emphasize that from a mathematical point of view we might have chosen equally well the U(3) basis.

In the complete SO(4) basis we have $n = 4$ in (5.188) with $G_1 =$ U(4), $G_2 =$ SO(4), $G_3 =$ SO(3), and $G_4 =$ SO(2). This represents a considerable complication compared with the U(2) $\supset$ SO(2) case discussed in Section 1.7, and in the interest of simplicity it is useful to introduce the associated tensor calculus in two stages. In the first we take only part of the chain, that is, $n = 3$ and $G_1 =$ SO(4), $G_2 =$ SO(3), and $G_3 =$ SO(2). The tensor calculus in this case can be fully worked out along the lines of Section 1.7. In the second stage we consider

5.8 Tensor calculus

the complete set of algebras occurring in the SO(4) limit, which thus amounts to finding tensor properties under U(4) in addition to those derived in stage 1. Although no extension of concepts is required, a general derivation of results in this case is prevented by technical complications.

We thus begin with a discussion of the tensor calculus defined by the chain of algebras

$$\begin{array}{ccc} \mathrm{SO}(4) \supset & \mathrm{SO}(3) \supset & \mathrm{SO}(2) \\ | & | & | \\ (\omega,\omega') & l & m \end{array} \qquad (5.189)$$

Note that for the applications to diatomic molecules discussed in this chapter a single index ω suffices to characterize the allowed SO(4) representations and $\omega' = 0$. For later applications discussed in the two following chapters it is necessary to consider the more general case with two labels. The angular momentum content of a general SO(4) representation (ω, ω') is obtained in the same way as in the specific case $\omega' = 0$ (see discussion after (5.129)) and reads

$$l = |\omega'|, |\omega'|+1, \ldots, \omega \qquad (5.190)$$

As a notational convention we denote the two indices ω and ω' as a single character $\vec{\omega}$, and thus the basis states (5.189) become $|\vec{\omega}lm\rangle$. The action on this basis state of any operator $\hat{R}$ belonging to SO(4) can now formally be written as

$$\hat{R}|\vec{\omega}lm\rangle = \sum_{l'm'} \langle \vec{\omega}l'm'|\hat{R}|\vec{\omega}lm\rangle |\vec{\omega}l'm'\rangle \qquad (5.191)$$

This relation is the generalization of (1.90) from three to four dimensions. The operator $\hat{R}$ can be thought of as inducing a rotation in four-dimensional space which depends on six generalized "Euler" angles θ_i. In analogy with the treatment in the U(2) case we may also define tensor operators $\hat{T}^{(\vec{\omega})}_{lm}$ that transform under these generalized rotations in the same way as the basis states,

$$\hat{R}\hat{T}^{(\vec{\omega})}_{lm}\hat{R}^{-1} = \sum_{l'm'} \langle \vec{\omega}l'm'|\hat{R}|\vec{\omega}lm\rangle \hat{T}^{(\vec{\omega})}_{l'm'} \qquad (5.192)$$

For infinitesimal rotations $\hat{R} = 1 + \epsilon_i \hat{X}_i$, with ϵ_i small, this relation reduces to

$$[\hat{X}_i, \hat{T}^{(\vec{\omega})}_{lm}] = \sum_{l'm'} \langle \vec{\omega} l' m' | \hat{X}_i | \vec{\omega} l m \rangle \hat{T}^{(\vec{\omega})}_{l'm'} \tag{5.193}$$

which again generalizes (1.92).

The next problem concerns the coupling of the basis states or tensors we have defined so far. Denoting by $A^{(\vec{\omega})}_{lm}$ both states and tensors, it is clear that the product $A^{(\vec{\omega}_1)}_{l_1 m_1} A^{(\vec{\omega}_2)}_{l_2 m_2}$ in general will *not* transform according to (5.191) or (5.192). The question is whether it is possible to form linear combinations that have the appropriate transformation properties, that is, whether we can find some coefficients such that the sum of products $A^{(\vec{\omega}_1)}_{l_1 m_1} A^{(\vec{\omega}_2)}_{l_2 m_2}$ multiplied with these coefficients gives rise to a state or tensor with the correct transformation properties,

$$A^{(\vec{\omega})}_{lm} = \sum_{l_1 m_1} \sum_{l_2 m_2} \left\langle \begin{array}{cc} \vec{\omega}_1 & \vec{\omega}_2 \\ l_1 m_1 & l_2 m_2 \end{array} \middle| \begin{array}{c} \vec{\omega} \\ lm \end{array} \right\rangle A^{(\vec{\omega}_1)}_{l_1 m_1} A^{(\vec{\omega}_2)}_{l_2 m_2} \tag{5.194}$$

The coefficient in the angle brackets is the generalization of the Clebsch–Gordan coefficient introduced in Section 1.7 and we refer to it as a *coupling coefficient*. When appearing in the text (as opposed to in equations) we denote it as $\langle \vec{\omega}_1 l_1 m_1 \otimes \vec{\omega}_2 l_2 m_2 | \vec{\omega} l m \rangle$. Note that we implicitly assume that the coupled tensor (5.194) is uniquely determined by $\vec{\omega}$ or, equivalently, that each coupled representation $\vec{\omega}$ is contained only once in the uncoupled one $\vec{\omega}_1 \otimes \vec{\omega}_2$. This assertion is true for product representations of particular algebras only—such as SU(2) and SO(4)—but not in general. Algebras satisfying this property are said to be *simply* reducible as opposed to those that are *multiply* reducible. The fact that SU(2) is simply reducible is well known from angular momentum theory and is tacitly assumed in Section 1.7; the analogous property for SO(4) follows from the isomorphism SO(4) $\simeq$ SU$_r$(2) $\otimes$ SU$_{r'}$(2) and is demonstrated below when explicit expressions for the coupling coefficients $\langle \vec{\omega}_1 l_1 m_1 \otimes \vec{\omega}_2 l_2 m_2 | \vec{\omega} l m \rangle$ are constructed. In contrast, U(4) is multiply reducible, and many of the technical complications associated with a tensor calculus in a complete U(4) $\supset$ SO(4) $\supset$ SO(3) $\supset$ SO(2) basis and alluded to at the beginning of this section arise because of this multiplicity problem.

5.8 Tensor calculus

Since the coupling coefficients in (5.194) are elements of a unitary transformation from a coupled to an uncoupled basis, they satisfy the usual orthonormality relations in rows and columns,

$$\sum_{\substack{l_1 m_1 \\ l_2 m_2}} \left\langle \begin{array}{cc} \vec{\omega}_1 & \vec{\omega}_2 \\ l_1 m_1 & l_2 m_2 \end{array} \Big| \begin{array}{c} \vec{\omega}_a \\ l_a m_a \end{array} \right\rangle^* \left\langle \begin{array}{cc} \vec{\omega}_1 & \vec{\omega}_2 \\ l_1 m_1 & l_2 m_2 \end{array} \Big| \begin{array}{c} \vec{\omega}_b \\ l_b m_b \end{array} \right\rangle$$
$$= \delta_{\vec{\omega}_a \vec{\omega}_b} \delta_{l_a l_b} \delta_{m_a m_b} \qquad (5.195)$$

$$\sum_{\vec{\omega} l m} \left\langle \begin{array}{cc} \vec{\omega}_1 & \vec{\omega}_2 \\ l_{1a} m_{1a} & l_{2a} m_{2a} \end{array} \Big| \begin{array}{c} \vec{\omega} \\ l m \end{array} \right\rangle^* \left\langle \begin{array}{cc} \vec{\omega}_1 & \vec{\omega}_2 \\ l_{1b} m_{1b} & l_{2b} m_{2b} \end{array} \Big| \begin{array}{c} \vec{\omega} \\ l m \end{array} \right\rangle$$
$$= \delta_{l_{1a} l_{1b}} \delta_{l_{2a} l_{2b}} \delta_{m_{1a} m_{1b}} \delta_{m_{2a} m_{2b}}$$

In addition to providing a way to construct composite states or tensors from simpler ones, coupling coefficients also facilitate the evaluation of the matrix elements of operators between basis states (5.189). This task is greatly simplified by the application of the (generalized) Wigner–Eckart theorem, which states that the dependence on SO(3) and SO(2) quantum numbers is contained in the coupling coefficient [5]:

$$\langle \vec{\omega}_a l_a m_a | \hat{T}^{(\vec{\omega})}_{lm} | \vec{\omega}_b l_b m_b \rangle = \left\langle \begin{array}{cc} \vec{\omega}_b & \vec{\omega} \\ l_b m_b & l m \end{array} \Big| \begin{array}{c} \vec{\omega}_a \\ l_a m_a \end{array} \right\rangle \langle \vec{\omega}_a \| \hat{T}^{(\vec{\omega})} \| \vec{\omega}_b \rangle \qquad (5.196)$$

The quantity $\langle \vec{\omega}_a \| \hat{T}^{(\vec{\omega})} \| \vec{\omega}_b \rangle$ depends on SO(4) quantum numbers only (compare with (1.96)); it is called a reduced matrix element or, more precisely, an SO(4)-reduced matrix element to indicate that the dependence on quantum numbers of subalgebras of SO(4) has been factored out.

For the practical application of these results (i.e., the coupling (5.194) of states or tensors and the Wigner–Eckart theorem (5.196)), explicit expressions for the coupling coefficients are required. To calculate $\langle \vec{\omega}_1 l_1 m_1 \otimes \vec{\omega}_2 l_2 m_2 | \vec{\omega} l m \rangle$, we must first determine what representations $\vec{\omega}$ are contained in the product $\vec{\omega}_1 \otimes \vec{\omega}_2$. This is easily accomplished using the isomorphism between SO(4) and $SU_r(2) \otimes SU_{r'}(2)$, which for representations implies

$$\vec{\omega}_i \equiv (\omega_i, \omega'_i) \simeq \{r_i, r'_i\} \qquad (5.197)$$

The labels between round brackets refer to the SO(4) notation while those between curly brackets are associated with $SU_r(2) \otimes SU_{r'}(2)$, the two being related by (5.122) or (5.123). The product of two $SU_r(2) \otimes SU_{r'}(2)$ representations is obtained by applying twice the usual angular momentum coupling rule:

$$\{r_1, r'_1\} \otimes \{r_2, r'_2\} = \oplus \sum_{r=|r_1-r_2|}^{r_1+r_2} \sum_{r'=|r'_1-r'_2|}^{r'_1+r'_2} \{r, r'\} \tag{5.198}$$

With the help of (5.122) and (5.123) this can be converted back into

$$(\omega_1, \omega'_1) \otimes (\omega_2, \omega'_2) = \oplus \sum_{\alpha=0}^{\omega_+} \sum_{\beta=0}^{\omega_-} (\omega_1+\omega_2-\alpha-\beta, \omega'_1+\omega'_2-\alpha+\beta) \tag{5.199}$$

with $\omega_\pm = \min(\omega_1 \pm \omega'_1, \omega_2 \pm \omega'_2)$. An example of this product rule is shown in Figure 5.10, for the SO(4) labels (ω, ω') as well as for the $SU_r(2) \otimes SU_{r'}(2)$ labels $\{r, r'\}$. Geometrically, both notations are related by a simple transformation of axes, and this is the easiest way to infer (5.199) from (5.198). Note that (5.199) shows that no SO(4)

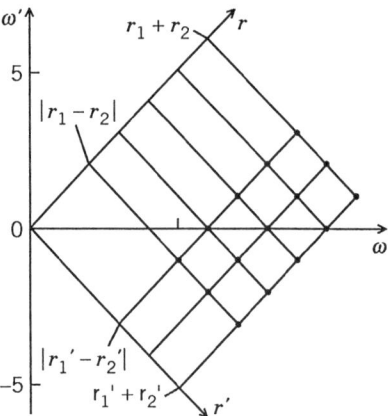

Figure 5.10: Illustration of the product rule for two SO(4) $\simeq SU_r(2) \otimes SU_{r'}(2)$ representations. The product $\{r_1, r'_1\} \otimes \{r_2, r'_2\} \simeq (\omega_1, \omega'_1) \otimes (\omega_2, \omega'_2)$ contains the representations $\{r, r'\} \simeq (\omega, \omega')$ indicated by a dot. The case shown corresponds to $\{4, 4\} \otimes \{2, 1\} \simeq (8, 0) \otimes (3, 1)$.

5.8 Tensor calculus

representation occurs more than once in any given product of SO(4) representations, and hence we have proven our earlier assertion that SO(4) is simply reducible.

We now proceed to derive explicit expressions for the coupling coefficients. We note that the uncoupled states $|\vec{\omega}_1 l_1 m_1\rangle|\vec{\omega}_2 l_2 m_2\rangle$ which appear on the right-hand side of (5.194) correspond to the following chain of algebras:

$$SO_1(4) \otimes SO_2(4) \supset SO_1(3) \otimes SO_2(3) \supset SO_1(2) \otimes SO_2(2) \supset SO(2) \quad (5.200)$$

where labels with a subscript i are associated with the algebra $SO_i(4)$. On the other hand, the state $|\vec{\omega} l m\rangle$ on the left-hand side of (5.194) is defined by the chain of coupled algebras

$$SO_1(4) \otimes SO_2(4) \supset SO(4) \supset SO(3) \supset SO(2) \quad (5.201)$$

where the coupled algebras G (i.e., those that have no subscript) are obtained by adding the generators of G_1 and G_2. It is clear that a third chain of algebras exists which is intermediate between the previous two and is of the form

$$SO_1(4) \otimes SO_2(4) \supset SO_1(3) \otimes SO_2(3) \supset SO(3) \supset SO(2) \quad (5.202)$$

In this way we have established the existence of three different coupling schemes, each with their associated labeling:

$$
\begin{array}{cccccc}
SO_1(4) & \otimes & SO_2(4) & \supset & SO(4) & \supset & SO(3) & \supset & SO(2) \\
| & & | & & | & & | & & | \\
\vec{\omega}_1 & & \vec{\omega}_2 & & \vec{\omega} & & l & & m
\end{array}
$$

$$
\begin{array}{cccccc}
SO_1(4) & \otimes & SO_2(4) & \supset & SO_1(3) & \otimes & SO_2(3) & \supset & SO(3) & \supset & SO(2) \\
| & & | & & | & & | & & | & & | \\
\vec{\omega}_1 & & \vec{\omega}_2 & & l_1 & & l_2 & & l & & m
\end{array}
$$

$$
\begin{array}{ccccccc}
SO_1(4) & \otimes & SO_2(4) & \supset & SO_1(3) & \otimes & SO_2(3) & \supset & SO_1(2) & \otimes & SO_2(2) & \supset & SO(2) \\
| & & | & & | & & | & & | & & | & & | \\
\vec{\omega}_1 & & \vec{\omega}_2 & & l_1 & & l_2 & & m_1 & & m_2 & & m
\end{array}
$$

$$(5.203)$$

These coupling schemes can conveniently be summarized in a *lattice of algebras* as

$$\begin{array}{ccccc}
SO_1(4) & & \otimes & & SO_2(4) \\
\downarrow & \searrow^a & & \swarrow^a & \downarrow \\
SO_1(3) & & SO(4) & & SO_2(3) \\
\downarrow & \searrow^b & \downarrow & \swarrow^b & \downarrow \\
SO_1(2) & & SO(3) & & SO_2(2) \\
& \searrow_c & \downarrow & \swarrow_c & \\
& & SO(2) & &
\end{array} \qquad (5.204)$$

where the different routes a, b, and c, indicated in the lattice, correspond to the first, second, and third reduction scheme in (5.203). Whenever we deal with a situation involving many different couplings, we shall often summarize the algebra chains with a lattice. With these three coupling schemes we associate the following basis states:

$$\begin{array}{ll}
(a) & |\vec{\omega}_1, \vec{\omega}_2; \vec{\omega} lm\rangle \\
(b) & |\vec{\omega}_1 l_1, \vec{\omega}_2 l_2; lm\rangle \\
(c) & |\vec{\omega}_1 l_1 m_1, \vec{\omega}_2 l_2 m_2; m\rangle \equiv |\vec{\omega}_1 l_1 m_1\rangle |\vec{\omega}_2 l_2 m_2\rangle
\end{array} \qquad (5.205)$$

The notation followed here is taken over from Part I: labels that are coupled with each other are separated by a colon and are separated by a semicolon from the label(s) to which they are coupled. Thus, for instance, in the state (c) only the SO(2) quantum numbers are coupled. Since the SO(2) coupling corresponds to the trivial addition $m_1 + m_2 = m$, this state is equivalent to a completely uncoupled one.

The transformation from (b) to (c) is readily established as it only requires the usual SO(3) $\supset$ SO(2) Clebsch–Gordan coefficient,

$$|\vec{\omega}_1 l_1, \vec{\omega}_2 l_2; lm\rangle = \sum_{m_1 m_2} \langle l_1 m_1 \, l_2 m_2 | lm \rangle |\vec{\omega}_1 l_1 m_1, \vec{\omega}_2 l_2 m_2; m\rangle \qquad (5.206)$$

The only remaining problem is finding the transformation matrix between the (a) and (b) bases, which can be obtained by rewriting the states in an $SU_r(2) \otimes SU_{r'}(2)$ notation as

$$\begin{array}{ll}
(a) & |\{r_1, r'_1\}, \{r_2, r'_2\}; \{r, r'\} lm\rangle \\
(b) & |\{r_1, r'_1\} l_1, \{r_2, r'_2\} l_2; lm\rangle
\end{array} \qquad (5.207)$$

5.8 Tensor calculus

In state (a), r_1 and r_2 are coupled to r, r'_1 and r'_2 are coupled to r', and r and r' are then combined to l; in state (b), r_i and r'_i are first coupled to l_i and then the coupling of l_1 and l_2 gives l. This shows that the transformation requires the recoupling of four angular momenta in SO(3), known from standard angular momentum theory, and can be written as

$$|\{r_1,r'_1\},\{r_2,r'_2\};\{r,r'\}lm\rangle$$
$$= \sum_{l_1 l_2} \begin{bmatrix} r_1 & r_2 & r \\ r'_1 & r'_2 & r' \\ l_1 & l_2 & l \end{bmatrix} |\{r_1,r'_1\}l_1,\{r_2,r'_2\}l_2;lm\rangle \qquad (5.208)$$

The symbol in square brackets is the overlap matrix encountered in the recoupling of four angular momenta and is related to the more commonly used $9j$ coefficient in the following way [21]:

$$\begin{bmatrix} r_1 & r_2 & r \\ r'_1 & r'_2 & r' \\ l_1 & l_2 & l \end{bmatrix} = \sqrt{(2r+1)(2r'+1)(2l_1+1)(2l_2+1)} \begin{Bmatrix} r_1 & r_2 & r \\ r'_1 & r'_2 & r' \\ l_1 & l_2 & l \end{Bmatrix}$$
$$(5.209)$$

We emphasize that the transformation (5.208), describing the recoupling of *four* angular momenta in SO(3), at the same time represents a coupling coefficient for *two* representations in SO(4). To bring out more clearly the latter coupling, we revert to the SO(4) notation and rewrite the relation (5.208) as

$$|\vec{\omega}_1,\vec{\omega}_2;\vec{\omega}lm\rangle = \sum_{l_1 l_2} \left\langle \begin{matrix} \vec{\omega}_1 & \vec{\omega}_2 \\ l_1 & l_2 \end{matrix} \middle| \begin{matrix} \vec{\omega} \\ l \end{matrix} \right\rangle |\vec{\omega}_1 l_1, \vec{\omega}_2 l_2; lm\rangle \qquad (5.210)$$

with

$$\left\langle \begin{matrix} \vec{\omega}_1 & \vec{\omega}_2 \\ l_1 & l_2 \end{matrix} \middle| \begin{matrix} \vec{\omega} \\ l \end{matrix} \right\rangle$$
$$= \left\langle \begin{matrix} (\omega_1,\omega'_1) & (\omega_2,\omega'_2) \\ l_1 & l_2 \end{matrix} \middle| \begin{matrix} (\omega,\omega') \\ l \end{matrix} \right\rangle \qquad (5.211)$$
$$= \begin{bmatrix} (\omega_1+\omega'_1)/2 & (\omega_2+\omega'_2)/2 & (\omega+\omega')/2 \\ (\omega_1-\omega'_1)/2 & (\omega_2-\omega'_2)/2 & (\omega-\omega')/2 \\ l_1 & l_2 & l \end{bmatrix}$$

The coefficient $\langle \vec{\omega}_1 l_1 \otimes \vec{\omega}_2 l_2 | \vec{\omega} l \rangle$ has the same structure as the familiar Clebsch–Gordan coefficient $\langle l_1 m_1 \, l_2 m_2 | lm \rangle$, but the latter pertains to the reduction $SO(3) \supset SO(2)$ while (5.211) is related to $SO(4) \supset SO(3)$. Its properties can be deduced from those of the $9j$ coefficient [21]. A particularly important one concerns the orthonormality of the $SO(4) \supset SO(3)$ coupling coefficients:

$$\sum_{l_1 l_2} \left\langle \begin{matrix} \vec{\omega}_1 & \vec{\omega}_2 \\ l_1 & l_2 \end{matrix} \middle| \begin{matrix} \vec{\omega}_a \\ l \end{matrix} \right\rangle \left\langle \begin{matrix} \vec{\omega}_1 & \vec{\omega}_2 \\ l_1 & l_2 \end{matrix} \middle| \begin{matrix} \vec{\omega}_b \\ l \end{matrix} \right\rangle = \delta_{\vec{\omega}_a \vec{\omega}_b} \qquad (5.212)$$

Alternatively, a relation expressing the orthonormality in l_1 and l_2 can be derived:

$$\sum_{\vec{\omega}} \left\langle \begin{matrix} \vec{\omega}_1 & \vec{\omega}_2 \\ l_{1a} & l_{2a} \end{matrix} \middle| \begin{matrix} \vec{\omega} \\ l \end{matrix} \right\rangle \left\langle \begin{matrix} \vec{\omega}_1 & \vec{\omega}_2 \\ l_{1b} & l_{2b} \end{matrix} \middle| \begin{matrix} \vec{\omega} \\ l \end{matrix} \right\rangle = \delta_{l_{1a} l_{1b}} \delta_{l_{2a} l_{2b}} \qquad (5.213)$$

Other useful properties are concerned with symmetries of the $SO(4) \supset SO(3)$ coupling coefficients under the exchange of its arguments:

$$\left\langle \begin{matrix} \vec{\omega}_2 & \vec{\omega}_1 \\ l_2 & l_1 \end{matrix} \middle| \begin{matrix} \vec{\omega} \\ l \end{matrix} \right\rangle = (-)^{\omega_1 + \omega_2 + \omega + l_1 + l_2 + l} \left\langle \begin{matrix} \vec{\omega}_1 & \vec{\omega}_2 \\ l_1 & l_2 \end{matrix} \middle| \begin{matrix} \vec{\omega} \\ l \end{matrix} \right\rangle \qquad (5.214)$$

and

$$\left\langle \begin{matrix} \vec{\omega}_2 & \vec{\omega} \\ l_2 & l \end{matrix} \middle| \begin{matrix} \vec{\omega}_1 \\ l_1 \end{matrix} \right\rangle = \left[\frac{\dim(\vec{\omega}_1) \dim(l)}{\dim(\vec{\omega}) \dim(l_1)} \right]^{1/2} \left\langle \begin{matrix} \vec{\omega}_1 & \vec{\omega}_2 \\ l_1 & l_2 \end{matrix} \middle| \begin{matrix} \vec{\omega} \\ l \end{matrix} \right\rangle \qquad (5.215)$$

where dim denotes the dimension of a representation. For an $SO(3)$ representation l we have $\dim(l) = 2l+1$ while $SO(4)$ representations have $\dim(\vec{\omega}) = (\omega + \omega' + 1)(\omega - \omega' + 1)$.

Finally, the transformation (5.194) (i.e., from (a) to (c) in (5.205)) can be found by combining (5.206) and (5.210), yielding an expression for the full $SO(4) \supset SO(3) \supset SO(2)$ coupling coefficient:

$$\left\langle \begin{matrix} \vec{\omega}_1 & \vec{\omega}_2 \\ l_1 m_1 & l_2 m_2 \end{matrix} \middle| \begin{matrix} \vec{\omega} \\ lm \end{matrix} \right\rangle = \left\langle \begin{matrix} \vec{\omega}_1 & \vec{\omega}_2 \\ l_1 & l_2 \end{matrix} \middle| \begin{matrix} \vec{\omega} \\ l \end{matrix} \right\rangle \langle l_1 m_1 \, l_2 m_2 | lm \rangle \qquad (5.216)$$

This coefficient can thus be written as the product of two factors, the first associated with $SO(4) \supset SO(3)$ and the second with $SO(3) \supset$

5.8 Tensor calculus

SO(2) (the usual Clebsch–Gordan coefficient). The coefficient with the SO(3) ⊃ SO(2) part factored out is sometimes referred to as an *isoscalar factor*; in this case, $\langle \vec{\omega}_1 l_1 \otimes \vec{\omega}_2 l_2 | \vec{\omega} l \rangle$ is the SO(4) ⊃ SO(3) isoscalar factor. The relation (5.216) provides an example of *Racah's factorization lemma* [22]. This lemma states that for a general set (5.188) of nested algebras the coupling coefficient can be written in terms of coefficients associated with the reductions $G_i \supset G_{i+1}$. If, in addition, all algebras in the chain are simply reducible, the full coupling coefficient is just the product of the $G_i \supset G_{i+1}$ coefficients, as in (5.216).

We have now completely determined the tensor calculus associated with SO(4) ⊃ SO(3) ⊃ SO(2): we know how to construct composite tensors from simple ones and can make use of the Wigner–Eckart theorem to calculate matrix elements of tensors in this basis. These computations can be carried out analytically since the SO(4) ⊃ SO(3) ⊃ SO(2) coupling coefficients are known in closed form.

We now turn our attention to the considerably more difficult problem of extending this tensor calculus to include the algebra U(4). The relevant chain of algebras then becomes

$$\begin{array}{cccc} \mathrm{U}(4) \supset & \mathrm{SO}(4) \supset & \mathrm{SO}(3) \supset & \mathrm{SO}(2) \\ | & | & | & | \\ [\vec{h}] & \gamma\vec{\omega} & l & m \end{array} \quad (5.217)$$

In general, four indices are needed to label a U(4) representation (see Appendix B), and we denote them with a single character as $[\vec{h}] \equiv [h, h', h'', h''']$. The branching rule from U(4) to SO(4) now becomes more problematic: whereas a given *symmetric* U(4) representation (i.e., $h' = h'' = h''' = 0$) never contains an SO(4) representation more than once, this is no longer the case for general, non-symmetric U(4) representations. As a consequence we must introduce extra labels to completely characterize the SO(4) representations contained in U(4). These are sometimes referred to as *missing labels* and are collectively denoted in (5.217) as γ. Because they are not associated with a particular algebra but rather "live" somewhere in between two algebras (here U(4) and SO(4)), these missing labels complicate the application of group-theoretical techniques.

The basis states (5.217) are fully characterized as $|[\vec{h}]\gamma\vec{\omega}lm\rangle$, and under the action of an operator $\hat{U}$ belonging to U(4), they transform as

$$\hat{U}|[\vec{h}]\gamma\vec{\omega}lm\rangle = \sum_{\gamma'\vec{\omega}'l'm'} \langle[\vec{h}]\gamma'\vec{\omega}'l'm'|\hat{U}|[\vec{h}]\gamma\vec{\omega}lm\rangle|[\vec{h}]\gamma'\vec{\omega}'l'm'\rangle \quad (5.218)$$

We can also define tensor operators with definite transformation properties under (5.217),

$$\hat{U}\hat{T}^{[\vec{h}]}_{\gamma\vec{\omega}lm}\hat{U}^{-1} = \sum_{\gamma'\vec{\omega}'l'm'} \langle[\vec{h}]\gamma'\vec{\omega}'l'm'|\hat{U}|[\vec{h}]\gamma\vec{\omega}lm\rangle \hat{T}^{[\vec{h}]}_{\gamma'\vec{\omega}'l'm'} \quad (5.219)$$

a relation which, for infinitesimal rotations $\hat{U} = 1 + \epsilon^i \hat{X}_i$, reduces to

$$[\hat{X}_i, \hat{T}^{[\vec{h}]}_{\gamma\vec{\omega}lm}] = \sum_{\gamma'\vec{\omega}'l'm'} \langle[\vec{h}]\gamma'\vec{\omega}'l'm'|\hat{X}_i|[\vec{h}]\gamma\vec{\omega}lm\rangle \hat{T}^{[\vec{h}]}_{\gamma'\vec{\omega}'l'm'} \quad (5.220)$$

Having already encountered the problem of the missing labels in the U(4) ⊃ SO(4) branching rule, there is another complication that arises because U(4) is not simply reducible. The general formula for the coupling of two U(4) ⊃ SO(4) ⊃ SO(3) ⊃ SO(2) states or tensors reads

$$A^{\lambda[\vec{h}]}_{\gamma\vec{\omega}lm} = \sum_{\substack{\gamma_1\vec{\omega}_1l_1m_1 \\ \gamma_2\vec{\omega}_2l_2m_2}} \left\langle \begin{array}{cc} [\vec{h}_1] & [\vec{h}_2] \\ \gamma_1\vec{\omega}_1l_1m_1 & \gamma_2\vec{\omega}_2l_2m_2 \end{array} \middle| \begin{array}{c} \lambda[\vec{h}] \\ \gamma\vec{\omega}lm \end{array} \right\rangle$$
$$\times A^{[\vec{h}_1]}_{\gamma_1\vec{\omega}_1l_1m_1} A^{[\vec{h}_2]}_{\gamma_2\vec{\omega}_2l_2m_2} \quad (5.221)$$

Because, in general, the product $[\vec{h}_1]\otimes[\vec{h}_2]$ may contain the representation $[\vec{h}]$ several times, an additional label λ appears in this equation; λ is thus a *multiplicity label* and should not be confused with the missing labels (here γ_1, γ_2, and γ).

As a result of the occurrence of missing as well as multiplicity labels, the explicit calculation of the U(4) ⊃ SO(4) ⊃ SO(3) ⊃ SO(2) coupling coefficients becomes a formidable task which, in fact, generally cannot be carried out analytically. In spite of these additional complications, many of the basic properties of coupling coefficients remain valid for $\langle[\vec{h}_1]\gamma_1\vec{\omega}_1l_1m_1 \otimes [\vec{h}_2]\gamma_2\vec{\omega}_2l_2m_2|\lambda[\vec{h}]\gamma\vec{\omega}lm\rangle$. For example, they still satisfy

5.8 Tensor calculus

orthonormality relations of the type

$$\sum_{\substack{\gamma_1\vec{\omega}_1l_1m_1 \\ \gamma_2\vec{\omega}_2l_2m_2}} \left\langle \begin{array}{cc|c} [\vec{h}_1] & [\vec{h}_2] & \lambda_a[\vec{h}_a] \\ \gamma_1\vec{\omega}_1l_1m_1 & \gamma_2\vec{\omega}_2l_2m_2 & \gamma_a\vec{\omega}_al_am_a \end{array} \right\rangle^*$$

$$\times \left\langle \begin{array}{cc|c} [\vec{h}_1] & [\vec{h}_2] & \lambda_b[\vec{h}_b] \\ \gamma_1\vec{\omega}_1l_1m_1 & \gamma_2\vec{\omega}_2l_2m_2 & \gamma_b\vec{\omega}_bl_bm_b \end{array} \right\rangle$$

$$= \delta_{\lambda_a\lambda_b}\delta_{\vec{h}_a\vec{h}_b}\delta_{\gamma_a\gamma_b}\delta_{\vec{\omega}_a\vec{\omega}_b}\delta_{l_al_b}\delta_{m_am_b}$$

$$\sum_{\lambda\vec{h}\gamma\vec{\omega}lm} \left\langle \begin{array}{cc|c} [\vec{h}_1] & [\vec{h}_2] & \lambda[\vec{h}] \\ \gamma_{1a}\vec{\omega}_{1a}l_{1a}m_{1a} & \gamma_{2a}\vec{\omega}_{2a}l_{2a}m_{2a} & \gamma\vec{\omega}lm \end{array} \right\rangle^* \quad (5.222)$$

$$\times \left\langle \begin{array}{cc|c} [\vec{h}_1] & [\vec{h}_2] & \lambda[\vec{h}] \\ \gamma_{1b}\vec{\omega}_{1b}l_{1b}m_{1b} & \gamma_{2b}\vec{\omega}_{2b}l_{2b}m_{2b} & \gamma\vec{\omega}lm \end{array} \right\rangle$$

$$= \delta_{\gamma_{1a}\gamma_{1b}}\delta_{\gamma_{2a}\gamma_{2b}}\delta_{\vec{\omega}_{1a}\vec{\omega}_{1b}}\delta_{\vec{\omega}_{2a}\vec{\omega}_{2b}}$$

$$\times \delta_{l_{1a}l_{1b}}\delta_{l_{2a}l_{2b}}\delta_{m_{1a}m_{1b}}\delta_{m_{2a}m_{2b}}$$

Also, in many cases of practical interest, missing and/or multiplicity labels are not needed and the expressions for the coupling coefficients simplify accordingly. If, for instance, none of the U(4) to SO(4) branching rules requires a missing label γ, the Wigner–Eckart theorem acquires the following form:

$$\langle [\vec{h}_a]\vec{\omega}_al_am_a|\hat{T}^{[\vec{h}]}_{\vec{\omega}lm}|[\vec{h}_b]\vec{\omega}_bl_bm_b\rangle$$
$$= \sum_{\lambda_a} \left\langle \begin{array}{cc|c} [\vec{h}_b] & [\vec{h}] & \lambda_a[\vec{h}_a] \\ \vec{\omega}_bl_bm_b & \vec{\omega}lm & \vec{\omega}_al_am_a \end{array} \right\rangle \langle \lambda_a[\vec{h}_a] \| \hat{T}^{[\vec{h}]} \| [\vec{h}_b]\rangle \quad (5.223)$$

The action of the operator $\hat{T}^{[\vec{h}]}_{\vec{\omega}lm}$ on a $[\vec{h}_b]$ state is equivalent to taking the product $[\vec{h}_b] \otimes [\vec{h}]$, and hence its result cannot be characterized by U(4) labels $[\vec{h}_a]$ only but needs an additional multiplicity label λ_a. In consequence, several U(4)-reduced matrix elements appear which must be determined separately. A further simplification occurs if $[\vec{h}_b] \otimes [\vec{h}]$ is simply reducible, in which case λ_a becomes a superfluous label and the sum in (5.223) reduces to one term. In addition, Racah's factorization lemma can be applied in its simplest form, and the coupling coefficient reduces to a product of two isoscalar factors (associated with U(4) ⊃

SO(4) and SO(4) $\supset$ SO(3)) and a Clebsch–Gordan coefficient:

$$\left\langle \begin{array}{cc} [\vec{h}_1] & [\vec{h}_2] \\ \vec{\omega}_1 l_1 m_1 & \vec{\omega}_2 l_2 m_2 \end{array} \middle| \begin{array}{c} [\vec{h}] \\ \vec{\omega} l m \end{array} \right\rangle$$
$$= \left\langle \begin{array}{cc} [\vec{h}_1] & [\vec{h}_2] \\ \vec{\omega}_1 & \vec{\omega}_2 \end{array} \middle| \begin{array}{c} [\vec{h}] \\ \vec{\omega} \end{array} \right\rangle \left\langle \begin{array}{cc} \vec{\omega}_1 & \vec{\omega}_2 \\ l_1 & l_2 \end{array} \middle| \begin{array}{c} \vec{\omega} \\ l \end{array} \right\rangle \langle l_1 m_1 \, l_2 m_2 | l m \rangle \quad (5.224)$$

From the orthonormality properties of the Clebsch–Gordan coefficients and of the SO(4) $\supset$ SO(3) isoscalar factors we may derive the corresponding properties for the U(4) $\supset$ SO(4) isoscalar factors which, in the absence of missing and multiplicity labels, reduce to

$$\sum_{\vec{\omega}_1 \vec{\omega}_2} \left\langle \begin{array}{cc} [\vec{h}_1] & [\vec{h}_2] \\ \vec{\omega}_1 & \vec{\omega}_2 \end{array} \middle| \begin{array}{c} [\vec{h}_a] \\ \vec{\omega}_a \end{array} \right\rangle \left\langle \begin{array}{cc} [\vec{h}_1] & [\vec{h}_2] \\ \vec{\omega}_1 & \vec{\omega}_2 \end{array} \middle| \begin{array}{c} [\vec{h}_b] \\ \vec{\omega}_b \end{array} \right\rangle = \delta_{\vec{h}_a \vec{h}_b} \delta_{\vec{\omega}_a \vec{\omega}_b}$$

$$\sum_{\vec{h}\vec{\omega}} \left\langle \begin{array}{cc} [\vec{h}_1] & [\vec{h}_2] \\ \vec{\omega}_{1a} & \vec{\omega}_{2a} \end{array} \middle| \begin{array}{c} [\vec{h}] \\ \vec{\omega} \end{array} \right\rangle \left\langle \begin{array}{cc} [\vec{h}_1] & [\vec{h}_2] \\ \vec{\omega}_{1b} & \vec{\omega}_{2b} \end{array} \middle| \begin{array}{c} [\vec{h}] \\ \vec{\omega} \end{array} \right\rangle = \delta_{\vec{\omega}_{1a} \vec{\omega}_{1b}} \delta_{\vec{\omega}_{2a} \vec{\omega}_{2b}}$$

(5.225)

To conclude this section, we illustrate with some examples of how the U(4) $\supset$ SO(4) $\supset$ SO(3) $\supset$ SO(2) tensor calculus can be used in the calculation of matrix elements, rederiving some of the results of Section 5.5 from a different perspective. As in Chapter 1 the first task at hand is to determine the tensor character of the various generators of U(4). The elementary building blocks of the U(4) boson model are the creation operators $p_m^\dagger$ and $s^\dagger$, which transform as $\hat{T}^{[1]}_{(1,0)1m}$ and $\hat{T}^{[1]}_{(1,0)00}$ tensors, respectively. (We follow the convention not to show U(4) labels that are zero; thus [1] denotes the U(4) representation $[1,0,0,0]$.) The modified annihilation operators $\tilde{p}_m$ and $\tilde{s}$ transform as $[1^3] \equiv [1,1,1,0]$, which is the representation *conjugate* to [1] (see Section 3.4); their tensor characters are $\hat{T}^{[1^3]}_{(1,0)1m}$ and $\hat{T}^{[1^3]}_{(1,0)00}$, respectively. The definition of elementary tensors is thus as follows:

$$\begin{aligned} \hat{T}^{[1]}_{(1,0)00} &= s^\dagger, & \hat{T}^{[1]}_{(1,0)1m} &= p_m^\dagger \\ \hat{T}^{[1^3]}_{(1,0)00} &= \tilde{s}, & \hat{T}^{[1^3]}_{(1,0)1m} &= \tilde{p}_m \end{aligned} \quad (5.226)$$

It will become clear below that this definition of elementary tensors is appropriate for U(4) $\supset$ $\overline{\text{SO}}(4)$ $\supset$ SO(3) $\supset$ SO(2). Application of the

5.8 Tensor calculus

canonical transformation (5.24) gives the corresponding definition for U(4) ⊃ SO(4) ⊃ SO(3) ⊃ SO(2) tensors,

$$\hat{T}^{[1]}_{(1,0)00} = s^\dagger, \quad \hat{T}^{[1]}_{(1,0)1m} = -i p^\dagger_m$$
$$\hat{T}^{[1^3]}_{(1,0)00} = \tilde{s}, \quad \hat{T}^{[1^3]}_{(1,0)1m} = i \tilde{p}_m \tag{5.227}$$

The generators of U(4) are bilinear products of creation times annihilation operators, and their properties under U(4) transformations are obtained by considering the product $[1] \otimes [1^3]$. From the Young tableau rules for U(4) (see Appendix B) we find

$$\square \otimes \begin{array}{c}\square\\\square\\\square\end{array} = \begin{array}{c}\square\\\square\\\square\\\square\end{array} \oplus \begin{array}{c}\square\square\\\square\\\square\end{array} \tag{5.228}$$

or

$$[1] \otimes [1^3] = [1^4] \oplus [2, 1^2]$$
$$4 \times 4 = 1 + 15 \tag{5.229}$$

where below each representation we indicate its dimensionality. The product of SO(4) representations can be worked out using the general rule (5.199):

$$(1,0) \otimes (1,0) = (2,0) \oplus (1,1) \oplus (1,-1) \oplus (0,0)$$
$$4 \times 4 = 9 + 3 + 3 + 1 \tag{5.230}$$

From dimensional considerations we conclude that $[1^4]$ contains $(0,0)$ while $[2, 1^2]$ contains all other SO(4) representations (i.e., $(2,0)$, $(1,1)$, and $(1,-1)$). The branching rule from SO(4) to SO(3) is known from (5.190) and leads to the enumeration of all possible tensors that can be formed from bilinear products $b^\dagger_{lm} \tilde{b}_{l'm'}$ (see Table 5.5).

Since there exists only one generator with $\lambda = 2$, we immediately deduce that the quadrupole generator $\hat{Q}_\mu$ transforms as a $\hat{T}^{[2,1^2]}_{(2,0)2\mu}$ tensor. Two (three) generators exist with multipolarity $\lambda = 0$ (1), and as a consequence, their tensor character cannot be deduced from a simple

Table 5.5: Tensor character of U(4) generators under U(4) ⊃ SO(4) ⊃ SO(3) ⊃ SO(2)

Tensor character				Operator
U(4) $[h, h', h'', h''']$	SO(4) (ω, ω')	SO(3) λ	SO(2) μ	
$[1,1,1,1]$	$(0,0)$	0	0	$\frac{1}{2}(\hat{n}_p + \hat{n}_s)$
$[2,1,1,0]$	$(1,1)$	1	μ	$\frac{1}{2}(\hat{L}_\mu - \hat{D}_\mu)$
	$(1,-1)$	1	μ	$\frac{1}{2}(\hat{L}_\mu + \hat{D}_\mu)$
	$(2,0)$	0	0	$\frac{1}{2}\left(-\sqrt{\frac{1}{3}}\hat{n}_p + \sqrt{3}\hat{n}_s\right)$
		1	μ	$-i\sqrt{\frac{1}{2}}\hat{D}'_\mu$
		2	μ	$\hat{Q}_\mu$

enumeration argument. Instead, they should be constructed explicitly according to

$$\hat{T}^{[\vec{h}]}_{(\omega,\omega')\lambda\mu} = \sum_{\substack{lm \\ l'm'}} \left\langle \begin{matrix} [1] \\ (1,0)lm \end{matrix} \begin{matrix} [1^3] \\ (1,0)l'm' \end{matrix} \middle| \begin{matrix} [\vec{h}] \\ (\omega,\omega')\lambda\mu \end{matrix} \right\rangle \hat{T}^{[1]}_{(1,0)lm} \hat{T}^{[1^3]}_{(1,0)l'm'}$$

(5.231)

where use is made of (5.221). Because no multiplicity or missing labels are required in the coupling coefficient, it can be factorized into two isoscalar factors and a Clebsch–Gordan coefficient:

$$\left\langle \begin{matrix} [1] \\ (1,0)lm \end{matrix} \begin{matrix} [1^3] \\ (1,0)l'm' \end{matrix} \middle| \begin{matrix} [\vec{h}] \\ (\omega,\omega')\lambda\mu \end{matrix} \right\rangle$$
$$= \left\langle \begin{matrix} [1] \\ (1,0) \end{matrix} \begin{matrix} [1^3] \\ (1,0) \end{matrix} \middle| \begin{matrix} [\vec{h}] \\ (\omega,\omega') \end{matrix} \right\rangle \left\langle \begin{matrix} (1,0) \\ l \end{matrix} \begin{matrix} (1,0) \\ l' \end{matrix} \middle| \begin{matrix} (\omega,\omega') \\ \lambda \end{matrix} \right\rangle \langle lm\, l'm'|\lambda\mu\rangle$$

(5.232)

Furthermore, the first of the orthonormality relations (5.225) implies that the modulus of the U(4) ⊃ SO(4) isoscalar factor is equal to 1 and thus we find

$$\hat{T}^{[\vec{h}]}_{(\omega,\omega')\lambda\mu} = \sum_{ll'} \left\langle \begin{matrix} (1,0) \\ l \end{matrix} \begin{matrix} (1,0) \\ l' \end{matrix} \middle| \begin{matrix} (\omega,\omega') \\ \lambda \end{matrix} \right\rangle (-i)^l (i)^{l'} \mathcal{B}^{(\lambda)}_\mu(l,l')$$
$$= \sum_{ll'} \sqrt{(\omega+\omega'+1)(\omega-\omega'+1)(2l+1)(2l'+1)}$$

5.8 Tensor calculus

$$\times \begin{Bmatrix} 1/2 & 1/2 & (\omega+\omega')/2 \\ 1/2 & 1/2 & (\omega-\omega')/2 \\ l & l' & \lambda \end{Bmatrix} (-i)^{l}(i)^{l'} \mathcal{B}_{\mu}^{(\lambda)}(l,l') \quad (5.233)$$

Inserting the appropriate values for the $9j$ coefficients (references to numerical tables are given in [3]), we find the different tensors given in Table 5.5 in terms of the generators $\hat{n}_{\mathrm{p}}$, $\hat{n}_{\mathrm{s}}$, $\hat{L}_\mu$, $\hat{D}_\mu$, $\hat{D}'_\mu$, and $\hat{Q}_\mu$ defined in Section 5.2. We note that the $\overline{\mathrm{SO}}(4)$ generator $\hat{D}'_\mu$ is a (2,0) tensor under SO(4) and hence obeys the selection rule $\Delta\omega = 0, \pm 2$, in agreement with the results of Table 5.4. This is a consequence of the definition (5.227) of elementary tensors. Had we followed the definition (5.226), appropriate for $\overline{\mathrm{SO}}(4)$, we would have found instead that $\hat{D}_\mu$ is the (2,0) tensor with non-zero $\Delta\omega = \pm 2$ matrix elements. This argument may be viewed as a justification of the $\overline{\mathrm{SO}}(4)$ and SO(4) associations proposed in (5.226) and (5.227).

Since we have determined the tensor character of all U(4) generators, we can, in principle, derive their matrix elements between SO(4) states using the Wigner–Eckart theorem. The application of (5.223) is complicated because (i) we need to determine the various reduced matrix elements and (ii) we have no general expression available for the U(4) $\supset$ SO(4) $\supset$ SO(3) $\supset$ SO(2) coupling coefficients. If we assume no multiplicity occurs in the U(4) products (which is always the case if bra and ket of the matrix element transform symmetrically under U(4)), we can, using the Wigner–Eckart theorem, readily establish many relations between matrix elements of generators, and hence easily deduce one from another. We illustrate this with two examples.

First, we can relate matrix elements of the same generator. For example, the ratio of quadrupole SO(3)-reduced matrix elements

$$\frac{\langle [N](\omega+2,0)L \| \hat{Q} \| [N](\omega,0)L \rangle}{\langle [N](\omega+2,0)L+2 \| \hat{Q} \| [N](\omega,0)L \rangle} \quad (5.234)$$

is equal to the following expression involving SO(4) $\supset$ SO(3) isoscalar factors:

$$\frac{\left\langle \begin{matrix} (\omega,0) & (2,0) \\ L & 2 \end{matrix} \middle| \begin{matrix} (\omega+2,0) \\ L \end{matrix} \right\rangle}{\left\langle \begin{matrix} (\omega,0) & (2,0) \\ L & 2 \end{matrix} \middle| \begin{matrix} (\omega+2,0) \\ L+2 \end{matrix} \right\rangle} \quad (5.235)$$

Other factors do not arise in (5.235): the U(4)-reduced matrix elements and U(4) ⊃ SO(4) isoscalar factors in numerator and denominator cancel and no Clebsch–Gordan coefficient occurs since we take the ratio of two SO(3)-reduced matrix elements. The SO(4) ⊃ SO(3) isoscalar factors are related to the $9j$ coefficients

$$\begin{Bmatrix} \omega/2 & 1 & (\omega+2)/2 \\ \omega/2 & 1 & (\omega+2)/2 \\ L & 2 & L \end{Bmatrix}$$
$$= -\frac{1}{(\omega+1)_3} \left[\frac{(\omega-L+1)_2(\omega+L+2)_2 L(L+1)}{30(2L-1)(2L+1)(2L+3)} \right]^{1/2} \quad (5.236)$$

and

$$\begin{Bmatrix} \omega/2 & 1 & (\omega+2)/2 \\ \omega/2 & 1 & (\omega+2)/2 \\ L & 2 & L+2 \end{Bmatrix} = \frac{1}{(\omega+1)_3} \left[\frac{(\omega+L+2)_4(L+1)(L+2)}{20(2L+1)(2L+3)(2L+5)} \right]^{1/2}$$
$$(5.237)$$

where $(a)_s$ denotes a Pochhammer symbol [10]. Taking the ratio of the two $9j$ coefficients, we find

$$\frac{\langle [N](\omega+2,0)L \| \hat{Q} \| [N](\omega,0)L \rangle}{\langle [N](\omega+2,0)L+2 \| \hat{Q} \| [N](\omega,0)L \rangle}$$
$$= -\left[\frac{2(\omega-L+1)_2 L(2L+5)}{3(\omega+L+4)_2(L+2)(2L-1)} \right]^{1/2} \quad (5.238)$$

This agrees up to a sign with the results of Table 5.4. Differences in sign may occur since the phase convention followed in the construction of the SO(4) states (5.161) does not necessarily coincide with the one used for the U(4) ⊃ SO(4) ⊃ SO(3) ⊃ SO(2) coupling coefficients.

Second, we can use the same procedure to relate matrix elements of different operators, for instance, $\hat{n}_p$ and $\hat{D}'_\mu$. From Table 5.5 we have

$$\hat{T}^{[2,1^2]}_{(2,0)00} = \frac{1}{2}\left(\sqrt{3}(\hat{N}-\hat{n}_p) - \sqrt{\tfrac{1}{3}}\hat{n}_p\right) = \sqrt{\tfrac{3}{4}}\hat{N} - \sqrt{\tfrac{4}{3}}\hat{n}_p \quad (5.239)$$

Since the number operator $\hat{N}$ does not contribute to off-diagonal matrix elements, we find that the ratio

$$\frac{\langle [N](\omega+2,0)L \| -2\hat{n}_p/\sqrt{3} \| [N](\omega,0)L \rangle}{\langle [N](\omega+2,0)L+1 \| -i\hat{D}'/\sqrt{2} \| [N](\omega,0)L \rangle} \quad (5.240)$$

5.8 Tensor calculus

of off-diagonal matrix elements is given by

$$\frac{\left\langle \begin{matrix} (\omega,0) & (2,0) \\ L & 0 \end{matrix} \middle| \begin{matrix} (\omega+2,0) \\ L \end{matrix} \right\rangle}{\left\langle \begin{matrix} (\omega,0) & (2,0) \\ L & 1 \end{matrix} \middle| \begin{matrix} (\omega+2,0) \\ L+1 \end{matrix} \right\rangle} \tag{5.241}$$

Using the expressions for the $9j$ coefficients

$$\left\{ \begin{matrix} \omega/2 & 1 & (\omega+2)/2 \\ \omega/2 & 1 & (\omega+2)/2 \\ L & 0 & L \end{matrix} \right\} = \frac{1}{(\omega+1)_3} \left[\frac{(\omega-L+1)_2(\omega+L+2)_2}{3(2L+1)} \right]^{1/2} \tag{5.242}$$

and

$$\left\{ \begin{matrix} \omega/2 & 1 & (\omega+2)/2 \\ \omega/2 & 1 & (\omega+2)/2 \\ L & 1 & L+1 \end{matrix} \right\}$$

$$= \frac{1}{(\omega+1)_3} \left[\frac{(\omega-L+1)(\omega+L+2)_3(L+1)}{6(2L+1)(2L+3)} \right]^{1/2} \tag{5.243}$$

we find the result

$$\frac{\langle [N](\omega+2,0)L \,\|\, \hat{n}_{\mathrm{p}} \,\|\, [N](\omega,0)L \rangle}{\langle [N](\omega+2,0)L+1 \,\|\, i\hat{D}' \,\|\, [N](\omega,0)L \rangle} = \left[\frac{(\omega-L+2)(2L+3)}{4(\omega+L+4)(L+1)} \right]^{1/2} \tag{5.244}$$

which—again up to a phase—agrees with Table 5.4.

We have shown in this section how many of the results, derived in Section 5.5 using the explicit boson representation of SO(4) states, can be obtained in a concise manner by exploiting the algebraic structure of the vibron model. Note, however, that not all matrix elements of U(4) generators can be derived algebraically: at least one of them is needed to determine the U(4)-reduced matrix element. In addition, the calculation is complicated because of our partial knowledge of the relevant coupling coefficients, with general expressions available for the SO(4) ⊃ SO(3) ⊃ SO(2) part only and not for the U(4) ⊃ SO(4) isoscalar factors. Nevertheless, these isoscalar factors have been obtained in closed form for many simple representations of interest [23]

Table 5.6: SO(4)-reduced matrix elements of U(4) generators

$$\langle [N]\omega \| \hat{T}^{(0,0)} \| [N]\omega \rangle = \tfrac{1}{2}N$$

$$\langle [N]\omega \| \hat{T}^{(1,1)} \| [N]\omega \rangle = \sqrt{\omega(\omega+2)}$$

$$\langle [N]\omega \| \hat{T}^{(1,-1)} \| [N]\omega \rangle = \sqrt{\omega(\omega+2)}$$

$$\langle [N]\omega \| \hat{T}^{(2,0)} \| [N]\omega \rangle = \tfrac{1}{2}(N+2)$$

$$\langle [N]\omega - 2 \| \hat{T}^{(2,0)} \| [N]\omega \rangle = -\left[\frac{(N-\omega+2)(N+\omega+2)(\omega+1)}{4(\omega-1)} \right]^{1/2}$$

$$\langle [N]\omega + 2 \| \hat{T}^{(2,0)} \| [N]\omega \rangle = -\left[\frac{(N-\omega)(N+\omega+4)(\omega+3)}{4(\omega+1)} \right]^{1/2}$$

which cover all cases listed in Table 5.4. We thus conclude that all matrix elements in Table 5.4 can be derived from a single one, which represents a considerable simplification over the method employed in Section 5.5.

A somewhat less ambitious scheme is to express all matrix elements in terms of several SO(4)-reduced matrix elements. This is readily possible by combining the results of Table 5.4 with the explicit expression (5.211) for the SO(4) isoscalar factor. In this way we can rewrite (see Table 5.6) all matrix elements in terms of a few SO(4)-reduced matrix elements of the tensors $\hat{T}^{(\omega,\omega')}_{\lambda\mu}$ defined in Table 5.5. We will demonstrate in the next two chapters that these results, obtained through the development of an SO(4) tensor calculus, are indispensable in the practical application of the vibron model to more complex molecular systems.

Summary

In this chapter the algebraic structure of the vibron model of molecular rotation–vibration spectra is reviewed. The basic building blocks of this model are p and s bosons resulting in a U(4) algebra which can be studied in close analogy with the U(2) case expounded in Chapter 1. The U(4) algebra arises through the boson commutators (5.1) and the definition of the generators (5.3) and their commutations relations (5.4). Apart from dimensionality these are all identical to the corresponding

Summary

expressions in U(2). An important additional ingredient in the U(4) case is the appearance of an angular momentum SO(3) subalgebra; this allows the definition of the U(4) generators (5.8) which have well-defined transformation properties under rotations. The most general vibron hamiltonian which includes up to two-body interactions acquires a considerably simplified form by imposing rotational invariance and is given in (5.26). The alternative form (5.50) of this hamiltonian involves Casimir invariants of the subalgebras of U(4) and is more useful in the study of its analytical solutions. Two such solutions exist, the U(3) limit and the SO(4) limit, which can be analyzed with techniques that are generalizations of those employed in Chapter 1. The U(3) limit is obtained with the hamiltonian (5.51) and corresponds to the classification (5.52). Its solutions can be found, first in terms of spherical polar coordinates, (5.73) and then by making use of Dragt's theorem, as polynomial functions of the boson operators, (5.87). In turn, the SO(4) limit is obtained with the hamiltonian (5.115) and corresponds to the classification (5.116). The SO(4) algebra is shown to be isomorphic to $SU(2) \otimes SU(2)$, (5.121), a result which is useful in the analysis of SO(4)-type symmetries throughout Part II of this book. Again, the SO(4) states are first obtained in a coordinate representation (5.149) and are then converted into the boson-operator polynomials (5.161). Section 5.7 deals with applications of the vibron model which show that the SO(4) limit gives a reasonable first-order description of rigid rotational diatomic molecules with scope for improvement by considering higher-order interactions and/or SO(4)-breaking terms. In the final section the SO(4) limit is analyzed from a different perspective using $SO(4) \supset SO(3) \supset SO(2)$ tensor calculus. By defining the infinitesimal operators inducing a rotation in four-dimensional space, it is possible to define tensor operators through the commutation relations (5.193). States or tensors can now be coupled using an expansion of the type (5.194). The coefficients in this expansion are generalized Clebsch–Gordan coefficients and satisfy the orthonormality relations (5.195). Another fundamental result of $SO(4) \supset SO(3) \supset SO(2)$ tensor calculus is the Wigner–Eckart theorem (5.196), which states that the dependence of matrix elements on the SO(3) and SO(2) quantum numbers is entirely contained in a generalized Clebsch–Gordan coefficient. A general expression (5.211) is derived for the latter. A tensor calculus can, in principle, be formulated for any chain of nested Lie algebras, as is illustrated in the latter part of Section 5.8 with $U(4) \supset SO(4) \supset SO(3) \supset SO(2)$. In this case, however, no closed, general expression can be obtained for the coupling coefficients.

References

1. F. Iachello, "Algebraic methods for molecular rotation–vibration spectra," Chem. Phys. Lett. **78** (1981) 581.
2. F. Iachello and R.D. Levine, "Algebraic approach to molecular rotation–vibration spectra. I. Diatomic molecules," J. Chem. Phys. **77** (1982) 3046.
3. A.R. Edmonds, *Angular Momentum in Quantum Mechanics*, Princeton University Press, Princeton, 1957.
4. G. Racah, "Sulla Caratterizzazione delle Rapresentazioni Irriducibili dei Gruppi Semisemplici di Lie," Lincei Rend. Sci. Fis. Mat. Nat. **8** (1950) 108.
5. B.G. Wybourne, *Classical Groups for Physicists*, Wiley-Interscience, New York, 1974.
6. G. Herzberg, *Molecular Spectra and Molecular Structure. I. Spectra of Diatomic Molecules*, van Nostrand, New York, 1950.
7. R.S. Berry, "A general phenomenology for small clusters, however floppy," in *Quantum Dynamics of Molecules*, edited by R.G. Woolley, Plenum, New York, 1980, p. 143.
8. M. Moshinsky, *The Harmonic Oscillator in Modern Physics: From Atoms to Quarks*, Gordon & Breach, New York, 1969.
9. M.E. Rose, *Elementary Theory of Angular Momentum*, Wiley, New York, 1957.
10. I.S. Gradshteyn and I.M. Ryzhik, *Table of Integrals, Series, and Products*, Academic, New York, 1965.
11. R. Gilmore, *Lie Groups, Lie Algebras, and Some of Their Applications*, Wiley-Interscience, New York, 1974.
12. A.K. Kerman, "Pairing forces and nuclear collective motion," Ann. Phys. (NY) **12** (1961) 300.

References

13. J.L. Dunham, "The energy levels of a rotating vibrator," Phys. Rev. **41** (1932) 721.
14. A. Frank and R. Lemus, "The O(4) wave functions in the vibron model for diatomic molecules," J. Chem. Phys. **84** (1986) 2698; *ibid.* **85** (1986) 642.
15. O.S. van Roosmalen, *Algebraic descriptions of nuclear and molecular rotation–vibration spectra*, doctoral dissertation, Rijksuniversiteit Groningen, 1982.
16. G. Herzberg and L.L. Howe, "The Lyman bands of molecular hydrogen," Can. J. Phys. **37** (1959) 636.
17. T.G. Waech and R.B. Bernstein, "Calculated spectrum of quasibound states for $H_2(^1\Sigma_g^+)$ and resonances in H + H scattering," J. Chem. Phys. **46** (1967) 4905.
18. R. Herman and R.F. Wallis, "Influence of vibration–rotation interaction on line intensities in vibration–rotation bands of diatomic molecules," J. Chem. Phys. **23** (1955) 637.
19. F. Iachello and S. Oss, "Overtone frequencies and intensities of bent XY_2 molecules in the vibron model," J. Mol. Spectrosc. **142** (1990) 85.
20. F. Iachello, A. Leviatan, and A. Mengoni, "Algebraic approach to molecular rotation–vibration spectra. III. Infrared intensities," J. Chem. Phys. **95** (1991) 1449.
21. L.C. Biedenharn and J.D. Louck, *Angular Momentum in Quantum Physics*, Addison-Wesley, Reading, MA, 1981.
22. G. Racah, "Theory of complex spectra. I," Phys. Rev. **61** (1942) 186; **62** (1942) 438; **63** (1943) 367; **76** (1949) 1352.
23. K.T. Hecht and S.C. Pang, "On the Wigner supermultiplet scheme," J. Math. Phys. **19** (1969) 1571.

Chapter 6

Triatomic Molecules

6.1 Introduction

In the previous chapter we analyzed in detail the properties of the U(4) vibron model which proposes an algebraic description of (heteronuclear) diatomic molecules in terms of a set of interacting p and s bosons. We saw that this model is capable of generating molecular spectra with rigid as well as non-rigid rotational characteristics and that in both cases the associated wave functions can be determined in terms of the constituent bosons, enabling the prediction of additional molecular properties such as electromagnetic transition intensities.

For diatomic molecules the algebraic approach gives results which are comparable in quality to those obtained with a Dunham expansion in terms of rotation–vibration quantum numbers [1], with the added advantage that, in contrast to the Dunham approach, the vibron model produces not only energies of molecular states but also the associated wave functions.

Besides the Dunham approach another traditional method—the *potential approach*—exists in molecular spectroscopy, which consists of solving the Schrödinger equation associated with an interatomic potential, determined either phenomenologically or from more fundamental *ab initio* calculations. The potential approach is fairly easy to apply to diatomic molecules and gives results of the same—if not better— quality than those obtained with the U(4) model. However, for more complex molecules the potential approach leads to a system of coupled differential equations whose solution quickly becomes intractable as the number of atoms increases. In contrast, the vibron model, as we shall see ahead, can be readily applied to polyatomic molecules and the ensuing extension can be solved using powerful group-theoretical techniques. It must be clear, therefore, that, although the U(4) model might be considered as an interesting curiosity when applied to diatomic molecules, its main domain of potential applications lies in the treatment of polyatomic molecules for which no simple alternative is available.

It is the purpose of this chapter to show how the U(4) model can in principle be extended to molecules with many atoms. We first discuss in detail the treatment of triatomic molecules since that is the case which has been mostly discussed in papers by Iachello and collaborators [2–5]. We then indicate, in the last section, how a generalization to molecules with more than three atoms can be achieved in a similar framework.

The idea behind the extension of the vibron model beyond diatomic molecules is a simple one. In Section 5.1 we argued that because many features of diatomic molecules can be characterized in terms of the distance vector **r** between the nuclei, the U(4) algebra, realized with p and s bosons, is suitable for the description of such molecules. In triatomic molecules two distance vectors $\mathbf{r}_1$ and $\mathbf{r}_2$ are needed to specify the relative positions of the atoms (see Figure 6.1), and by analogy, we propose a description based on the algebra $U_1(4) \otimes U_2(4)$, the subscripts $\rho = 1, 2$ referring to the respective bonds in the molecule. Note also the similarity between this approach and the one followed in Section 2.7 where a $U_1(2) \otimes U_2(2)$ algebra is proposed for the description of stretching vibrations in X–Y–X molecules, each of the $U_\rho(2)$ algebras corresponding to an X–Y bond and its associated interatomic potential. The $U_1(4) \otimes U_2(4)$ model is similar in philosophy but has the advantage that it is not restricted to stretching modes but allows for a simultaneous treatment of bending modes and rotational degrees of freedom as well.

This procedure is easily extended from triatomic to polyatomic molecules: each interatomic bond is associated with a U(4) algebra, and we arrive at a description in terms of the product algebra $U_1(4) \otimes U_2(4) \otimes \cdots \otimes U_m(4)$, where m is the number of bonds in the molecule. For the

6.2 The $U_1(4) \otimes U_2(4)$ algebra

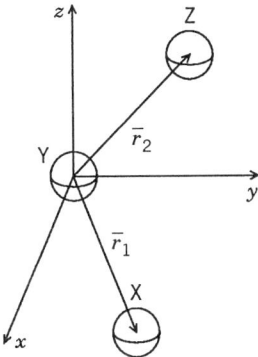

Figure 6.1: Schematic illustration of the geometric structure of a triatomic molecule X–Y–Z and the two distance vectors $\mathbf{r}_1$ and $\mathbf{r}_2$ to characterize the relative positions of the nuclei in the molecule.

moment we investigate the algebraic structure of $U_1(4) \otimes U_2(4)$, the topic of the following sections.

6.2 The $U_1(4) \otimes U_2(4)$ algebra

The $U_1(4) \otimes U_2(4)$ algebra is defined starting from the operators $p^\dagger_{\rho m}$ and $s^\dagger_\rho$ and their hermitian conjugates $p_{\rho m}$ and s_ρ, with $\rho = 1, 2$ referring to the two bonds of the triatomic molecule. These creation and annihilation operators satisfy the commutation relations

$$[b_{\rho i}, b^\dagger_{\rho' j}] = \delta_{\rho \rho'} \delta_{ij}, \qquad \rho, \rho' = 1, 2, \quad i, j = 1, 2, 3, 4 \qquad (6.1)$$

with all other commutators being zero and where the following notation is introduced:

$$\begin{aligned} b^\dagger_{\rho 1} &\equiv p^\dagger_{\rho -1}, & b^\dagger_{\rho 2} &\equiv p^\dagger_{\rho 0}, & b^\dagger_{\rho 3} &\equiv p^\dagger_{\rho +1}, & b^\dagger_{\rho 4} &\equiv s^\dagger_\rho \\ b_{\rho 1} &\equiv p_{\rho -1}, & b_{\rho 2} &\equiv p_{\rho 0}, & b_{\rho 3} &\equiv p_{\rho +1}, & b_{\rho 4} &\equiv s_\rho \end{aligned} \qquad (6.2)$$

for $\rho = 1, 2$. Defining the bilinear operators

$$\mathcal{G}^{\rho j}_{\rho i} \equiv b^\dagger_{\rho i} b_{\rho j}, \qquad i, j = 1, 2, 3, 4 \qquad (6.3)$$

we find them to satisfy the relations

$$[\mathcal{G}^{\rho j}_{\rho i}, \mathcal{G}^{\rho' l}_{\rho' k}] = (\mathcal{G}^{\rho l}_{\rho i}\delta_{jk} - \mathcal{G}^{\rho j}_{\rho k}\delta_{il})\delta_{\rho\rho'}, \qquad i,j,k,l = 1,2,3,4 \qquad (6.4)$$

This shows, as in Section 2.1, that the $\mathcal{G}^{\rho j}_{\rho i}$ generate a direct-product algebra, in this case $U_1(4) \otimes U_2(4)$.

The p_ρ and s_ρ bosons have the same rotation–reflection properties as the p and s bosons of the U(4) model. For the creation operators we have, for rotations $\hat{R}$ characterized by the Euler angles θ_i,

$$\hat{R}^{-1} p^\dagger_{\rho m} \hat{R} = \sum_{m'} D^1_{m'm}(\theta_1, \theta_2, \theta_3) p^\dagger_{\rho m'}, \qquad \hat{R}^{-1} s^\dagger_\rho \hat{R} = s^\dagger_\rho \qquad (6.5)$$

and, for reflections $\hat{P}$,

$$\hat{P}^{-1} p^\dagger_{\rho m} \hat{P} = -p^\dagger_{\rho m}, \qquad \hat{P}^{-1} s^\dagger_\rho \hat{P} = s^\dagger_\rho \qquad (6.6)$$

The modified annihilation operators

$$\tilde{p}_{\rho m} \equiv (-)^{1+m} p_{\rho -m}, \qquad \tilde{s}_\rho \equiv s_\rho \qquad (6.7)$$

satisfy the same rotation–reflection properties.

The (sub)structure of each of the $U_\rho(4)$ algebras is entirely analogous to that of the U(4) algebra, and hence much of the discussion can be taken from the previous chapter. In particular, we find the $U_\rho(3)$ subalgebras with generators

$$\hat{n}_{p_\rho} \equiv \sqrt{3}[p^\dagger_\rho \times \tilde{p}_\rho]^{(0)}_0, \quad \hat{L}_{\rho\mu} \equiv \sqrt{2}[p^\dagger_\rho \times \tilde{p}_\rho]^{(1)}_\mu, \quad \hat{Q}_{\rho\mu} \equiv [p^\dagger_\rho \times \tilde{p}_\rho]^{(2)}_\mu \quad (6.8)$$

the $SO_\rho(4)$ subalgebras consisting of

$$\hat{L}_{\rho\mu} \equiv \sqrt{2}[p^\dagger_\rho \times \tilde{p}_\rho]^{(1)}_\mu, \qquad \hat{D}_{\rho\mu} \equiv i[p^\dagger_\rho \times \tilde{s}_\rho + s^\dagger_\rho \times \tilde{p}_\rho]^{(1)}_\mu \qquad (6.9)$$

and the $\overline{SO}_\rho(4)$ subalgebras generated by

$$\hat{L}_{\rho\mu} \equiv \sqrt{2}[p^\dagger_\rho \times \tilde{p}_\rho]^{(1)}_\mu, \qquad \hat{D}'_{\rho\mu} \equiv [p^\dagger_\rho \times \tilde{s}_\rho - s^\dagger_\rho \times \tilde{p}_\rho]^{(1)}_\mu \qquad (6.10)$$

leading for $U_1(4)$ as well as $U_2(4)$ to the three possible chains of algebras discussed in Section 5.2. Although the classification of all possible subalgebra chains for the *separate* $U_\rho(4)$ algebras is straightforward, the corresponding problem for the *direct product* $U_1(4) \otimes U_2(4)$ leads to a bewildering number of different possibilities, as shown in the next section.

6.3 Symmetry limits

We begin this section with a brief discussion of the general form of the $U_1(4) \otimes U_2(4)$ hamiltonian. For a hermitian, rotation–reflection-invariant hamiltonian which includes up to two-body interactions we have

$$\hat{H} = \hat{H}_1 + \hat{H}_2 + \hat{V}_{12} \tag{6.11}$$

where $\hat{H}_1$ and $\hat{H}_2$ are the hamiltonians associated with bond-1 and bond-2 bosons (see (5.26) under the appropriate substitution of p and s by p_ρ and s_ρ bosons) and $\hat{V}_{12}$ describes the interaction between the two bonds,

$$\begin{aligned}
\hat{V}_{12} =\ & v_1[[p_1^\dagger \times p_2^\dagger]^{(0)} \times [\tilde{p}_1 \times \tilde{p}_2]^{(0)}]_0^{(0)} + v_2[[p_1^\dagger \times p_2^\dagger]^{(1)} \times [\tilde{p}_1 \times \tilde{p}_2]^{(1)}]_0^{(0)} \\
& + v_3[[p_1^\dagger \times p_2^\dagger]^{(2)} \times [\tilde{p}_1 \times \tilde{p}_2]^{(2)}]_0^{(0)} + v_4[[p_1^\dagger \times s_2^\dagger]^{(1)} \times [\tilde{p}_1 \times \tilde{s}_2]^{(1)}]_0^{(0)} \\
& + v_5[[p_1^\dagger \times p_2^\dagger]^{(0)} \times [\tilde{s}_1 \times \tilde{s}_2]^{(0)} + [s_1^\dagger \times s_2^\dagger]^{(0)} \times [\tilde{p}_1 \times \tilde{p}_2]^{(0)}]_0^{(0)} \\
& + v_6[[p_1^\dagger \times s_2^\dagger]^{(1)} \times [\tilde{s}_1 \times \tilde{p}_2]^{(1)} + [s_1^\dagger \times p_2^\dagger]^{(1)} \times [\tilde{p}_1 \times \tilde{s}_2]^{(1)}]_0^{(0)} \\
& + v_7[[s_1^\dagger \times p_2^\dagger]^{(1)} \times [\tilde{s}_1 \times \tilde{p}_2]^{(1)}]_0^{(0)} + v_8[[s_1^\dagger \times s_2^\dagger]^{(0)} \times [\tilde{s}_1 \times \tilde{s}_2]^{(0)}]_0^{(0)}
\end{aligned} \tag{6.12}$$

To simplify (6.11) we can now proceed as in Section 5.3 and use the property that the boson numbers N_1 and N_2 are conserved quantities of the model. Carrying out the replacements

$$\hat{n}_{s_\rho} = \hat{N}_\rho - \hat{n}_{p_\rho} \mapsto N_\rho - \hat{n}_{p_\rho} \tag{6.13}$$

for $\rho = 1, 2$, we reduce the number of independent terms (excluding a constant overall shift) to 13: four in $\hat{H}_1$, four in $\hat{H}_2$, and five in $\hat{V}_{12}$. The parameters in front of these terms can be related to those appearing in the original hamiltonian (6.11).

To enumerate all possible subalgebra chains of $U_1(4) \otimes U_2(4)$, we start with the trivial ones which have wave functions that are products of two U(4) states, associated with the first and second bond, respectively. Requiring the two U(4) states to be classified by either U(3) or

SO(4), we have the following two possibilities:

$$\begin{array}{cccccc} U_1(4) \otimes U_2(4) & \supset & U_1(3) \otimes U_2(3) & \supset & SO_1(3) \otimes SO_2(3) \\ | \quad\quad | & & | \quad\quad | & & | \quad\quad | \\ [N_1] \quad [N_2] & & n_{p_1} \quad n_{p_2} & & L_1 \quad L_2 \end{array}$$

$$\supset SO_1(2) \otimes SO_2(2)$$
$$\begin{array}{cc} | & | \\ M_1 & M_2 \end{array}$$

(6.14)

and

$$\begin{array}{cccccc} U_1(4) \otimes U_2(4) & \supset & SO_1(4) \otimes SO_2(4) & \supset & SO_1(3) \otimes SO_2(3) \\ | \quad\quad | & & | \quad\quad | & & | \quad\quad | \\ [N_1] \quad [N_2] & & \omega_1 \quad \omega_2 & & L_1 \quad L_2 \end{array}$$

$$\supset SO_1(2) \otimes SO_2(2)$$
$$\begin{array}{cc} | & | \\ M_1 & M_2 \end{array}$$

(6.15)

with associated basis states

$$|[N_1]n_{p_1}L_1M_1, [N_2]n_{p_2}L_2M_2\rangle \equiv |[N_1]n_{p_1}L_1M_1\rangle|[N_2]n_{p_2}L_2M_2\rangle \quad (6.16)$$

and

$$|[N_1]\omega_1 L_1 M_1, [N_2]\omega_2 L_2 M_2\rangle \equiv |[N_1]\omega_1 L_1 M_1\rangle|[N_2]\omega_2 L_2 M_2\rangle \quad (6.17)$$

Although these are perfectly respectable bases from a mathematical point of view, they are not of much use in the context of the vibron model. This is because the hamiltonian (6.11), which by construction is rotationally invariant, leads to eigenstates with definite values for the total angular momentum L—a quantum number manifestly absent from the classifications (6.14) and (6.15). It is more useful to work in a basis with good L since in that case each L value can be treated separately, given that states with different L are not mixed by a rotationally invariant hamiltonian Such bases are readily constructed from (6.14) and (6.15) by angular momentum coupling, leading to

$$|[N_1]n_{p_1}L_1, [N_2]n_{p_2}L_2; LM_L\rangle$$
$$= \sum_{M_1 M_2} \langle L_1 M_1\, L_2 M_2 | LM_L\rangle |[N_1]n_{p_1}L_1 M_1\rangle |[N_2]n_{p_2}L_2 M_2\rangle$$

(6.18)

6.3 Symmetry limits

and

$$|[N_1]\omega_1 L_1, [N_2]\omega_2 L_2; LM_L\rangle = \sum_{M_1 M_2} \langle L_1 M_1\, L_2 M_2 | LM_L \rangle |[N_1]\omega_1 L_1 M_1\rangle |[N_2]\omega_2 L_2 M_2\rangle$$
(6.19)

with the associated chains of algebras

$$
\begin{array}{ccccccc}
U_1(4) \otimes U_2(4) & \supset & U_1(3) \otimes U_2(3) & \supset & SO_1(3) \otimes SO_2(3) & \supset & SO_{12}(3) \\
| & | & | & | & | & | & | \\
[N_1] & [N_2] & n_{p_1} & n_{p_2} & L_1 & L_2 & L
\end{array}
$$
(6.20)

and

$$
\begin{array}{ccccccc}
U_1(4) \otimes U_2(4) & \supset & SO_1(4) \otimes SO_2(4) & \supset & SO_1(3) \otimes SO_2(3) & \supset & SO_{12}(3) \\
| & | & | & | & | & | & | \\
[N_1] & [N_2] & \omega_1 & \omega_2 & L_1 & L_2 & L
\end{array}
$$
(6.21)

respectively. The coupled algebra $SO_{12}(3)$ consists of the sum of generators of $SO_1(3)$ and $SO_2(3)$ (i.e., $\hat{L}_{1\mu} + \hat{L}_{2\mu}$), as implied by the subscript 12. Henceforth we only deal with chains of algebras that have the total angular momentum L as a good quantum number, and as in (6.20) and (6.21), we do not show the reduction from L to M_L since it is common to all such labeling schemes.

Clearly, it is possible to form coupled algebras other than $SO_{12}(3)$ by adding the corresponding generators of subalgebras of $U_1(4)$ and $U_2(4)$. In the notation introduced in Section 5.8 the resulting algebra chains can be summarized with lattices of algebras which in this case become

$$
\begin{array}{ccccc}
U_1(4) & \otimes & & & U_2(4) \\
\downarrow {\scriptstyle a} & \searrow & & \swarrow {\scriptstyle a} & \downarrow \\
U_1(3) & & U_{12}(4) & & U_2(3) \\
\downarrow & \searrow {\scriptstyle b} & \downarrow & \swarrow {\scriptstyle b} & \downarrow \\
SO_1(3) & & U_{12}(3) & & SO_2(3) \\
& \searrow {\scriptstyle c} & \downarrow & \swarrow {\scriptstyle c} & \\
& & SO_{12}(3) & &
\end{array}
$$
(6.22)

and

$$
\begin{array}{ccccc}
U_1(4) & & \otimes & & U_2(4) \\
\downarrow & \searrow^{a} & & \swarrow^{a} & \downarrow \\
SO_1(4) & & U_{12}(4) & & SO_2(4) \\
\downarrow & \searrow^{b} & \downarrow & \swarrow^{b} & \downarrow \\
SO_1(3) & & SO_{12}(4) & & SO_2(3) \\
& \searrow^{c} & \downarrow & \swarrow^{c} & \\
& & SO_{12}(3) & &
\end{array}
\qquad (6.23)
$$

The classifications (6.20) and (6.21) correspond to routes c in the above lattices. Routes a and b give rise to chains of algebras where the coupling occurs at the level of $U(4)$, $U(3)$, or $SO(4)$; these are of most practical value in the description of triatomic molecules (both rigid and non-rigid rotational) and are analyzed in detail in Sections 6.4 and 6.5. The coupled algebra $U_{12}(4)$ is simply obtained by adding the generators of $U_1(4)$ and $U_2(4)$,

$$\mathcal{G}_i^j \equiv \mathcal{G}_{1i}^{1j} + \mathcal{G}_{2i}^{2j} \qquad (6.24)$$

which can be shown to satisfy $U(4)$ commutation relations. In the same manner the subalgebras of $U_{12}(4)$ are formed; specifically, we have $U_{12}(3)$ with generators

$$\hat{n}_p \equiv \hat{n}_{p_1} + \hat{n}_{p_2}, \qquad \hat{L}_\mu \equiv \hat{L}_{1\mu} + \hat{L}_{2\mu}, \qquad \hat{Q}_\mu \equiv \hat{Q}_{1\mu} + \hat{Q}_{2\mu} \qquad (6.25)$$

and $SO_{12}(4)$ generated by

$$\hat{L}_\mu \equiv \hat{L}_{1\mu} + \hat{L}_{2\mu}, \qquad \hat{D}_\mu \equiv \hat{D}_{1\mu} + \hat{D}_{2\mu} \qquad (6.26)$$

The complete enumeration of all symmetry limits of $U_1(4) \otimes U_2(4)$ leads to many more schemes than the six given in (6.22) and (6.23). Although they are of less practical value in the analysis of triatomic molecules, we give here a brief overview of all additional bases, mainly with the purpose to illustrate the richness of the algebraic structure of $U_1(4) \otimes U_2(4)$.

First, we note the existence of the "mixed" limit

$$U_1(4) \otimes U_2(4) \supset U_1(3) \otimes SO_2(4) \supset SO_1(3) \otimes SO_2(3) \supset SO_{12}(3) \quad (6.27)$$

6.3 Symmetry limits

This limit is appropriate for molecules where the first bond has non-rigid (or U(3)) characteristics whereas the second bond is rigid rotational (or SO(4)). (The situation corresponding to a rigid first and a non-rigid second bond is completely equivalent and is not separately enumerated here.) Although a mixed basis might be of practical value for the description of some molecules, the specific basis (6.27) is rather limited in scope since the corresponding hamiltonian contains only one interaction term between the two bonds, namely the one associated with the coupled algebra $SO_{12}(3)$. This property is shared by all mixed limits of the vibron model. It thus appears that a proper description of these mixed, rigid–non-rigid situations cannot be achieved in the context of a symmetry limit of $U_1(4) \otimes U_2(4)$ but requires a numerical approach (i.e., the numerical diagonalization of a general vibron hamiltonian of the type (6.11) in an appropriate basis).

Second, we may exploit the existence of the alternative algebras $\overline{SO}_1(4)$ and $\overline{SO}_2(4)$ to construct a batch of additional algebra chains. From (6.23) we construct the following two alternatives:

$$
\begin{array}{ccccc}
U_1(4) & & \otimes & & U_2(4) \\
\downarrow & \searrow^{a} & & \swarrow^{a} & \downarrow \\
\overline{SO}_1(4) & & U_{12}(4) & & \overline{SO}_2(4) \\
\downarrow & \searrow^{b} & \downarrow & \swarrow^{b} & \downarrow \\
SO_1(3) & & \overline{SO}_{12}(4) & & SO_2(3) \\
& \searrow^{c} & \downarrow & \swarrow^{c} & \\
& & SO_{12}(3) & &
\end{array}
\qquad (6.28)
$$

and

$$
\begin{array}{ccccc}
U_1(4) & & \otimes & & U_2(4) \\
\downarrow & & & & \downarrow \\
SO_1(4) & & & & \overline{SO}_2(4) \\
\downarrow & \searrow^{a} & & \swarrow^{a} & \downarrow \\
SO_1(3) & & \widetilde{SO}_{12}(4) & & SO_2(3) \\
& \searrow^{b} & \downarrow & \swarrow^{b} & \\
& & SO_{12}(3) & &
\end{array}
\qquad (6.29)
$$

Note that the coupled SO(4) algebras have generators different from those in (6.23), and a notation is introduced in (6.28) and (6.29) to stress this fact. The coupled algebras in question are defined as follows: $\overline{SO}_{12}(4)$ consists of the generators

$$\hat{L}_{1\mu} + \hat{L}_{2\mu}, \qquad \hat{D}'_{1\mu} + \hat{D}'_{2\mu} \qquad (6.30)$$

while the algebra we denote as $\widetilde{SO}_{12}(4)$ is generated by

$$\hat{L}_{1\mu} + \hat{L}_{2\mu}, \qquad \hat{D}_{1\mu} + \hat{D}'_{2\mu} \qquad (6.31)$$

Whereas the first set of operators consists of summed generators of $\overline{SO}_1(4)$ and $\overline{SO}_2(4)$ and hence trivially closes under commutation to form $\overline{SO}_{12}(4)$, the closure property is a little more delicate for the second set of operators. It follows from a direct computation of the commutation relations:

$$\begin{aligned}
[\hat{L}_{1\mu} + \hat{L}_{2\mu}, \hat{L}_{1\mu'} + \hat{L}_{2\mu'}] &= a_{\mu\mu'} \left(\hat{L}_{1\mu+\mu'} + \hat{L}_{2\mu+\mu'} \right) \\
[\hat{L}_{1\mu} + \hat{L}_{2\mu}, \hat{D}_{1\mu'} + \hat{D}'_{2\mu'}] &= a_{\mu\mu'} \left(\hat{D}_{1\mu+\mu'} + \hat{D}'_{2\mu+\mu'} \right) \\
[\hat{D}_{1\mu} + \hat{D}'_{2\mu}, \hat{D}_{1\mu'} + \hat{D}'_{2\mu'}] &= a_{\mu\mu'} \left(\hat{L}_{1\mu+\mu'} + \hat{L}_{2\mu+\mu'} \right)
\end{aligned} \qquad (6.32)$$

with $a_{\mu\mu'} = -\sqrt{2}\langle 1\mu\, 1\mu'|1\mu+\mu'\rangle$. This indeed establishes the existence of $\widetilde{SO}_{12}(4)$ as a Lie algebra.

Obviously, the chain of algebras (6.27) can be given the same treatment, and after replacing $SO_2(4)$ by $\overline{SO}_2(4)$, we obtain

$$U_1(4) \otimes U_2(4) \supset U_1(3) \otimes \overline{SO}_2(4) \supset SO_1(3) \otimes SO_2(3) \supset SO_{12}(3) \quad (6.33)$$

So far we have constructed subalgebras of $U_1(4) \otimes U_2(4)$ by *adding* generators of $U_1(4)$ and $U_2(4)$ since this trivially leads to closed coupled algebras. This procedure is not the only possible one, and coupled algebras can also be formed by *substracting* generators. To see why this comes about, we consider a $U(4)$ algebra and apply to it the transformation

$$b^\dagger_{lm} \to \tilde{b}_{lm}, \qquad \tilde{b}_{lm} \to -b^\dagger_{lm} \qquad (6.34)$$

which replaces boson particles by boson holes and *vice versa* and is sometimes referred to as a *particle–hole conjugation*. It is readily shown that the boson commutation relations (5.1) remain invariant under this transformation:

$$[b_{lm}, b^\dagger_{l'm'}] \to [-(-)^{l-m} b^\dagger_{l-m}, (-)^{l'+m'} b_{l'-m'}] = \delta_{ll'}\delta_{mm'} \qquad (6.35)$$

6.3 Symmetry limits

The U(4) generators $\mathcal{B}^{(\lambda)}_\mu(l,l') = [b^\dagger_l \times \tilde{b}_{l'}]^{(\lambda)}_\mu$ transform into

$$\mathcal{B}^{(\lambda)}_\mu(l,l') \to (-)^{l+l'-\lambda+1} \mathcal{B}^{(\lambda)}_\mu(l',l) - \sqrt{2l+1}\,\delta_{ll'}\delta_{\lambda 0}\delta_{\mu 0} \qquad (6.36)$$

or explicitly,

$$\begin{array}{lll}
\hat{n}_s \to -\hat{n}_s - 1, & \hat{n}_p \to -\hat{n}_p - \sqrt{3}, & \hat{L}_\mu \to \hat{L}_\mu \\
\hat{D}_\mu \to -\hat{D}_\mu, & \hat{D}'_\mu \to \hat{D}'_\mu, & \hat{Q}_\mu \to -\hat{Q}_\mu
\end{array} \qquad (6.37)$$

These equations can be used as a guide in the construction of additional coupled algebras: adding corresponding generators *after* the application of (6.34) to one of the algebras in $U_1(4) \otimes U_2(4)$ (say $U_2(4)$) leads to

$$\begin{array}{lll}
\hat{n}_{s_1} - \hat{n}_{s_2}, & \hat{n}_{p_1} - \hat{n}_{p_2}, & \hat{L}_{1\mu} + \hat{L}_{2\mu} \\
\hat{D}_{1\mu} - \hat{D}_{2\mu}, & \hat{D}'_{1\mu} + \hat{D}'_{2\mu}, & \hat{Q}_{1\mu} - \hat{Q}_{2\mu}
\end{array} \qquad (6.38)$$

By virtue of the invariance property (6.35) this set of operators must close—as indeed can be verified by direct computation of the commutation relations—and we refer to this algebra as $U^*_{12}(4)$. (Note that the constants -1 and $-\sqrt{3}$, appearing in the number operators in (6.37), are not included in the definition of $U^*_{12}(4)$ since they have no effect on the commutation relations and are inconsequential for our discussion here.) By slightly modifying the dipole operators, yet another subalgebra of $U_1(4) \otimes U_2(4)$ is established, namely,

$$\begin{array}{lll}
\hat{n}_{s_1} - \hat{n}_{s_2}, & \hat{n}_{p_1} - \hat{n}_{p_2}, & \hat{L}_{1\mu} + \hat{L}_{2\mu} \\
\hat{D}_{1\mu} + \hat{D}_{2\mu}, & \hat{D}'_{1\mu} - \hat{D}'_{2\mu}, & \hat{Q}_{1\mu} - \hat{Q}_{2\mu}
\end{array} \qquad (6.39)$$

which we refer to as $U^{**}_{12}(4)$.

A new set of symmetry limits can now be built on the basis of the coupled algebras $U^*_{12}(4)$ and $U^{**}_{12}(4)$. Starting with symmetry limits of the U(3) type, we can construct the lattice of algebras

$$\begin{array}{ccccc}
U_1(4) & & \otimes & & U_2(4) \\
\downarrow & \searrow^a & & \swarrow^a & \downarrow \\
U_1(3) & & U^*_{12}(4) & & U_2(3) \\
& \searrow_b & \downarrow^b & \swarrow & \\
& & U^*_{12}(3) & & \\
& & \downarrow & & \\
& & SO_{12}(3) & &
\end{array} \qquad (6.40)$$

where we have omitted the third route c since it would be identical to the one in (6.22). The $U_{12}^*(3)$ algebra in (6.40) is obtained by removing from $U_{12}^*(4)$ the generators containing an s_ρ boson,

$$\hat{n}_{p_1} - \hat{n}_{p_2}, \qquad \hat{L}_{1\mu} + \hat{L}_{2\mu}, \qquad \hat{Q}_{1\mu} - \hat{Q}_{2\mu} \qquad (6.41)$$

Clearly, a lattice similar to (6.40) can be constructed with $U_{12}^{**}(4)$. Care must be taken to avoid double counting and enumerating the same algebra chain twice. Since both $U_{12}^*(4)$ and $U_{12}^{**}(4)$ have the same $U_{12}^*(3)$ subalgebra, this means that only one additional chain of algebras is obtained:

$$U_1(4) \otimes U_2(4) \supset U_{12}^{**}(4) \supset U_{12}^*(3) \supset SO_{12}(3) \qquad (6.42)$$

We can go through the same exercise and establish all possible lattices based on SO(4)-type algebras, with in this case the added complication of choosing between SO(4) algebras constructed with the dipole operators $\hat{D}_{\rho\mu}$ or those obtained with $\hat{D}'_{\rho\mu}$. In the first class we find the classification schemes

$$\begin{array}{ccccc}
U_1(4) & & \otimes & & U_2(4) \\
\downarrow & \searrow^a & & \swarrow^a & \downarrow \\
SO_1(4) & & U_{12}^*(4) & & SO_2(4) \\
& \searrow^b & \downarrow^b & \swarrow^b & \\
& & SO_{12}^*(4) & & \\
& & \downarrow & & \\
& & SO_{12}(3) & &
\end{array} \qquad (6.43)$$

and

$$U_1(4) \otimes U_2(4) \supset U_{12}^{**}(4) \supset SO_{12}(4) \supset SO_{12}(3) \qquad (6.44)$$

where $SO_{12}^*(4)$ is formed by

$$\hat{L}_{1\mu} + \hat{L}_{2\mu}, \qquad \hat{D}_{1\mu} - \hat{D}_{2\mu} \qquad (6.45)$$

In the second class we obtain the algebra chains

$$U_1(4) \otimes U_2(4) \supset U_{12}^*(4) \supset \overline{SO}_{12}(4) \supset SO_{12}(3) \qquad (6.46)$$

6.3 Symmetry limits

and

$$\begin{array}{ccccc}
U_1(4) & & \otimes & & U_2(4) \\
\downarrow & \searrow^a & & \swarrow^a & \downarrow \\
\overline{SO}_1(4) & & U_{12}^{**}(4) & & \overline{SO}_2(4) \\
& \searrow_b & \downarrow & \swarrow_b & \\
& & \overline{SO}_{12}^*(4) & & \\
& & \downarrow & & \\
& & SO_{12}(3) & &
\end{array} \qquad (6.47)$$

with the $\overline{SO}_{12}^*(4)$ algebra generated by

$$\hat{L}_{1\mu} + \hat{L}_{2\mu}, \qquad \hat{D}'_{1\mu} - \hat{D}'_{2\mu} \qquad (6.48)$$

Finally, there is one more hybrid algebra chain based on route a in the lattice (6.29), which after application of the transformation (6.34) on $U_2(4)$ leads to

$$U_1(4) \otimes U_2(4) \supset SO_1(4) \otimes \overline{SO}_2(4) \supset \widetilde{SO}_{12}^*(4) \supset SO_{12}(3) \qquad (6.49)$$

where $\widetilde{SO}_{12}^*(4)$ has the (admittedly peculiar) generators

$$\hat{L}_{1\mu} + \hat{L}_{2\mu}, \qquad \hat{D}_{1\mu} - \hat{D}'_{2\mu} \qquad (6.50)$$

We have by now established the existence of 23 different coupling schemes in the $U_1(4) \otimes U_2(4)$ model! Note that they all share the property of ending up in $SO_{12}(3)$ generated by the total angular momentum operator $\hat{L}_{1\mu} + \hat{L}_{2\mu}$, implying that the associated hamiltonians, by construction, are rotationally invariant. We have carried out the preceding analysis to illustrate the amazing proliferation of subalgebras in the vibron model once coupled U(4) algebras are considered, and we certainly do not expect all bases to be of importance in the classification of triatomic rotation–vibration spectra. To decide which are the algebra chains of relevance, worthy of further analysis, we may proceed via a process of elimination.

First, we discard the mixed bases (6.27) and (6.33) on the grounds that they contain only one interaction term (see discussion after (6.27)).

Second, we do not separately investigate those cases that involve $\overline{SO}_\rho(4)$ algebras (i.e., the schemes (6.28), (6.29), (6.46), (6.47), and

(6.49)). Our argumentation is the same as in Chapter 5, where it is shown that a consistent treatment (use of the same dipole operator in the hamiltonian and in the E1 transition operator) leads to identical results in the SO(4) and $\overline{\text{SO}}(4)$ limits.

A third class of bases, namely those obtained through the transformation (6.34) and given in (6.40), (6.42), (6.43), and (6.44), will not be analyzed in detail either. These cases, however, are *not* trivially related to (6.22) or (6.23), as we now proceed to show. Let us first consider the effect of the transformation (6.34) on a single-boson state and on its U(4)-representation character in particular. In Section 5.8 it is shown that the creation operators $b^\dagger_{lm}$ transform as the U(4) representation $[1] \equiv [1,0,0,0]$ and the modified annihilation operators $\tilde{b}_{lm}$ as $[1^3] \equiv [1,1,1,0]$. Thus the transformation (6.34) converts a single-boson representation into its conjugate. This property persists for a many-boson state. For example, all two-boson hole states must transform according to one of the two U(4) representations (or a mixture) on the right-hand side of

$$\square\square \otimes \square\square = \begin{array}{c}\square\square\\\square\square\\\square\end{array} \oplus \begin{array}{c}\square\square\\\square\square\\\square\square\end{array} \qquad (6.51)$$

The first U(4) representation on the right-hand side, $[2,2,1,1]$, is identical to $[1,1]$ and contains all *antisymmetric* two-boson states. Since a system of identical bosons must have *symmetric* states, these necessarily are contained in the second U(4) representation $[2,2,2]$. This argument can be generalized to an arbitrary number of bosons, and it follows that the effect of the particle–hole conjugation (6.34) is to convert the completely symmetric representation $[N]$ for a system of identical boson particles into the conjugate representation $[N^3] \equiv [N, N, N, 0]$, appropriate for boson holes. It can be further generalized to an arbitrary U(n) algebra, in which case a symmetric representation $[N]$ goes over into $[N^{n-1}]$. Note that for the schematic models considered in Part I ($n = 2$) the particle–hole conjugation transforms a symmetric representation into itself and as a consequence has a trivial effect.

If one considers a single U(4) algebra, the same results (energy spectra, transition properties, etc.) are obtained irrespective of whether

6.3 Symmetry limits

one starts from the symmetric representation $[N]$ or its conjugate $[N^3]$. This must be the case since the particle–hole conjugation transforms the U(4) algebra into itself—see (6.37). (The constants in the number operators in (6.37) give rise to an overall shift in energy, but all other properties remain unaffected.) Therefore there is no need to consider the hole case separately in Chapter 5. The situation is different, however, for $U_1(4) \otimes U_2(4)$ when the particle–hole conjugation is applied to one of the U(4) algebras only. The representations for the resulting coupled algebra $U_{12}^*(4)$ are determined in this case from the multiplication

$$[N_1] \otimes [N_2, N_2, N_2] \tag{6.52}$$

and are completely different from those obtained for $U_{12}(4)$, which follow from

$$[N_1] \otimes [N_2] \tag{6.53}$$

The energy spectra and other properties are different in the two cases. Another way of understanding this result is to note that the application of (6.34) to bond-2 bosons does *not* transform $U_{12}(4)$ into itself but rather into $U_{12}^*(4)$.

The particle–hole bases cannot be dismissed on the grounds that they are trivially related to particle–particle bases and give identical results. In fact, in the U(6) models discussed in Part III such particle–hole bases have been frequently proposed to describe certain nuclei. The difference between the molecular and nuclear models is that in the latter a clear interpretation exists for the bosons (i.e., they are pairs of nucleons which can have a particle or a hole character) whereas this interpretation is lacking in the former. As a result it is not at all clear at present what is the physical significance of particle–hole conjugation in the vibron model, and for this reason they have been less well studied. (In [2] a discussion is given of the mathematical aspects of some particle–hole bases but no applications are presented.)

We conclude that it is best to concentrate our efforts on the study of the lattices (6.22) and (6.23), corresponding to U(3) and SO(4) limits of $U_1(4) \otimes U_2(4)$, respectively. Although, judging from the results for diatomic molecules, we expect SO(4)-type symmetries to be of more relevance in actual applications, we nevertheless give in the next section a brief overview of the U(3) limits.

6.4 The U(3) limits

Our starting point in this discussion is the route c in the lattice (6.22), which defines the basis

$$\begin{array}{cccccccc}
U_1(4) & \otimes & U_2(4) & \supset U_1(3) \otimes U_2(3) \supset SO_1(3) \otimes SO_2(3) \supset SO_{12}(3) \\
\mid & & \mid & \mid \quad\quad\quad \mid \quad\quad\quad \mid \quad\quad\quad \mid \quad\quad\quad \mid \\
[N_1] & & [N_2] & n_{p_1} \quad\quad n_{p_2} \quad\quad L_1 \quad\quad L_2 \quad\quad L
\end{array} \quad (6.54)$$

The branching rules in (6.54) are known from Section 5.4,

$$n_{p_\rho} = 0, 1, \ldots, N_\rho \quad (6.55)$$

and

$$L_\rho = n_{p_\rho}, n_{p_\rho} - 2, \ldots, 1 \text{ or } 0 \quad (6.56)$$

and the total angular momentum L is obtained with the usual coupling rule:

$$L = |L_1 - L_2|, |L_1 - L_2| + 1, \ldots, L_1 + L_2 \quad (6.57)$$

The hamiltonian corresponding to (6.54) contains only one interaction term between the first and second bond. This feature limits its applicability, but the basis nevertheless has some importance because (i) it is a convenient one to perform numerical calculations and (ii) its branching rules are known and can be used to derive those in the other bases.

Route b in the lattice (6.22) defines the basis

$$\begin{array}{|cccccc|}
\hline
U_1(4) \otimes U_2(4) \supset U_1(3) \otimes U_2(3) \supset U_{12}(3) \supset SO_{12}(3) \\
\mid \quad\quad\quad \mid \quad\quad\quad \mid \quad\quad\quad \mid \quad\quad\quad \mid \quad\quad\quad \mid \\
[N_1] \quad\quad [N_2] \quad\quad n_{p_1} \quad\quad n_{p_2} \quad\quad (n_p, n'_p) \quad\quad \kappa L \\
\hline
\end{array} \quad (6.58)$$

which we refer to as the $U_1(3) \otimes U_2(3)$ limit. As in the previous chapters the hamiltonian diagonal in this basis is defined in terms of Casimir invariants associated with the different algebras in (6.58). Formally, it is of the form

$$\begin{aligned}
\hat{H}_{Ib} &= \epsilon_1 \mathcal{C}_1[U_1(3)] + \epsilon_2 \mathcal{C}_1[U_2(3)] + \alpha_1 \mathcal{C}_2[U_1(3)] + \alpha_2 \mathcal{C}_2[U_2(3)] \\
&\quad + \alpha \mathcal{C}_2[U_{12}(3)] + \gamma \mathcal{C}_2[SO_{12}(3)]
\end{aligned} \quad (6.59)$$

6.4 The U(3) limits

where the Casimir invariants of $U_\rho(4)$ and of $U_{12}(3)$ are not included since they reduce to constants or can be expressed in terms of other invariants. Introducing the explicit expressions for the various terms (see Table 5.2), we rewrite this hamiltonian as

$$\hat{H}_{Ib} = \tilde{\epsilon}_1 \hat{n}_{p_1} + \tilde{\epsilon}_2 \hat{n}_{p_2} + \alpha_1 \hat{n}_{p_1}^2 + \alpha_2 \hat{n}_{p_2}^2 + \alpha \left(\tfrac{1}{3}\hat{n}_p^2 + \tfrac{1}{2}\hat{L}^2 + \hat{Q}^2\right) + \gamma \hat{L}^2 \quad (6.60)$$

with $\tilde{\epsilon}_\rho = \epsilon_\rho + 3\alpha_\rho$. Note that in this chapter the operators $\hat{n}_p$, $\hat{L}_\mu$, and $\hat{Q}_\mu$ have a realization different from that in Chapter 5 since they correspond to *sums* of bond-1 and bond-2 operators, as defined in (6.25) and (6.26). The eigenvalues of $\hat{H}_{Ib}$ are

$$\begin{aligned}E_{Ib}&(n_{p_1}, n_{p_2}, n_p, n_p', L) \\ &= \tilde{\epsilon}_1 n_{p_1} + \tilde{\epsilon}_2 n_{p_2} + \alpha_1 n_{p_1}^2 + \alpha_2 n_{p_2}^2 + \alpha[n_p(n_p + 2) + n_p'^2] \\ &\quad + \gamma L(L+1)\end{aligned} \quad (6.61)$$

The various terms in this equation also appear in the U(4) expression (5.53) with the exception of $n_p(n_p + 2) + n_p'^2$, the eigenvalue associated with $C_2[U_{12}(3)]$. This contribution is found by applying the general expressions for eigenvalues of Casimir invariants of $U(n)$, given in Appendix B. Note also the similarity with the U(2) result (2.65).

To obtain the spectrum defined by $\hat{H}_{Ib}$, we need the various branching and multiplication rules in (6.58). The U(4) ⊃ U(3) branching rule is identical to the one in (6.54); the U(3) multiplication can be obtained from the Young tableaux rules given in Appendix B and leads to (see also (2.42))

$$(n_{p_1}, 0) \otimes (n_{p_2}, 0) = (n_{p_1} + n_{p_2}, 0) \oplus (n_{p_1} + n_{p_2} - 1, 1) \oplus \cdots$$
$$\cdots \oplus \begin{cases} (n_{p_1}, n_{p_2}) & \text{if } n_{p_1} \geq n_{p_2} \\ (n_{p_2}, n_{p_1}) & \text{if } n_{p_1} \leq n_{p_2} \end{cases} \quad (6.62)$$

Note that $U_{12}(3)$ is characterized by *two* labels n_p and n_p', and as a consequence we must generalize the U(3) ⊃ SO(3) branching rule (6.56). This more general branching rule can be derived by noting that the allowed L values in the two bases (6.54) and (6.58) must be the same. Since we know the L values in the former from (6.55), (6.56), and (6.57), we can deduce the U(3) ⊃ SO(3) branching rule in the latter. To illustrate the procedure, we consider first the simple case of $n_{p_1} > n_{p_2} = 1$. This implies $L_2 = 1$ in (6.54) and hence the allowed L values are

$$L = n_{p_1}+1, n_{p_1}, (n_{p_1}-1)^2, n_{p_1}-2, (n_{p_1}-3)^2, \ldots, \begin{cases} 2^2, 1, 0 & \text{odd } n_{p_1} \\ 3^2, 2, 1 & \text{even } n_{p_1} \end{cases} \quad (6.63)$$

where a power denotes a multiplicity (i.e., L^q implies that L occurs q times). On the other hand, the $U_{12}(3)$ representation (n_p, n'_p) in (6.58) can be $(n_{p_1}+1, 0)$ or $(n_{p_1}, 1)$. Substracting the L values occurring in $(n_{p_1}+1, 0)$, we find those contained in $(n_{p_1}, 1)$:

$$L = n_{p_1}, n_{p_1}-1, \ldots, 1 \quad (6.64)$$

It is instructive to work out the example of $n_{p_2} = 2$. We then have $L_2 = 0, 2$ and the allowed values for L are

$$L = n_{p_1}+2, n_{p_1}+1, (n_{p_1})^3, (n_{p_1}-1)^2, (n_{p_1}-2)^4, (n_{p_1}-3)^2,$$
$$(n_{p_1}-4)^2, \ldots, \begin{cases} 4^2, 3^4, 2^2, 1^3 & \text{odd } n_{p_1} \\ 3^2, 2^4, 1, 0^2 & \text{even } n_{p_1} \end{cases} \quad (6.65)$$

The allowed $U_{12}(3)$ representations are $(n_{p_1}+2, 0)$, $(n_{p_1}+1, 1)$, and $(n_{p_1}, 2)$, and since we now know the angular momentum content of the first two representations, we can derive the L values contained in $(n_{p_1}, 2)$:

$$L = n_{p_1}, n_{p_1}-1, (n_{p_1}-2)^2, n_{p_1}-3, (n_{p_1}-4)^2, \ldots, \begin{cases} 3^2, 2, 1 & \text{odd } n_{p_1} \\ 3, 2^2, 0 & \text{even } n_{p_1} \end{cases} \quad (6.66)$$

From this last example it becomes clear that a general $U_{12}(3)$ representation of the form (n_p, n'_p) may contain certain L values more than once. As discussed in Section 5.8 (after equation (5.217)), in such cases one must introduce one or several multiplicity labels. Here only one is needed, denoted in (6.58) as κ. The procedure outlined above for $n_{p_2} = 1$ and $n_{p_2} = 2$ can be extended indefinitely, and a general rule emerges from it which may be expressed as follows [7]. Introducing the Elliott labels (λ, μ) defined as

$$\lambda = n_p - n'_p, \qquad \mu = n'_p \quad (6.67)$$

6.4 The U(3) limits

one finds that κ can take the values

$$\kappa = \min(\lambda, \mu), \min(\lambda, \mu) - 2, \ldots, 1 \text{ or } 0 \tag{6.68}$$

and that the L values contained in (λ, μ) for a given κ are

$$L = \kappa, \kappa + 1, \ldots, \kappa + \max(\lambda, \mu) \tag{6.69}$$

for $\kappa \neq 0$ and

$$L = \max(\lambda, \mu), \max(\lambda, \mu) - 2, \ldots, 1 \text{ or } 0 \tag{6.70}$$

for $\kappa = 0$. This general rule reduces for $\mu = 0, 1, 2$ to the ones given above. Using the expression (6.61) and the various branching and multiplication rules, the energy spectrum of $\hat{H}_{Ib}$ can now be derived. An example is shown in Figure 6.2 for $N_1 = N_2 = 2$ bosons.

The analysis of route a in (6.22) proceeds along the same lines. The

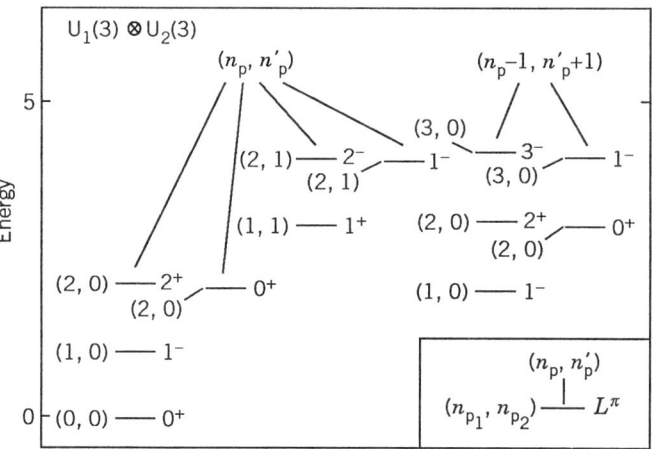

Figure 6.2: The $U_1(3) \otimes U_2(3)$ eigenspectrum of the hamiltonian $\hat{H}_{Ib}$ with $\tilde{\epsilon}_1 = 1$, $\tilde{\epsilon}_2 = 2$, $\alpha_1 = \alpha_2 = 0$, $\alpha = 0.01$, and $\gamma = 0.01$ for $N_1 = N_2 = 2$ bosons. Levels are labeled by n_{p_1}, n_{p_2}, (n_p, n'_p), total angular momentum L, and parity $\pi = (-)^{n_{p_1} + n_{p_2}} = (-)^{n_p + n'_p}$. Energies are in arbitrary units.

corresponding basis is labeled as follows:

$$
\begin{array}{ccccc}
U_1(4) \otimes U_2(4) \supset & U_{12}(4) \supset & U_{12}(3) \supset & SO_{12}(3) \\
\downarrow \quad \downarrow & \downarrow & \downarrow & \downarrow \\
[N_1] \quad [N_2] & [h,h'] & (n_p, n'_p) & \kappa L
\end{array}
\tag{6.71}
$$

with the associated hamiltonian

$$\hat{H}_{\mathrm{Ia}} = \eta \mathcal{C}_2[U_{12}(4)] + \epsilon \mathcal{C}_1[U_{12}(3)] + \alpha \mathcal{C}_2[U_{12}(3)] + \gamma \mathcal{C}_2[SO_{12}(3)] \tag{6.72}$$

or explicitly,

$$\begin{aligned}
\hat{H}_{\mathrm{Ia}} &= \eta \left(\tfrac{1}{3}\hat{n}_p^2 + \tfrac{1}{2}(\hat{L}^2 + \hat{D}^2 + \hat{D}'^2) + \hat{Q}^2 + \hat{n}_s^2\right) + \epsilon \hat{n}_p \\
&\quad + \alpha \left(\tfrac{1}{3}\hat{n}_p^2 + \tfrac{1}{2}\hat{L}^2 + \hat{Q}^2\right) + \gamma \hat{L}^2
\end{aligned} \tag{6.73}$$

We refer to the basis (6.71) as the $U_{12}(4)$ limit. The eigenvalues of $\hat{H}_{\mathrm{Ia}}$ are

$$\begin{aligned}
E_{\mathrm{Ia}}&(h, h', n_p, n'_p, L) \\
&= \eta[h(h+3) + h'(h'+1)] + \epsilon(n_p + n'_p) \\
&\quad + \alpha[n_p(n_p+2) + n'^2_p] + \gamma L(L+1)
\end{aligned} \tag{6.74}$$

The U(4) multiplication rule can be obtained as usual with Young tableaux and determines the $U_{12}(4)$ representations $[h, h']$:

$$\begin{aligned}
[N_1] \otimes [N_2] &= [N_1+N_2, 0] \oplus [N_1+N_2-1, 1] \oplus \cdots \\
&\quad \cdots \oplus \begin{cases} [N_1, N_2] & \text{if } N_1 \geq N_2 \\ [N_2, N_1] & \text{if } N_1 \leq N_2 \end{cases}
\end{aligned} \tag{6.75}$$

Note that this expression is formally identical to (2.42), but here $[N_\rho]$ and $[h, h']$ denote U(4) representations, whereas in Chapter 2 they refer to U(2). Since the U(3) $\supset$ SO(3) branching rule is known from our previous discussion of the $U_1(3) \otimes U_2(3)$ limit, the only remaining task is to find the labels (n_p, n'_p) in a given $U_{12}(4)$ representation $[h, h']$. This again can be achieved by requiring the allowed values of (n_p, n'_p) to be the same in both bases (6.58) and (6.71). For instance, for $N_1 > N_2 = 1$

6.4 The U(3) limits

we can have the values $n_{p_2} = 0$ and $n_{p_2} = 1$ in (6.58), and hence we find the (n_p, n_p') labels

$$\begin{aligned}(n_p, n_p') &= (N_1, 0), (N_1 - 1, 0), \ldots, (0, 0), \\ &\quad (N_1 + 1, 0), (N_1, 0), \ldots, (1, 0), \\ &\quad (N_1, 1), (N_1 - 1, 1), \ldots, (1, 1)\end{aligned} \quad (6.76)$$

On the other hand, the $U_{12}(4)$ representation $[h, h']$ can be $[N_1+1, 0]$ or $[N_1, 1]$. Substracting the $U_{12}(3)$ representations contained in $[N_1+1, 0]$ (known from (6.55)), we obtain those contained in $[N_1, 1]$:

$$\begin{aligned}(n_p, n_p') &= (N_1, 0), (N_1 - 1, 0), \ldots, (1, 0), \\ &\quad (N_1, 1), (N_1 - 1, 1), \ldots, (1, 1)\end{aligned} \quad (6.77)$$

Continuing in this way for higher values for N_2, the following simple $U(4) \supset U(3)$ branching rule from $[h, h']$ to (n_p, n_p') emerges:

$$h \geq n_p \geq h', \qquad h' \geq n_p' \geq 0 \quad (6.78)$$

This is a special case of the general $U(n) \supset U(n-1)$ result given in Appendix B. This derivation, together with (6.74), determines the energy spectrum of $\hat{H}_{Ia}$, an example of which is shown in Figure 6.3 for $N_1 = N_2 = 2$ bosons.

The derivation of $U_{12}(3)$ properties presented in this section has been rather dense, leaving out the discussion of topics such as the construction of wave functions in the different limits and the calculation of all possible matrix elements. This would require the computation of generalized coupling coefficients associated with $U(4) \supset U(3) \supset SO(3) \supset SO(2)$ (or a subchain) and the development of the corresponding tensor calculus. We will not pursue these matters here since, ultimately, we are primarily interested in the applications of the $U_1(4) \otimes U_2(4)$ model to triatomic rotation–vibration spectra, which so far are confined to SO(4)-type limits. Properties of limits of this type are studied in detail in Section 6.5 and applications are discussed in Section 6.6.

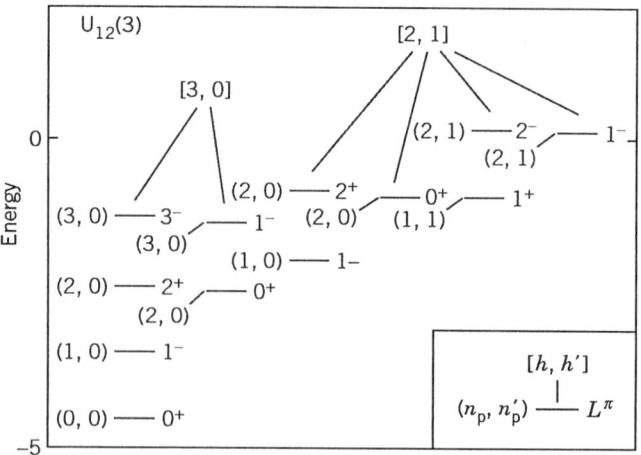

Figure 6.3: The $U_{12}(3)$ eigenspectrum of the hamiltonian $\hat{H}_{Ia}$ with $\eta = -0.25$, $\epsilon = 1$, $\alpha = 0.01$, and $\gamma = 0.01$ for $N_1 = N_2 = 2$ bosons. Levels are labeled by $[h, h']$, (n_p, n'_p), total angular momentum L, and parity $\pi = (-)^{n_p + n'_p}$. Energies are in arbitrary units.

6.5 The SO(4) limits

As in the previous section we begin by discussing the uncoupled basis corresponding to route c in the lattice (6.23). This limit is characterized by the algebra chain

$$
\begin{array}{ccccccc}
U_1(4) \otimes U_2(4) & \supset & SO_1(4) \otimes SO_2(4) & \supset & SO_1(3) \otimes SO_2(3) & \supset & SO_{12}(3) \\
| & & | \quad\quad\quad | & & | \quad\quad\quad | & & | \\
[N_1] \quad [N_2] & & \omega_1 \quad\quad \omega_2 & & L_1 \quad\quad L_2 & & L
\end{array}
\tag{6.79}
$$

with branching rules known from Section 5.5:

$$\omega_\rho = N_\rho, N_\rho - 2, \ldots, 1 \text{ or } 0 \tag{6.80}$$

and

$$L_\rho = 0, 1, \ldots, \omega_\rho \tag{6.81}$$

6.5 The SO(4) limits

and the total angular momentum L determined from (6.57). The states associated with (6.79),

$$|[N_1]\omega_1 L_1, [N_2]\omega_2 L_2; LM_L\rangle \qquad (6.82)$$

provide a useful basis to perform numerical calculations, and moreover, the branching rules (6.80) and (6.81) can be used to derive those in the bases considered below. The hamiltonian diagonal in (6.79) is of the form

$$\hat{H}_{\text{IIc}} = \beta_1 \mathcal{C}_2[\text{SO}_1(4)] + \beta_2 \mathcal{C}_2[\text{SO}_2(4)] + \gamma_1 \mathcal{C}_2[\text{SO}_1(3)] + \gamma_2 \mathcal{C}_2[\text{SO}_2(3)]$$
$$+ \gamma \mathcal{C}_2[\text{SO}_{12}(3)] \qquad (6.83)$$

where, as usual in this chapter, operators that give constant contributions to the energy are omitted. Written in terms of generators, this hamiltonian becomes

$$\hat{H}_{\text{IIc}} = \beta_1(\hat{L}_1^2 + \hat{D}_1^2) + \beta_2(\hat{L}_2^2 + \hat{D}_2^2) + \gamma_1 \hat{L}_1^2 + \gamma_2 \hat{L}_2^2 + \gamma \hat{L}^2 \qquad (6.84)$$

and it has the eigenvalues

$$E_{\text{IIc}}(\omega_1, \omega_2, L_1, L_2, L)$$
$$= \beta_1 \omega_1(\omega_1 + 2) + \beta_2 \omega_2(\omega_2 + 2) + \gamma_1 L_1(L_1 + 1) + \gamma_2 L_2(L_2 + 1)$$
$$+ \gamma L(L + 1) \qquad (6.85)$$

where the eigenvalue expression of the Casimir invariant $\mathcal{C}_2[\text{SO}_\rho(4)]$ in the symmetric representation of $\text{SO}_\rho(4)$ is taken from Section 5.5. The vibron hamiltonian in this limit contains only one interaction term between the two bonds of the triatomic molecule (of the form $2\gamma \hat{L}_1 \cdot \hat{L}_2$), and this is found to be insufficient to generate realistic molecular spectra. The hamiltonian $\hat{H}_{\text{IIc}}$ thus appears to be of limited use and we do not discuss it in further detail here.

Routes a and b in the lattice (6.23) define limits that are of more practical value in the application to rigid molecular spectra. A hamiltonian exists which is intermediate between these two limits and which can reproduce a wide spectrum of triatomic molecular structures, with characteristics ranging from local-mode to normal-mode vibrations. The two limiting dynamical symmetries corresponding to routes a and b

can thus be viewed as "benchmarks" which enclose the solutions to the more general hamiltonian. In addition, it is possible to define two different hamiltonians—the first appropriate for linear and the second for bent molecules—and it is necessary to treat these two cases separately. In the following we present a detailed analysis of the various possible combinations which encompasses a local-mode as well as a normal-mode description of linear as well as bent triatomic molecules, and we show that several of these cases can be treated in an analytic manner in the vibron model. It should also be emphasized that from this procedure a more general hamiltonian emerges that deals in a continuous way with intermediate situations (i.e., from local to normal modes of quasi-linear to quasi-bent molecules). Examples of this approach are discussed in Section 6.6.

6.5.1 Linear molecules

We begin with route b in the lattice (6.23), which defines the basis

$$\begin{array}{|cccccc|} \hline U_1(4) \otimes U_2(4) \supset SO_1(4) \otimes SO_2(4) \supset SO_{12}(4) \supset SO_{12}(3) \\ | & | & | & | & | & | \\ [N_1] & [N_2] & \omega_1 & \omega_2 & (\omega,\omega') & L \\ \hline \end{array} \quad (6.86)$$

which we refer to as the $SO_1(4) \otimes SO_2(4)$ limit. All branching and multiplication rules in (6.86) have been derived previously. The $U_\rho(4) \supset SO_\rho(4)$ branching rule is given in (6.80). The multiplication of $SO(4)$ representations is known in the general case (see (5.199)) and reduces here to

$$(\omega_1, 0) \otimes (\omega_2, 0) = \oplus \sum_{\alpha\beta=0}^{\omega_{\min}} (\omega_1 + \omega_2 - \alpha - \beta, -\alpha + \beta) \quad (6.87)$$

with $\omega_{\min} = \min(\omega_1, \omega_2)$. The $SO_{12}(4) \supset SO_{12}(3)$ branching rule is also known from Chapter 5 (see (5.190)) and reads, in the present notation,

$$L = |\omega'|, |\omega'| + 1, \ldots, \omega \quad (6.88)$$

The hamiltonian diagonal in this basis is of the form

$$\begin{aligned} \hat{H}_{\text{IIb}}^{\text{lin}} &= \beta_1 \mathcal{C}_2[SO_1(4)] + \beta_2 \mathcal{C}_2[SO_2(4)] + \beta \mathcal{C}_2[SO_{12}(4)] + \gamma \mathcal{C}_2[SO_{12}(3)] \\ &= \beta_1(\hat{L}_1^2 + \hat{D}_1^2) + \beta_2(\hat{L}_2^2 + \hat{D}_2^2) + \beta(\hat{L}^2 + \hat{D}^2) + \gamma \hat{L}^2 \end{aligned} \quad (6.89)$$

6.5 The SO(4) limits

and has the eigenvalues

$$
\begin{aligned}
E_{\text{IIb}}^{\text{lin}}&(\omega_1,\omega_2,\omega,\omega',L) \\
&= \beta_1\omega_1(\omega_1+2) + \beta_2\omega_2(\omega_2+2) + \beta[\omega(\omega+2)+\omega'^2] \\
&\quad + \gamma L(L+1)
\end{aligned}
\qquad (6.90)
$$

where the eigenvalue expression of $C_2[SO_{12}(4)]$ in the general representation (ω,ω') is taken from (5.126). According to the discussion in Section 5.8, the basis states (6.86) can be written in terms of (6.82):

$$
\boxed{\begin{aligned}
&|[N_1]\omega_1,[N_2]\omega_2;(\omega,\omega')LM_L\rangle \\
&= \sum_{L_1 L_2} \left\langle \begin{matrix}(\omega_1,0) & (\omega_2,0) \\ L_1 & L_2\end{matrix} \,\bigg|\, \begin{matrix}(\omega,\omega') \\ L\end{matrix} \right\rangle \\
&\quad \times |[N_1]\omega_1 L_1,[N_2]\omega_2 L_2;LM_L\rangle
\end{aligned}}
\qquad (6.91)
$$

The coefficient $\langle(\omega_1,0)L_1 \otimes (\omega_2,0)L_2|(\omega,\omega')L\rangle$ is an isoscalar factor associated with $SO(4) \supset SO(3)$, for which a closed expression is given in (5.211).

The expansion (6.91) enables us to study the transformation properties of the states $|[N_1]\omega_1,[N_2]\omega_2;(\omega,\omega')LM_L\rangle$ under the inversion operation $\hat{P}$. We first note that the parity of the states (6.82) is determined from

$$
\hat{P}|[N_1]\omega_1 L_1,[N_2]\omega_2 L_2;LM_L\rangle = (-)^{L_1+L_2}|[N_1]\omega_1 L_1,[N_2]\omega_2 L_2;LM_L\rangle
\qquad (6.92)
$$

which is valid since the parity of each state $|[N_\rho]\omega_\rho L_\rho M_\rho\rangle$ separately is $(-)^{L_\rho}$. Acting with the parity operator on (6.91) then gives

$$
\begin{aligned}
&\hat{P}|[N_1]\omega_1,[N_2]\omega_2;(\omega,\omega')LM_L\rangle \\
&= \sum_{L_1 L_2} (-)^{L_1+L_2} \left\langle \begin{matrix}(\omega_1,0) & (\omega_2,0) \\ L_1 & L_2\end{matrix} \,\bigg|\, \begin{matrix}(\omega,\omega') \\ L\end{matrix} \right\rangle \\
&\quad \times |[N_1]\omega_1 L_1,[N_2]\omega_2 L_2;LM_L\rangle
\end{aligned}
\qquad (6.93)
$$

which shows that, in general, the states (6.91) do not have a well-defined parity. From (5.211) and the symmetry properties of the $9j$ coefficient

[8] we deduce

$$\left\langle \begin{matrix} (\omega_1,0) & (\omega_2,0) \\ L_1 & L_2 \end{matrix} \middle| \begin{matrix} (\omega,\omega') \\ L \end{matrix} \right\rangle$$
$$= (-)^{\omega_1+\omega_2+\omega+L_1+L_2+L} \left\langle \begin{matrix} (\omega_1,0) & (\omega_2,0) \\ L_1 & L_2 \end{matrix} \middle| \begin{matrix} (\omega,-\omega') \\ L \end{matrix} \right\rangle$$
(6.94)

which shows that, for the specific case $\omega' = 0$, the isoscalar factor vanishes unless $\omega_1 + \omega_2 + \omega + L_1 + L_2 + L$ is even. Hence we find

$$\hat{P}|[N_1]\omega_1,[N_2]\omega_2;(\omega,0)LM_L\rangle$$
$$= (-)^{\omega_1+\omega_2+\omega+L}|[N_1]\omega_1,[N_2]\omega_2;(\omega,0)LM_L\rangle$$
$$= (-)^{L}|[N_1]\omega_1,[N_2]\omega_2;(\omega,0)LM_L\rangle \quad (6.95)$$

where the last equality follows from (6.87) with $-\alpha + \beta = 0$. We conclude that for $\omega' = 0$ the state has a well-defined parity $(-)^L$. For $\omega' \neq 0$ we obtain

$$\hat{P}|[N_1]\omega_1,[N_2]\omega_2;(\omega,\omega')LM_L\rangle$$
$$= (-)^{\omega_1+\omega_2+\omega+L}|[N_1]\omega_1,[N_2]\omega_2;(\omega,-\omega')LM_L\rangle \quad (6.96)$$

We may now define the linear combinations (assuming $\omega' > 0$ to avoid double counting)

$$|[N_1]\omega_1,[N_2]\omega_2;(\omega,\omega')LM_L\rangle^{\pm}$$
$$= \sqrt{\tfrac{1}{2}}\Big(|[N_1]\omega_1,[N_2]\omega_2;(\omega,\omega')LM_L\rangle$$
$$\pm(-)^{\omega_1+\omega_2+\omega+L}|[N_1]\omega_1,[N_2]\omega_2;(\omega,-\omega')LM_L\rangle\Big) \quad (6.97)$$

which transform properly under parity,

$$\hat{P}|[N_1]\omega_1,[N_2]\omega_2;(\omega,\omega')LM_L\rangle^{\pm} = \pm|[N_1]\omega_1,[N_2]\omega_2;(\omega,\omega')LM_L\rangle^{\pm}$$
(6.98)

Since, according to (6.90), states with $\pm\omega'$ are degenerate in energy, the combinations are still eigenstates of the hamiltonian $\hat{H}_{\text{IIb}}^{\text{lin}}$ but now have a well-defined parity. Strictly speaking they are no longer classified by

6.5 The SO(4) limits

$SO_{12}(4)$ but rather by $O_{12}(4)$ (see the discussions in Sections 1.5 and 2.5).

The preceding discussion enables us to cast the branching and multiplication rules in this limit in a slightly different but equivalent way [3] which includes the parity quantum number π. In the $SO_1(4) \otimes SO_2(4)$ multiplication we restrict ω' to non-negative values, that is,

$$\omega = \omega_1 + \omega_2 - \alpha - \beta, \qquad \omega' = -\alpha + \beta, \qquad 0 \leq \alpha \leq \beta \leq \omega_{\min} \quad (6.99)$$

It then follows that the $SO_{12}(4)$ representation (ω, ω') contains the following angular momentum–parity states:

$$L^\pi = 0^+, 1^-, 2^+, \ldots, \begin{cases} \omega^+ & \text{even } \omega \\ \omega^- & \text{odd } \omega \end{cases} \quad (6.100)$$

for $\omega' = 0$ and

$$L^\pi = \omega'^\pm, (\omega'+1)^\pm, \ldots, \omega^\pm \quad (6.101)$$

for $\omega' \neq 0$. This is the convention we henceforth adopt, that is, we take $\omega' \geq 0$ and label states by their parity.

For the interpretation of the energy spectrum it is essential that we relate the quantum numbers ω_1, ω_2, ω, and ω' to those traditionally used to describe linear triatomic molecules. The assumption of forces between nuclei that vary quadratically with the deviation from equilibrium position (harmonic-oscillator interactions) leads to normal vibrations [9] and to the concept of a normal-mode basis. In a linear XYZ molecule one finds two stretching vibrations v_1 and v_3 and two (degenerate) bending vibrations v_{2a} and v_{2b}. The stretching vibrations v_1 and v_3 do not generate any angular momentum, and the total (rotational and vibrational) angular momentum $\mathbf{L}$ is perpendicular to the axis of symmetry (see Figure 6.4). The bending vibrations may combine to give a composite harmonic motion consisting of each nucleus swinging around the axis of symmetry. A single v_2 quantum gives rise to a vibrational angular momentum $\mathbf{l}$ with magnitude $l = 1$ pointing along the axis of symmetry; this state is doubly degenerate on account of the two possible alignments of $\mathbf{l}$. The combination of several $v_2 = v_{2a} + v_{2b}$ quanta leads to various species of vibrational levels, denoted as $\Lambda = \Sigma, \Pi, \Delta, \Phi, \ldots$ and corresponding to vibrational

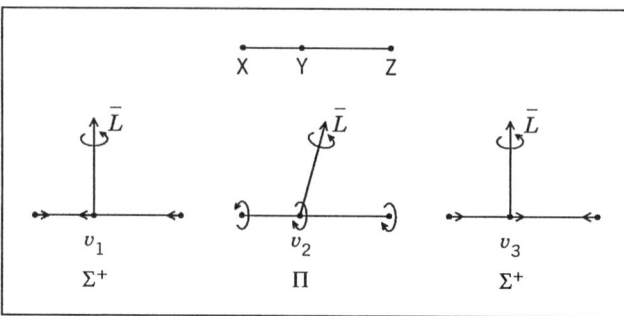

Figure 6.4: Stretching (v_1 and v_3) and bending (v_2) normal vibrations in linear XYZ molecules. Also shown are the total angular momentum **L**, the vibrational angular momentum l (of v_2 vibrations), and the vibrational species designation Σ^+ and $\Pi^\pm$.

angular momenta $l = 0, 1, 2, 3, \ldots$. Except for the Σ species these are all doubly degenerate. A general vibrational state in the normal-mode basis consists of simultaneous v_1, v_2, and v_3 vibrations (overtone frequencies) and is thus characterized by (v_1, v_2^l, v_3). The combination of rotational motion with the various species of vibrational levels leads to the rotational bands schematically depicted in Figure 6.5. We point out that the degeneracy of $\Pi, \Delta, \Phi, \ldots$ levels is removed as the rotation increases (l-type doubling [9]), but this effect is relatively small and neglected here.

In more recent years the limitations of the normal-mode picture have become clear, and the extension of the observation range to higher overtone levels has shown the advantages of the local-mode description, particularly for X–H vibrations [10]. The local-mode basis appears more appropriate for asymmetrical molecules. A linear XYZ molecule has two local stretching vibrations, denoted here as v_a and v_c, and two (degenerate) bent vibrations, v_{b1} and v_{b2}. The latter are identical to the normal vibrations, but the former are different from their normal counterparts (see Figure 6.6). In the local-mode basis a state is thus characterized by (v_a, v_b^l, v_c).

From the analogous problem discussed in Section 2.7 in the context

6.5 The SO(4) limits

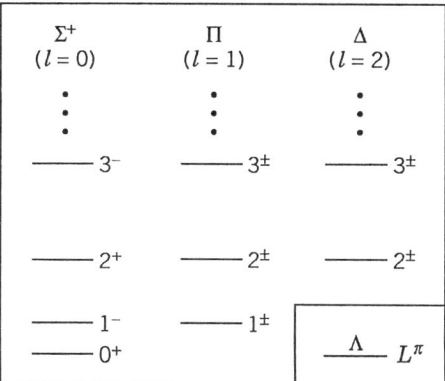

Figure 6.5: Rotational levels in various species of vibrational levels of rigid linear molecules. Levels are labeled by the total angular momentum L, parity π, and the vibrational species Λ.

of the $U_1(2) \otimes U_2(2)$ model, we expect the basis (6.86) to correspond to local-mode states whereas route a in the lattice (6.23) defines a normal-mode basis. We thus seek to establish a connection between $(\omega_1, \omega_2, \omega, \omega')$ and the local-mode quantum numbers (v_a, v_b^l, v_c). The labels v_a and v_c count the vibrational quanta for the bonds X–Y and Y–Z separately, and hence, because of the relation found between v and ω for diatomic molecules (see (5.131)), it is natural to propose the correspondence $v_a = (N_1 - \omega_1)/2$ and $v_c = (N_2 - \omega_2)/2$. A comparison of the angular momenta and parities (6.100) and (6.101) contained in an $SO_{12}(4)$ representation (ω, ω') with those in a rotational band of a given vibrational species (see Figure 6.5) leads to the association $l = \omega'$. A last connection between the algebraic and the local-mode quantum numbers can be proposed using the relations established in Section 2.7 which are appropriate for stretching vibrations of triatomic molecules (i.e., for $v_b = 0$). It is found there (see (2.93)) that the sum $v_a + v_c$ is given by $(N_1 + N_2 - m_1 - m_2)/2$ or $(N_1 + N_2 - m)/2$, where m_ρ is the $SO_\rho(2)$ label and m is associated with the combined algebra $SO_{12}(2)$. It is natural to identify the $SO_{12}(4)$ label ω with m and consequently we find the relation $\omega = N_1 + N_2 - 2v_a - 2v_c$, valid for pure stretching vibrations. On account of the double degeneracy of

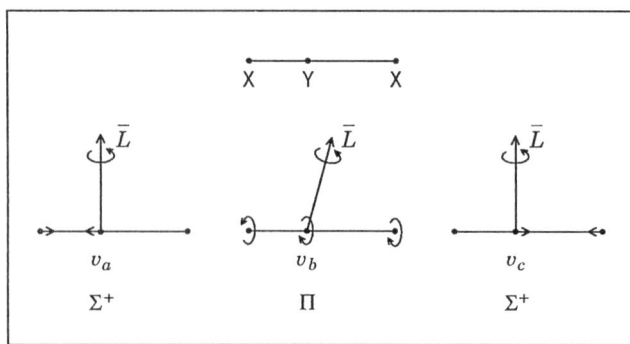

Figure 6.6: Stretching (v_a and v_c) and bending (v_b) local vibrations in linear XYZ molecules. Also shown are the total angular momentum **L**, the vibrational angular momentum **l** (of v_b vibrations), and the vibrational species designation Σ^+ and $\Pi^\pm$.

the bending vibration this is generalized to an arbitrary vibration as $\omega = N_1 + N_2 - 2v_a - v_b - 2v_c$. Summing up, we propose the relations

$$\begin{aligned} \omega_1 &= N_1 - 2v_a, & \omega &= N_1 + N_2 - 2v_a - v_b - 2v_c \\ \omega_2 &= N_2 - 2v_c, & \omega' &= l \end{aligned} \quad (6.102)$$

which inserted in the energy expression (6.90) lead to

$$\begin{aligned} E_{\text{IIb}}^{\text{lin}}&(v_a, v_b^l, v_c, L) \\ =& \beta_1(N_1+1)(N_1+3) + \beta_2(N_2+1)(N_2+3) + \beta(N+3)(N+5) \\ & -4[\beta_1(N_1+2) + \beta(N+4)]\left(v_a + \tfrac{1}{2}\right) - 2\beta(N+4)(v_b+1) \\ & -4[\beta_2(N_2+2) + \beta(N+4)]\left(v_c + \tfrac{1}{2}\right) \\ & +4(\beta_1+\beta)\left(v_a+\tfrac{1}{2}\right)^2 + \beta(v_b+1)^2 + 4(\beta_2+\beta)\left(v_c+\tfrac{1}{2}\right)^2 \\ & +4\beta\left(v_a+\tfrac{1}{2}\right)(v_b+1) + 8\beta\left(v_a+\tfrac{1}{2}\right)\left(v_c+\tfrac{1}{2}\right) \\ & +4\beta(v_b+1)\left(v_c+\tfrac{1}{2}\right) \\ & +\beta l^2 + \gamma L(L+1) \end{aligned} \quad (6.103)$$

where $N = N_1 + N_2$. We see that (6.103) is a specific example of a

6.5 The SO(4) limits

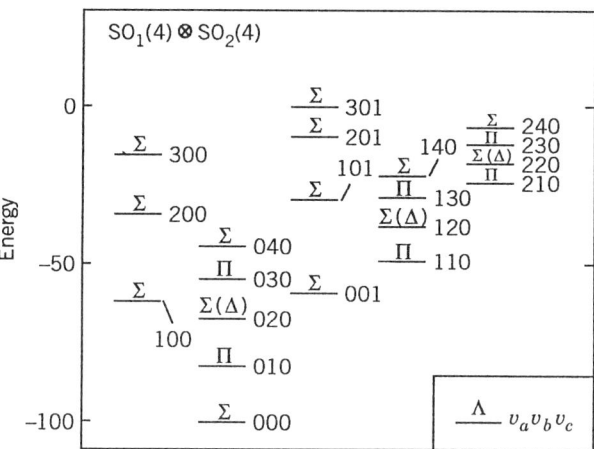

Figure 6.7: Schematic $SO_1(4) \otimes SO_2(4)$ eigenspectrum of the hamiltonian $\hat{H}^{\text{lin}}_{\text{IIb}}$ with $\beta_1 = -1$, $\beta_2 = -0.25$, and $\beta = -1$ for $N_1 = 6$ and $N_2 = 2$ bosons. Levels are labeled by (v_a, v_b, v_c) and the species Σ, Π, and Δ ($l = 0, 1, 2$). Energies are in arbitrary units. Only the energy of the lowest species Σ or Π is shown for each level. Eventual higher species are shown in parentheses, but their energy, which is slightly different from that of the Σ or Π species, is not plotted. Not shown in the figure are the rotational bands on top of each vibrational level.

Dunham-like expansion in terms of the local vibrational quanta v_a, v_b, v_c, and l. An example of an $SO_1(4) \otimes SO_2(4)$ spectrum, constructed from the branching rules (6.80), (6.99), (6.100), and (6.101) and the energy expression (6.103), is shown in Figure 6.7. Due to the relations (6.102) we find that the stretching modes are determined by the $SO_1(4) \otimes SO_2(4)$ representation and all bending modes on top of a fixed stretching excitation are contained in a single $SO_1(4) \otimes SO_2(4)$ representation. The rotational band on top of each vibrational level belongs to a single $SO_{12}(4)$ representation. (The individual rotational levels are not shown in Figure 6.7 but are as in Figure 6.5.) One point that should be emphasized is that the species $\Sigma, \Pi, \Delta, \ldots$ appropriate to the doubly degenerate nature of the v_b bending vibrations are *automatically* reproduced by the rule (6.99).

The second possibility of interest for linear triatomic molecules corresponds to route a in the lattice (6.23) with basis states defined by

$$\begin{array}{|ccccc|}
\hline
U_1(4) \otimes U_2(4) & \supset U_{12}(4) & \supset SO_{12}(4) & \supset SO_{12}(3) \\
| & | & | & | & | \\
{[N_1]} & [N_2] & [h,h'] & \gamma(\omega,\omega') & L \\
\hline
\end{array} \quad (6.104)$$

We refer to this dynamical symmetry as the $SO_{12}(4)$ limit. The $U(4) \supset SO(4)$ branching rule is the only one yet to be determined in (6.104) and can be derived with the technique explained in Section 6.4 which consists of deriving it for increasing values of h'. We start with the simplest (non-trivial) case of $N_1 > N_2 = 1$. In the basis (6.86) this implies $\omega_2 = 1$, and hence the allowed (ω, ω') values are obtained from (6.87) with $\omega_1 = N_1, N_1 - 2, \ldots, 1$ or 0, and $\omega_2 = 1$,

$$\begin{aligned}
(\omega,\omega') &= (N_1+1,0), (N_1-1,0), \ldots, \left\{ \begin{array}{c} (2,0) \\ (1,0) \end{array} \right\}, \\
&\quad (N_1,\pm 1), (N_1-2,\pm 1), \ldots, \left\{ \begin{array}{c} (1,\pm 1) \\ (2,\pm 1) \end{array} \right\}, \\
&\quad (N_1-1,0), (N_1-3,0), \ldots, \left\{ \begin{array}{c} (0,0) \\ (1,0) \end{array} \right\} \quad (6.105)
\end{aligned}$$

where the upper (lower) labels in the curly brackets should be taken for odd (even) N_1. On the other hand, the $U_{12}(4)$ representation $[h,h']$ can be $[N_1+1,0]$ or $[N_1,1]$. Substracting the $SO_{12}(4)$ representations contained in $[N_1+1,0]$ (known from (6.80)), we obtain those contained in $[N_1,1]$:

$$\begin{aligned}
(\omega,\omega') &= (N_1-1,0), (N_1-3,0), \ldots, \left\{ \begin{array}{c} (2,0) \\ (1,0) \end{array} \right\}, \\
&\quad (N_1,\pm 1), (N_1-2,\pm 1), \ldots, \left\{ \begin{array}{c} (1,\pm 1) \\ (2,\pm 1) \end{array} \right\} \quad (6.106)
\end{aligned}$$

In contrast to the cases considered in Section 6.4, it is more difficult to give a general rule for the $U(4) \supset SO(4)$ reduction, but it is clear that the procedure illustrated here for $h' = 1$ can be extended indefinitely to higher values of h' since the $SO(4)$ multiplication (6.87) is known

6.5 The SO(4) limits

in general. Table C-2 of [11] includes all cases up to $h' \leq h \leq 9$. We also note that the reduction $U(4) \supset SO(4)$ is not simply reducible, and hence a missing label γ is needed in (6.104).

The hamiltonian in this limit is, for linear molecules,

$$\begin{aligned}
\hat{H}_{\text{IIa}}^{\text{lin}} &= \eta \mathcal{C}_2[U_{12}(4)] + \beta \mathcal{C}_2[SO_{12}(4)] + \gamma \mathcal{C}_2[SO_{12}(3)] \\
&= \eta \left(\tfrac{1}{3}\hat{n}_{\text{p}}^2 + \tfrac{1}{2}(\hat{L}^2 + \hat{D}^2 + \hat{D}'^2) + \hat{Q}^2 + \hat{n}_{\text{s}}^2 \right) \\
&\quad + \beta(\hat{L}^2 + \hat{D}^2) + \gamma \hat{L}^2
\end{aligned} \quad (6.107)$$

and has the eigenvalues

$$\begin{aligned}
E_{\text{IIa}}^{\text{lin}}&(h, h', \omega, \omega', L) \\
&= \eta[h(h+3) + h'(h'+1)] + \beta[\omega(\omega+2) + \omega'^2] \\
&\quad + \gamma L(L+1)
\end{aligned} \quad (6.108)$$

The eigenstates of $\hat{H}_{\text{IIa}}^{\text{lin}}$ can be written as

$$\boxed{\begin{aligned}
&|[N_1], [N_2]; [h, h']\gamma(\omega, \omega') L M_L\rangle \\
&= \sum_{\omega_1 \omega_2} \sum_{L_1 L_2} \left\langle \begin{array}{cc} [N_1] & [N_2] \\ (\omega_1, 0) & (\omega_2, 0) \end{array} \bigg| \begin{array}{c} [h, h'] \\ \gamma(\omega, \omega') \end{array} \right\rangle \\
&\quad \times \left\langle \begin{array}{cc} (\omega_1, 0) & (\omega_2, 0) \\ L_1 & L_2 \end{array} \bigg| \begin{array}{c} (\omega, \omega') \\ L \end{array} \right\rangle \\
&\quad \times |[N_1]\omega_1 L_1, [N_2]\omega_2 L_2; L M_L\rangle \\
&= \sum_{\omega_1 \omega_2} \left\langle \begin{array}{cc} [N_1] & [N_2] \\ (\omega_1, 0) & (\omega_2, 0) \end{array} \bigg| \begin{array}{c} [h, h'] \\ \gamma(\omega, \omega') \end{array} \right\rangle \\
&\quad \times |[N_1]\omega_1, [N_2]\omega_2; (\omega, \omega') L M_L\rangle
\end{aligned}} \quad (6.109)$$

The expansion now involves two kinds of isoscalar factors associated with $U(4) \supset SO(4)$ and $SO(4) \supset SO(3)$, respectively. General expressions for the former coefficient are not known, and hence (6.109) can only be regarded as a formal expansion.

The states (6.109) in general do not have a definite parity, but as in the $SO_1(4) \otimes SO_2(4)$ limit, it is possible to define linear combinations that remain eigenstates of the hamiltonian $\hat{H}_{\text{IIa}}^{\text{lin}}$ and do transform properly under the inversion operator $\hat{P}$. From (6.95) and the second expansion given in (6.109) it follows that for $\omega' = 0$ these states have parity $(-)^L$. In general, we have the property

$$\hat{P}|[N_1],[N_2];[h,h']\gamma(\omega,\omega')LM_L\rangle$$
$$= (-)^{N_1+N_2+\omega+L}|[N_1],[N_2];[h,h']\gamma(\omega,-\omega')LM_L\rangle \quad (6.110)$$

which follows from (6.96) and the fact that $N_\rho - \omega_\rho$ is even. As a consequence, for $\omega' \neq 0$, the linear combinations

$$|[N_1],[N_2];[h,h']\gamma(\omega,\omega')LM_L\rangle^\pm$$
$$= \sqrt{\tfrac{1}{2}}\Big(|[N_1],[N_2];[h,h']\gamma(\omega,\omega')LM_L\rangle$$
$$\pm(-)^{N_1+N_2+\omega+L}|[N_1],[N_2];[h,h']\gamma(\omega,-\omega')LM_L\rangle\Big) \quad (6.111)$$

have definite parity $\pm$. As an alternative (but equivalent) labeling scheme we may again restrict ω' to non-negative values and characterize eigenstates by their parity π.

Our next problem is to find the relation between the quantum numbers appearing in (6.104) and vibrational quantum numbers commonly used in the interpretation of triatomic molecular spectra. While we argued that the labels defined in (6.86) are connected to the local-mode vibrations, we expect, in analogy with the discussion in Section 2.7, that (h, h', ω, ω') are related to the normal-mode quantum numbers (v_1, v_2^l, v_3). Note, however, that algebraically only three independent labels exist (since $h + h' = N_1 + N_2$ is constant), and we cannot hope to find a one-to-one correspondence between the two sets. A fourth independent quantum number appears in (6.104) in the form of the missing label γ. However, because it is not connected to any algebra, its interpretation is difficult. Also, being absent from the eigenvalue expression (6.108), it is of no relevance to the energy spectrum. To find the connection between (h, h', ω, ω') and (v_1, v_2^l, v_3), we observe that the bending vibrations are identical in local- and normal-mode bases, $v_b = v_2$. Consequently, we must adopt the same relation between (ω, ω') and (v_2^l) as in (6.102). A third connection between the algebraic and normal-mode quantum numbers can be established on the basis of relations derived in the context of the $U_1(2) \otimes U_2(2)$ model of Section 2.7. We note that the sum of the stretching vibrational quantum numbers is the same in the local- and normal-mode basis; in the $U_1(2) \otimes U_2(2)$

6.5 The SO(4) limits

model it equals $v_1 + v_3 = v_a + v_c = (N_1 + N_2 - m)/2$ where m is the $SO_{12}(2)$ label. Furthermore, the number of quanta in the symmetric mode, v_1, is given by $v_1 = j - m/2$ (see (2.101)), which can be rewritten as $v_1 = (N_1 + N_2 - m - 2h')/2$ (see (2.38)), where h' is the second $U_{12}(2)$ label. The combination of these results leads to $h' = v_3$. In the $SO_{12}(4)$ limit of the $U_1(4) \otimes U_2(4)$ model we adopt, for pure stretching vibrations, the same relation $h' = v_3$, where h' is now the second $U_{12}(4)$ label. For a general vibration van Roosmalen et al. [3] propose the relation $h' = v_2 + v_3$. In summary, the following correspondence is established:

$$h = N_1 + N_2 - v_2 - v_3, \qquad \omega = N_1 + N_2 - 2v_1 - v_2 - 2v_3$$
$$h' = v_2 + v_3, \qquad \omega' = l \tag{6.112}$$

Inserted in the energy expression (6.108), this leads to

$$\begin{aligned}
E^{\text{lin}}_{\text{IIa}}&(v_1, v_2^l, v_3, L) \\
&= \tfrac{15}{2}(\eta + 2\beta) + 2(3\eta + 4\beta)N + (\eta + \beta)N^2 \\
&\quad -4\beta(N+4)\left(v_1 + \tfrac{1}{2}\right) - 2(\eta + \beta)(N+4)(v_2 + 1) \\
&\quad -2(\eta + 2\beta)(N+4)\left(v_3 + \tfrac{1}{2}\right) \\
&\quad +4\beta\left(v_1 + \tfrac{1}{2}\right)^2 + (2\eta + \beta)(v_2 + 1)^2 + 2(\eta + 2\beta)\left(v_3 + \tfrac{1}{2}\right)^2 \\
&\quad +4\beta\left(v_1 + \tfrac{1}{2}\right)(v_2 + 1) + 8\beta\left(v_1 + \tfrac{1}{2}\right)\left(v_3 + \tfrac{1}{2}\right) \\
&\quad +4(\eta + \beta)(v_2 + 1)\left(v_3 + \tfrac{1}{2}\right) \\
&\quad +\beta l^2 + \gamma L(L+1)
\end{aligned} \tag{6.113}$$

where $N = N_1 + N_2$. Again, we find a Dunham expansion, but this time in terms of the normal vibrational quanta v_1, v_2, v_3, and l. In Figure 6.8 an example is shown of a $SO_{12}(4)$ spectrum calculated with (6.113). Note that in (6.113) it is not possible to adjust independently the energies of the three vibrational modes since the coefficients of $(v_1 + 1/2)$, $(v_2 + 1)$, and $(v_3 + 1/2)$ depend on just two parameters η and β. Also, the energies of the vibrational levels do not depend on N_1 and N_2 separately (as is the case in (6.103)) but only on the sum $N_1 + N_2$. We conclude that the $SO_{12}(4)$ limit leads to a normal-mode

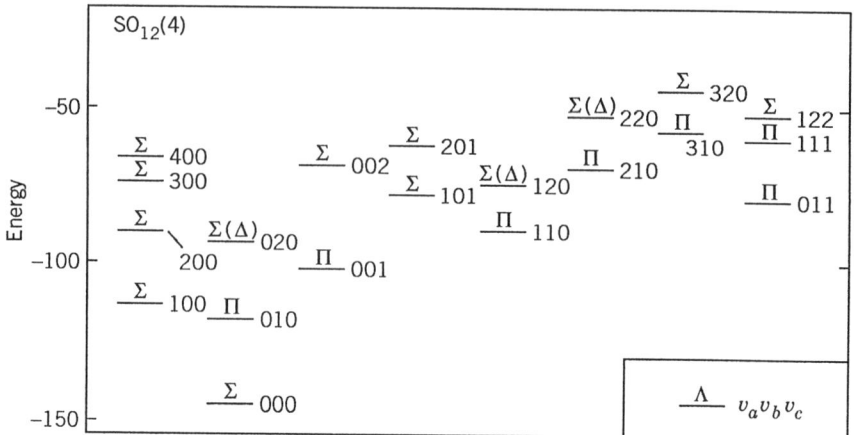

Figure 6.8: Schematic $SO_{12}(4)$ eigenspectrum of the hamiltonian $\hat{H}^{\text{lin}}_{\text{IIa}}$ with $\eta = -0.75$ and $\beta = -1$ for $N_1 = 6$ and $N_2 = 2$ bosons. Levels are labeled by (v_1, v_2, v_3) and the species Σ, Π, and Δ ($l = 0, 1, 2$), as in Figure 6.7. Energies are in arbitrary units.

basis but does not give rise to the most general hamiltonian for rigid linear triatomic molecules.

6.5.2 Bent molecules

In a linear triatomic molecule the projection l of the angular momentum along the molecular axis can, for a given number v_2 of bending quanta, take the values

$$l = v_2, v_2 - 2, \ldots, 1 \text{ or } 0 \qquad (6.114)$$

In the previous section this rule automatically followed as a consequence of the branching and multiplication rules (6.99) together with the relations

$$v_2 = \omega_1 + \omega_2 - \omega = \alpha + \beta \qquad (6.115)$$

and

$$l = \omega' = -\alpha + \beta = v_2 - 2\alpha \qquad (6.116)$$

6.5 The SO(4) limits

The most important difference between linear and non-linear molecules is that in the latter the values of the projection quantum number (usually referred to as K) are unrestricted because the molecule is permanently bent.

The hamiltonians proposed in the previous section, which are special cases of the general hamiltonian (6.11), are not capable of generating spectra with the degeneracies appropriate for bent triatomic molecules. This objective can be achieved, however, by considering the effect of terms of the type $\hat{L} \cdot \hat{D}$ in the hamiltonian (6.11). More specifically, we may add to (6.11) the following interaction:

$$\hat{V}'_{12} = v_9[[p_1^\dagger \times p_2^\dagger]^{(1)} \times [\tilde{p}_1 \times \tilde{s}_2]^{(1)} + [p_1^\dagger \times s_2^\dagger]^{(1)} \times [\tilde{p}_1 \times \tilde{p}_2]^{(1)}]_0^{(0)}$$
$$+ v_{10}[[p_1^\dagger \times p_2^\dagger]^{(1)} \times [\tilde{s}_1 \times \tilde{p}_2]^{(1)} + [s_1^\dagger \times p_2^\dagger]^{(1)} \times [\tilde{p}_1 \times \tilde{p}_2]^{(1)}]_0^{(0)}$$
(6.117)

Alternatively, this interaction can be written in terms of $\hat{L} \cdot \hat{D}$ and $\hat{L} \cdot \hat{D}'$. As usual we study the properties of the interaction $\hat{L} \cdot \hat{D}$, involving the generator of $SO_{12}(4)$, since the other interaction $\hat{L} \cdot \hat{D}'$ can be dealt with in a similar fashion in the context of the $\overline{SO}_{12}(4)$ limit. Note that $\hat{L} \cdot \hat{D}$ contains terms of the type $\hat{L}_\rho \cdot \hat{D}_\rho$; these vanish identically in the symmetric $U_\rho(4)$ representations, and consequently $\hat{L} \cdot \hat{D}$ is a two-body interaction term.

The interaction $\hat{V}'_{12}$ is rotationally invariant (or scalar) but violates parity; it is therefore sometimes referred to as *pseudo-scalar*. In the context of the vibron model pseudo-scalar interactions generally are not of much use since they do not lead to eigenstates that carry a definite parity. For this reason the interactions $\hat{L} \cdot \hat{D}$ and $\hat{L} \cdot \hat{D}'$ are not included in the general hamiltonian (6.11). However, a slight modification of the interaction $\hat{L} \cdot \hat{D}$, added to a hamiltonian with an $SO_{12}(4)$ dynamical symmetry, leads to the description of bent triatomic molecules while still maintaining the property of parity invariance, as we now proceed to show.

We first deduce the matrix elements of $\hat{L} \cdot \hat{D}$ in a basis with good $SO_{12}(4)$ quantum numbers (ω, ω'), such as (6.86) or (6.104). In terms of the operators $\hat{R}_\mu$ and $\hat{R}'_\mu$ introduced in (5.118), $\hat{L} \cdot \hat{D}$ can be written as $\hat{R}^2 - \hat{R}'^2$, and hence its diagonal matrix elements are given by

$$r(r+1) - r'(r'+1) = \omega'(\omega+1) \tag{6.118}$$

where use is made of the correspondence (5.123). This also shows that $\hat{L} \cdot \hat{D}$ is diagonal in the bases (6.86) and (6.104), and as we demonstrated in Section 6.5.1, the associated wave functions do not carry good parity but appropriate linear combinations of $+\omega'$ and $-\omega'$ do. For the particular hamiltonians considered in the previous section with eigenvalues (6.90) and (6.108), states with $\pm\omega'$ are degenerate in energy and hence such linear combinations remain eigenstates of the hamiltonian. If, however, the term $\hat{L} \cdot \hat{D}$ is added to the hamiltonian, the states with $\pm\omega'$ are no longer degenerate and the construction of eigenstates with good parity fails. This argument constitutes an explicit proof of the pseudo-scalar nature of the interaction $\hat{L} \cdot \hat{D}$. A parity-invariant modification of this interaction can be defined as

$$\mathcal{C}'_2[SO_{12}(4)] = |\hat{L} \cdot \hat{D}| \qquad (6.119)$$

which clearly has the same eigenvalue $|\omega'(\omega+1)|$ for $+\omega'$ and $-\omega'$, and consequently the construction of eigenstates with good parity becomes possible again. Once good-parity eigenstates are defined, we may again, as in Section 6.5.1, label states by a non-negative ω' and a parity π (instead of positive and negative ω' values), and this is the convention we henceforth follow.

The definition of an absolute value is meaningful for diagonal operators (such as $\mathcal{C}'_2[SO_{12}(4)]$ in a basis with $SO_{12}(4)$ symmetry) since it then simply corresponds to taking the absolute value of the eigenvalue. It cannot, however, be extended to operators with off-diagonal matrix elements. We conclude that the operator (6.119) is properly defined only in bases with $SO_{12}(4)$ symmetry and we intend to use it in such bases only.

To arrive at a description of bent triatomic molecules, we add the operator $\mathcal{C}'_2[SO_{12}(4)]$ to the hamiltonian $\hat{H}^{\text{lin}}_{\text{II}b}$ proposed for linear molecules and define the most general (two-body) hamiltonian associated with route b in the lattice (6.23):

$$\begin{aligned}
\hat{H}_{\text{II}b} &= \beta_1 \mathcal{C}_2[SO_1(4)] + \beta_2 \mathcal{C}_2[SO_2(4)] + \beta \mathcal{C}_2[SO_{12}(4)] \\
&\quad + \beta' \mathcal{C}'_2[SO_{12}(4)] + \gamma \mathcal{C}_2[SO_{12}(3)] \\
&= \beta_1(\hat{L}_1^2 + \hat{D}_1^2) + \beta_2(\hat{L}_2^2 + \hat{D}_2^2) + \beta(\hat{L}^2 + \hat{D}^2) + \beta'|\hat{L} \cdot \hat{D}| + \gamma \hat{L}^2
\end{aligned} \qquad (6.120)$$

6.5 The SO(4) limits

For $\beta' = 2\beta$ this hamiltonian reduces to one appropriate for bent molecules,

$$\hat{H}_{\text{IIb}}^{\text{bent}} = \beta_1 \mathcal{C}_2[SO_1(4)] + \beta_2 \mathcal{C}_2[SO_2(4)] + \beta \mathcal{C}_2[SO_{12}(4)]$$
$$+ 2\beta \mathcal{C}_2'[SO_{12}(4)] + \gamma \mathcal{C}_2[SO_{12}(3)] \quad (6.121)$$

with eigenvalues

$$E_{\text{IIb}}^{\text{bent}}(\omega_1, \omega_2, \omega, \omega', L)$$
$$= \beta_1 \omega_1(\omega_1 + 2) + \beta_2 \omega_2(\omega_2 + 2) + \beta(\omega + \omega')(\omega + \omega' + 2)$$
$$+ \gamma L(L + 1) \quad (6.122)$$

Note that the eigenstates of $\hat{H}_{\text{IIb}}^{\text{bent}}$ (or indeed of the more general hamiltonian $\hat{H}_{\text{IIb}}$) are identical to the ones found for linear molecules, as are the associated branching and multiplication rules, and the relevant results obtained in the previous section are also valid for bent molecules. Differences between the two cases only arise due to the different eigenvalue expressions (6.90) and (6.122) and also because the labels ω and ω' have a different meaning. This different interpretation of the algebraic quantum numbers is suggested by the correlation diagram [12] which relates energy levels in linear and bent molecules (see Figure 6.9). This correlation diagram shows that the appropriate correspondence is

$$v_b^{\text{lin}} \mapsto 2v_b^{\text{bent}} + l, \qquad l \mapsto K \quad (6.123)$$

and after inserting these relations into (6.102), we find

$$\omega_1 = N_1 - 2v_a, \qquad \omega = N_1 + N_2 - 2v_a - 2v_b - 2v_c - K$$
$$\omega_2 = N_2 - 2v_c, \qquad \omega' = K$$
$$(6.124)$$

Note the difference between the relations (6.102) and (6.124) with regard to the bending mode v_b. This originates from the doubly degenerate nature of this mode in linear molecules as opposed to it being non-degenerate in bent molecules. Inserting (6.124) into the energy expression (6.122), we obtain

$$E_{\text{IIb}}^{\text{bent}}(v_a, v_b, v_c, K, L)$$
$$= \beta_1(N_1 + 1)(N_1 + 3) + \beta_2(N_2 + 1)(N_2 + 3) + \beta(N + 3)(N + 5)$$

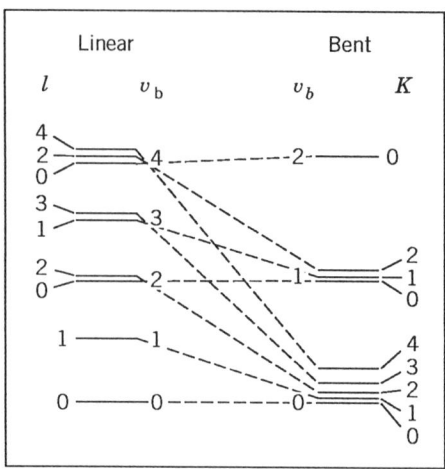

Figure 6.9: Correlation diagram for the rotation–vibration spectra of linear and bent triatomic molecules.

$$-4[\beta_1(N_1+2)+\beta(N+4)]\left(v_a+\tfrac{1}{2}\right)-4\beta(N+4)\left(v_b+\tfrac{1}{2}\right)$$
$$-4[\beta_2(N_2+2)+\beta(N+4)]\left(v_c+\tfrac{1}{2}\right)$$
$$+4(\beta_1+\beta)\left(v_a+\tfrac{1}{2}\right)^2+4\beta\left(v_b+\tfrac{1}{2}\right)^2+4(\beta_2+\beta)\left(v_c+\tfrac{1}{2}\right)^2$$
$$+8\beta\left(v_a+\tfrac{1}{2}\right)\left(v_b+\tfrac{1}{2}\right)+8\beta\left(v_a+\tfrac{1}{2}\right)\left(v_c+\tfrac{1}{2}\right)$$
$$+8\beta\left(v_b+\tfrac{1}{2}\right)\left(v_c+\tfrac{1}{2}\right)$$
$$+\gamma L(L+1) \tag{6.125}$$

Since (6.122) depends only on the sum of ω and ω' and not on their separate values, relations (6.124) lead to a K-independent energy expression. This is illustrated in Figure 6.10, constructed from (6.125) together with the branching and multiplication rules given in the previous section.

The absence of a K quantum number from (6.125) indicates that this energy expression is, strictly speaking, valid for a spherical top only. Symmetric and asymmetric tops can be obtained by relaxing the condition $\beta'=2\beta$ and by introducing higher-order terms in the hamiltonian $\hat{H}_{\mathrm{IIb}}$. In this way one arrives at a more realistic description

6.5 The SO(4) limits

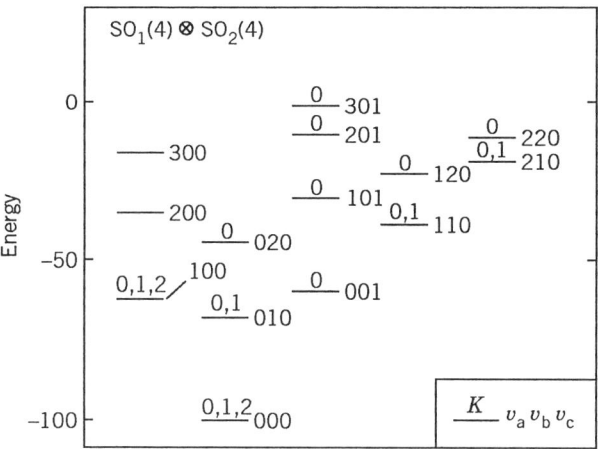

Figure 6.10: Schematic $SO_1(4) \otimes SO_2(4)$ eigenspectrum of the hamiltonian $\hat{H}_{\text{IIb}}^{\text{bent}}$ with $\beta_1 = -0.25$, $\beta_2 = -1$, and $\beta = -1$ for $N_1 = 6$ and $N_2 = 2$ bosons. Levels are labeled by (v_a, v_b, v_c) and the rotational quantum number K. Energies are in arbitrary units. Not shown in the figure are the rotational bands on top of each vibrational level.

of bent triatomic molecules.

The K independence of the energy expressions is a necessary feature of a first-order description of non-linear molecules. We have seen that in the $SO_1(4) \otimes SO_2(4)$ limit this K independence comes about by adding an $|\hat{L} \cdot \hat{D}|$ term with strength $\beta' = 2\beta$ to the hamiltonian for linear molecules. This leads to a constant energy for levels with constant $\omega + \omega'$ (and all other labels the same) and, because of (6.124), to K independence. It is not possible to repeat the same procedure for the $SO_{12}(4)$ limit corresponding to route a in the lattice (6.23), since in this limit the $SO_{12}(4)$ representations (ω, ω') with constant $\omega + \omega'$ are contained in different $U_{12}(4)$ representations $[h, h']$ and hence are, in general, not degenerate in energy. We thus find that it is not possible to describe bent triatomic molecules in the $SO_{12}(4)$ limit of the vibron model.

This concludes our review of analytically solvable hamiltonians that describe various types of rigid rotational triatomic molecules, ranging from linear to bent with vibrations exhibiting local-mode as well as normal-mode behavior. In the process we have collected all ingredients for constructing a more general hamiltonian (to be treated numerically) that deals with intermediate situations in a continuous way. This more general hamiltonian is applied in the next section to the calculation of triatomic molecular spectra.

6.6 Examples of triatomic spectra

From the previous section it is clear that a versatile vibron hamiltonian that deals with a wide variety of different types of triatomic spectra is of the form

$$\begin{aligned}\hat{H}_{\text{II}} &= \eta \mathcal{C}_2[\text{U}_{12}(4)] + \beta_1 \mathcal{C}_2[\text{SO}_1(4)] + \beta_2 \mathcal{C}_2[\text{SO}_2(4)] \\ &\quad + \beta \mathcal{C}_2[\text{SO}_{12}(4)] + \beta' \mathcal{C}'_2[\text{SO}_{12}(4)] + \gamma \mathcal{C}_2[\text{SO}_{12}(3)] \\ &= \eta \left(\tfrac{1}{3}\hat{n}_p^2 + \tfrac{1}{2}(\hat{L}^2 + \hat{D}^2 + \hat{D}'^2) + \hat{Q}^2 + \hat{n}_s^2 \right) \\ &\quad + \beta_1 (\hat{L}_1^2 + \hat{D}_1^2) + \beta_2 (\hat{L}_2^2 + \hat{D}_2^2) \\ &\quad + \beta (\hat{L}^2 + \hat{D}^2) + \beta' |\hat{L} \cdot \hat{D}| + \gamma \hat{L}^2 \end{aligned} \quad (6.126)$$

This is not the most general vibron hamiltonian for a triatomic molecule since no Casimir invariant related to any of the U(3) algebras appears in (6.126) nor do we include the operators $\mathcal{C}_2[\text{SO}_1(3)]$ or $\mathcal{C}_2[\text{SO}_2(3)]$. Our choice is guided by the requirement for the hamiltonian to exhibit an $\text{SO}_{12}(4)$ symmetry, that is, to have eigenstates with good ω and ω'. This, according to our discussion of Section 6.5.2, is a necessary condition for the operator $|\hat{L} \cdot \hat{D}|$—crucial for the description of bent molecules—to be well defined. Since the operators $\mathcal{C}_n[\text{U}_\rho(3)]$ ($n = 1, 2$ and $\rho = 1, 2, 12$) as well as $\mathcal{C}_2[\text{SO}_\rho(3)]$ ($\rho = 1, 2$) break the $\text{SO}_{12}(4)$ symmetry (i.e., have off-diagonal matrix elements in an (ω, ω') basis), they are not included in $\hat{H}_{\text{II}}$. Nevertheless, we note that the hamiltonian (6.126) includes all the important dynamical symmetries discussed in Section 6.5. For $\beta' = 0$ we describe linear triatomic molecules while $\beta' = 2\beta$ covers the case of bent molecules. In addition, the Casimir invariant $\mathcal{C}_2[\text{U}_{12}(4)]$ drives the eigensolutions towards the normal-mode basis whereas the invariants $\mathcal{C}_2[\text{SO}_\rho(4)]$ ($\rho = 1, 2$) induce local-mode

6.6 Examples of triatomic spectra

characteristics. We may thus expect that the hamiltonian (6.126) is sufficiently versatile to be of practical value.

The hamiltonian $\hat{H}_{\text{II}}$ can be diagonalized numerically in the bases discussed in the previous sections which have well-defined $\text{SO}_{12}(4)$ symmetry, but most convenient for our purpose here is the one in (6.86). All terms in (6.126), with the exception of the first, are diagonal in this basis with eigenvalues given in Sections 6.5.1 and 6.5.2. Hence the only task remaining is to evaluate the matrix elements of $\mathcal{C}_2[\text{U}_{12}(4)]$ in the basis (6.86). We do this by rewriting this operator as follows:

$$\begin{aligned}
\mathcal{C}_2[\text{U}_{12}(4)] &= \tfrac{1}{3}\hat{n}_{\text{p}}^2 + \tfrac{1}{2}(\hat{L}^2 + \hat{D}^2 + \hat{D}'^2) + \hat{Q}^2 + \hat{n}_{\text{s}}^2 \\
&= \mathcal{C}_2[\text{U}_1(4)] + \mathcal{C}_2[\text{U}_2(4)] + 2\left(\tfrac{1}{3}\hat{n}_{\text{p}_1}\hat{n}_{\text{p}_2} + \hat{n}_{\text{s}_1}\hat{n}_{\text{s}_2} + \hat{Q}_1 \cdot \hat{Q}_2\right) \\
&\quad + \hat{L}_1 \cdot \hat{L}_2 + \hat{D}_1 \cdot \hat{D}_2 + \hat{D}'_1 \cdot \hat{D}'_2 \\
&= \mathcal{C}_2[\text{U}_1(4)] + \mathcal{C}_2[\text{U}_2(4)] \\
&\quad + \tfrac{1}{2}(\mathcal{C}_2[\text{SO}_{12}(4)] - \mathcal{C}_2[\text{SO}_1(4)] - \mathcal{C}_2[\text{SO}_2(4)]) \\
&\quad + 2\left(\tfrac{1}{3}\hat{n}_{\text{p}_1}\hat{n}_{\text{p}_2} + \tfrac{1}{2}\hat{D}'_1 \cdot \hat{D}'_2 + \hat{Q}_1 \cdot \hat{Q}_2 + \hat{n}_{\text{s}_1}\hat{n}_{\text{s}_2}\right) \quad (6.127)
\end{aligned}$$

In the last expression all Casimir invariants have diagonal matrix elements known from the previous section. Furthermore, it is clear that the matrix elements of the other terms $\hat{n}_{\text{p}_1}\hat{n}_{\text{p}_2}$, $\hat{D}'_1 \cdot \hat{D}'_2$, $\hat{Q}_1 \cdot \hat{Q}_2$, and $\hat{n}_{\text{s}_1}\hat{n}_{\text{s}_2}$ can be obtained by using the expansion (6.91) of the basis states and, through a summation over intermediate states, can be expressed in terms of matrix elements of bond-1 generators times matrix elements of bond-2 generators. It is equally clear that although this is a perfectly straightforward procedure (since the $\text{SO}(4) \supset \text{SO}(3)$ isoscalar factors are known explicitly and all necessary matrix elements are given in Table 5.4), it is quite cumbersome. Moreover, would we want to compute other quantities (e.g., matrix elements of higher-order interactions or of transition operators), we would have to carry out a similar calculation all over again. Therefore we wish to follow here a different approach which makes use of the $\text{SO}(4)$ tensor properties of $\mathcal{C}_2[\text{U}_{12}(4)]$. This derivation is also lengthy (but more elegant than the brute-force method outlined above) because we need to develop some new results in the $\text{SO}(4)$ tensor calculus. Once they are obtained, however, the matrix elements of $\mathcal{C}_2[\text{U}_{12}(4)]$ (as well as of other operators) are derived almost immediately.

We begin our derivation by noting that the sum of operators in the last line of (6.127) must be scalar under $SO_{12}(4)$ and related to

$$[\hat{T}_1^{(2,0)} \times \hat{T}_2^{(2,0)}]_{00}^{(0,0)} \tag{6.128}$$

where the generators of $SO_\rho(4)$ are denoted by their tensor character (see Table 5.5) and an additional subscript $\rho = 1$ or $\rho = 2$. (Since we are not interested in the tensor character under the $U(4)$ algebras, the corresponding labels are omitted and a tensor is denoted as $\hat{T}_{\rho\lambda\mu}^{(\omega,\omega')}$.) This is confirmed by working out the tensor product:

$$[\hat{T}_1^{(2,0)} \times \hat{T}_2^{(2,0)}]_{00}^{(0,0)}$$

$$= \sum_{\lambda\mu} \left\langle \begin{matrix} (2,0) & (2,0) \\ \lambda & \lambda \end{matrix} \middle| \begin{matrix} (0,0) \\ 0 \end{matrix} \right\rangle \langle \lambda\mu\, \lambda - \mu | 00 \rangle \hat{T}_{1\lambda\mu}^{(2,0)} \hat{T}_{2\lambda-\mu}^{(2,0)}$$

$$= \tfrac{1}{3}\left(\tfrac{1}{12}\hat{n}_{p_1}\hat{n}_{p_2} + \tfrac{3}{4}\hat{n}_{s_1}\hat{n}_{s_2} - \tfrac{1}{4}\hat{n}_{p_1}\hat{n}_{s_2} - \tfrac{1}{4}\hat{n}_{s_1}\hat{n}_{p_2} + \tfrac{1}{2}\hat{D}_1' \cdot \hat{D}_2' + \hat{Q}_1 \cdot \hat{Q}_2\right)$$

$$= \tfrac{1}{3}\left(-\tfrac{1}{4}\hat{N}_1\hat{N}_2 + \tfrac{1}{3}\hat{n}_{p_1}\hat{n}_{p_2} + \tfrac{1}{2}\hat{D}_1' \cdot \hat{D}_2' + \hat{Q}_1 \cdot \hat{Q}_2 + \hat{n}_{s_1}\hat{n}_{s_2}\right) \tag{6.129}$$

where use is made of the expression (5.211) for the $SO(4) \supset SO(3)$ isoscalar factor as well as the explicit form of the tensors given in Table 5.5. Introducing this result in (6.126), we find

$$\begin{aligned}\mathcal{C}_2[U_{12}(4)] &= \tfrac{1}{2}\mathcal{C}_2[U_1(4)]\mathcal{C}_2[U_2(4)] + \mathcal{C}_2[U_1(4)] + \mathcal{C}_2[U_2(4)] \\ &+ \tfrac{1}{2}(\mathcal{C}_2[SO_{12}(4)] - \mathcal{C}_2[SO_1(4)] - \mathcal{C}_2[SO_2(4)]) \\ &+ 6[\hat{T}_1^{(2,0)} \times \hat{T}_2^{(2,0)}]_{00}^{(0,0)}\end{aligned} \tag{6.130}$$

Once again all Casimir invariants are diagonal in the basis (6.86) and our task is reduced to finding the matrix element of the last term.

The operator expression (6.128) is a special case of the general tensor product

$$[\hat{T}_1^{(\vec{\omega}_1)} \times \hat{T}_2^{(\vec{\omega}_2)}]_{lm}^{(\vec{\omega})} \tag{6.131}$$

where the two $SO(4)$ indices are denoted, as in Section 5.8, by a single boldface character $\vec{\omega}_x$. We will derive a general expression for the matrix element of (6.131) in the basis (6.86). Since any operator that is at most quadratic in the generators of the vibron model can always be expressed as a sum of tensor products of the type (6.131), this expression will be useful in a variety of cases.

6.6 Examples of triatomic spectra

We note that (6.131) is a generalization to SO(4) of the SO(3) tensor product

$$[\hat{T}_1^{(l_1)} \times \hat{T}_2^{(l_2)}]_m^{(l)} = \sum_{m_1 m_2} \langle l_1 m_1 \, l_2 m_2 | lm \rangle \hat{T}_{1m_1}^{(l_1)} \hat{T}_{2m_2}^{(l_2)} \quad (6.132)$$

For the purpose of calculating the matrix element of (6.131) in a coupled SO(4) basis, it is instructive to give the corresponding SO(3) result which can be found in many textbooks (see, e.g., Section 3.17 of [13]):

$$\langle l_{1a}, l_{2a}; l_a \| [\hat{T}_1^{(l_1)} \times \hat{T}_2^{(l_2)}]^{(l)} \| l_{1b}, l_{2b}; l_b \rangle$$
$$= \begin{bmatrix} l_{1b} & l_{2b} & l_b \\ l_1 & l_2 & l \\ l_{1a} & l_{2a} & l_a \end{bmatrix} \langle l_{1a} \| \hat{T}_1^{(l_1)} \| l_{1b} \rangle \langle l_{2a} \| \hat{T}_2^{(l_2)} \| l_{2b} \rangle \quad (6.133)$$

where the symbol in square brackets is related to the $9j$ coefficient (see (5.209)). This equation is written in terms of matrix elements that are reduced with respect to SO(3) and are related to the matrix elements proper by a Clebsch–Gordan coefficient, as discussed in Section 1.7. The derivation of (6.133) does not involve anything more than the use of the definition of coupled states and coupled tensors, the application of the Wigner–Eckart theorem, and the orthonormality properties of the Clebsch–Gordan coefficients. It may be expected, therefore, that this result can be generalized to SO(4). The main problem in this generalization is the derivation of the relevant *re*coupling coefficients which involve the recoupling of SO(4) rather than the usual SO(3) representations.

So, our first objective is to generalize angular momentum recoupling theory from three to four dimensions, that is, from SO(3) to SO(4). Although the coefficient in (6.133) pertains to the recoupling of four SO(3) representations, which eventually must be generalized to SO(4), we begin our discussion with the simpler case of three representations. Three SO(4) representations $\vec{\omega}_1$, $\vec{\omega}_2$, and $\vec{\omega}_3$ can be combined to a total SO(4) representation $\vec{\omega}$ in the following two different coupling orders. In the first, $\vec{\omega}_1$ and $\vec{\omega}_2$ are coupled to an intermediate representation $\vec{\omega}_{12}$, which in turn is coupled to $\vec{\omega}_3$ to give the total $\vec{\omega}$. This state can be denoted as

$$|(\vec{\omega}_1, \vec{\omega}_2)\vec{\omega}_{12}, \vec{\omega}_3; \vec{\omega} lm\rangle \quad (6.134)$$

and its explicit expression in terms of uncoupled SO(4) states is

$$\sum_{\substack{l_1 l_2 l_{12} l_3 \\ m_1 m_2 m_{12} m_3}} \left\langle \begin{matrix} \vec{\omega}_1 & \vec{\omega}_2 \\ l_1 m_1 & l_2 m_2 \end{matrix} \middle| \begin{matrix} \vec{\omega}_{12} \\ l_{12} m_{12} \end{matrix} \right\rangle \left\langle \begin{matrix} \vec{\omega}_{12} & \vec{\omega}_3 \\ l_{12} m_{12} & l_3 m_3 \end{matrix} \middle| \begin{matrix} \vec{\omega} \\ lm \end{matrix} \right\rangle$$
$$\times |\vec{\omega}_1 l_1 m_1\rangle |\vec{\omega}_2 l_2 m_2\rangle |\vec{\omega}_3 l_3 m_3\rangle \tag{6.135}$$

Using Racah's factorization lemma (5.216) for the SO(4) ⊃ SO(3) ⊃ SO(2) coupling coefficient, this sum can be rewritten as

$$\sum_{l_1 l_2 l_3 l_{12}} \left\langle \begin{matrix} \vec{\omega}_1 & \vec{\omega}_2 \\ l_1 & l_2 \end{matrix} \middle| \begin{matrix} \vec{\omega}_{12} \\ l_{12} \end{matrix} \right\rangle \left\langle \begin{matrix} \vec{\omega}_{12} & \vec{\omega}_3 \\ l_{12} & l_3 \end{matrix} \middle| \begin{matrix} \vec{\omega} \\ l \end{matrix} \right\rangle |(\vec{\omega}_1 l_1, \vec{\omega}_2 l_2) l_{12}, \vec{\omega}_3 l_3; lm\rangle \tag{6.136}$$

in terms of SO(4) ⊃ SO(3) isoscalar factors and SO(3)-coupled states. In the second coupling order, the final SO(4) representation $\vec{\omega}$ is obtained from $\vec{\omega}_1$ and $\vec{\omega}_{23}$, the latter arising from the coupling of $\vec{\omega}_2$ and $\vec{\omega}_3$. The resulting state is

$$|\vec{\omega}_1, (\vec{\omega}_2, \vec{\omega}_3)\vec{\omega}_{23}; \vec{\omega} lm\rangle \tag{6.137}$$

and has the following explicit expression:

$$\sum_{l_1 l_2 l_3 l_{23}} \left\langle \begin{matrix} \vec{\omega}_1 & \vec{\omega}_{23} \\ l_1 & l_{23} \end{matrix} \middle| \begin{matrix} \vec{\omega} \\ l \end{matrix} \right\rangle \left\langle \begin{matrix} \vec{\omega}_2 & \vec{\omega}_3 \\ l_2 & l_3 \end{matrix} \middle| \begin{matrix} \vec{\omega}_{23} \\ l_{23} \end{matrix} \right\rangle |\vec{\omega}_1 l_1, (\vec{\omega}_2 l_2, \vec{\omega}_3 l_3), l_{23}; lm\rangle \tag{6.138}$$

The states (6.134) and (6.137) are connected through a unitary transformation which can be formally written as

$$|(\vec{\omega}_1, \vec{\omega}_2)\vec{\omega}_{12}, \vec{\omega}_3; \vec{\omega} lm\rangle$$
$$= \sum_{\vec{\omega}_{23}} U(\vec{\omega}_1 \vec{\omega}_2 \vec{\omega} \vec{\omega}_3; \vec{\omega}_{12} \vec{\omega}_{23}) |\vec{\omega}_1, (\vec{\omega}_2, \vec{\omega}_3)\vec{\omega}_{23}; \vec{\omega} lm\rangle \tag{6.139}$$

The U coefficient represents the overlap matrix for the recoupling of three SO(4) representations and is a direct generalization of the corresponding SO(3) U coefficient which is defined through the transformation

$$|(l_1, l_2) l_{12}, l_3; lm\rangle = \sum_{l_{23}} U(l_1 l_2 l l_3; l_{12} l_{23}) |l_1, (l_2, l_3) l_{23}; lm\rangle \tag{6.140}$$

6.6 Examples of triatomic spectra

and related to the $6j$ coefficient in the following way [13]:

$$U(l_1 l_2 l l_3; l_{12} l_{23}) = (-)^{l_1+l_2+l_3+l}\sqrt{(2l_{12}+1)(2l_{23}+1)}\begin{Bmatrix} l_1 & l_2 & l_{12} \\ l_3 & l & l_{23} \end{Bmatrix}$$
(6.141)

Note that the SO(3) U coefficient in (6.140) is independent of the magnetic quantum number m, a result well known from angular momentum theory. In the same way, we may expect the SO(4) U coefficient to be independent of the angular momentum l and its projection m. (This is shown below and is anticipated in the notation in (6.139).) Introducing the expansions (6.136) and (6.138) in the transformation (6.139) and using the orthonormality of the SO(4) $\supset$ SO(3) isoscalar factors, we obtain the following expression for the SO(4) U coefficient:

$$\begin{aligned} & U(\vec{\omega}_1 \vec{\omega}_2 \vec{\omega} \vec{\omega}_3; \vec{\omega}_{12} \vec{\omega}_{23}) \\ & \quad - \sum_{l_1 l_2 l_3 l_{12} l_{23}} U(l_1 l_2 l l_3; l_{12} l_{23}) \\ & \quad \times \left\langle \begin{matrix} \vec{\omega}_1 & \vec{\omega}_2 \\ l_1 & l_2 \end{matrix} \middle| \begin{matrix} \vec{\omega}_{12} \\ l_{12} \end{matrix} \right\rangle \left\langle \begin{matrix} \vec{\omega}_{12} & \vec{\omega}_3 \\ l_{12} & l_3 \end{matrix} \middle| \begin{matrix} \vec{\omega} \\ l \end{matrix} \right\rangle \\ & \quad \times \left\langle \begin{matrix} \vec{\omega}_1 & \vec{\omega}_{23} \\ l_1 & l_{23} \end{matrix} \middle| \begin{matrix} \vec{\omega} \\ l \end{matrix} \right\rangle \left\langle \begin{matrix} \vec{\omega}_2 & \vec{\omega}_3 \\ l_2 & l_3 \end{matrix} \middle| \begin{matrix} \vec{\omega}_{23} \\ l_{23} \end{matrix} \right\rangle \end{aligned}$$
(6.142)

The SO(4) U coefficient can thus be written as a sum of products of four $9j$ coefficients and one U coefficient in SO(3). It is still not clear from (6.142) that $U(\vec{\omega}_1 \vec{\omega}_2 \vec{\omega} \vec{\omega}_3; \vec{\omega}_{12} \vec{\omega}_{23})$ is independent of l and that leads us to suspect that a simpler expression must be available. This is indeed the case, and the easiest way to see this is not by working out explicitly the summation (6.142), but by converting to the $SU_r(2) \otimes SU_{r'}(2)$ notation introduced in Section 5.5. In this notation the state (6.134) becomes

$$|(\{r_1, r'_1\}, \{r_2, r'_2\})\{r_{12}, r'_{12}\}, \{r_3, r'_3\}; \{r, r'\} lm\rangle \qquad (6.143)$$

This shows that the unprimed and primed angular momenta are coupled *independently* to r and r', respectively, and only then coupled to total angular momentum l. Consequently, (6.134) can be rewritten as

$$\sum_{m_r m_{r'}} \langle r m_r \, r' m_{r'} | lm \rangle |(r_1, r_2) r_{12}, r_3; r m_r \rangle |(r'_1, r'_2) r'_{12}, r'_3; r' m_{r'} \rangle \qquad (6.144)$$

and similarly, the state (6.137) becomes

$$\sum_{m_r m_{r'}} \langle r m_r\, r' m_{r'} | lm \rangle |r_1, (r_2, r_3) r_{23}; r m_r\rangle |r'_1, (r'_2, r'_3) r'_{23}; r' m_{r'}\rangle \quad (6.145)$$

In this notation it is clear that the transformation between the two states is the product of two SO(3) U coefficients, and hence we find

$$U(\vec{\omega}_1 \vec{\omega}_2 \vec{\omega} \vec{\omega}_3; \vec{\omega}_{12} \vec{\omega}_{23}) = U(r_1 r_2 r r_3; r_{12} r_{23}) U(r'_1 r'_2 r' r'_3; r'_{12} r'_{23}) \quad (6.146)$$

always assuming the relations $r_x = (\omega_x + \omega'_x)/2$ and $r'_x = (\omega_x - \omega'_x)/2$, where x represents any subscript. It is convenient for later purposes to define, in analogy with the SO(3) case, the $6\vec{\omega}$ coefficient via the relation

$$U(\vec{\omega}_1 \vec{\omega}_2 \vec{\omega} \vec{\omega}_3; \vec{\omega}_{12} \vec{\omega}_{23})$$
$$= (-)^{\omega_1 + \omega_2 + \omega_3 + \omega} \sqrt{\dim(\vec{\omega}_{12}) \dim(\vec{\omega}_{23})} \begin{Bmatrix} \vec{\omega}_1 & \vec{\omega}_2 & \vec{\omega}_{12} \\ \vec{\omega}_3 & \vec{\omega} & \vec{\omega}_{23} \end{Bmatrix}$$
$$(6.147)$$

where $\dim(\vec{\omega}_x) = (\omega_x + \omega'_x + 1)(\omega_x - \omega'_x + 1)$ is the dimension of the SO(4) representation $\vec{\omega}_x$. Combining this relation with the corresponding one for SO(3) U coefficients, we see from (6.146) that the $6\vec{\omega}$ coefficient can be written as the product of two $6j$ coefficients:

$$\begin{Bmatrix} \vec{\omega}_1 & \vec{\omega}_2 & \vec{\omega}_{12} \\ \vec{\omega}_3 & \vec{\omega} & \vec{\omega}_{23} \end{Bmatrix} = \begin{Bmatrix} r_1 & r_2 & r_{12} \\ r_3 & r & r_{23} \end{Bmatrix} \begin{Bmatrix} r'_1 & r'_2 & r'_{12} \\ r'_3 & r' & r'_{23} \end{Bmatrix} \quad (6.148)$$

We conclude that the recoupling coefficients of three SO(4) representations are known explicitly since they can be expressed in terms of two corresponding coefficients in SO(3).

The recoupling of four SO(4) representations can be treated similarly in close analogy with the SO(3) case. We now need the transformation matrix between the SO(4)-coupled states

$$|(\vec{\omega}_1, \vec{\omega}_2)\vec{\omega}_{12}, (\vec{\omega}_3, \vec{\omega}_4)\vec{\omega}_{34}; \vec{\omega} l m\rangle \quad (6.149)$$

and

$$|(\vec{\omega}_1, \vec{\omega}_3)\vec{\omega}_{13}, (\vec{\omega}_2, \vec{\omega}_4)\vec{\omega}_{24}; \vec{\omega} l m\rangle \quad (6.150)$$

6.6 Examples of triatomic spectra

The two bases are connected through a unitary transformation which we write as

$$|(\vec{\omega}_1,\vec{\omega}_2)\vec{\omega}_{12},(\vec{\omega}_3,\vec{\omega}_4)\vec{\omega}_{34};\vec{\omega}lm\rangle$$
$$= \sum_{\vec{\omega}_{12}\vec{\omega}_{34}} \begin{bmatrix} \vec{\omega}_1 & \vec{\omega}_2 & \vec{\omega}_{12} \\ \vec{\omega}_3 & \vec{\omega}_4 & \vec{\omega}_{34} \\ \vec{\omega}_{13} & \vec{\omega}_{24} & \vec{\omega} \end{bmatrix} |(\vec{\omega}_1,\vec{\omega}_3)\vec{\omega}_{13},(\vec{\omega}_2,\vec{\omega}_4)\vec{\omega}_{24};\vec{\omega}lm\rangle$$
(6.151)

Using the expansion of (6.149) and (6.150) in terms of SO(3)-coupled states, that is,

$$|(\vec{\omega}_1,\vec{\omega}_2)\vec{\omega}_{12},(\vec{\omega}_3,\vec{\omega}_4)\vec{\omega}_{34};\vec{\omega}lm\rangle$$
$$= \sum_{l_1l_2l_3l_4l_{12}l_{34}} \left\langle \begin{matrix}\vec{\omega}_1 & \vec{\omega}_2 \\ l_1 & l_2\end{matrix} \Bigg| \begin{matrix}\vec{\omega}_{12} \\ l_{12}\end{matrix} \right\rangle \left\langle \begin{matrix}\vec{\omega}_3 & \vec{\omega}_4 \\ l_3 & l_4\end{matrix} \Bigg| \begin{matrix}\vec{\omega}_{34} \\ l_{34}\end{matrix} \right\rangle \left\langle \begin{matrix}\vec{\omega}_{12} & \vec{\omega}_{34} \\ l_{12} & l_{34}\end{matrix} \Bigg| \begin{matrix}\vec{\omega} \\ l\end{matrix} \right\rangle$$
$$\times |(\vec{\omega}_1l_1,\vec{\omega}_2l_2)l_{12},(\vec{\omega}_3l_3,\vec{\omega}_4l_4)l_{34};lm\rangle$$
(6.152)

and

$$|(\vec{\omega}_1,\vec{\omega}_3)\vec{\omega}_{13},(\vec{\omega}_2,\vec{\omega}_4)\vec{\omega}_{24};\vec{\omega}lm\rangle$$
$$= \sum_{l_1l_3l_2l_4l_{13}l_{24}} \left\langle \begin{matrix}\vec{\omega}_1 & \vec{\omega}_3 \\ l_1 & l_3\end{matrix} \Bigg| \begin{matrix}\vec{\omega}_{13} \\ l_{13}\end{matrix} \right\rangle \left\langle \begin{matrix}\vec{\omega}_2 & \vec{\omega}_4 \\ l_2 & l_4\end{matrix} \Bigg| \begin{matrix}\vec{\omega}_{24} \\ l_{24}\end{matrix} \right\rangle \left\langle \begin{matrix}\vec{\omega}_{13} & \vec{\omega}_{24} \\ l_{13} & l_{24}\end{matrix} \Bigg| \begin{matrix}\vec{\omega} \\ l\end{matrix} \right\rangle$$
$$\times |(\vec{\omega}_1l_1,\vec{\omega}_3l_3)l_{13},(\vec{\omega}_2l_2,\vec{\omega}_4l_4)l_{24};lm\rangle$$
(6.153)

and with the help of the orthonormality properties of the SO(4) $\supset$ SO(3) isoscalar factors, we deduce the following expression for the SO(4) recoupling coefficient:

$$\begin{bmatrix} \vec{\omega}_1 & \vec{\omega}_2 & \vec{\omega}_{12} \\ \vec{\omega}_3 & \vec{\omega}_4 & \vec{\omega}_{34} \\ \vec{\omega}_{13} & \vec{\omega}_{24} & \vec{\omega} \end{bmatrix}$$
$$= \sum_{\substack{l_1l_2l_3l_4 \\ l_{12}l_{34}l_{13}l_{24}}} \begin{bmatrix} l_1 & l_2 & l_{12} \\ l_3 & l_4 & l_{34} \\ l_{13} & l_{24} & l \end{bmatrix}$$
$$\times \left\langle \begin{matrix}\vec{\omega}_1 & \vec{\omega}_2 \\ l_1 & l_2\end{matrix} \Bigg| \begin{matrix}\vec{\omega}_{12} \\ l_{12}\end{matrix} \right\rangle \left\langle \begin{matrix}\vec{\omega}_3 & \vec{\omega}_4 \\ l_3 & l_4\end{matrix} \Bigg| \begin{matrix}\vec{\omega}_{34} \\ l_{34}\end{matrix} \right\rangle \left\langle \begin{matrix}\vec{\omega}_{12} & \vec{\omega}_{34} \\ l_{12} & l_{34}\end{matrix} \Bigg| \begin{matrix}\vec{\omega} \\ l\end{matrix} \right\rangle$$

$$\times \left\langle \begin{array}{cc|c} \vec{\omega}_1 & \vec{\omega}_3 & \vec{\omega}_{13} \\ l_1 & l_3 & l_{13} \end{array} \right\rangle \left\langle \begin{array}{cc|c} \vec{\omega}_2 & \vec{\omega}_4 & \vec{\omega}_{24} \\ l_2 & l_4 & l_{24} \end{array} \right\rangle \left\langle \begin{array}{cc|c} \vec{\omega}_{13} & \vec{\omega}_{24} & \vec{\omega} \\ l_{13} & l_{24} & l \end{array} \right\rangle$$

(6.154)

A much simpler expression is obtained by reverting to the $SU_r(2) \otimes SU_{r'}(2)$ notation. A similar argument as in the case of the recoupling of three $SO(4)$ representations now leads to

$$\begin{bmatrix} \vec{\omega}_1 & \vec{\omega}_2 & \vec{\omega}_{12} \\ \vec{\omega}_3 & \vec{\omega}_4 & \vec{\omega}_{34} \\ \vec{\omega}_{13} & \vec{\omega}_{24} & \vec{\omega} \end{bmatrix} = \begin{bmatrix} r_1 & r_2 & r_{12} \\ r_3 & r_4 & r_{34} \\ r_{13} & r_{24} & r \end{bmatrix} \begin{bmatrix} r'_1 & r'_2 & r'_{12} \\ r'_3 & r'_4 & r'_{34} \\ r'_{13} & r'_{24} & r' \end{bmatrix} \quad (6.155)$$

that is, the $SO(4)$ recoupling coefficient again factors out into the product of two corresponding $SO(3)$ coefficients. It is useful to introduce the $9\vec{\omega}$ *coefficient* defined as

$$\left\{ \begin{array}{ccc} \vec{\omega}_1 & \vec{\omega}_2 & \vec{\omega}_{12} \\ \vec{\omega}_3 & \vec{\omega}_4 & \vec{\omega}_{34} \\ \vec{\omega}_{13} & \vec{\omega}_{24} & \vec{\omega} \end{array} \right\} = \left\{ \begin{array}{ccc} r_1 & r_2 & r_{12} \\ r_3 & r_4 & r_{34} \\ r_{13} & r_{24} & r \end{array} \right\} \left\{ \begin{array}{ccc} r'_1 & r'_2 & r'_{12} \\ r'_3 & r'_4 & r'_{34} \\ r'_{13} & r'_{24} & r' \end{array} \right\} \quad (6.156)$$

which has symmetry properties that can be deduced from those of the $9j$ coefficients.

Although the previous derivations have been somewhat lengthy, the result of it all is surprisingly simple and can be stated as follows. Any $3n\vec{\omega}$ coefficient in $SO(4)$ with arguments $\vec{\omega}_i$ ($i = 1, 2, \ldots, 3n$) can be written as the product of two $3nj$ coefficients in $SO(3)$ with arguments $r_i = (\omega_i + \omega'_i)/2$ and $r'_i = (\omega_i - \omega'_i)/2$, respectively. The relations (6.148) and (6.156) provide illustrations of this general result. Note that $n = 1$ is an exception to this rule, that is, a coupling or $3\vec{\omega}$ coefficient in $SO(4)$ is *not* simply the product of two coupling or $3j$ coefficients in $SO(3)$. This is because the $SO(4) \supset SO(3) \supset SO(2)$ coupling coefficients depend on the $SO(4)$ representations $\vec{\omega}_i$ as well as the angular momenta l_i and their projections m_i, in contrast to the *re*coupling coefficients ($n > 1$), which only depend on $\vec{\omega}_i$.

We can now return to the calculation of the matrix element of (6.131), which by virtue of the generalized Wigner–Eckart theorem can be written in terms of an $SO(4)$-reduced matrix element,

$$\langle \vec{\omega}_{1a}, \vec{\omega}_{2a}; \vec{\omega}_a L_a M_a | [\hat{T}_1^{(\vec{\omega}_1)} \times \hat{T}_2^{(\vec{\omega}_2)}]_{lm}^{(\vec{\omega})} | \vec{\omega}_{1b}, \vec{\omega}_{2b}; \vec{\omega}_b L_b M_b \rangle$$

6.6 Examples of triatomic spectra

$$= \left\langle \begin{array}{cc|c} \vec{\omega}_b & \vec{\omega} & \vec{\omega}_a \\ L_b & l & L_a \end{array} \right\rangle \langle L_b M_b \, lm | L_a M_a \rangle$$
$$\times \langle \vec{\omega}_{1a}, \vec{\omega}_{2a}; \vec{\omega}_a \| [\hat{T}_1^{(\vec{\omega}_1)} \times \hat{T}_2^{(\vec{\omega}_2)}]^{(\vec{\omega})} \| \vec{\omega}_{1b}, \vec{\omega}_{2b}; \vec{\omega}_b \rangle \quad (6.157)$$

The reduced matrix element in this equation is the direct generalization of the SO(3)-reduced matrix element on the left-hand side of (6.133). We already noted that the derivation of the SO(3) result (6.133) only involves elementary properties of the coupling coefficients (such as their orthonormality). These properties remain valid for the SO(4) ⊃ SO(3) isoscalar factors, and we may readily generalize (6.133) to SO(4):

$$\langle \vec{\omega}_{1a}, \vec{\omega}_{2a}; \vec{\omega}_a \| [\hat{T}_1^{(\vec{\omega}_1)} \times \hat{T}_2^{(\vec{\omega}_2)}]^{(\vec{\omega})} \| \vec{\omega}_{1b}, \vec{\omega}_{2b}; \vec{\omega}_b \rangle$$
$$= \begin{bmatrix} \vec{\omega}_{1b} & \vec{\omega}_{2b} & \vec{\omega}_b \\ \vec{\omega}_1 & \vec{\omega}_2 & \vec{\omega} \\ \vec{\omega}_{1a} & \vec{\omega}_{2a} & \vec{\omega}_a \end{bmatrix} \langle \vec{\omega}_{1a} \| \hat{T}_1^{(\vec{\omega}_1)} \| \vec{\omega}_{1b} \rangle \langle \vec{\omega}_{2a} \| \hat{T}_2^{(\vec{\omega}_2)} \| \vec{\omega}_{2b} \rangle$$

$$(6.158)$$

with the recoupling coefficient known in terms of the product of two $9j$ coefficients. The matrix element of (6.131) is now completely determined from (6.157) and (6.158).

Let us now apply this result to the tensor product (6.128). Because it is a scalar tensor product, the coupling coefficients in (6.157) reduce to unity, and we find

$$\langle [N_1]\omega_{1a}, [N_2]\omega_{2a}; \vec{\omega} L M_L | [\hat{T}_1^{(2,0)} \times \hat{T}_2^{(2,0)}]_{00}^{(0,0)} | [N_1]\omega_{1b}, [N_2]\omega_{2b}; \vec{\omega} L M_L \rangle$$
$$= \begin{bmatrix} (\omega_{1b},0) & (\omega_{2b},0) & (\omega,\omega') \\ (2,0) & (2,0) & (0,0) \\ (\omega_{1a},0) & (\omega_{2a},0) & (\omega,\omega') \end{bmatrix}$$
$$\times \langle [N_1]\omega_{1a} \| \hat{T}_1^{(2,0)} \| [N_1]\omega_{1b} \rangle \langle [N_2]\omega_{2a} \| \hat{T}_2^{(2,0)} \| [N_2]\omega_{2b} \rangle \quad (6.159)$$

where, for clarity, we use the full notation including the $U_\rho(4)$ labels N_ρ. The SO(4)-reduced matrix elements are known from Chapter 5 (see Table 5.6) and the SO(4) recoupling coefficient can be written as the product of two SO(3) recoupling coefficients,

$$\begin{bmatrix} (\omega_{1b},0) & (\omega_{2b},0) & (\omega,\omega') \\ (2,0) & (2,0) & (0,0) \\ (\omega_{1a},0) & (\omega_{2a},0) & (\omega,\omega') \end{bmatrix}$$

$$= \begin{bmatrix} \omega_{1b}/2 & \omega_{2b}/2 & (\omega+\omega')/2 \\ 1 & 1 & 0 \\ \omega_{1a}/2 & \omega_{2a}/2 & (\omega+\omega')/2 \end{bmatrix} \begin{bmatrix} \omega_{1b}/2 & \omega_{2b}/2 & (\omega-\omega')/2 \\ 1 & 1 & 0 \\ \omega_{1a}/2 & \omega_{2a}/2 & (\omega-\omega')/2 \end{bmatrix}$$
(6.160)

which can be further simplified to

$$\tfrac{1}{3}(-)^{\omega_{1b}+\omega_{2a}+\omega}(\omega_{1a}+1)(\omega_{2a}+1)\left\{\begin{array}{ccc}(\omega_{1a},0) & (\omega_{2a},0) & (\omega,\omega') \\ (\omega_{2b},0) & (\omega_{1b},0) & (2,0)\end{array}\right\}$$ (6.161)

in terms of the $6\vec{\omega}$ coefficient defined in (6.148). This result shows that the tensor product (6.128) (and consequently the Casimir invariant $C_2[U_{12}(4)]$) in general has non-zero matrix elements (diagonal as well as off-diagonal) for $|\omega_{1a}-\omega_{1b}|=0,2$ and $|\omega_{2a}-\omega_{2b}|=0,2$. The preceding derivation also illustrates how the properties of SO(4) recoupling coefficients (such as symmetries under the permutation of the arguments or the reduction of a particular $9\vec{\omega}$ coefficient to a $6\vec{\omega}$ coefficient) can be deduced from the corresponding properties of $3nj$ coefficients. As a result the SO(4) tensor formalism described here is as easy to handle and as powerful as conventional angular momentum theory.

It should be emphasized that (6.157) and (6.158) are general expressions that can be used to calculate the matrix elements, not just of $C_2[U_{12}(4)]$, but of a wide variety of operators in the vibron model. A case which frequently occurs is that of an operator depending on one of the bond variables only. For instance, in lowest order the dipole operator of the vibron model can be written [4–6] as a sum of bond dipole operators, $\hat{T}=\hat{T}_1+\hat{T}_2$. The calculation of the matrix elements of such operators is considerably simplified by the application of (6.157) and (6.158). For an operator depending on bond-1 variables the latter equation reduces further to

$$\langle [N_1]\omega_{1a},[N_2]\omega_{2a};\vec{\omega}_a\|\hat{T}_1^{(\vec{\omega}_1)}\|[N_1]\omega_{1b},[N_2]\omega_{2b};\vec{\omega}_b\rangle$$
$$= (-)^{\omega_{1a}+\omega_{2b}+\omega_1+\omega_b}(\omega_{1a}+1)\sqrt{(\omega_b+\omega'_b+1)(\omega_b-\omega'_b+1)}\delta_{\omega_{2a}\omega_{2b}}$$
$$\times \left\{\begin{array}{ccc}(\omega_a,\omega'_a) & (\omega_{2a},0) & (\omega_{1a},0) \\ (\omega_{1b},0) & (\omega_1,\omega'_1) & (\omega_b,\omega'_b)\end{array}\right\}\langle[N_1]\vec{\omega}_{1a}\|\hat{T}_1^{(\vec{\omega}_1)}\|[N_1]\vec{\omega}_{1b}\rangle$$

(6.162)

6.6 Examples of triatomic spectra

The corresponding equation for a bond-2 operator is

$$\langle [N_1]\omega_{1a}, [N_2]\omega_{2a}; \vec{\omega}_a \| \hat{T}_2^{(\vec{\omega}_2)} \| [N_1]\omega_{1b}, [N_2]\omega_{2b}; \vec{\omega}_b \rangle$$
$$= (-)^{\omega_{1b}+\omega_{2b}+\omega_2+\omega_a}(\omega_{2a}+1)\sqrt{(\omega_b+\omega_b'+1)(\omega_b-\omega_b'+1)}\delta_{\omega_{1a}\omega_{1b}}$$
$$\times \left\{ \begin{array}{ccc} (\omega_a,\omega_a') & (\omega_{1a},0) & (\omega_{2a},0) \\ (\omega_{2b},0) & (\omega_2,\omega_2') & (\omega_b,\omega_b') \end{array} \right\} \langle [N_2]\vec{\omega}_{2a} \| \hat{T}_2^{(\vec{\omega}_2)} \| [N_2]\vec{\omega}_{2b} \rangle$$

(6.163)

We are now in a position to proceed with a discussion of the results that can be obtained with the hamiltonian (6.126). We have available closed, analytic expressions for the diagonal and off-diagonal matrix elements of the Casimir invariant $\mathcal{C}_2[U_{12}(4)]$ in the basis (6.86), and all other terms in $\hat{H}_{II}$ are diagonal in this basis with eigenvalues known from Sections 6.5.1 and 6.5.2. The hamiltonian (6.126) can thus be diagonalized and its eigenvalues and eigenvectors obtained numerically. This approach has been followed by Iachello and collaborators to calculate overtone frequencies in the linear molecules CO_2, $C^{13}O_2$, OCS, HCN, and N_2O (where a superscript indicates atomic mass) as well as the bent molecules H_2O, H_2O^{18}, D_2O, H_2S, and SO_2. We show here some typical examples of these results referring the reader to the original papers [4–6] for more details.

Although it is possible to give a unified description of linear and bent triatomic molecules with the hamiltonian $\hat{H}_{II}$ (including quasi-linear and quasi-bent structures), their treatment is kept separate in [4–6]. We follow this approach and assume $\beta' = 0$ for linear and $\beta' = 2\beta$ for bent molecules without considering intermediate cases. For non-symmetric molecules XYZ (or also X–X–Y molecules such as N_2O) the independent parameters in $\hat{H}_{II}$ are β_1, β_2, β, and η if we confine ourselves to vibrational excitations, in which case the $\hat{L}^2$ term is irrelevant. One also adjusts the boson numbers N_1 and N_2, although these are to some extent constrained by the values used in corresponding diatomic molecules. Thus in the simplest approach the description of non-symmetric molecules requires the adjustment of $4+2$ parameters. For symmetric bent molecules X–Y–X we have the additional constraints $\beta_1 = \beta_2$ and $N_1 = N_2$ because of the equality of the first and second bonds. This reduces the number of parameters to $3+1$. The same constraints apply to symmetric linear molecules, but in that case it is found necessary to

Table 6.1: Summary of results for linear molecules

	CO_2	$C^{13}O_2$	OCS	HCN	N_2O
$N_1{}^a$	153	150	159	47	165
$N_2{}^a$	153	150	190	140	136
Number of levels	47	19	64	34	25
$\Delta_{\rm rms}\ ({\rm cm}^{-1})^a$	11.3	7.3	12.4	9.7	11.7
$\Delta_{\rm rms}\ ({\rm cm}^{-1})^b$	5.3	2.6	2.9	5.0	0.4
$\Delta_{\rm rms}\ ({\rm cm}^{-1})^c$	4.2	0.4	1.8	2.6	0.3
ξ^a	0.92	0.92	0.40	0.24	0.87

[a] Fit 1 with $4+1$ (CO_2) or $4+2$ parameters.
[b] Fit 2 with $8+1$ (CO_2) or $10+2$ parameters.
[c] Fit 3 with $10+1$ (CO_2) or $13+2$ parameters.

introduce, even in lowest order, an operator that properly describes the phenomenon of Fermi resonances [9], increasing the number of parameters by 1. The form of this operator (which is higher than quadratic in the generators of the vibron model) is specified in [5] in terms of its matrix elements in the basis (6.86). In the simplest approach, which we refer to as fit 1, based on the hamiltonian (6.126) with in some cases an additional operator to account for Fermi resonances, the number of parameters ranges from 4 to 6. Iachello and collaborators also consider more elaborate hamiltonians which involve products of the Casimir invariants considered so far. Two refinements of the simplest calculation are considered (fits 2 and 3) with numbers of parameters summarized in Tables 6.1 and 6.2 for linear and bent molecules, respectively.

We also show in these tables the number of levels included in the fit as well as the boson numbers N_1 and N_2. For symmetric molecules these are equal: $N_1 = N_2$ for CO_2 in Table 6.1 and for all bent molecules in Table 6.2. We note that N_1 and N_2 are obtained through a least-squares fit to the spectra of triatomic molecules, and we should expect these boson numbers to be close to those found for corresponding diatomic molecules. As discussed in Section 5.7, a simple estimate of the boson number in a diatomic molecule follows from its relation to parameters in the Dunham expansion, $Y_{10}/Y_{20} = -(N+2)$. In CH one finds $N = 43$, close to the value of 47 associated with the H–C bond in HCN (N_1 for HCN in Table 6.1). A comparable agreement is found for other bonds:

6.6 Examples of triatomic spectra

Table 6.2: Summary of results for bent molecules

	H_2O	H_2O^{18}	D_2O	H_2S	SO_2
$N_1 = N_2{}^a$	36	40	56	41	215
Number of levels	52	12	14	20	40
Δ_{rms} (cm^{-1})^a	44.3	11.9	8.4	16.4	14.5
Δ_{rms} (cm^{-1})^b	6.7	4.0	2.7	2.0	12.7
Δ_{rms} (cm^{-1})^c	6.5	3.1	1.8	1.1	5.6
ξ^a	0.46	0.41	0.56	0.23	0.81

^{a} Fit 1 with $3+1$ parameters.
^{b} Fit 2 with $6+1$ parameters.
^{c} Fit 3 with $8+1$ parameters.

$N = 155$ for CN compared to $N_2 = 140$ for the C–N bond in HCN, $N = 159$ for CO compared to $N_1 = N_2 = 153$ in CO_2 and $N_1 = 159$ for the O–C bond in OCS, $N = 196$ for CS compared to $N_2 = 190$ for the C–S bond in OCS, $N = 134$ for NO compared to $N_2 = 136$ for the N–O bond in N_2O, and $N = 161$ for N_2 compared to $N_1 = 165$ for the N–N bond in N_2O.

Tables 6.1 and 6.2 also list the root-mean-square deviations Δ_{rms} in cm^{-1} for the three different fits. In the simplest approach (fit 1, based on the hamiltonian $\hat{H}_{\text{II}}$) one finds an average deviation of the order of 10 cm^{-1}, with the exception of H_2O, where it is significantly higher. In the more elaborate fits this average deviation is reduced to 1–5 cm^{-1}. In Table 6.3 we show more detailed results of this approach for a typical example (HCN). Note that v_1, v_2^l, v_3 are only approximate quantum numbers and the values listed in the table are those associated with the biggest normal-mode component of the eigenstate. Similar tables for the other molecules can be found in [4–6]. Because of the accuracy 1–5 cm^{-1} of the best fits, it is possible to make reliable predictions concerning the location of overtones not yet measured. In addition, in some cases this accuracy provides a criterion to reassign levels that deviate considerably from uniformly good fits.

One of the most appealing aspects of these calculations is that they provide a unified framework which smoothly connects the local-mode with the normal-mode basis. This feature can be quantified by intro-

Table 6.3: Vibrational energy levels of HCN[a]

v_1	v_2^l	v_3	Expt	Fit 1	Δ_1	Fit 2	Δ_2	Fit 3	Δ_3
0	2^0	0	1411.4	1406.5	−4.9	1415.3	3.9	1412.4	0.9
1	0^0	0	2096.9	2105.4	8.5	2095.2	−1.6	2095.8	−1.1
0	0^0	1	3311.5	3298.4	−13.1	3307.9	−3.6	3312.6	1.1
0	4^0	0	2802.9	2799.8	−3.1	2801.2	−1.6	2804.2	1.4
1	2^0	0	3501.1	3496.5	−4.6	3502.9	1.8	3502.8	1.6
2	0^0	0	4173.1	4186.0	12.9	4174.3	1.2	4173.8	0.7
0	2^0	1	4684.3	4688.4	4.1	4688.4	4.1	4686.1	1.8
1	0^0	1	5393.7	5388.4	−5.3	5384.3	−9.4	5391.6	−2.1
0	0^0	2	6519.6	6502.2	−17.4	6518.7	−0.9	6521.2	1.6
2	2^0	0	5571.9	5561.5	−10.4	5573.9	2.0	5574.8	2.9
1	2^0	1	6761.3	6763.0	1.7	6757.5	−3.8	6758.9	−2.4
1	0^0	2	8585.6	8576.8	−8.8	8577.2	−8.4	8583.6	−2.0
0	0^0	3	9627.0	9611.5	−15.5	9631.1	4.1	9628.6	1.6
1	2^0	2	9914.4	9934.9	20.5	9916.5	2.1	9910.9	−3.5
1	0^0	3	11674.5	11670.7	−3.8	11672.3	−2.2	11674.2	−0.3
0	0^0	4	12635.9	12626.3	−9.6	12643.5	7.6	12637.4	1.5
0	0^0	5	15551.9	15546.6	−5.3	15554.5	2.6	15550.7	−1.2
3	0^0	3	15710.5	15714.3	3.8	15704.0	−6.5	15705.9	−4.6
2	0^0	4	16674.2	16688.9	14.7	16675.9	1.7	16673.7	−0.5
1	0^0	5	17550.4	17574.9	24.5	17563.5	13.1	17561.1	10.7
0	0^0	6	18377.0	18372.3	−4.7	18362.9	−14.1	18371.6	−5.4
5	0^0	2	16640.3	16626.1	−14.2	16642.2	1.9	16641.2	0.9
0	1^1	0	712.0	709.5	−2.5	714.6	2.6	712.4	0.4
0	3^1	0	2113.5	2109.3	−4.2	2115.1	1.6	2113.9	0.4
1	1^1	0	2805.6	2807.1	1.5	2805.8	0.2	2805.3	−0.3
0	1^1	1	4004.2	3999.5	−4.7	4004.9	0.7	4005.2	1.0
1	3^1	0	4201.3	4191.4	−9.9	4198.8	−2.5	4201.5	0.2
2	1^1	0	4878.3	4879.7	1.4	4880.7	2.4	4880.1	1.8
0	3^1	1	5366.9	5382.9	16.0	5371.0	4.1	5368.2	1.3
1	1^1	1	6083.4	6081.6	−1.8	6077.5	−5.9	6080.9	−2.5
0	1^1	2	7194.8	7194.9	0.1	7198.6	3.8	7193.9	−0.9
0	2^2	0	1426.5	1424.7	−1.8	1428.0	1.5	1425.9	−0.6
0	4^2	0	2818.2	2817.4	−0.8	2813.6	−4.6	2816.9	−1.3
1	2^2	0	3516.9	3514.1	−2.8	3515.1	−1.8	3515.5	−1.4
0	2^2	1	4699.2	4706.0	6.8	4700.8	1.6	4698.5	−0.7
Δ_{rms}					9.7		5.0		2.6

[a] All energies in cm^{-1}.

6.7 Polyatomic molecules

ducing the transition parameters

$$\xi_\rho = \frac{2}{\pi} \arctan\left|\frac{4\eta}{\beta_\rho}\right|, \qquad \rho = 1, 2 \qquad (6.164)$$

in terms of the parameters η and β_ρ in the hamiltonian $\hat{H}_{\mathrm{II}}$. A local-mode basis is obtained for $\eta \ll \beta_\rho$ or $\xi_\rho \approx 0$, while a normal-mode basis occurs for $\eta \gg \beta_\rho$ or $\xi_\rho \approx 1$. Note that each of the bonds in a triatomic molecule has a transition parameter ξ_ρ associated with it. In a symmetric molecule X–Y–X these are equal and we can define $\xi = \xi_1 = \xi_2$; for non-symmetric molecules we may define the geometric mean $\xi = \sqrt{\xi_1 \xi_2}$. This transition parameter ξ is shown in the last row of Tables 6.1 and 6.2 for the different molecules considered in the fit. The results are consistent with those of Child and Halonen [10], who introduce a similar transition parameter. The value of the local-to-normal transition parameter associated with a bond X–Y is predominantly determined by the difference in mass of the X and the Y atoms. A big mass difference results in a local-mode basis, while the case of equal masses displays normal-mode characteristics.

Using the wave functions obtained by numerically diagonalizing the hamiltonian (6.126), one can now compute the matrix elements of the vibron dipole operator and evaluate the intensities of dipole transitions. In lowest order the transition operator can be written as a sum of the dipole operators belonging to each bond, $\hat{T} = \hat{T}_1 + \hat{T}_2$; higher-order contributions involve certain cross terms. A full account of the form of the bond dipole operator as well as the results obtained with it is given in [4–6], and we refer the interested reader to these references for more details.

6.7 Polyatomic molecules

With the formalism developed in the previous sections we can now extend the vibron model to polyatomic molecules. Formally this extension can be carried out with relative ease using the SO(4) tensor calculus discussed in Sections 5.8 and 6.6. We note, however, that very few applications have been discussed in the literature, and as a consequence, many questions concerning the interpretation and the validity of the vibron model, as applied to polyatomic molecules, remain unanswered.

We begin our discussion with a formal outline of the extension of the vibron model to polyatomic molecules and then illustrate the procedure with the example of acetylene (C_2H_2). To each of the m bonds in the molecule we associate a $U(4)$ algebra, denoted as $U_\rho(4)$, $\rho = 1, 2, \ldots, m$. The dynamical algebra associated with the system then becomes the product algebra

$$U_1(4) \otimes U_2(4) \otimes \cdots \otimes U_m(4) \tag{6.165}$$

The number of different bases that can be constructed for this dynamical algebra increases explosively with m, and it is clearly not worthwhile to investigate or even list all of them. The discussion in the preceding sections makes it clear, though, that applications are confined to bases with an overall $SO(4)$ symmetry. Furthermore, we want to develop the model by extending the ideas of the previous section and also by exploiting the analogy with the $U(2)$ application to polyatomic molecules presented in Section 2.8. The ensuing extension essentially relies on the (numerical) diagonalization of an appropriate vibron hamiltonian (to be discussed below) in some bases which result from the coupling of the $U_\rho(4)$ algebras. A possible basis is of the form

$$\boxed{\begin{aligned} &U_1(4) \otimes U_2(4) \otimes \cdots \otimes U_m(4) \\ &\supset SO_1(4) \otimes SO_2(4) \otimes \cdots \otimes SO_m(4) \\ &\supset SO_{12}(4) \otimes SO_3(4) \otimes \cdots \otimes SO_m(4) \\ &\supset SO_{123}(4) \otimes SO_4(4) \otimes \cdots \otimes SO_m(4) \\ &\cdots \\ &\supset SO_{12\ldots m}(4) \supset SO_{12\ldots m}(3) \end{aligned}} \tag{6.166}$$

where a subscript $\rho\sigma\ldots\tau$ in an $SO(4)$ algebra indicates that it is obtained by adding the corresponding generators of $SO_\rho(4)$, $SO_\sigma(4)$, $\ldots$, and $SO_\tau(4)$. Note that the basis (6.166) has an overall $SO(4)$ symmetry since the final $SO(4)$ algebra, $SO_{12\ldots m}(4)$, is obtained by adding the corresponding generators of all bonds. Since we intend to propose a vibron hamiltonian that preserves the overall $SO(4)$ symmetry, the use of the basis (6.166) leads to considerable computational simplifications. For notational simplicity we henceforth denote $SO_{12\ldots m}(n)$ as $SO(n)$.

6.7 Polyatomic molecules

Clearly, coupling orders other than the one in (6.166) can be proposed. For example, we might choose to first combine the SO(4) algebras in pairs $(12), (34), \ldots$ and subsequently couple the pair representations in some convenient order. Which coupling scheme is best depends on the particular application (i.e., on the molecule); it should be carefully selected since great computational simplifications might result from an appropriate choice.

States in the basis (6.166) are written as

$$\left|(([N_1]\omega_1, [N_2]\omega_2)\vec{\omega}_{12}, [N_3]\omega_3)\vec{\omega}_{123}\ldots;\vec{\omega}LM_L\right\rangle \qquad (6.167)$$

for which we introduce the more compact notation

$$\left|((\omega_1, \omega_2)\vec{\omega}_{12}, \omega_3)\vec{\omega}_{123}\ldots;\vec{\omega}LM_L\right\rangle \qquad (6.168)$$

in the usual convention that a boldface $\vec{\omega}_x$ represents two labels (ω_x, ω'_x).

We propose to diagonalize the following vibron hamiltonian in the basis (6.166):

$$\begin{aligned} \hat{H}_{\mathrm{II}} &= \sum_{\rho<\sigma}^{m} \eta_{\rho\sigma}\mathcal{C}_2[\mathrm{U}_{\rho\sigma}(4)] + \sum_{\rho=1}^{m} \beta_\rho \mathcal{C}_2[\mathrm{SO}_\rho(4)] \\ &+ \sum_{\rho<\sigma}^{m} \beta_{\rho\sigma}\mathcal{C}_2[\mathrm{SO}_{\rho\sigma}(4)] + \beta \mathcal{C}_2[\mathrm{SO}(4)] \\ &+ \beta' \mathcal{C}'_2[\mathrm{SO}(4)] + \gamma \mathcal{C}_2[\mathrm{SO}(3)] \end{aligned} \qquad (6.169)$$

The various terms in $\hat{H}_{\mathrm{II}}$ are immediate generalizations of those appearing in the triatomic hamiltonian (6.126). The Casimir invariants $\mathcal{C}_2[\mathrm{SO}_\rho(4)]$ are associated with a single bond ρ, while the other operators in (6.169) represent bond–bond interactions. Apart from the three last terms, there are, as in Section 6.6, two types of bond–bond interactions in (6.169): (i) the $\mathcal{C}_2[\mathrm{SO}_{\rho\sigma}(4)]$ invariants which preserve the local nature of the spectrum and (ii) the $\mathcal{C}_2[\mathrm{U}_{\rho\sigma}(4)]$ invariants inducing normal-mode characteristics. Note that the $\mathcal{C}'_2[\mathrm{SO}_{\rho\sigma}(4)]$ invariants cannot be included since generally they are not diagonal—and consequently not well-defined—in the basis (6.167) (see the discussion after (6.119)). Concerning the introduction of the appropriate discrete symmetries into (6.169), we refer to Section 2.8.

General expressions can be derived for the matrix elements of the different Casimir invariants in (6.169) in the basis (6.167). The operators $\mathcal{C}_2[SO_\rho(4)]$, $\mathcal{C}_2[SO(4)]$, $\mathcal{C}'_2[SO(4)]$, and $\mathcal{C}_2[SO(3)]$ are diagonal in this basis with eigenvalues $\omega_\rho(\omega_\rho + 2)$, $\omega(\omega + 2) + \omega'^2$, $|\omega'(\omega + 1)|$, and $L(L+1)$, respectively. Matrix elements of the other operators in (6.169), which we collectively denote as $\hat{T}_{\rho\sigma}$, are obtained by a change of SO(4)-coupling order,

$$|((\omega_\rho, \omega_\sigma)\vec{\omega}_{\rho\sigma}, \omega_3)\vec{\omega}_{\rho\sigma 3} \ldots ; \vec{\omega} L M_L\rangle$$
$$= \sum_{\vec{\omega}_x} \langle((\omega_1, \omega_2)\vec{\omega}_{12}, \omega_3)\vec{\omega}_{123}\ldots ; \vec{\omega}|((\omega_\rho, \omega_\sigma)\vec{\omega}_{\rho\sigma}, \omega_3)\vec{\omega}_{\rho\sigma 3}\ldots ; \vec{\omega}\rangle$$
$$\times |((\omega_1, \omega_2)\vec{\omega}_{12}, \omega_3)\vec{\omega}_{123}\ldots ; \vec{\omega} L M_L\rangle \quad (6.170)$$

where the summation runs over intermediate SO(4) representations, $x = 12, 123, \ldots$. The original states are thus transformed to a basis where the first and second bonds are interchanged with the ρ and σ bonds, respectively. According to the discussion of Section 6.6, the SO(4) transformation matrix in (6.170) equals the product of two similar matrices in SO(3):

$$\langle((\omega_1, \omega_2)\vec{\omega}_{12}, \omega_3)\vec{\omega}_{123}\ldots ; \vec{\omega}|((\omega_\rho, \omega_\sigma)\vec{\omega}_{\rho\sigma}, \omega_3)\vec{\omega}_{\rho\sigma 3}\ldots ; \vec{\omega}\rangle$$
$$= \langle((r_1, r_2)r_{12}, r_3)r_{123}\ldots ; r|((r_\rho, r_\sigma)r_{\rho\sigma}, r_3)r_{\rho\sigma 3}\ldots ; r\rangle$$
$$\times \langle((r'_1, r'_2)r'_{12}, r'_3)r'_{123}\ldots ; r'|((r'_\rho, r'_\sigma)r'_{\rho\sigma}, r'_3)r'_{\rho\sigma 3}\ldots ; r'\rangle$$
$$(6.171)$$

where, as before, $r_x = (\omega_x + \omega'_x)/2$ and $r'_x = (\omega_x - \omega'_x)/2$. The two SO(3) recoupling coefficients can, in principle, be written in terms of $3nj$ coefficients. Whether their calculation is feasible in practical terms depends on the actual coupling orders involved. Note that over the years very powerful techniques have been developed to deal with recoupling in SO(3). A detailed account of this problem is given by Jucys *et al.* [14], who consider $3nj$ coefficients up to and including $n = 6$.

The matrix elements of $\hat{T}_{\rho\sigma}$ are easily obtained in the recoupled basis (6.170) since

$$\langle((\omega_{\rho a}, \omega_{\sigma a})\vec{\omega}_{\rho\sigma}, \omega_3)\vec{\omega}_{\rho\sigma 3}\ldots ; \vec{\omega}\|\hat{T}_{\rho\sigma}\|((\omega_{\rho b}, \omega_{\sigma b})\vec{\omega}_{\rho\sigma}, \omega_3)\vec{\omega}_{\rho\sigma 3}\ldots ; \vec{\omega}\rangle$$
$$= \langle[N_\rho]\omega_{\rho a}, [N_\sigma]\omega_{\sigma a}; \vec{\omega}_{\rho\sigma}\|\hat{T}_{\rho\sigma}\|[N_\rho]\omega_{\rho b}, [N_\sigma]\omega_{\sigma b}; \vec{\omega}_{\rho\sigma}\rangle \quad (6.172)$$

6.7 Polyatomic molecules

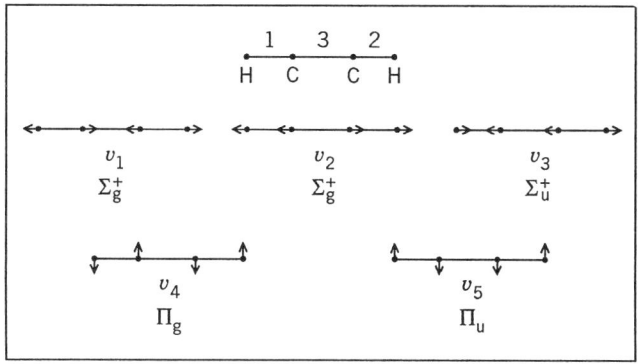

Figure 6.11: The bond coordinates of linear acetylene. Also shown are the stretching (v_1, v_2, and v_3) and bending (v_4 and v_5) normal vibrations and the vibrational species designation Σ^+ and $\Pi^\pm$. The displacements are not drawn to scale.

where the reduced matrix elements on the right-hand side are known from the previous section. This relation is valid for any operator $\hat{T}_{\rho\sigma}$ that is an SO(4) scalar, and hence it is valid for the two operators $\mathcal{C}_2[\mathrm{U}_{\rho\sigma}(4)]$ and $\mathcal{C}_2[\mathrm{SO}_{\rho\sigma}(4)]$. Note that (6.172) is diagonal in all SO(4) labels except ω_ρ and ω_σ, which, for $\hat{T}_{\rho\sigma} = \mathcal{C}_2[\mathrm{U}_{\rho\sigma}(4)]$, can take different values in bra and ket.

We illustrate the procedure outlined above with the specific example of the molecule acetylene (C_2H_2) as discussed by Iachello et al. [15]. We label the bonds as in Figure 6.11, which also illustrates the five normal-mode vibrations exhibited by linear C_2H_2.

The basis (6.166) now reduces to

$$\begin{aligned} &\mathrm{U}_1(4) \otimes \mathrm{U}_2(4) \otimes \mathrm{U}_3(4) \\ &\supset \mathrm{SO}_1(4) \otimes \mathrm{SO}_2(4) \otimes \mathrm{SO}_3(4) \\ &\supset \mathrm{SO}_{12}(4) \otimes \mathrm{SO}_3(4) \\ &\supset \mathrm{SO}(4) \supset \mathrm{SO}(3) \end{aligned} \quad (6.173)$$

with the states labeled as

$$|([N_1]\omega_1, [N_2]\omega_2)\vec{\omega}_{12}, [N_3]\omega_3; \vec{\omega} L M_L\rangle \equiv |(\omega_1, \omega_2)\vec{\omega}_{12}, \omega_3; \vec{\omega} L M_L\rangle \quad (6.174)$$

The particular coupling order (12)3 results in computational simplifications, as is discussed below.

One can investigate the properties of the states (6.166) under the inversion operation $\hat{P}$, which has the effect of interchanging the first and the third bonds. The procedure to determine the action of $\hat{P}$ was illustrated in the preceding sections in the case of triatomic molecules. Similar arguments lead to the conclusion that for acetylene the proper form for wave functions carrying a well-defined parity is (for $\omega' \neq 0$)

$$\begin{aligned}&|(\omega_1,\omega_2)\vec{\omega}_{12},\omega_3;\vec{\omega}LM_L\rangle^{\pm} \\ &= \sqrt{\tfrac{1}{2}}\Big(|(\omega_1,\omega_2)\vec{\omega}_{12},\omega_3;\vec{\omega}LM_L\rangle \\ &\quad \pm(-)^{\omega_1+\omega_2+\omega+L}|(\omega_1,\omega_2)\vec{\omega}_{12}^{\,-},\omega_3;\vec{\omega}^{\,-}LM_L\rangle\Big)\end{aligned} \quad (6.175)$$

where we use the notation $\vec{\omega}_x^{\,-} = (\omega_x, -\omega'_x)$. Note that there are two second-row SO(4) labels, ω'_{13} and ω', which can have the same or the opposite sign, corresponding to the alignment or antialignment of the two vibrational angular momenta l_4 and l_5. We may restrict one of the second labels, say ω', to non-negative values and label the states by their parity π instead.

The most general hamiltonian is of the form (6.169) with $m = 3$. Its matrix elements can be related to those obtained for triatomic molecules. For the bond–bond interaction terms (i.e, for $\mathcal{C}_2[U_{\rho\sigma}(4)]$ and $\mathcal{C}_2[SO_{\rho\sigma}(4)]$, denoted as $\hat{T}_{\rho\sigma}$) we find

$$\begin{aligned}&\langle(\omega_{1a},\omega_{2a})\vec{\omega}_{12a},\omega_{3a};\vec{\omega}\,\|\,\hat{T}_{12}\,\|\,(\omega_{1b},\omega_{2b})\vec{\omega}_{12b},\omega_{3b};\vec{\omega}\rangle \\ &= \langle[N_1]\omega_{1a},[N_2]\omega_{2a};\vec{\omega}_{12a}\,\|\,\hat{T}_{12}\,\|\,[N_1]\omega_{1b},[N_2]\omega_{2b};\vec{\omega}_{12b}\rangle\delta_{\omega_{3a}\omega_{3b}}\delta_{\vec{\omega}_{12a}\vec{\omega}_{12b}} \\[4pt]
&\langle(\omega_{1a},\omega_{2a})\vec{\omega}_{12a},\omega_{3a};\vec{\omega}\,\|\,\hat{T}_{13}\,\|\,(\omega_{1b},\omega_{2b})\vec{\omega}_{12b},\omega_{3b};\vec{\omega}\rangle \\ &= \sum_{\vec{\omega}_{13}}(-)^{\omega_{1a}+\omega_{1b}}U(\omega_{1a}\omega_{2a}\omega_{3a}\vec{\omega};\vec{\omega}_{12}\vec{\omega}_{13a})U(\omega_{1b}\omega_{2b}\omega_{3b}\vec{\omega};\vec{\omega}_{12}\vec{\omega}_{13b}) \\ &\qquad\times\langle[N_1]\omega_{1a},[N_2]\omega_{2a};\vec{\omega}_{12}\,\|\,\hat{T}_{12}\,\|\,[N_1]\omega_{1b},[N_2]\omega_{2b};\vec{\omega}_{12}\rangle\delta_{\omega_{2a}\omega_{2b}} \\[4pt]
&\langle(\omega_{1a},\omega_{2a})\vec{\omega}_{12a},\omega_{3a};\vec{\omega}\,\|\,\hat{T}_{23}\,\|\,(\omega_{1b},\omega_{2b})\vec{\omega}_{12b},\omega_{3b};\vec{\omega}\rangle \\ &= \sum_{\vec{\omega}_{23}}U(\omega_{1a}\omega_{3a}\vec{\omega}\omega_{2a};\vec{\omega}_{13a}\vec{\omega}_{23})U(\omega_{1b}\omega_{3b}\vec{\omega}\omega_{2b};\vec{\omega}_{13b}\vec{\omega}_{23}) \\ &\qquad\times\langle[N_2]\omega_{2a},[N_3]\omega_{3a};\vec{\omega}_{23}\,\|\,\hat{T}_{23}\,\|\,[N_2]\omega_{2b},[N_3]\omega_{3b};\vec{\omega}_{23}\rangle\delta_{\omega_{1a}\omega_{1b}}\end{aligned}$$

$$(6.176)$$

6.7 Polyatomic molecules 257

The matrix element of $\hat{T}_{12}$ reduces to a single triatomic matrix element. and the expressions for $\hat{T}_{13}$ and $\hat{T}_{23}$ involve a summation of products of two SO(4) U coefficients.

In [15] Iachello et al. use this approach to calculate the vibrational levels of the acetylene molecule C_2H_2 and its isotopic variations C_2D_2 and C_2HD. For C_2H_2 they report the results of two fits: one with a hamiltonian containing 8 parameters (fit 1) and a second with an extended hamiltonian with 16 parameters (fit 2). These parameter numbers exclude the boson numbers, which are chosen as $N_1 = N_2 = 43$ and $N_3 = 137$. It is found that the bond–bond interactions $\hat{T}_{13}$ and $\hat{T}_{23}$ are unimportant and can be neglected in first approximation. As a consequence, only the simplest matrix element in (6.176) needs to be evaluated, which explains why the coupling order (12)3 is the most convenient. Neglecting the $\hat{T}_{13}$ and $\hat{T}_{23}$ interactions, the calculation of vibrational levels with the general hamiltonian (6.169) involves six parameters (η_{12}, $\beta_1 = \beta_2$, β_3, β_{12}, β, and β'). Two additional, higher-order interactions are needed to account for the splitting of l degeneracy and to obtain a description of 96 acetylene levels with a root-mean-square deviation $\Delta_{\rm rms}$ of 16.8 cm^{-1}. This deviation can be further reduced by including eight more higher-order interactions (fit 2). In Table 6.4 we show the results of these two fits for lowest vibrational levels in C_2H_2. Again we note that the v_i's given in the table are only approximate quantum numbers and the values listed are those associated with the largest normal-mode component of the eigenstate. Similar results for the isotopic variations of acetylene are given in [15].

This result for acetylene, combined with similar results for triatomic molecules, raises the hope that the vibron model offers a relatively simple and comprehensive computational scheme that can be used to predict overtone frequencies of complex molecules with an overall accuracy of the order of 10 cm^{-1}.

This concludes our analysis of the vibron model as applied to triatomic and polyatomic molecules. It is perhaps appropriate to reiterate the two main conclusions that have emerged from our review. First, we have repeatedly stressed that polyatomic molecules provide the ideal testing ground of the algebraic approach as proposed in the vibron model, since this approach is typically geared to the analysis of complex

Table 6.4: Vibrational energy levels of C_2H_2 [a]

v_1	v_2	v_3	$v_4^{l_4}$	$v_5^{l_5}$		Expt	Fit 1	Δ_1	Fit 2	Δ_2
0	0	0	2^0	0^0	Σ_g^+	1230.4	1216.3	14.1	1227.6	2.8
0	0	0	1^1	1^1	Σ_u^+	1328.1	1328.1	0.0	1333.1	5.0
0	0	0	1^1	1^1	Σ_u^-	1340.6	1341.5	-0.9	1343.5	-3.0
0	0	0	0^0	2^0	Σ_g^+	1449.1	1446.6	2.5	1443.3	5.8
0	1	0	0^0	0^0	Σ_g^+	1974.3	1970.8	3.5	1975.8	-1.5
0	0	0	3^1	1^1	Σ_u^+	2560.6	2533.5	27.1	2549.7	10.9
0	0	0	3^1	1^1	Σ_u^-	2583.9	2559.7	24.2	2570.1	13.7
0	0	0	2^2	2^2	Σ_g^+	2648.0	2646.7	1.3	2656.4	-8.4
1	0	0	2^2	2^2	Σ_g^-	2661.2	2665.7	-4.6	2671.9	-10.7
0	0	0	1^1	3^1	Σ_u^+	2757.8	2764.2	-6.4	2766.2	-8.4
0	0	0	1^1	3^1	Σ_u^-	2783.7	2791.1	-7.4	2787.1	-3.4
0	0	0	0^0	4^0	Σ_u^-	2880.2	2883.8	-3.6	2877.7	2.5
0	0	0	1^1	0^0	Π_g	611.7	612.1	-0.4	617.1	-5.4
0	0	0	0^0	1^1	Π_u	729.2	726.5	2.7	724.5	4.7
0	0	0	3^1	0^0	Π_u	1854.5	1822.7	31.9	1839.1	15.5
0	0	0	2^0	1^1	Π_{uII}	1940.0	1935.2	4.8	1945.0	-5.0
0	0	0	2^0	1^1	Π_{uI}	1959.7	1954.0	5.7	1959.7	0.0
0	0	0	1^1	2^0	Π_{gII}	2047.9	2049.8	-1.9	2052.9	-5.0
0	0	0	1^1	2^0	Π_{gI}	2065.8	2068.8	-3.0	2067.7	-1.9
0	0	0	0^0	3^1	Π_u	2169.2	2168.2	0.9	2163.3	5.9
0	1	0	1^1	0^0	Π_g	2573.2	2577.3	-4.2	2579.9	-6.7
0	1	0	0^0	1^1	Π_u	2701.9	2691.7	10.2	2696.7	5.2
0	0	0	2^0	0^0	Δ_g	1228.8	1226.4	2.4	1235.2	-6.4
0	0	0	1^1	1^1	Δ_u	1342.8	1343.3	-0.5	1344.5	-1.7
0	0	0	0^0	2^2	Δ_g	1458.3	1454.3	4.0	1450.1	8.2
0	0	0	3^1	1^1	Δ_{uII}	2556.8	2544.9	11.9	2558.2	-1.4
0	0	0	3^1	1^1	Δ_{uI}	2585.0	2567.8	17.2	2576.0	8.9
0	0	0	2^2	2^2	Δ_{gII}	2661.4	2660.4	1.0	2666.7	-5.3
0	0	0	1^1	3^1	Δ_{uII}	2768.5	2773.3	-4.9	2774.0	-5.6
0	0	0	1^1	3^1	Δ_{uI}	2790.8	2796.7	-5.9	2792.2	-1.4
0	0	0	0^0	4^2	Δ_g	2889.4	2890.9	-1.6	2883.9	5.4
0	0	0	3^3	0^0	Φ_g	1851.3	1842.8	8.5	1853.9	-2.6
0	0	0	2^2	1^1	Φ_u	1961.9	1961.9	0.0	1965.0	3.1
0	0	0	1^1	2^2	Φ_g	2074.2	2076.0	-1.8	2073.0	1.2
0	0	0	0^0	3^3	Φ_u	2187.5	2183.1	4.4	2176.3	11.2
			Δ_{rms}					16.8		11.4

[a] All energies in cm^{-1}.

spectroscopic properties and a clear and simple alternative description is lacking for such systems. Second, we have shown, in particular in this section, that there are, in principle, no technical difficulties in extending the model to complex molecules. However, few examples of such calculations are available at present and more are needed to test their predictive power as well as to enable an unambiguous interpretation of the results.

Summary

This chapter is concerned with the coupling of several U(4) algebras which is of relevance to a vibron description of polyatomic molecules. It primarily deals with the simplest of such cases, namely the coupling of two U(4) algebras. The corresponding algebraic structure is $U_1(4) \otimes U_2(4)$ and is defined by relations (6.1), (6.3), and (6.4). This product algebra contains a large number of possible chains of subalgebras which are enumerated in Section 6.3. Only a small subset of them can be used in applications to molecular spectra, and these can be summarized by means of the two lattices of algebras (6.22) and (6.23). Again two types of symmetry limits arise, according to whether they pertain to U(3) or SO(4). The former have only a restricted applicability and are briefly summarized in Section 6.4. The various U(3) limits differ in the coupling scheme used: in (6.58) the algebras are coupled at the level of U(3), and in (6.71) the coupling occurs at the level of U(4). The SO(4) limits are of more interest. Again various coupling schemes exist, two of which are given in (6.86) and (6.104), corresponding to SO(4) and U(4) coupling, respectively. The associated wave functions (6.91) and (6.109) can be written in an uncoupled basis with expansion coefficients that are generalized coupling coefficients associated with $SO(4) \supset SO(3) \supset SO(2)$ and $U(4) \supset SO(4) \supset SO(3) \supset SO(2)$, respectively. The physical interpretation of these two classifications is that the first describes local vibrations of the molecule whereas the second is used for normal vibrations. The discussion is further subdivided to deal with linear and bent molecules separately, which require different combinations of Casimir invariants to arrive at a proper classification of their rotation–vibration levels. In Section 6.6 examples of triatomic spectra are discussed. The hamiltonian (6.126) is transitional between the different SO(4) limits of Section 6.5 and can thus be used to calculate a wide range of molecular spectra. Its matrix elements in one of the SO(4) bases can be evaluated with use of the tensor calculus developed in Section 5.8. Detailed applications to several triatomic molecules are presented. In the final section this procedure is generalized to polyatomic molecules. The basis (6.166), the wave functions (6.167), and the hamiltonian (6.169) are straightforward generalizations of the concepts introduced previously, and hamiltonian matrix elements can be calculated in an analogous way. The chapter concludes with an application to linear acetylene.

References

1. J.L. Dunham, "The energy levels of a rotating vibrator," Phys. Rev. **41** (1932) 721.
2. O.S. van Roosmalen, A.E.L. Dieperink, and F. Iachello, "A dynamic algebra for rotation–vibration spectra of complex molecules," Chem. Phys. Lett. **85** (1982) 32.
3. O.S. van Roosmalen, F. Iachello, R.D. Levine, and A.E.L. Dieperink, "Algebraic approach to molecular rotation–vibration spectra. II. Triatomic molecules," J. Chem. Phys. **79** (1983) 2515.
4. F. Iachello and S. Oss, "Overtone frequencies and intensities of bent XY_2 molecules in the vibron model," J. Mol. Spectrosc. **142** (1990) 85.
5. F. Iachello, S. Oss, and R. Lemus, "Vibrational spectra of linear triatomic molecules in the vibron model," J. Mol. Spectrosc. **146** (1991) 56.
6. F. Iachello, A. Leviatan, and A. Mengoni, "Algebraic approach to molecular rotation–vibration spectra. III. Infrared intensities," J. Chem. Phys. **95** (1991) 1449.
7. J.P. Elliott, "Collective motion in the nuclear shell model. I. Classification schemes for states of mixed configurations," Proc. Roy. Soc. A **245** (1958) 128.
8. A.R. Edmonds, *Angular Momentum in Quantum Mechanics*, Princeton University Press, Princeton, 1957.

References

9. G. Herzberg, *Molecular Spectra and Molecular Structure. II. Infrared and Raman Spectra of Polyatomic Molecules*, van Nostrand, New York, 1945.
10. M.S. Child and L. Halonen, "Overtone frequencies and intensities in the local mode picture," Adv. Chem. Phys. **54** (1984) 1.
11. B.G. Wybourne, *Symmetry Principles and Atomic Spectroscopy*, Wiley-Interscience, New York, 1970.
12. G. Herzberg, *Molecular Spectra and Molecular Structure. III. Electronic Spectra of Polyatomic Molecules*, van Nostrand, New York, 1950.
13. L.C. Biedenharn and J.D. Louck, *Angular Momentum in Quantum Physics*, Addison-Wesley, Reading, MA, 1981.
14. A.P. Jucys, I.B. Levinson, and V.V. Vanagas, *Mathematical Apparatus of the Theory of Angular Momentum*, Israel Program for Scientific Translations, Jerusalem, 1962.
15. F. Iachello, S. Oss, and R. Lemus, "Linear four-atom molecules in the vibron model," J. Mol. Spectrosc. **149** (1991) 132.

Chapter 7

Bose–Fermi Symmetries and Molecular Electronic Spectra

7.1 Introduction

In the previous chapters we have analyzed the structure of the vibron model and shown how group-theoretical methods give rise to an elegant and compact framework for the description of rotation–vibration spectra of linear molecules. This procedure, however, does not explicitly consider the different electronic configurations possible in the ground state of diatomic molecules or the electronic excitations to higher-lying states [1]. The applications of this algebraic formalism are thus restricted to the description of the lowest electronic state, with the additional requirement that the total electronic spin is coupled to zero, that is, to Σ^+ ground configurations in linear molecules [2]. In order to attempt a complete description of the observed spectra, the electronic degrees of freedom must be considered explicitly, and the present chapter is devoted to showing how this can be accomplished [3]. Although we shall restrict our discussion to the case of diatomic molecules, it has been shown [1] that the introduction of electronic degrees of freedom permits a distinction between heteronuclear and homonuclear molecules. Mathematically this is equivalent for this case to the introduction of point-group properties into the model, and thus it is feasible to extend the method to include the description of polyatomic molecules where discrete symmetries are essential. However, no explicit calculations have been carried out as yet.

We start our analysis by writing the total molecular wave function as

$$\psi = \sum_n \psi_n^{\text{rv}} \psi_n^{\text{e}} \qquad (7.1)$$

where ψ_n^{rv} denotes a rotation–vibration state and ψ_n^{e} an electronic state. In Chapters 5 and 6 it is shown that the rotation–vibration states may be described in terms of an algebra, G, to which we now attach the superscript rv. For diatomic molecules this algebra is $U(4)$ while for triatomic molecules it is $U(4) \otimes U(4)$. In analogy with the analysis in Chapter 4, where fermion degrees of freedom are introduced through an algebra $U^{\text{F}}(2)$, we show that the electronic states can also be described by means of a unitary algebra, G, to which we attach a superscript e. To define this algebra, we consider here the "united-atoms" limit, that is, the hypothetical case in which the atoms are placed in a common center, a device commonly used in traditional treatments of diatomic molecules [2]. This limit is useful for the definition of a complete set of states for the electrons and, in particular, for the description of very light molecules. Other choices, such as the "separated-atoms" limit, may be more appropiate for heavier molecules where the basis functions are those pertaining to the separated, non-interacting atoms [2]. For simplicity we consider hydrogenic levels, and for the classification of electronic levels in the united-atoms limit we start with the case where the electrons do not interact among themselves. This initial assumption gives rise to the definition of G^{e} which for the electrons plays the role of dynamical algebra, in an analogous fashion as $U(4)$ (or $U(4) \otimes U(4) \cdots$) does for rotation–vibration spectra. Following this choice of basis, the algebra G^{e} may be taken as $U(2n^2)$, where n is the hydrogenic principal quantum number, $n = 1, 2, \ldots$, and the factor 2 takes into account the spin of the electrons. This selection for G^{e} implies in addition that we only consider electronic states originating from a single shell in the united-atoms limit. This restriction can be removed by extending G^{e}, or, in a simpler fashion, by introducing effective interactions to take into account the effect of other shells.

As a specific example we consider in this chapter the case of electronic states originating from the 2s–2p hydrogenic levels, for which

7.2 Algebra

$G^{\text{e}} = \text{U}(8)$, in a heteronuclear diatomic molecule, for which $G^{\text{rv}} = \text{U}(4)$ [4,5].

7.2 Algebra

With each electronic level we associate the creation and annihilation operators $a^{\dagger}_{nlm,s\sigma}$ and $a_{nlm,s\sigma}$ where the quantum numbers n, l, m, s, and σ label the united-atoms states [6]. We consider the case $n = 2$ and henceforth we do not write this quantum number explicitly. The orbital angular momentum takes the values $l = 0, 1$, which are associated with the 2s–2p states, respectively. These operators satisfy the usual Fermi anticommutation relations:

$$\begin{aligned}
\{a_{lm,1/2\sigma}, a^{\dagger}_{l'm',1/2\sigma'}\} &= \delta_{ll'}\delta_{mm'}\delta_{\sigma\sigma'} \\
\{a_{lm,1/2\sigma}, a_{l'm',1/2\sigma'})\} &= \{a^{\dagger}_{lm,1/2\sigma}, a^{\dagger}_{l'm',1/2\sigma'}\} = 0
\end{aligned} \quad (7.2)$$

The 64 bilinear products of the form $a^{\dagger}_{lm,1/2\sigma} a_{l'm',1/2\sigma'}$ generate the unitary algebra $\text{U}^{\text{e}}(8)$. The hermitian generators for this algebra may be written as

$$\begin{array}{ll}
[a^{\dagger}_{0,1/2} \times \tilde{a}_{0,1/2}]^{(0,S)}_{0,M_S} & 4 \\
[a^{\dagger}_{1,1/2} \times \tilde{a}_{1,1/2}]^{(L,S)}_{M_L,M_S} & 36 \\
[a^{\dagger}_{1,1/2} \times \tilde{a}_{0,1/2} - a^{\dagger}_{0,1/2} \times \tilde{a}_{1,1/2}]^{(1,S)}_{M_L,M_S} & 12 \\
i[a^{\dagger}_{1,1/2} \times \tilde{a}_{0,1/2} + a^{\dagger}_{0,1/2} \times \tilde{a}_{1,1/2}]^{(1,S)}_{M_L,M_S} & 12
\end{array} \quad (7.3)$$

where the notation $[a^{\dagger}_{l,1/2} \times \tilde{a}_{l',1/2}]^{(L,S)}_{M_L,M_S}$ implies the coupling of l and l' to total L and projection M_L and the coupling of the spins to S and projection M_S. In analogy with the boson case we introduce the covariant operator

$$\tilde{a}_{lm,1/2\sigma} = (-)^{l+m+1/2+\sigma} a_{l-m,1/2-\sigma} \quad (7.4)$$

which transforms as $a^{\dagger}_{lm,1/2\sigma}$ under both the (orbital) SO(3) and the (spin) SU(2) algebras. By restricting the generators (7.3) to those with total spin $S = 0$, we arrive at the subalgebra $U^e(4)$ with the 16 generators associated with the orbital part of the electronic wave functions:

$$
\begin{aligned}
&[a^\dagger_{0,1/2} \times \tilde{a}_{0,1/2}]^{(0,0)}_{0,0} & 1 \\
&[a^\dagger_{1,1/2} \times \tilde{a}_{1,1/2}]^{(L,0)}_{M_L,0} & 9 \\
&[a^\dagger_{1,1/2} \times \tilde{a}_{0,1/2} - a^\dagger_{0,1/2} \times \tilde{a}_{1,1/2}]^{(1,0)}_{M_L,0} & 3 \\
&i[a^\dagger_{1,1/2} \times \tilde{a}_{0,1/2} + a^\dagger_{0,1/2} \times \tilde{a}_{1,1/2}]^{(1,0)}_{M_L,0} & 3
\end{aligned}
\qquad (7.5)
$$

Likewise, the three operators

$$
\sum_l \sqrt{\tfrac{1}{2}(2l+1)} [a^\dagger_{l,1/2} \times \tilde{a}_{l,1/2}]^{(0,1)}_{0,M_S} \qquad (7.6)
$$

generate the total-spin subalgebra $SU^e(2)$ of $U^e(8)$. It is straightforward to show that the spin operators (7.6) commute with the orbital ones in (7.5). We thus arrive at the decomposition [6]

$$
U^e(8) \supset U^e(4) \otimes SU^e(2) \qquad (7.7)
$$

which corresponds to a decomposition of the electronic wave function of the form $\psi^e = \psi^e_{\text{orbital}} \psi^e_{\text{spin}}$.

7.3 Hamiltonian

We assume that the full hamiltonian for diatomic molecules can be written in the form

$$
\hat{H} = \hat{H}^{\text{rv}} + \hat{H}^e + \hat{V}^{\text{rv}-e} \qquad (7.8)
$$

where the first term is a pure boson contribution and corresponds to the vibron hamiltonian discussed in Chapter 5, $\hat{H}^e$ is a one- and two-body electronic hamiltonian which for the $n = 2$ case under consideration can be expressed in terms of the $U^e(8)$ dynamical algebra (7.3), and $\hat{V}^{\text{rv}-e}$ is the boson–fermion interaction which is essential for the description of electronic excitations in the molecule. The most general expression (up to and including two-body terms) for $\hat{H}^e$ is of the form

$$
\hat{H}^e = \eta_0 [a^\dagger_{0,1/2} \times \tilde{a}_{0,1/2}]^{(0,0)0}
$$

7.3 Hamiltonian

$$+\eta_1[a^\dagger_{1,1/2}\times \tilde{a}_{1,1/2}]^{(0,0)0} + \rho_1[a^\dagger_{1,1/2}\times \tilde{a}_{1,1/2}]^{(1,1)0}$$

$$+\sum_{l_i LS}\eta^{LS}_{l_1l_2l_3l_4}[[a^\dagger_{l_1,1/2}\times a^\dagger_{l_2,1/2}]^{(L,S)}\times [\tilde{a}_{l_3,1/2}\times \tilde{a}_{l_4,1/2}]^{(L,S)}]^{(0,0)0}$$

$$+\sum_{l_i L_i S_i}\rho^{L_1 S_1 L_2 S_2 S_3}_{l_1l_2l_3l_4}[[a^\dagger_{l_1,1/2}\times a^\dagger_{l_2,1/2}]^{(L_1,S_1)}$$
$$\times [\tilde{a}_{l_3,1/2}\times \tilde{a}_{l_4,1/2}]^{(L_2,S_2)}]^{(S_3,S_3)0} \tag{7.9}$$

where the notation is such that the final L and S are coupled to $J=0$ and the η parameters are associated with interactions where L and S are both 0. In the model hamiltonian discussed in this chapter we ignore all terms associated with the ρ parameters since they are known to be small [2], particularly for light molecules where the spin–orbit coupling is very weak. The $\hat{V}^{\text{rv}-\text{e}}$ interaction can be written in terms of the boson and fermion operators (again up to two-body terms) as

$$\hat{V}^{\text{rv}-\text{e}} = \sum_{l_i L}\alpha^L_{l_1l_2l_3l_4}[[b^\dagger_{l_1}\times a^\dagger_{l_2,1/2}]^{(L,1/2)}\times [\tilde{b}_{l_3}\times \tilde{a}_{l_4,1/2}]^{(L,1/2)}]^{(0,0)0}$$
$$+\sum_{l_i L}\beta^L_{l_1l_2l_3l_4}[[b^\dagger_{l_1}\times a^\dagger_{l_2,1/2}]^{(L,1/2)}\times [\tilde{b}_{l_3}\times \tilde{a}_{l_4,1/2}]^{(L,1/2)}]^{(1,1)0}$$

$$\tag{7.10}$$

Again, we ignore the β terms as they contribute essentially to the finer details of the spectrum [2].

The general form (7.8)–(7.10) arises from considerations of hermiticity and rotation–inversion invariance plus the assumption of separate conservation of boson and fermion number, which generalizes the boson-number conservation of the vibron model.

An alternative form for the boson–fermion interaction provides a more useful and physically meaningful parametrization. This is the multipole expansion

$$\hat{V}^{\text{rv}-\text{e}} = \sum_k \alpha_k \hat{T}^{(k)}_{\text{rv}}\cdot \hat{T}^{(k)}_{\text{e}} \tag{7.11}$$

This simple form of the interaction is valid when the $\beta^L_{l_1l_2l_3l_4}$ terms vanish and thus the α_k are linear combinations of the $\alpha^L_{l_1l_2l_3l_4}$ alone. The structure of the spectrum turns out to be dominated by the dipole ($k=1$) and quadrupole ($k=2$) contributions to (7.11),

$$\alpha_1\hat{D}_{\text{rv}}\cdot \hat{D}_{\text{e}} + \alpha'_1 \hat{L}_{\text{rv}}\cdot \hat{L}_{\text{e}} \tag{7.12}$$

Table 7.1: Generators of fermion and boson–fermion algebras in the $U^{rv}(4) \otimes U^e(8)$ model

Algebra	Generators
	Fermion
$U^e(8)$	$\mathcal{A}_{\mu,\nu}^{(\lambda,\sigma)}(l,l') \equiv [a_{l,1/2}^\dagger \times \tilde{a}_{l',1/2}]_{\mu,\nu}^{(\lambda,\sigma)}$
$U^e(4)$	$\hat{n}_\sigma \equiv \sqrt{2}[a_{0,1/2}^\dagger \times \tilde{a}_{0,1/2}]_{0,0}^{(0,0)}$
	$\hat{n}_\pi \equiv \sqrt{6}[a_{1,1/2}^\dagger \times \tilde{a}_{1,1/2}]_{0,0}^{(0,0)}$
	$\hat{L}_{e\mu} \equiv 2[a_{1,1/2}^\dagger \times \tilde{a}_{1,1/2}]_{\mu,0}^{(1,0)}$
	$\hat{Q}_{e\mu} \equiv \sqrt{2}[a_{1,1/2}^\dagger \times \tilde{a}_{1,1/2}]_{\mu,0}^{(2,0)}$
	$\hat{D}_{e\mu} \equiv \sqrt{2}[a_{1,1/2}^\dagger \times \tilde{a}_{0,1/2} - a_{0,1/2}^\dagger \times \tilde{a}_{1,1/2}]_{\mu,0}^{(1,0)}$
	$\hat{D}'_{e\mu} \equiv i\sqrt{2}[a_{1,1/2}^\dagger \times \tilde{a}_{0,1/2} + a_{0,1/2}^\dagger \times \tilde{a}_{1,1/2}]_{\mu,0}^{(1,0)}$
$SU^e(2)$	$\hat{S}_\nu \equiv \sum_l \sqrt{\frac{1}{2}(2l+1)}[a_{l,1/2}^\dagger \times \tilde{a}_{l,1/2}]_{0,\nu}^{(0,1)}$
$SO^e(4)$	$\hat{L}_{e\mu}, \hat{D}_{e\mu}$
$U^e(3)$	$\hat{n}_\pi, \hat{L}_{e\mu}, \hat{Q}_{e\mu}$
$SO^e(3)$	$\hat{L}_{e\mu}$
	Boson–Fermion
$U(4)$	$\mathcal{B}_\mu^{(\lambda)}(l,l') + \sqrt{2}\mathcal{A}_{\mu,0}^{(\lambda,0)}(l,l'), \quad l,l' = 0,1$
$SO(4)$	$\hat{L}_{rv\mu} + \hat{L}_{e\mu}, \hat{D}_{rv\mu} + \hat{D}_{e\mu}$
$U(3)$	$\hat{n}_p + \hat{n}_\pi, \hat{L}_{rv\mu} + \hat{L}_{e\mu}, \hat{Q}_{rv\mu} + \hat{Q}_{e\mu}$
$SO(3)$	$\hat{L}_{rv\mu} + \hat{L}_{e\mu}$
$SU(2)$	$\hat{L}_{rv\mu} + \hat{L}_{e\mu} + \hat{S}_{e\mu}$

and

$$\alpha_2 \hat{Q}_{rv} \cdot \hat{Q}_e \tag{7.13}$$

where the operators $\hat{D}_{rv}$, $\hat{L}_{rv}$, and $\hat{Q}_{rv}$ are defined in (5.18) and (5.20) and $\hat{D}_e$, $\hat{L}_e$, and $\hat{Q}_e$ are given in Table 7.1.

To find the eigensolutions of the hamiltonian (7.8), we first need to construct a set of basis states classified according to the representations of the dynamical algebra for the molecular system. This dynamical algebra is given by the direct product of the boson and fermion dynamical

algebras:
$$G = U^{rv}(4) \otimes U^e(8) \tag{7.14}$$

The associated wave functions will thus be labeled by the fully symmetric irreducible representations $[N]$ of the vibron model and the fully antisymmetric ones $[1^M]$ for the electronic algebra $U^e(8)$ [6], where M is the number of electrons present in the 2s–2p shell of the united-atoms limit. In the next section we construct a complete basis for the hamiltonian (7.8) by considering a convenient chain of subalgebras of G.

7.4 Wave functions

In order to find an appropriate coupling scheme for the total wave function (7.1), we require a complete classification for the electronic states. We start from the LS decomposition (7.7) which constitutes a reasonable approximation in light molecules of the kind we intend to describe here. In addition, we know that the SO(4) symmetry in the vibron model is essential for the description of rotation–vibration spectra of rigid molecules. When searching for a basis for the electronic configurations, we consider the limit of M non-interacting electrons in the united-atoms limit which also provides us with an SO(4) algebra [4], namely the well-known symmetry algebra associated with hydrogenic atoms [7]. This $SO^e(4)$ algebra contains the electronic orbital angular momentum algebra $SO^e(3)$ as a subalgebra. We arrive at the chain of algebras

$$\begin{array}{cccc}
U^e(8) \supset & U^e(4) \otimes SU^e(2) \supset & SO^e(4) \otimes SU^e(2) \\
| & | & | & | \\
[1^M] & S & (\tau_e, \tau'_e) & S \\
\\
\supset SO^e(3) \otimes SU^e(2) \supset & SO^e(2) \otimes U^e(1) & & \\
| & | & | & | \\
L_e & S & M_e & M_S
\end{array} \tag{7.15}$$

where S is the total spin associated with the operator (7.6). Here again we encounter general $SO^e(4)$ representations with $\tau'_e \neq 0$, as is the case for the $U(4) \otimes U(4)$ model for triatomic molecules. The chain $U^e(4) \supset SO^e(4) \supset SO^e(3)$ is isomorphic to the $SO(4)$ chain of the vibron model. However, more general representations of $U^e(4)$ may occur than just the symmetric one $[N]$. In (7.15) we do not indicate the $U^e(4)$ representations, as they are completely determined by the spin S. This follows from the fact that the $U^e(8)$ representation is the fully antisymmetric one $[1^M]$. As shown in Chapter 2, (2.38), an $SU(2)$ algebra with representation j and associated with N particles arises from a $U(2)$ algebra with representation $[h, h']$ with $N = h + h'$ and $j = (h - h')/2$. Thus, for this case,

$$M = h + h', \qquad S = \tfrac{1}{2}(h - h') \tag{7.16}$$

In the decomposition

$$U^e(8) \supset U^e(4) \otimes U^e(2) \tag{7.17}$$

the representations of $U^e(4)$ should be conjugated to the $U^e(2)$ representations in order for the inner product [8] to be contained in the fully antisymmetric representation $[1^M]$ of $U^e(8)$. The conjugate representation is defined by the replacement of rows by columns in the Young tableau:

$$\begin{array}{c} \overset{h}{\square\square\square\cdots\square} \\ \underset{h'}{\square\square\cdots\square} \end{array} \quad \rightarrow \quad \left.\begin{array}{c}\square\\\square\\\square\\\vdots\\\square\end{array}\right\}h \quad \left.\begin{array}{c}\square\\\square\\\vdots\\\square\end{array}\right\}h' \tag{7.18}$$

This result generalizes the well-known fact that a fully antisymmetric state can be built from a product of symmetric and antisymmetric states. From this discussion we conclude that the $U^e(4)$ representations are given by

$$[2^{h'}, 1^{h-h'}, 0, 0] = [2^{M/2-S}, 1^{2S}, 0, 0] \tag{7.19}$$

The $U^e(4)$ and $U^e(2)$ (or $SU^e(2)$) algebras are said to be *complementary* within the $[1^M]$ representation of $U^e(8)$. In Table 7.1 the generators of the algebras in (7.15) are shown while in Table 7.2 we list the reduction rules for the labels for M running from one to eight, the maximum number of electrons in the 2s–2p hydrogenic shell.

7.4 Wave functions

Table 7.2: Electronic states in the 2s–2p shell

$U^e(8)$	$U^e(4)$	$SU^e(2)$	$SO^e(4)$
$[1^M]$	$[h, h', h'', h''']$	S	(τ_e, τ'_e)
[1]	[1]	1/2	(1,0)
$[1^2]$	[1,1]	1	(1,±1)
	[2]	0	(0,0)
			(2,0)
$[1^3]$	[1,1,1]	3/2	(1,0)
	[2,1]	1/2	(1,0)
			(2,±1)
$[1^4]$	[1,1,1,1]	2	(0,0)
	[2,1,1]	1	(1,±1)
			(2,0)
	[2,2]	0	(0,0)
			(2,0)
			(2,±2)
$[1^5]$	[2,1,1,1]	3/2	(1,0)
	[2,1,1]	1/2	(1,0)
			(2,±1)
$[1^6]$	[2,1,1,1]	1	(1,±1)
	[2,2,2]	0	(0,0)
			(2,0)
$[1^7]$	[2,2,2,1]	1/2	(1,0)
$[1^8]$	[2,2,2,2]	0	(0,0)

The electronic states $|[1^M](\tau_e, \tau'_e)L_e M_e, S M_S\rangle$ were constructed explicitly in [6]. Here we illustrate the procedure for $M = 1$ and $M = 2$. We use the isomorphism $SO(4) \simeq SU_r(2) \otimes SU_{r'}(2)$ to find the necessary representations and coupling coefficients. The single-electron states are—as far as their orbital character is concerned—equivalent to the one-boson states of Chapter 5 and transform as the (1,0) representation of SO(4),

$$|[1](1,0)l_e m_e, 1/2\sigma\rangle = a^\dagger_{l_e m_e, 1/2\sigma}|0\rangle \quad (7.20)$$

as is clear from the allowed values of l_e (0 and 1) in (1,0). The two-electron states can be most easily found using the relations (5.122),

$$\tau_e = j + j', \qquad \tau'_e = j - j' \qquad (7.21)$$

from which $(1,0) \simeq \{1/2, 1/2\}$, where we use the notation $\{j, j'\}$ to label the $SU(2) \otimes SU(2)$ representations. Since the coupling of two (independent) spin $1/2$ states gives spin 0 and 1, the two-electron states must correspond to one of the $\{0,0\}$, $\{1,0\}$, $\{0,1\}$, or $\{1,1\}$ representations. Using (7.21) again, we find

$$(1,0) \otimes (1,0) = (0,0) \oplus (1,1) \oplus (1,-1) \oplus (2,0) \qquad (7.22)$$

The two-electron states can be constructed by coupling two one-electron states (7.20) to one of the $SO^e(4)$ representations (7.22) and by coupling the spins $1/2$ to a total spin S:

$$|[1^2](\tau_e, \tau'_e) L_e M_e, S M_S\rangle$$
$$= \sum_{l_1 l_2} \sum_{m_1 m_2} \sum_{\sigma_1 \sigma_2} \left\langle \begin{matrix} (1,0) & (1,0) \\ l_1 & l_2 \end{matrix} \middle| \begin{matrix} (\tau_e, \tau'_e) \\ L_e \end{matrix} \right\rangle \langle l_1 m_1\, l_2 m_2 | L_e M_e \rangle$$
$$\times \langle 1/2\sigma_1\, 1/2\sigma_2 | S M_S \rangle a^\dagger_{l_1 m_1, 1/2\sigma_1} a^\dagger_{l_2 m_2, 1/2\sigma_2} |0\rangle \qquad (7.23)$$

The first factor under the summation sign is the $SO(4) \supset SO(3)$ isoscalar factor, discussed in Section 5.8, for which a general expression is available (see (5.211)). Here it reduces to

$$\sqrt{(2l_1+1)(2l_2+1)(\tau_e + \tau'_e + 1)(\tau_e - \tau'_e + 1)} \begin{Bmatrix} 1/2 & 1/2 & (\tau_e + \tau'_e)/2 \\ 1/2 & 1/2 & (\tau_e - \tau'_e)/2 \\ l_1 & l_2 & L_e \end{Bmatrix}$$
$$(7.24)$$

The $9j$ coefficients in (7.24) can be evaluated in general [9] and thus lead to the complete determination of the two-electron states [1]. The $SO(4)$ states $(1,1)$ and $(1,-1)$ are associated with $S = 1$ and the $(0,0)$ and $(2,0)$ representations with $S = 0$, as indicated in Table 7.2. This fact can be understood from the direct product of $U(4)$ representations, as discussed in Appendix B,

$$[1] \otimes [1] = [1,1] \oplus [2,0] \qquad (7.25)$$

7.4 Wave functions

and the subsequent $U(4) \supset SO(4)$ reduction (see Chapter 5), $[1,1] \to (1,1)$ and $[2,0] \to (2,0) \oplus (0,0)$. The complementarity of $U(4)$ and $SU(2)$ representations then leads to the corresponding relation with the spin S. Successive couplings analogous to (7.23) allow the construction of the states for $M > 2$ electrons [6].

Once we know both the boson and the fermion states, we may label the complete rotation–vibration–electronic wave functions via the chain

$$\begin{array}{c}
U^{rv}(4) \otimes U^e(8) \supset U^{rv}(4) \otimes U^e(4) \otimes SU^e(2) \\
\hspace{0.3em}|\hspace{3em}|\hspace{4em}|\hspace{4em}|\hspace{4em}| \\
[N]\hspace{1em}[1^M]\hspace{2em}[N]\hspace{5em}S \\[0.5em]
\supset SO^{rv}(4) \otimes SO^e(4) \otimes SU^e(2) \supset SO(4) \otimes SU^e(2) \\
\hspace{0.3em}|\hspace{3em}|\hspace{4em}|\hspace{4em}|\hspace{3em}| \\
(\omega,0)\hspace{1em}(\tau_e,\tau_e')\hspace{2em}S\hspace{3em}(\tau,\tau')\hspace{2em}S \\[0.5em]
\supset SO(3) \otimes SU^e(2) \supset SU(2) \\
\hspace{0.3em}|\hspace{3em}|\hspace{4em}| \\
L\hspace{2em}S\hspace{3em}J
\end{array} \quad (7.26)$$

The generators of the electronic and rotation–vibration–electronic algebras in (7.26) are given in Table 7.1, and below each algebra we indicate the quantum numbers required to label them. The coupling of L and S leads to the total angular momentum J. States classified according to (7.26) are written as

$$\begin{aligned}
&|[N](\omega,0),[1^M](\tau_e,\tau_e');(\tau,\tau')LSJM_J\rangle \\
&= \sum_{L_{rv} L_e} \sum_{M_{rv} M_e} \sum_{M_L M_S} \left\langle \begin{array}{cc|c} (\omega,0) & (\tau_e,\tau_e') & (\tau,\tau') \\ L_{rv} & L_e & L \end{array} \right\rangle \\
&\quad \times \langle L_{rv} M_{rv}\, L_e M_e | L M_L\rangle \langle L M_L\, S M_S | J M_J\rangle \\
&\quad \times |[N]\omega l m\rangle |[1^M](\tau_e,\tau_e')l_e m_e; S M_S\rangle
\end{aligned} \quad (7.27)$$

which can be explicitly constructed for all cases since the $SO(4) \supset SO(3)$ isoscalar factors are known through the relation (5.211). The allowed values for (τ,τ') and L can be given in closed form [10]:

$$\begin{aligned}
\tau &= \tfrac{1}{2}|\omega - \tau_e - \tau_e'| + \tfrac{1}{2}|\omega - \tau_e + \tau_e'| + \alpha + \beta \\
\tau' &= \tfrac{1}{2}|\omega - \tau_e - \tau_e'| + \tfrac{1}{2}|\omega - \tau_e + \tau_e'| + \alpha - \beta
\end{aligned} \quad (7.28)$$

with
$$\alpha = 0, 1, \ldots, \min(\omega, \tau_e + \tau'_e)$$
$$\beta = 0, 1, \ldots, \min(\omega, \tau_e - \tau'_e) \quad (7.29)$$
and
$$L = |\tau'|, |\tau'| + 1, \ldots, \tau \quad (7.30)$$

For example, we find that for $(1,0) \otimes (1,0)$ we recover the result (7.22).

To conclude our discussion about the wave functions (7.27), we note that the model hamiltonian (7.8) conserves angular momentum and parity. Parity is in general not well defined for the molecular states (7.27) as a consequence of the lack of definite parity of the $SO^e(4)$ electronic states. The parity of the electronic states is well defined in the alternative chain [1,6,10]

$$\begin{array}{ccccccc}
U^e(8) \supset & U^e(4) \otimes SU^e(2) \supset & U^e(3) & \otimes\, SU^e(2) \supset & SO^e(3) \otimes SU^e(2) \\
| & | & | & | & | & | \\
[1^M] & S & [h, h', h''] & S & L_e & S
\end{array}$$
(7.31)

which is analogous to the U(3) chain in the vibron model, discussed in Chapter 5. As shown in Appendix B, in this case the $U^e(3)$ representations require in general three labels, denoted by $[h, h', h'']$ above, which form a partition of an integer n_π counting the number of electrons in the p orbital of the united-atoms limit,

$$n_\pi = h + h' + h'' \quad (7.32)$$

Since electronic parity is positive for s orbitals and negative for p orbitals, the total parity for the states is simply $(-)^{n_\pi} = (-)^{h+h'+h''}$. With this result it is possible to find linear combinations of the SO(4) wave functions (7.27) with good parity. They are given by [10]

$$\psi^\pm = \sqrt{\tfrac{1}{2}}\Big(|[N](\omega,0), [1^M](\tau_e, \tau'_e); (\tau, \tau') LSJM_J\rangle$$
$$\pm (-)^{\omega+\tau+L+\delta} |[N](\omega,0), [1^M](\tau_e, -\tau'_e); (\tau, -\tau') LSJM_J\rangle\Big)$$
(7.33)

where
$$\delta = \begin{cases} 0 & \text{for } \tau'_e \neq 0 \\ (-)^{\tau_e + \lfloor \frac{2S+1}{3} \rfloor} & \text{for } \tau'_e = 0 \end{cases} \quad (7.34)$$

7.5 Symmetry limits

and $\lfloor x \rfloor$ indicates the integral part of x. A short-hand notation for the states (7.33) is

$$|[N](\omega,0),[1^M](\tau_e,\tau'_e);(\tau,\tau')LSJ^\pi M_J\rangle \tag{7.35}$$

where $\pi = \pm$ is the parity. The states (7.27) (or (7.35)) are eigenstates of a dynamical-symmetry hamiltonian which we discuss in the next section.

7.5 Symmetry limits

The general algebraic hamiltonian (7.8) includes a large number of interactions and parameters, as seen from (7.9) and (7.10). It is possible, however, to find particular combinations of these interactions that give rise to simple solutions, as was the case for the pure boson hamiltonians of Chapters 5 and 6. Indeed, we may consider a linear combination of invariant operators in the chain (7.26):

$$\begin{aligned}\hat{H}_{\text{II}} &= \beta_{\text{rv}}\mathcal{C}_2[\text{SO}^{\text{rv}}(4)] + \beta_e\mathcal{C}_2[\text{SO}^e(4)] + \beta\mathcal{C}_2[\text{SO}(4)] \\ &\quad + \gamma_L\hat{L}^2 + \gamma_S\hat{S}^2 + \gamma_J\hat{J}^2\end{aligned} \tag{7.36}$$

where the interactions are Casimir invariants of the indicated algebras. This hamiltonian is diagonal in the basis (7.26) (or (7.35)) with eigenvalues [7]

$$\begin{aligned}&E_{\text{II}}(\omega,\tau_e,\tau'_e,\tau,\tau',L,S,J) \\ &= \beta_{\text{rv}}\omega(\omega+2) + \beta_e[\tau_e(\tau_e+2)+\tau'^2_e] + \beta[\tau(\tau+2)+\tau'^2] \\ &\quad + \gamma_L L(L+1) + \gamma_S S(S+1) + \gamma_J J(J+1)\end{aligned} \tag{7.37}$$

When the hamiltonian is of the form (7.36), it corresponds to the dynamical symmetry SO(4). In Figure 7.1 we show a typical SO(4) spectrum obtained from (7.37).

The explicit form of the Casimir invariants is given in Table 7.3. The eigenvalue of any SO(4) Casimir invariant acting on a state (τ,τ') is of the form $\tau(\tau+2) + \tau'^2$, as was shown in Chapter 5 using the SO(4) $\simeq$ SU$_r$(2) $\otimes$ SU$_{r'}$(2) isomorphism (7.21).

Before considering other possible limits, we first discuss the physical significance of the classification scheme defined by (7.36). We note

Table 7.3: Casimir invariants in the $U^{rv}(4) \otimes U^e(8)$ model[a]

Algebra	Order	Casimir invariant	Eigenvalue
		Fermion	
$U^e(8)$	1	$\hat{M} \equiv \hat{n}_\sigma + \hat{n}_\pi$	M
	2	$\sum_{ll'\lambda\sigma}(-)^{l+l'}\mathcal{A}^{(\lambda,\sigma)}(l,l'):\mathcal{A}^{(\lambda,\sigma)}(l,l')$	$M(9-M)$
$U^e(4)$	1	$\hat{M}$	M
	2	$2\sum_{ll'\lambda}(-)^{l+l'}\mathcal{A}^{(\lambda,0)}(l,l')\cdot\mathcal{A}^{(\lambda,0)}(l,l')$	$6M - \frac{1}{2}M^2 - 2S(S+1)$
$U^e(3)$	1	$\hat{n}_\pi$	n_π
	2	$2\sum_\lambda \mathcal{A}^{(\lambda,0)}(1,1)\cdot\mathcal{A}^{(\lambda,0)}(1,1)$	$h^2 + h'^2 + h''^2$
		$= \frac{1}{3}\hat{n}_\pi^2 + \frac{1}{2}\hat{L}_e^2 + \hat{Q}_e^2$	$-2h - 2h''$
$SO^e(4)$	2	$\hat{L}_e^2 + \hat{D}_e^2$	$\tau_e(\tau_e+2) + \tau_e'^2$
$SO^e(3)$	2	$\hat{L}_e^2$	$l_e(l_e+1)$
$SU^e(2)$	2	$\hat{S}^2$	$S(S+1)$
		Boson–Fermion	
$U(4)$	1	$\hat{M} + \hat{N}$	$M+N$
	2	$\mathcal{C}_2[U^{rv}(4)] + \mathcal{C}_2[U^e(4)]$	$h^2 + h'^2 + h''^2 + h'''^2$
		$-\sqrt{8}\sum_{ll'\lambda}(-)^{l+l'}\mathcal{B}^{(\lambda)}(l,l')\cdot\mathcal{A}^{(\lambda,0)}(l',l)$	$+3h + h' - h'' - 3h'''$
$SO(4)$	2	$(\hat{L}_{rv} + \hat{L}_e)^2 + (\hat{D}_{rv} + \hat{D}_e)^2$	$\tau(\tau+2) + \tau'^2$
$U(3)$	1	$\hat{n}_p + \hat{n}_\pi$	$n_p + n_\pi$
	2	$\frac{1}{3}(\hat{n}_p + \hat{n}_\pi)^2 + \frac{1}{2}(\hat{L}_{rv} + \hat{L}_e)^2$	$h(h+2) + h'^2$
		$+ (\hat{Q}_{rv} + \hat{Q}_e)^2$	$+h''(h''-2)$
$SO(3)$	2	$\hat{L}^2 \equiv (\hat{L}_{rv} + \hat{L}_e)^2$	$L(L+1)$
$SU(2)$	2	$\hat{J}^2 \equiv (\hat{L} + \hat{S})^2$	$J(J+1)$

[a] The double scalar product is defined as
$$\mathcal{A}^{(\lambda,\sigma)}:\mathcal{A}^{(\lambda,\sigma)} = \sum_{\mu\nu}(-)^{\mu+\nu}\mathcal{A}^{(\lambda,\sigma)}_{\mu,\nu}\mathcal{A}^{(\lambda,\sigma)}_{-\mu,-\nu}$$

7.5 Symmetry limits

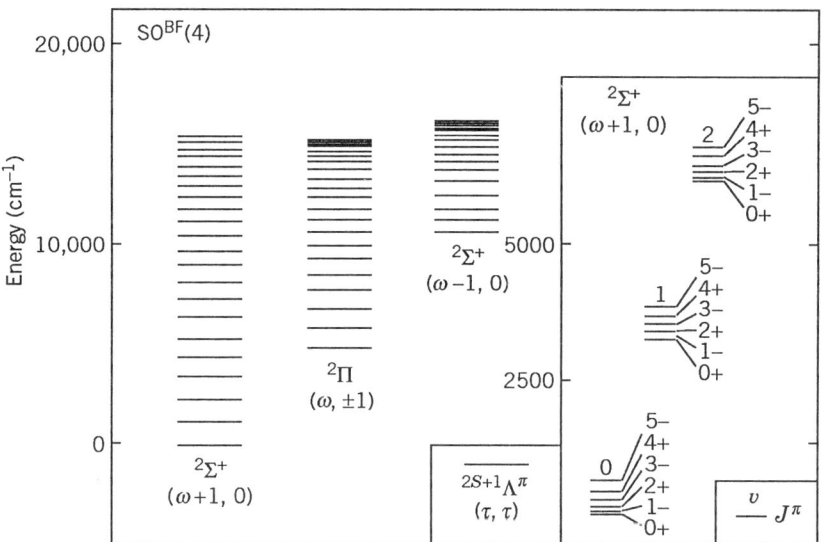

Figure 7.1: Partial $SO^{BF}(4)$ eigenspectrum of the hamiltonian $\hat{H}_{II}$ for $N = 42$ bosons and $M = 1$ or $M = 7$ electrons. The figure shows the electronic structure obtained with the parameters (in cm^{-1}) $\beta_{rv} = 157$ and $\beta = -174$. The inset shows the rotational structure of the electronic state $^2\Sigma^+(\omega+1,0)$ obtained with the parameters (in cm^{-1}) $\gamma_L = 18$ and $\gamma_J = 1$.

from Table 7.1 that the interactions in $\hat{H}_{II}$ contribute mostly to the $k = 1$ term in the multipole expansion (7.11), which indicates that this symmetry is dominated by dipole interactions. To establish a connection with experimental data, it is convenient to relate the group labels in (7.35), which do form a complete set, to quantum numbers commonly used in molecular physics [2]. These are based on the Born–Oppenheimer approximation [11,12]. Because of axial symmetry in the diatomic molecule, the projection of the orbital angular momentum on this axis is conserved, and thus the electronic states are classified according to the value of this projection. Its absolute value is denoted by Λ and takes the values $0, 1, 2, \ldots$, denoted by $\Sigma, \Pi, \Delta, \ldots$, respectively. The Σ states are either Σ^+ or Σ^- according to their behavior under reflection on a plane containing the symmetry axis. The states with $\Lambda \neq 0$, such as $\Pi^\pm, \Delta^\pm, \ldots$, are doubly degenerate to a very good approximation. Since Λ is a projection, we have $L \geq \Lambda$. Denoting the vibrational quantum number by v, the usual labeling of states is then given by

$$|v\Lambda LSJM_J\rangle \tag{7.38}$$

In Chapter 5 it is shown that in the vibron model

$$v = \tfrac{1}{2}(N - \omega) \tag{7.39}$$

so the boson number N fixes the maximum number of vibrational states present in an electronic state. The quantum numbers L, S, J, and M_J correspond to the ones in (7.35), and thus we are left with finding the algebraic equivalent to Λ. We note that, from (7.30), $\tau \geq L \geq |\tau'|$, and consequently we are able to make the identification $\Lambda = |\tau'|$. This implies the correlations

$$(\tau, 0) \leftrightarrow \Sigma, \qquad (\tau, \pm 1) \leftrightarrow \Pi, \qquad (\tau, \pm 2) \leftrightarrow \Delta, \qquad \ldots \tag{7.40}$$

To find the relation between the σ label in Λ^σ and the parity quantum number π, we reason as follows. Since reflection on a plane is equivalent to an inversion followed by a rotation by π along an axis perpendicular to the symmetry axis, we find the relation

$$\sigma = \pi(-)^L \tag{7.41}$$

since the effect of the rotation on the states (7.35) is just a phase factor $(-)^L$. Using this relation, we find the following classification of the states (7.35):

1. Σ states.

    (a) If $\tau'_e = \tau' = 0$, we conclude that

    $$|[N](\omega, 0), [1^M](\tau_e, 0); (\tau, 0)LSJM_J\rangle = \begin{cases} {}^{2S+1}\Sigma^+, & \alpha = \text{even} \\ {}^{2S+1}\Sigma^-, & \alpha = \text{odd} \end{cases}$$

    where $\alpha = \omega + \tau + \delta$ (see (7.34)) and we have included the multiplicity index $2S + 1$, as is usually done.

7.5 Symmetry limits

(b) If $\tau_e' \neq 0$ and $\tau' = 0$, we construct the states

$$\psi_\Sigma^\pm = \sqrt{\tfrac{1}{2}}\Big(|[N](\omega,0),[1^M](\tau_e,\tau_e');(\tau,0)LSJM_J\rangle$$
$$\pm |[N](\omega,0),[1^M](\tau_e,-\tau_e');(\tau,0)LSJM_J\rangle\Big)$$

and we find that

$$\psi_\Sigma^\pm = \begin{cases} {}^{2S+1}\Sigma^\pm, & \alpha = \text{even} \\ {}^{2S+1}\Sigma^\mp, & \alpha = \text{odd} \end{cases}$$

2. $\Lambda \neq 0$ states ($\tau' \neq 0$). We again consider the combinations

$$\psi_\Lambda^\pm = \sqrt{\tfrac{1}{2}}\Big(|[N](\omega,0),[1^M](\tau_e,\tau_e');(\tau,\tau')LSJM_J\rangle$$
$$\pm |[N](\omega,0),[1^M](\tau_e,-\tau_e');(\tau,\tau')LSJM_J\rangle\Big)$$

and find the pairs of degenerate states

$$\psi_\Lambda^\pm = \begin{cases} {}^{2S+1}\Lambda^\pm, & \alpha = \text{even} \\ {}^{2S+1}\Lambda^\mp, & \alpha = \text{odd} \end{cases}$$

This labeling of states is complete and coincides with the standard classification. The indices τ, τ_e, and τ_e' are additional indices supplied by group theory to uniquely specify the states, although as seen from (7.34) and (7.41), they are related to σ.

We now return to the question of whether there are other dynamical symmetries present in the model. It has been shown [10] that there are six possible limits that involve the initial reduction $U^{rv}(4) \otimes U^e(8) \supset U^{rv}(4) \otimes U^e(4) \otimes SU^e(2)$, including the one associated with $\hat{H}_{II}$ in (7.36). Here we mention only one other symmetry which has some physical interest. This is the U(3) limit, resulting from the coupling of the rotation–vibration U(3) chain (5.52) and the electronic U(3) chain (7.31):

$$\begin{aligned}
U^{rv}(4) \otimes U^e(8) &\supset U^{rv}(4) \otimes U^e(4) \otimes SU^e(2) \\
&\supset U^{rv}(3) \otimes U^e(3) \otimes SU^e(2) \\
&\supset U(3) \otimes SU^e(2) \\
&\supset SO(3) \otimes SU^e(2) \supset SU(2) \quad (7.42)
\end{aligned}$$

which is of importance for the description of homonuclear molecules [1,11,10]. In Table 7.3 we give the explicit form of the Casimir invariants of the algebras in (7.42) together with their eigenvalues [7].

In the next section we show an application of the techniques developed so far.

7.6 Application to the hydride molecules

From Figure 7.1 we conclude that the SO(4) limit incorporates the main characteristics found in diatomic molecular spectra. An inspection of spectroscopic data for molecules arising from the 2s–2p configurations (LiH, BeH, BH, CH, NH, and OH), however, reveals that the electronic level ordering implicit in the SO(4) limit is not displayed by them. This fact implies that the SO(4) limit lacks some important interactions which are necessary for the correct description of these molecules, which leads us to a discussion of SO(4) symmetry breaking.

As mentioned before, the $\hat{V}^{\mathrm{rv-e}}$ interaction (7.10) plays a major role in determining the relative position of the electronic states. The SO(4) limit incorporates dipole ($k = 1$) interactions only, so it is natural to study the influence of the $k = 0$ and $k = 2$ interactions in (7.11), in order to have a complete description up to two-body terms. Some of the monopole terms, such as the $\hat{n}_\mathrm{p}$ operator of the $\mathrm{U}^{\mathrm{rv}}(3)$ vibron chain, will modify exclusively the vibron levels, as described in Chapter 5. This operator will be important to fine-tune the fit to vibrational levels but will have no influence on the order of electronic states. On the other hand, the $\mathrm{U}^{\mathrm{e}}(3)$ invariants $\hat{n}_\pi$ and $\hat{Q}_\mathrm{e} \cdot \hat{Q}_\mathrm{e}$ have an important effect, since they contribute to the 2s–2p splitting and to the electron interactions in the molecule. The $\hat{S}^2$ and $\hat{J}^2$ operators in (7.36) may be neglected in a first approximation, as they correspond to spin–spin and spin–orbit interactions which only contribute to the fine structure of the spectrum.

The rotation–vibration–electronic states are then obtained by diagonalizing the molecular hamiltonian (7.8)

$$\hat{H} = \hat{H}^{\mathrm{rv}} + \hat{H}^{\mathrm{e}} + \hat{V}^{\mathrm{rv-e}} \qquad (7.43)$$

7.6 Application to the hydride molecules

where

$$\begin{aligned}
\hat{H}^{\rm rv} &= \beta_{\rm rv}\mathcal{C}_2[{\rm SO}^{\rm rv}(4)] + \gamma_L \hat{L}^2 \\
\hat{H}^{\rm e} &= \epsilon_e \hat{n}_\pi + \beta_e \mathcal{C}_2[{\rm SO}^{\rm e}(4)] + \kappa_e \hat{Q}_e^2 + \gamma_e \hat{L}_e^2 \\
\hat{H}^{\rm rv-e} &= \beta \mathcal{C}_2[{\rm SO}(4)] + \kappa \hat{Q}_{\rm rv}\hat{Q}_e + \gamma \hat{L}_{\rm rv}\hat{L}_e
\end{aligned} \quad (7.44)$$

Additional interactions, such as those involving the $\hat{D}_{{\rm rv}\mu}$ of ${\rm SO}^{\rm rv}(4)$ (and the corresponding ${\rm SO}^{\rm e}(4)$ operator $\hat{D}_{e\mu}$) may be included in (7.44), but it is found that it is not necessary to do so. The matrix elements of the molecular hamiltonian are evaluated in the basis (7.35). Some of the interactions are already diagonal in the SO(4) basis, but others are non-diagonal and it is necessary to evaluate their matrix elements. This involves the same kind of steps followed in the previous chapters:

1. We write all operators involved as SO(4) tensor operators $\hat{T}_{LM_L}^{(\tau,\tau')}$. In this case we find, for example,

$$\begin{aligned}
\hat{n}_\pi &= -\sqrt{\tfrac{9}{2}}\hat{T}_{00}^{(0,0)} + \sqrt{\tfrac{3}{2}}\hat{T}_{00}^{(2,0)} \\
\hat{Q}_{e\mu} &= \hat{T}_{2\mu}^{(2,0)}
\end{aligned} \quad (7.45)$$

where the tensors $\hat{T}_{00}^{(0,0)}$ and $\hat{T}_{00}^{(2,0)}$ are given by

$$\begin{aligned}
\hat{T}_{00}^{(0,0)} &= \Sigma_{lm\sigma} \left\langle \begin{matrix}(1,0) & (1,0) \\ l & l\end{matrix} \middle| \begin{matrix}(0,0) \\ 0\end{matrix} \right\rangle \langle lm\, l-m|00\rangle \\
&\quad \times \langle 1/2\sigma\, 1/2-\sigma|00\rangle a^\dagger_{lm,1/2\sigma}\tilde{a}_{l-m,1/2-\sigma} \\
&= \sqrt{\tfrac{1}{8}}(\hat{n}_\sigma + \hat{n}_\pi) \\
\hat{T}_{00}^{(2,0)} &= \Sigma_{lm\sigma} \left\langle \begin{matrix}(1,0) & (1,0) \\ l & l\end{matrix} \middle| \begin{matrix}(2,0) \\ 0\end{matrix} \right\rangle \langle lm\, l-m|00\rangle \\
&\quad \times \langle 1/2\sigma\, 1/2-\sigma|00\rangle a^\dagger_{lm,1/2\sigma}\tilde{a}_{l-m,1/2-\sigma} \\
&= \sqrt{\tfrac{3}{8}}(\hat{n}_\sigma - 3\hat{n}_\pi)
\end{aligned} \quad (7.46)$$

Since $\hat{L}_{\rm rv}\cdot\hat{L}_e = (\hat{L}^2 - \hat{L}_{\rm rv}^2 - \hat{L}_e^2)/2$, the matrix elements of this term can be easily computed in the basis (7.27).

2. To evaluate the matrix elements of a given tensor operator, we apply the generalized Wigner–Eckart theorem. To illustrate this

point, we calculate the matrix element of $\hat{n}_\pi$. Since this operator acts only on the fermion part of the states (7.27), the calculation is simplified. Using the expansion (7.27), we find

$$\langle [N](\omega,0), [1^M](\bar{\tau}_e, \bar{\tau}'_e); (\bar{\tau}, \bar{\tau}') LSJM_J |$$
$$\hat{n}_\pi | [N](\omega,0), [1^M](\tau_e, \tau'_e); (\tau, \tau') LSJM_J \rangle$$
$$= \sum_{ll_e} \left\langle \begin{array}{cc} (\omega,0) & (\bar{\tau}_e, \bar{\tau}'_e) \\ l & l_e \end{array} \middle| \begin{array}{c} (\bar{\tau}, \bar{\tau}') \\ L \end{array} \right\rangle \left\langle \begin{array}{cc} (\omega,0) & (\tau_e, \tau'_e) \\ l & l_e \end{array} \middle| \begin{array}{c} (\tau, \tau') \\ L \end{array} \right\rangle$$
$$\times \langle [1^M](\bar{\tau}_e, \bar{\tau}'_e) l_e m_e | \hat{n}_\pi | [1^M](\tau_e, \tau'_e) l_e m_e \rangle \qquad (7.47)$$

In (7.47) we use the fact that $\hat{n}_\pi$ is an SO(3) scalar, and consequently, the Clebsch–Gordan coefficients occurring in the expansion (7.27) sum up to 1. In addition, on the right-hand side of (7.47), we have suppressed the spin quantum numbers since there is no dependence on them. We now apply the Wigner–Eckart theorem to the electronic matrix element by substituting the first of relations (7.45) in (7.47):

$$\langle [1^M](\bar{\tau}_e, \bar{\tau}'_e) l_e m_e | \hat{n}_\pi | [1^M](\tau_e, \tau'_e) l_e m_e \rangle$$
$$= -\sqrt{\tfrac{9}{2}} \langle [1^M](\tau_e, \tau'_e) \| \hat{T}^{(0,0)} \| [1^M](\tau_e, \tau'_e) \rangle \delta_{\bar{\tau}_e,\tau_e} \delta_{\bar{\tau}'_e,\tau'_e}$$
$$+ \sqrt{\tfrac{3}{2}} \left\langle \begin{array}{cc} (\tau_e, \tau'_e) & (2,0) \\ l_e & 0 \end{array} \middle| \begin{array}{c} (\bar{\tau}_e, \bar{\tau}'_e) \\ l_e \end{array} \right\rangle$$
$$\times \langle [1^M](\bar{\tau}_e, \bar{\tau}'_e) \| \hat{T}^{(2,0)} \| [1^M](\tau_e, \tau'_e) \rangle \qquad (7.48)$$

where $\langle \cdots \| \hat{T}^{(\tau,\tau')} \| \cdots \rangle$ denotes the SO(4)-reduced matrix element introduced in Section 5.8.

3. We evaluate the reduced matrix elements by using the explicit forms (7.46) for the tensor operators and simple (highest-weight) states for the electronic wave functions [1,6], analogous to the evaluation of reduced matrix elements of boson operators discussed in Chapters 5 and 6.

The evaluation of the other matrix elements in (7.44) is carried out in a similar fashion and may be found in [10]. We again see that the computation of matrix elements involves simple manipulations and the

7.6 Application to the hydride molecules

knowledge of the coupling coefficients associated with particular Lie algebras.

Although in principle all interactions in (7.44) are involved in a least-squares fit, they are not all equally relevant. For example, the $\hat{L}_{\text{rv}}^2$ and $\hat{L}_{\text{rv}} \cdot \hat{L}_{\text{e}}$ have importance for the rotational spectrum, but hardly any for the electronic and vibrational features. The electronic hamiltonian, on the other hand, can be determined (or at least constrained) by carrying out *atomic* calculations, as we now discuss [13,14].

An atomic hamiltonian in the 2s–2p shell, including only Coulomb interactions, can be written in second quantized form in terms of the operators $\hat{n}_\pi$, $\hat{n}_\pi^2$, $\hat{D}_{\text{e}}^2$, and $\hat{Q}_{\text{e}}^2$ [6]. The contribution of $\hat{n}_\pi^2$, however, turns out to be of lesser importance than the others [6], which suggests that the operators $\hat{n}_\pi$, $\hat{D}_{\text{e}}^2$, and $\hat{Q}_{\text{e}}^2$ may be sufficient to obtain a reasonable fit for the excited atomic states of the 2s–2p atoms ^{3}Li, ^{4}Be, ^{5}B, ^{6}C, ^{7}N, ^{8}O, and ^{9}F. We rewrite $\hat{D}_{\text{e}}^2$ in terms of $\mathcal{C}_2[\text{SO}^{\text{e}}(4)]$ by including at the same time the operator $\hat{L}_{\text{e}}^2$,

$$\hat{H}^{\text{e}} = \epsilon_{\text{e}} \hat{n}_\pi + \beta_{\text{e}} \mathcal{C}_2[\text{SO}^{\text{e}}(4)] + \kappa_{\text{e}} \hat{Q}_{\text{e}}^2 + \gamma_{\text{e}} \hat{L}_{\text{e}}^2 \qquad (7.49)$$

If the hamiltonian $\hat{H}^{\text{e}}$ contained all the interactions appearing in the second quantization expression including $\hat{n}_\pi^2$, the parameters associated to $\mathcal{C}_2[\text{SO}^{\text{e}}(4)]$ and $\hat{L}_{\text{e}}^2$ would be constrained to satisfy $\gamma_{\text{e}} = -\beta_{\text{e}}$. If this condition is not imposed, we expect an almost exact cancellation, and this is indeed found to be the case. The hamiltonian $\hat{H}^{\text{e}}$ can be applied to the series of atoms ^{4}Be ($M = 2$), ^{5}B ($M = 3$), ^{6}C ($M = 4$), ^{7}N ($M = 5$), and ^{8}O ($M = 6$). The atoms ^{3}Li and ^{9}F are not included in the fit since they both have a single experimental state arising from the 2s–2p shell. The parameters in the hamiltonian $\hat{H}^{\text{e}}$ are determined from a least-squares fit to the energy levels of the five atoms. The root-mean-square deviation from the observed energies turns out to be a few percent of typical energy differences between states. In Figure 7.2 we plot the four parameters ϵ_{e}, β_{e}, κ_{e}, and γ_{e} as a function of the number of valence electrons M. There is a striking uniformity in the parameter values as well as a nearly exact cancellation of the $\hat{L}_{\text{e}}^2$ and $\mathcal{C}_2[\text{SO}^{\text{e}}(4)]$ interactions, as expected.

We now turn to the determination of the molecular hamiltonian appropriate for the hydride molecules. The molecular hamiltonian (7.44)

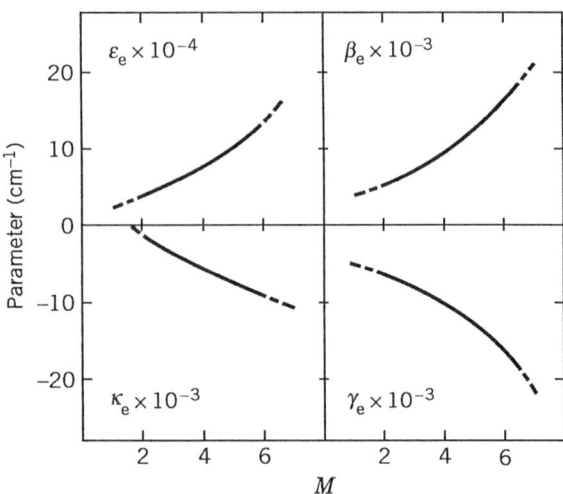

Figure 7.2: The parameters ϵ_e, κ_e, β_e, and γ_e in the hamiltonian $\hat{H}^e$ as a function of the electron number M.

includes fermion (electronic), boson, and boson–fermion parameters. We adopt the following prescription for the determination of the electronic parameters in this hamiltonian, which takes advantage of the uniformity of the atomic parameters in Figure 7.2. We start with the interactions in the united-atoms limit, that is, Be for LiH, B for BeH, and so on, and displace the parameter values simultaneously on top of the interpolated curves, moving in the direction of the neighboring lighter atom, for example, towards Be in BeH, until we achieve the best fit in the molecule. In this way we constrain the electronic parameters, effectively reducing their number to 1 and linking them to the atomic calculation. Note that this implies that we fix the electronic interactions "in between" the united- and separated-atoms limits.

For the vibron hamiltonian $\hat{H}^{rv}$ we take the SO(4) limit as a first approximation since we are mainly concerned here with the electronic structure of the molecules. As explained in the last two chapters, additional vibron interactions can be included, which would improve the fit to the vibrational levels. Lastly, for the electron–vibron interaction $\hat{V}^{rv-e}$, only the dipole and quadrupole terms are needed. The other

7.6 Application to the hydride molecules

interactions turn out not to influence the structure of the excited electronic states. To simplify matters we also take $\gamma_L = \gamma = 0$ in (7.44) since the corresponding operators only influence the rotational structure of the spectrum.

The molecular hamiltonian for the hydride molecules is thus given by [13,14]

$$\begin{aligned}\hat{H} &= \beta_{\mathrm{rv}}\mathcal{C}_2[\mathrm{SO}^{\mathrm{rv}}(4)] + \epsilon_e \hat{n}_\pi + \beta_e \mathcal{C}_2[\mathrm{SO}^e(4)] + \kappa_e \hat{Q}_e^2 \\ &\quad + \gamma_e \hat{L}_e^2 + \beta \mathcal{C}_2[\mathrm{SO}(4)] + \kappa \hat{Q}_{\mathrm{rv}} \cdot \hat{Q}_e\end{aligned} \qquad (7.50)$$

By means of the constraining procedure explained above, the energy fit involves only four parameters, associated with one vibrational, one electronic, and two boson–fermion degrees of freedom. Furthermore, the parameter β_{rv} is only related to the vibrational structure and can be fixed in the SO(4) dynamical symmetry from the vibrational behavior of the electronic ground states. The fermion operators do not affect the vibrational structure while the quadrupole interaction $\hat{Q}_e \cdot \hat{Q}_{\mathrm{rv}}$ has only a minor effect. Hence the vibrational parameters can be determined independently of the other parameters.

The boson–fermion parameter κ turns out to be essential for the correct ordering of the excited electronic states, producing the necessary level crossings when starting from the SO(4) spectrum, while β and β_e determine the distance between electronic states. We again point out that the hamiltonian (7.50) does not contain interactions generating the rotational structure, but the introduction of these spectroscopic characteristics is not difficult. The parameter values obtained in this way are summarized in Table 7.4. Figure 7.3 illustrates the smooth variation of parameters as a function of M with the example of the quadrupole interaction parameter κ.

In Figures 7.4 to 7.6 we show the best fits obtained with this procedure to the known electronic states of the first series of hydride molecules. We have included the vibrational levels, which are known experimentally according to [15]. In Table 7.5 we compare the calculated and experimental energies in the example of OH, where for each electronic level we include four excited vibrational states. Similar results are obtained for the other hydride molecules [13,14]. We remark that since our simplified vibrational calculation depends only on the

Table 7.4: Parameters (in cm^{-1}) used in the calculation of energy levels[a]

	LiH	BeH	BH	CH	NH	OH
β_{rv}	104	159	64	131	120	59
ϵ_e	42 000	38 421	55 269	86 000	100 000	44 000
β_e	5 500	5 236	7 208	10 800	12 500	5 650
κ_e	1 300	858	3 382	6 700	7 500	1 600
γ_e	−6 800	−6 458	−8 198	−11 200	−1 300	−7 000
β	−110	−170	−80	−148	−140	−80
κ	−800	−900	−1 000	−1 190	−1 400	−1 640

[a] The vibron number $N = 42$ for all molecules.

parameter β_{rv}, substantial improvement can be made by the inclusion of more terms in the hamiltonian.

The root-mean-square deviation for the electronic states in the six molecules is of the order of 4000 cm^{-1}. Although current *ab initio* calculations [16] are capable of accuracies much better than 4000 cm^{-1}, it should be emphasized that the present calculation includes 25 excited electronic states (of six molecules) using a minimal basis set (2s–2p shell) in the united-atoms limit. This is in contrast to *ab initio* calculations where it is common to use bases including orbitals up to the ninth–tenth shell in the separated-atoms limit.

The model produces naturally all electronic states originating from

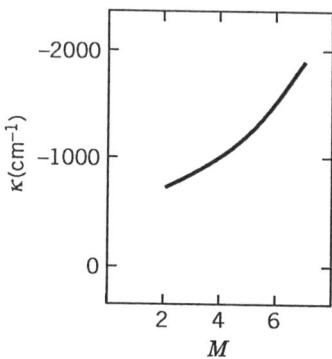

Figure 7.3: The parameter κ as a function of the electron number M.

7.6 Application to the hydride molecules

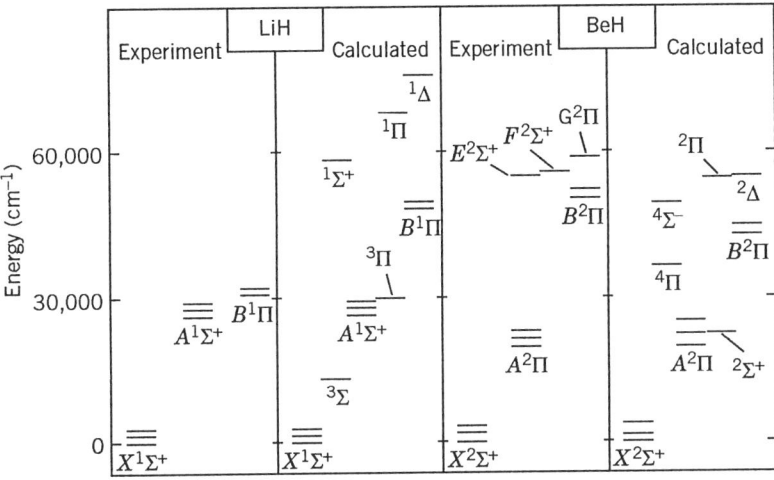

Figure 7.4: Calculated eigenspectra for $M = 2$ and $M = 3$ electrons and their comparison to the observed levels of the molecules LiH and BeH. The calculated spectrum is obtained with the parameters shown in Table 7.4 with the energy of the ground state normalized to zero.

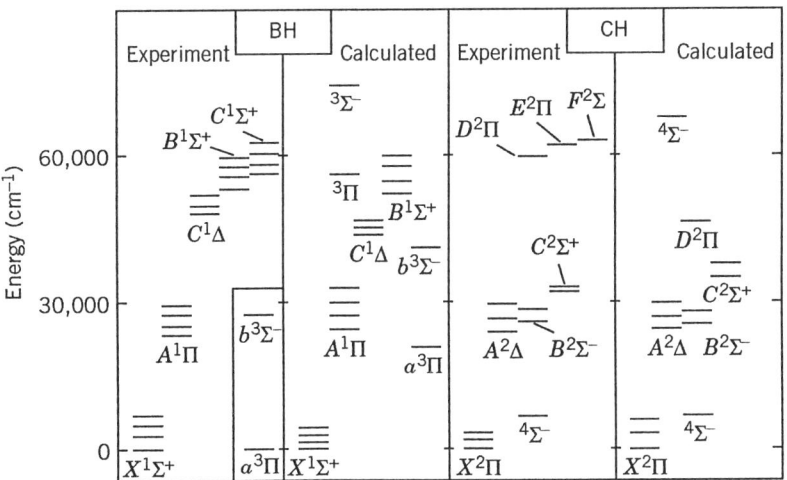

Figure 7.5: Same as Figure 7.4 but for $M = 4$ (BH) and $M = 5$ (CH).

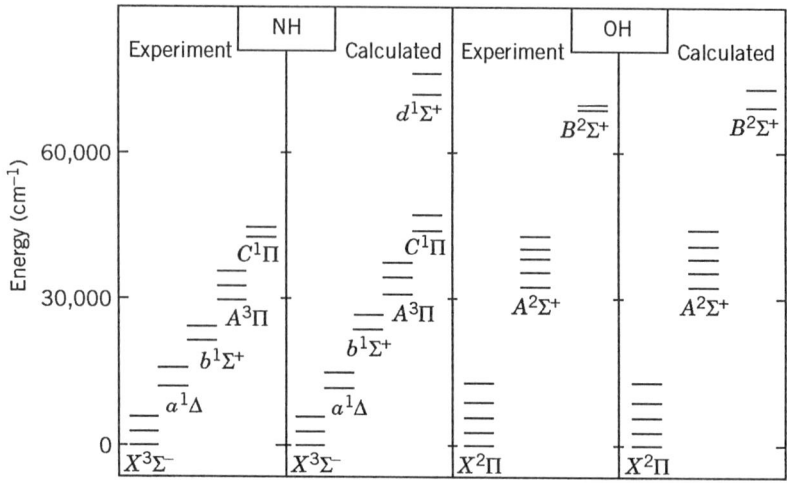

Figure 7.6: Same as Figure 7.4 but for $M = 6$ (NH) and $M = 7$ (OH).

Table 7.5: Electronic and vibrational energy levels of HO[a]

State	v	Experiment	Calculation	Error
$X^2\Pi$	0	0	0	—
	1	3 465	3 441	
	2	6 616	6 709	
	3	9 921	9 801	
	4	13 070	12 714	
$A^2\Sigma^+$	0	32 637	32 373	0.8%
	1	35 575	35 767	
	2	38 329	38 963	
	3	40 917	41 954	
	4	43 194	44 729	
$B^2\Sigma^+$	0	69 074	69 565	0.7%
	1	69 736	72 958	
	2		76 186	
	3		79 248	
	4		82 146	

[a] All energies in cm^{-1}.

the 2s–2p united-atoms levels, including states not known experimentally. A case in point is the BH molecule where the position of the $S = 1$ states $^3\Pi$ and $^3\Sigma^-$ relative to the ground-state configuration $^1\Sigma^+$ is fixed by the calculation. Another interesting result is the reproduction of the $^4\Sigma^-$ electronic state in CH where the experimental figure has not been confirmed [16–18]. The wave functions in the algebraic model are of manageable size and may be of considerable interest for the description of more complicated situations, such as those arising in molecular dynamics [19] or the scattering of electrons off molecules [20,21].

Summary

In this chapter boson–fermion symmetries, as introduced in Chapter 3 in the context of the schematic $U(2)$ model, are applied to the vibron model with the purpose to arrive at an algebraic description of molecular electronic spectra. Although the procedure can be defined in general terms, the specific example of electronic states originating from the 2s–2p hydrogenic levels is considered throughout the chapter. In addition, a united-atoms limit is taken for the electrons. As a result, the appropriate algebra to describe the fermion degrees of freedom is $U^e(8)$. The basic building blocks are $a^\dagger_{nlm,s\sigma}$ fermions which satisfy the anticommutation relations (7.2); the bilinear combinations (7.3) of the fermion creation and annihilation operators generate $U^e(8)$. Once the fermion algebra is defined, the classification of many-electron states in the 2s–2p shell can be carried out in the basis (7.15). Its essential ingredients are an LS decomposition, known to be a reasonable approximation in light molecules, and the occurrence of an $SO^e(4)$ algebra, which is a well-known symmetry associated with hydrogenic atoms. The next step is to couple these fermion wave functions to the boson states of the vibron model and the complete classification (7.26) of the rotation–vibration–electronic states results; the corresponding expression (7.27) for the wave function involves $SO(4) \supset SO(3)$ isoscalar factors which are known from Section 5.8. As in the previous chapters it is possible to define symmetry limits which correspond to a particular hamiltonian that can be solved analytically. Some of these limits are discussed in Section 7.5. They are useful since they provide an insight into the relation between the algebraic quantum numbers and the labels traditionally used to classify electronic molecular levels. For realistic applications to the hydride molecules a more general hamiltonian is needed which is specified in (7.50). It is found that with smoothly varying parameters the model is able to reproduce all electronic states originating from the 2s–2p united-atoms levels.

References

1. R. Lemus, *Descripción Espectroscópica de Moléculas Diatómicas Mediante Métodos Algebraicos*, doctoral dissertation, Universidad Nacional Autónoma de México, 1988.
2. G. Herzberg, *Molecular Spectra and Molecular Structure. I. Spectra of Diatomic Molecules*, van Nostrand, New York, 1950.
3. R. Lemus and A. Frank, "Approximate dynamical symmetry in the hydride molecules," Phys. Rev. Lett. **66** (1991) 2863.
4. A. Frank, F. Iachello, and R. Lemus, "Algebraic methods for molecular electronic spectra," Chem. Phys. Lett. **131** (1986) 380.
5. A. Frank, R. Lemus, and F. Iachello, "Algebraic approach to molecular electronic spectra. I. Energy levels," J. Chem. Phys. **91** (1989) 29.
6. E. Chacón, M. Moshinsky, O. Novaro, and C. Wulfman, "O(4) and U(3) symmetry breaking in the 2s–2p shell," Phys. Rev. A **3** (1971) 166.
7. B.G. Wybourne, *Classical Groups for Physicists*, Wiley-Interscience, New York, 1974.
8. M. Hamermesh, *Group Theory and Its Application to Physical Problems*, Addison-Wesley, Reading, MA, 1962.
9. G. Racah, "Theory of complex spectra. I," Phys. Rev. **61** (1942) 186; **62** (1942) 438; **63** (1943) 367; **76** (1949) 1352.
10. R. Lemus and A. Frank, "An algebraic model for molecular electronic excitations in diatomic molecules," Ann. Phys. (NY) **206** (1991) 122.
11. M. Born and R. Oppenheimer, "Zur Quantentheorie der Molekeln," Ann. Phys. **84** (1927) 457.

References

12. M. Born and K. Huang, *Dynamical Theory of Cristal Lattices*, Oxford University Press, Oxford, 1955.
13. A. Frank, R. Lemus, and F. Iachello, "Algebraic model for molecular electronic spectra," in *Symmetries in Science V*, edited by B. Gruber, L.C. Biedenharn, and H.D. Doebner, Plenum, New York, 1991, p. 173.
14. R. Lemus and A. Frank, "Constrained calculations in the electron–vibron model," Phys. Rev. A **47** (1993) 4920.
15. R.W.B. Pearse, *The Identification of Molecular Spectra*, Chapman and Hall, London, 1976.
16. C.W. Bauschlicher and S.R. Langhoff, "The study of molecular spectroscopy by *ab initio* methods," Chem. Rev. **91** (1991) 701.
17. S.N. Suchard, *Spectroscopic Data. I. Heteronuclear Diatomic Molccules*, Plenum, New York, 1975.
18. D.R. Yarkong, "On the electronic structure of the NH radical," J. Chem. Phys. **91** (1989) 4745.
19. A. Frank, R. Lemus, J. Recamier, and A. Amaya, "An algebraic appraoch to the study of three-dimensional atom–diatom collisions," Chem. Phys. Lett. **193** (1992) 176.
20. R. Bijker, R.D. Amado, and D.A. Sparrow, "Algebraic–eikonal approach to electron–molecule scattering: Diatomic molecules," Phys. Rev. A **33** (1986) 871.
21. R. Bijker and R.D. Amado, "Algebraic–eikonal approach to electron–molecule scattering. II. Rotational and vibrational excitation of LiF and KI," Phys. Rev. A **34** (1986) 71.

Part III

Nuclear Models

Chapter 8

The Interacting Boson Model

8.1 Introduction

In the third part of this book we turn our attention to the application of algebraic methods to nuclear structure physics. From the point of view of these applications, the interacting boson model (IBM) has had a major impact in the field and has established itself as a standard theoretical framework for the study of nuclear structure phenomena. The model was proposed in 1974 by Arima and Iachello [1], drawing some inspiration from earlier work, notably by Feshbach and Iachello [2,3] and Janssen et al. [4], with the basic aim to describe collective nuclear excitations in an algebraic framework. The starting point was to consider a realization in terms of a scalar (rank-0) operator $s^\dagger$ and quadrupole (rank-2) operators $d_m^\dagger$. The latter were first introduced by Bohr and Mottelson [5,6] in connection with surface vibrations of the nucleus, of which those of quadrupole nature ($l = 2$) were shown to be the most important ones in collective nuclear excitations. Soon after the proposal of the IBM it became apparent that a microscopic interpretation of the $s^\dagger$ and $d_m^\dagger$ bosons exists, relating them to valence nucleon pairs coupled to angular momenta $l = 0$ and $l = 2$, respectively [7]. Thus the IBM provided a framework for a microscopic description of collective quadrupole excitations and stimulated a large number of theoretical studies investigating its connection to the nuclear shell model. We do not pursue these matters here but refer the reader to the bibliography, concentrating in what follows on the algebraic aspects of the model.

Besides the two basic volumes devoted to the model [8,9], in the last years a large number of review articles have been written [10–14], covering a broad range of subjects. Thus our aim here is not to attempt a review of the many applications that have sprung in the last decade, but rather to concentrate on the algebraic techniques involved by presenting some selected examples and emphasizing the similarities with the simpler U(2) and U(4) models discussed in the previous chapters. As pointed out throughout Parts I and II, the techniques required in Part III are essentially identical to the ones developed for the U(2) and U(4) models, except for the somewhat more complicated manipulations arising in connection with the U(6) algebra.

In this chapter we review the algebraic framework for the IBM following the approach of Castaños et al. [15], which is not as well known as the one considered originally by Arima and Iachello [16–18] but which leads to a very complete analysis of the algebraic structure of U(6) and is very similar to the one presented in Chapter 5 for U(4).

8.2 The U(6) algebra

As exemplified in Chapters 1 and 5, the unitary algebras in any number of dimensions are connected with the harmonic oscillator. As in lower dimensions we may use either a coordinate or a bosonic realization to study the U(6) algebra. Both will be needed in the subsequent discussion.

The U(6) algebra may be realized in terms of six creation and six annihilation operators $b_i^\dagger$ and b_i ($i = 1, 2, \ldots, 6$). The form of these operators depends on physical considerations, the basic one being the conservation of angular momentum L. Since the fundamental degrees of freedom in collective nuclear excitations are of quadrupole ($l = 2$) nature, we consider in this case a realization in terms of a scalar (rank-0) operator $s^\dagger$ and quadrupole (rank-2) operators $d_m^\dagger$ [1].

The creation and annihilation operators of the IBM satisfy the commutation relations (5.1), with i and j now running from 1 to 6 and

8.2 The U(6) algebra

where one uses the notation

$$b_1^\dagger \equiv d_{-2}^\dagger, \quad b_2^\dagger \equiv d_{-1}^\dagger, \quad \ldots, \quad b_5^\dagger \equiv d_2^\dagger, \quad b_6^\dagger \equiv s^\dagger$$
$$b_1 \equiv d_{-2}, \quad b_2 \equiv d_{-1}, \quad \ldots, \quad b_5 \equiv d_2, \quad b_6 \equiv s \tag{8.1}$$

The bilinear operators

$$\boxed{\mathcal{G}_i^j \equiv b_i^\dagger b_j, \qquad i,j = 1, 2, \ldots, 6} \tag{8.2}$$

then satisfy the U(6) commutation relations

$$\boxed{[\mathcal{G}_i^j, \mathcal{G}_k^l] = \mathcal{G}_i^l \delta_{jk} - \mathcal{G}_k^j \delta_{il}, \qquad i,j,k,l = 1, 2, \ldots, 6} \tag{8.3}$$

which are identical to those of the U(2) and U(4) algebras.

The $s^\dagger$ and $d_m^\dagger$ operators transform under rotations $\hat{R}(\theta_1, \theta_2, \theta_3)$, characterized by the three Euler angles θ_i, by means of the **D** matrices of corresponding angular momentum

$$\hat{R}^{-1} d_m^\dagger \hat{R} = \sum_{m'} D_{m'm}^2(\theta_1, \theta_2, \theta_3) d_{m'}^\dagger, \qquad \hat{R}^{-1} s^\dagger \hat{R} = s^\dagger \tag{8.4}$$

and carry positive parity, that is,

$$\hat{P}^{-1} d_m^\dagger \hat{P} = d_m^\dagger, \qquad \hat{P}^{-1} s^\dagger \hat{P} = s^\dagger \tag{8.5}$$

where $\hat{P}$ is the reflection operator. As discussed in Sections 1.7 and 5.1, the annihilation operators do not transform as spherical tensors, but rather the modified operators

$$\tilde{d}_m \equiv (-)^{2+m} d_{-m}, \qquad \tilde{s} \equiv s \tag{8.6}$$

are the ones that transform appropriately. We may also define the generators in coupled form,

$$\boxed{\mathcal{B}_\mu^{(\lambda)}(l, l') \equiv [b_l^\dagger \times \tilde{b}_{l'}]_\mu^{(\lambda)} = \sum_{mm'} \langle lm\, l'm' | \lambda \mu \rangle b_{lm}^\dagger \tilde{b}_{l'm'}} \tag{8.7}$$

where $b_{2m}^\dagger \equiv d_m^\dagger$ and $b_{00}^\dagger \equiv s^\dagger$. This form is useful for their physical interpretation. For example, the d-boson number operator

$$\hat{n}_d = \sum_m d_m^\dagger d_m \tag{8.8}$$

may be related to the $\mathcal{B}_0^{(0)}(2,2)$ generator through

$$\mathcal{B}_0^{(0)}(2,2) = [d^\dagger \times \tilde{d}]_0^{(0)} = \sum_m \langle 2m\, 2-m|00\rangle d_m^\dagger \tilde{d}_{-m} = \sqrt{\tfrac{1}{5}}\hat{n}_d \qquad (8.9)$$

Proceeding as in Chapter 5, we write the 36 U(6) generators as

$$\begin{aligned}
\mathcal{B}_0^{(0)}(2,2) &= [d^\dagger \times \tilde{d}]_0^{(0)} & 1 \\
\mathcal{B}_\mu^{(1)}(2,2) &= [d^\dagger \times \tilde{d}]_\mu^{(1)} & 3 \\
\mathcal{B}_\mu^{(2)}(2,2) &= [d^\dagger \times \tilde{d}]_\mu^{(2)} & 5 \\
\mathcal{B}_\mu^{(3)}(2,2) &= [d^\dagger \times \tilde{d}]_\mu^{(3)} & 7 \\
\mathcal{B}_\mu^{(4)}(2,2) &= [d^\dagger \times \tilde{d}]_\mu^{(4)} & 9 \\
\mathcal{B}_\mu^{(2)}(2,0) &= [d^\dagger \times \tilde{s}]_\mu^{(2)} & 5 \\
\mathcal{B}_\mu^{(2)}(0,2) &= [s^\dagger \times \tilde{d}]_\mu^{(2)} & 5 \\
\mathcal{B}_0^{(0)}(0,0) &= [s^\dagger \times \tilde{s}]_0^{(0)} & 1
\end{aligned} \qquad (8.10)$$

Of the many possible chains of subalgebras, we restrict our attention to those that contain the angular momentum subalgebra. From (8.10) we see that the angular momentum operators must be proportional to the rank-1 generators $\mathcal{B}_\mu^{(1)}(2,2)$. To find the proportionality constant, we compute one of the commutators involving these generators, for instance,

$$[\mathcal{B}_0^{(1)}(2,2), \mathcal{B}_1^{(1)}(2,2)] = \sqrt{\tfrac{1}{10}}\mathcal{B}_1^{(1)}(2,2) \qquad (8.11)$$

and compare with the SO(3) commutators. This leads to the result

$$\hat{L}_\mu = \sqrt{10}\mathcal{B}_\mu^{(1)}(2,2) = \sqrt{10}[d^\dagger \times \tilde{d}]_\mu^{(1)} \qquad (8.12)$$

The commutation relations for U(n), in coupled notation and for arbitrary dimension n, are given in (5.16). This formula is useful in the search for possible algebra chains. In the present case we are looking for all algebras G that satisfy

$$U(6) \supset G \supset SO(3) \qquad (8.13)$$

where U(6) is generated by the 36 operators (8.10) and SO(3) by the $\hat{L}_\mu$ operators (8.12). As we shall see, in some cases G itself can have

8.2 The U(6) algebra

subalgebras G' which may be inserted between G and the angular momentum algebra.

We may readily find a subalgebra satisfying (8.13) by noting that the first 25 generators in (8.10) contain only d bosons and should close under commutation. It is easily verified that they generate a U(5) algebra, analogous to the U(3) subalgebra in chain (5.19), which is also obtained by deleting all generators containing s-boson operators. In this case there is an additional subalgebra of U(5) containing the SO(3) algebra. To prove this fact, we return to the explicit form of the U(5) generators,

$$\mathcal{B}_\mu^{(\lambda)}(2,2) = \sum_{mm'} \langle 2m\, 2m'|\lambda\mu\rangle d_m^\dagger \tilde{d}_{m'} \qquad (8.14)$$

and use the symmetry property of the Clebsch–Gordan coefficients [20]

$$\langle 2m'\, 2m|\lambda\mu\rangle = (-)^\lambda \langle 2m\, 2m'|\lambda\mu\rangle \qquad (8.15)$$

For odd values of λ we may rewrite $\mathcal{B}_\mu^{(\lambda)}(2,2)$ in the form

$$\mathcal{B}_\mu^{(\lambda)}(2,2) = \tfrac{1}{2} \sum_{mm'} \langle 2m\, 2m'|\lambda\mu\rangle (d_m^\dagger \tilde{d}_{m'} - d_{m'}^\dagger \tilde{d}_m), \quad \lambda = 1,3 \qquad (8.16)$$

and thus find that they are linear combinations of antisymmetric tensors.

As pointed out in Chapter 5 in connection with relation (5.22), the antisymmetric combinations of U(n) generators close under commutation to the SO(n) algebra. For $n = 5$ we may prove this directly through equation (5.16) by substituting the appropriate $6j$ coefficients [19]. It is, however, easier to revert to the uncoupled notation and prove that the antisymmetric combinations

$$\Lambda_{ij} \equiv b_i^\dagger b_j - b_j^\dagger b_i = \mathcal{G}_i^j - \mathcal{G}_j^i, \qquad i,j = 1,2,\ldots,5 \qquad (8.17)$$

indeed close under commutation. Using (8.3) we find

$$\begin{aligned}
{[\Lambda_{ij}, \Lambda_{kl}]} &= [\mathcal{G}_i^j - \mathcal{G}_j^i, \mathcal{G}_k^l - \mathcal{G}_l^k] \\
&= \mathcal{G}_i^l \delta_{jk} - \mathcal{G}_k^j \delta_{il} - \mathcal{G}_j^l \delta_{ik} + \mathcal{G}_k^i \delta_{jl} \\
&\quad - \mathcal{G}_i^k \delta_{jl} + \mathcal{G}_l^j \delta_{ik} + \mathcal{G}_j^k \delta_{il} - \mathcal{G}_l^i \delta_{jk} \\
&= \Lambda_{il}\delta_{jk} + \Lambda_{jk}\delta_{il} + \Lambda_{lj}\delta_{ik} + \Lambda_{ki}\delta_{jl}
\end{aligned} \qquad (8.18)$$

Table 8.1: Generators of algebras in the U(5) limit

Algebra	Generators		
U(6)	$[b_l^\dagger \times \tilde{b}_{l'}]_\mu^{(\lambda)}$,	$l, l' = 0, 2,$	$\lambda = 0, 1, \ldots, 4$
U(5)	$[d^\dagger \times \tilde{d}]_\mu^{(\lambda)}$,	$\lambda = 0, 1, \ldots, 4$	
SO(5)	$[d^\dagger \times \tilde{d}]_\mu^{(\lambda)}$,	$\lambda = 1, 3$	
SO(3)	$[d^\dagger \times \tilde{d}]_\mu^{(1)}$		

which are the SO(5) commutation relations (see Appendix B). In fact, both (8.3) and (8.18) are valid for index values ranging from 1 to n, and this proves that the general reduction

$$U(n) \supset SO(n) \tag{8.19}$$

is accomplished by means of the antisymmetric combinations (8.17). From (8.16) we see that the SO(5) subalgebra of U(5) indeed contains SO(3). We have thus found the first IBM chain

$$\boxed{\begin{array}{cccc} U(6) \supset & U(5) \supset & SO(5) \supset & SO(3) \\ | & | & | & | \\ {[N]} & n_d & v & n_\Delta L \end{array}} \tag{8.20}$$

where we indicate below each algebra the quantum numbers that label their representations. These are discussed in some detail in Section 8.4.

Note that we also indicate in (8.20) an additional quantum number n_Δ, which is the missing label in the SO(5) $\supset$ SO(3) reduction. As pointed out before, these are needed whenever a given representation of an algebra (in this case v) may contain more than one representation of the subalgebra (in this case L). The missing labels then distinguish between the different representations of the subalgebra. In Section 8.4 it is shown that n_Δ is related to the maximum number of d-boson triplets coupled to zero angular momentum in the U(5) $\supset$ SO(5) $\supset$ SO(3) wave functions. In Table 8.1 we present a summary of the generators in the U(5) chain.

Another IBM chain may be found by applying the $U(n) \supset SO(n)$ reduction for $n = 6$. That is, we consider the antisymmetric combinations

8.2 The U(6) algebra

Table 8.2: Generators of algebras in the SO(6) limit

Algebra	Generators		
U(6)	$[b_l^\dagger \times \tilde{b}_{l'}]_\mu^{(\lambda)}$,	$l, l' = 0, 2$,	$\lambda = 0, 1, \ldots, 4$
SO(6)	$i[d^\dagger \times \tilde{s} - s^\dagger \times \tilde{d}]_\mu^{(2)}$		
	$[d^\dagger \times \tilde{d}]_\mu^{(\lambda)}$,	$\lambda = 1, 3$	
SO(5)	$[d^\dagger \times \tilde{d}]_\mu^{(\lambda)}$,	$\lambda = 1, 3$	
SO(3)	$[d^\dagger \times \tilde{d}]_\mu^{(1)}$		

(8.17) of the complete set of U(6) generators, with $i, j = 1, 2, \ldots, 6$. We then restrict the latter to $i, j = 1, 2, \ldots, 5$ (i.e., omitting the operators containing s bosons) arriving at the chain

$$\boxed{\begin{array}{cccc} U(6) & \supset SO(6) & \supset SO(5) & \supset SO(3) \\ | & | & | & | \\ [N] & \sigma & v & n_\Delta L \end{array}} \quad (8.21)$$

Note that this basis differs from (8.20) only through the σ label, associated with the SO(6) subalgebra. The basis (8.21) is analyzed in Section 8.5. In Table 8.2 we summarize the generators associated with the reduction (8.21). Note that we have included a factor i in the definition of the quadrupole operator of SO(6), to make its $\mu = 0$ component hermitian, as explained in Section 5.2 after equation (5.21).

Before continuing our discussion, we point out that, in analogy with the situation for SO(4) in Chapter 5, there is a different realization for SO(6) (which we denote by $\overline{SO(6)}$) arising from the transformation

$$d_m^\dagger \to -id_m^\dagger, \qquad \tilde{d}_m \to i\tilde{d}_m \quad (8.22)$$

which is similar to (5.24). The SO(6) commutation relations (8.18) are preserved by this canonical transformation, but the explicit realization of the generators involving s bosons is altered. Although Arima and Iachello [18] use the $\overline{SO(6)}$ realization, we prefer the SO(6) form as given in Table 8.2 because it leads to a simpler treatment in coordinate space, as is discussed in Section 8.5. It should be stressed that the two bases are equivalent and that a consistent treatment gives the same results for all matrix elements. The $\overline{SO}(6)$ form of the generators, Casimir invariants, and basis states can always be recovered using the transformation (8.22).

From the analysis of their generators we see that bases (8.20) and (8.21) are the higher-dimensional analogs of the U(3) and SO(4) bases, respectively, of Chapter 5. There is in the U(6) case one additional algebra satisfying (8.13) with no mathematical counterpart in the U(4) model. This is the SU(3) algebra. Note that SU(3) is also present as a subalgebra of U(4) in (5.19); its realization in Chapter 5, however, is very different from the one we present below.

The SU(3) algebra we now discuss was originally studied by Elliott [21] in connection with the s–d nuclear shell model. The mathematical structure of that system is very similar to that of the IBM since the nucleons are allowed to occupy orbits with orbital angular momenta $l = 0$ and $l = 2$. They thus generate a U(6) orbital space, although being fermions, the representations allowed are not restricted to the fully symmetric ones.

The SU(3) generators involve, besides the three angular momentum operators (8.12), a quadrupole tensor which may be defined in second quantized form as [17,21]

$$\begin{aligned} \hat{Q}_\mu &= -\left(\frac{2\pi}{5}\right)^{1/2} \sum_{lm}\sum_{l'm'} \langle 2lm|r^2 Y_\mu^2(\theta,\phi)|2l'm'\rangle b_{lm}^\dagger b_{l'm'} \\ &= [d^\dagger \times \tilde{s} + s^\dagger \times \tilde{d}]_\mu^{(2)} - \sqrt{\tfrac{7}{4}}[d^\dagger \times \tilde{d}]_\mu^{(2)} \end{aligned} \qquad (8.23)$$

The kets $|2lm\rangle$ are harmonic-oscillator states of two quanta, which can have angular momenta $l = 0$ and $l = 2$. The matrix elements of the quadrupole operator $r^2 Y_m^2(\theta,\phi)$ are well known in this basis [22].

We can compute the SU(3) commutation relations using the general formula (5.16). We find, after some algebra and the substitution of the appropriate $6j$ coefficients [19],

$$\begin{aligned} [\hat{L}_\mu, \hat{Q}_{\mu'}] &= -\sqrt{6}\langle 1\mu\, 2\mu'|2\mu+\mu'\rangle \hat{Q}_{\mu+\mu'} \\ [\hat{Q}_\mu, \hat{Q}_{\mu'}] &= \tfrac{3}{4}\sqrt{\tfrac{5}{2}}\langle 2\mu\, 2\mu'|1\mu+\mu'\rangle \hat{L}_{\mu+\mu'} \end{aligned} \qquad (8.24)$$

which together with the SO(3) commutation relations

$$[\hat{L}_\mu, \hat{L}_{\mu'}] = -\sqrt{2}\langle 1\mu\, 1\mu'|1\mu+\mu'\rangle \hat{L}_{\mu+\mu'} \qquad (8.25)$$

8.3 Symmetry limits

Table 8.3: Generators of algebras in the SU(3) limit

Algebra	Generators
U(6)	$[b_l^\dagger \times \tilde{b}_{l'}]_\mu^{(\lambda)}$, $l, l' = 0, 2$, $\lambda = 0, 1, \ldots, 4$
SU(3)	$[d^\dagger \times \tilde{s} + s^\dagger \times \tilde{d}]_\mu^{(2)} - \sqrt{\frac{7}{4}}[d^\dagger \times \tilde{d}]_\mu^{(2)}$
	$[d^\dagger \times \tilde{d}]_\mu^{(1)}$
SO(3)	$[d^\dagger \times \tilde{d}]_\mu^{(1)}$

constitute the SU(3) algebra. Note that (8.25) is a compact form of writing the usual angular momentum relations (5.13). We have thus arrived at the third IBM chain,

$$\boxed{\begin{array}{ccc} U(6) & \supset SU(3) & \supset SO(3) \\ | & | & | \\ [N] & (\lambda, \mu) & \kappa L \end{array}} \quad (8.26)$$

Again a missing label κ is needed in the reduction SU(3) $\supset$ SO(3), as pointed out in Chapter 6. In Section 8.7 the basis (8.26) is analyzed and in Table 8.3 the results concerning the SU(3) limit obtained so far are summarized. No other algebras satisfying condition (8.13) exist.

In the next section we introduce the IBM hamiltonian and discuss its relation to the three dynamical-symmetry chains.

8.3 Symmetry limits

The most general hamiltonian in the s–d-boson space, containing one- and two-body interactions, is of the same form as in the U(4) model,

$$\hat{H} = E_0 + \sum_{ij} \epsilon_{ij} b_i^\dagger b_j + \sum_{ijkl} e_{ijkl} b_i^\dagger b_j^\dagger b_l b_k \quad (8.27)$$

except that the indices run from 1 to 6 in this case. The hermiticity condition and rotational invariance of the hamiltonian substantially reduce the number of independent parameters. After imposing these conditions, we arrive at the IBM hamiltonian [8,15]

$$\begin{aligned}
\hat{H} &= E_0 + \epsilon_s \hat{n}_s + \epsilon_d \hat{n}_d + e_1 [[d^\dagger \times d^\dagger]^{(0)} \times [\tilde{d} \times \tilde{d}]^{(0)}]_0^{(0)} \\
&\quad + e_2 [[d^\dagger \times d^\dagger]^{(2)} \times [\tilde{d} \times \tilde{d}]^{(2)}]_0^{(0)} + e_3 [[d^\dagger \times d^\dagger]^{(4)} \times [\tilde{d} \times \tilde{d}]^{(4)}]_0^{(0)} \\
&\quad + e_4 [[d^\dagger \times d^\dagger]^{(0)} \times [\tilde{s} \times \tilde{s}]^{(0)} + [s^\dagger \times s^\dagger]^{(0)} \times [\tilde{d} \times \tilde{d}]^{(0)}]_0^{(0)} \\
&\quad + e_5 [[d^\dagger \times s^\dagger]^{(2)} \times [\tilde{d} \times \tilde{s}]^{(2)}]_0^{(0)} + e_6 [[s^\dagger \times s^\dagger]^{(0)} \times [\tilde{s} \times \tilde{s}]^{(0)}]_0^{(0)} \\
&\quad + e_7 [[d^\dagger \times d^\dagger]^{(2)} \times [\tilde{d} \times \tilde{s}]^{(2)} + [d^\dagger \times s^\dagger]^{(2)} \times [\tilde{d} \times \tilde{d}]^{(2)}]_0^{(0)}
\end{aligned}$$

(8.28)

where $\hat{n}_s$ and $\hat{n}_d$ are the number operators for s and d bosons while ϵ_s, ϵ_d, and e_i are parameters related to the one- and two-body matrix elements. There are thus nine parameters, although for a fixed boson number N not all are independent. For the moment we retain the general expression (8.28) since in this form it allows the possibility of a simultaneous fit to *several* nuclei (i.e., allowing N to vary) [23]. For each even–even nucleus the boson number N is determined as half the number of valence nucleons, that is, half the number of protons and neutrons outside closed shells [1]. Note that this number is counted from the nearest closed shell: if more than half of the valence shell is occupied, the bosons are treated as pairs of *holes* coupled to $l = 0$ or $l = 2$.

As in the case of U(4) the hamiltonian (8.28) can be written equivalently as an expansion in terms of invariant operators of U(6) and its subalgebras. We thus need to obtain the different Casimir invariants of the algebras in chains (8.20), (8.21), and (8.26), which can be done using the general expressions derived in Section 5.3. The approach followed in Section 5.3 is to convert the expressions for the Casimir invariants to a coupled notation and subsequently rewrite them in terms of scalar products of the various generators (see, e.g., (5.37)). Here we follow a simpler approach based on manipulations of uncoupled generators.

We first consider the linear U(6) invariant

$$\mathcal{C}_1[\text{U}(6)] = \sum_{i=1}^{6} \mathcal{G}_i^i = s^\dagger s + \sum_m d_m^\dagger d_m = \hat{n}_s + \hat{n}_d = \hat{N} \qquad (8.29)$$

which is just the total boson number operator. The quadratic Casimir invariant of U(6) is most simply given in terms of the uncoupled gen-

8.3 Symmetry limits

erators:

$$\begin{aligned}
\mathcal{C}_2[\mathrm{U}(6)] &\equiv \sum_{ij=1}^{6} \mathcal{G}_i^j \mathcal{G}_j^i = \sum_{ij=1}^{6} b_i^\dagger b_j b_j^\dagger b_i \\
&= \sum_{ij=1}^{6} (b_i^\dagger [b_j, b_j^\dagger] b_i + b_i^\dagger b_j^\dagger b_j b_i) \\
&= 6\sum_{i=1}^{6} b_i^\dagger b_i + \sum_{ij=1}^{6} b_i^\dagger [b_j^\dagger, b_i] b_j + \sum_{i=1}^{6} b_i^\dagger b_i \sum_{j=1}^{6} b_j^\dagger b_j \\
&= \hat{N}(\hat{N}+5)
\end{aligned} \qquad (8.30)$$

Note that this operator is again a function of the total boson number $\hat{N}$. This only occurs for particular U(6) representations, namely the fully symmetric ones, labeled by the Young tableau

$$\overbrace{\square\square\cdots\square}^{N} \qquad (8.31)$$

These are the representations that label the boson wave functions of the IBM. In general, the quadratic Casimir invariants of U(n) cannot be written solely in terms of the linear invariants, as exemplified for the U(2) model in Section 2.4.

The linear and quadratic Casimir invariants of U(5) are computed in the same way, except that the indices now range from 1 to 5. We find

$$\mathcal{C}_1[\mathrm{U}(5)] = \sum_m d_m^\dagger d_m = \hat{n}_d \qquad (8.32)$$

and

$$\mathcal{C}_2[\mathrm{U}(5)] = \hat{n}_d (\hat{n}_d + 4) \qquad (8.33)$$

As shown in Section 5.3, the linear SO(n) invariants vanish identically. Using the notation of (8.17) for the SO(n) generators,

$$\Lambda_{ij} \equiv b_i^\dagger b_j - b_j^\dagger b_i = \mathcal{G}_i^j - \mathcal{G}_j^i, \qquad i,j = 1, 2, \ldots, n \qquad (8.34)$$

we may compute the SO(6) quadratic Casimir invariant as follows:

$$\mathcal{C}_2[\mathrm{SO}(6)] \equiv \tfrac{1}{2} \sum_{ij=1}^{6} \Lambda_{ij} \Lambda_{ji} = \tfrac{1}{2} \sum_{ij=1}^{6} (\mathcal{G}_i^j - \mathcal{G}_j^i)(\mathcal{G}_j^i - \mathcal{G}_i^j)$$

$$= \sum_{ij=1}^{6}(\mathcal{G}_i^j\mathcal{G}_j^i - \mathcal{G}_i^j\mathcal{G}_i^j) = \mathcal{C}_2[\text{U}(6)] - \sum_{ij=1}^{6} b_i^\dagger b_j b_i^\dagger b_j$$

$$= \hat{N}(\hat{N}+5) - \sum_{ij=1}^{6} b_i^\dagger[b_j,b_i^\dagger]b_j - \sum_{i=1}^{6} b_i^\dagger b_i^\dagger \sum_{j=1}^{6} b_j b_j$$

$$= \hat{N}(\hat{N}+4) - \sum_{i=1}^{6} b_i^\dagger b_i^\dagger \sum_{j=1}^{6} b_j b_j \quad (8.35)$$

Converting back to the coupled notation we find

$$\mathcal{C}_2[\text{SO}(6)] = \hat{N}(\hat{N}+4) - (d^\dagger \cdot d^\dagger + s^\dagger s^\dagger)(\tilde{d} \cdot \tilde{d} + \tilde{s}\tilde{s}) \quad (8.36)$$

where the dot denotes a scalar product, $d^\dagger \cdot d^\dagger = \sum_m (-)^m d_m^\dagger d_{-m}^\dagger$. Proceeding in the same way for SO(5) and using the expression for $\mathcal{C}_2[\text{U}(5)]$, we obtain

$$\mathcal{C}_2[\text{SO}(5)] = \hat{n}_d(\hat{n}_d + 3) - (d^\dagger \cdot d^\dagger)(\tilde{d} \cdot \tilde{d}) \quad (8.37)$$

while the SO(3) Casimir invariant is computed directly from the SO(3) generators (8.12)

$$\mathcal{C}_2[\text{SO}(3)] = \hat{L}^2 = \sum_\mu (-)^\mu \hat{L}_\mu \hat{L}_{-\mu} = \sum_\mu (-)^\mu [d^\dagger \times \tilde{d}]^{(1)}_\mu [d^\dagger \times \tilde{d}]^{(1)}_{-\mu} \quad (8.38)$$

Lastly, the SU(3) Casimir invariant turns out to be given by [21]

$$\mathcal{C}_2[\text{SU}(3)] = \hat{Q}^2 + \tfrac{1}{2}\hat{L}^2 \quad (8.39)$$

which involves the quadrupole operators $\hat{Q}_\mu$ defined in (8.23). This expression for $\mathcal{C}_2[\text{SU}(3)]$ is identical to the one for the U(3) Casimir invariant given in Section 5.3, except for the term $\hat{n}_p^2/3$ involving the boson number operator. The difference will be explained in Section 8.7.

We are now in a position to express the IBM hamiltonian (8.28) in terms of the invariant operators introduced above. We find

$$\boxed{\begin{aligned}\hat{H} = {} & E_0' + \eta_1 \hat{N} + \eta_2 \hat{N}^2 \\ & + \epsilon \hat{n}_d + \alpha \hat{n}_d^2 + \eta_3 \hat{N}\hat{n}_d + \xi \mathcal{C}_2[\text{SU}(3)] \\ & + \beta \mathcal{C}_2[\text{SO}(6)] + \zeta \mathcal{C}_2[\text{SO}(5)] + \gamma \mathcal{C}_2[\text{SO}(3)]\end{aligned}} \quad (8.40)$$

8.4 Vibrational nuclei: The U(5) limit

where the nine parameters are linear combinations of those appearing in (8.28) [8,15]. For a single nucleus N is fixed and the term $\hat{N}\hat{n}_d$ is not independent of $\hat{n}_d$ while the $\hat{N}$ and $\hat{N}^2$ interactions can be included in E'_0 and deleted from (8.40).

Some of the two-body interactions in (8.28) are readily expressed in terms of invariants. Thus, for example, using the expression for the SO(5) invariant,

$$[[d^\dagger \times d^\dagger]^{(0)} \times [\tilde{d} \times \tilde{d}]^{(0)}]_0^{(0)} = \tfrac{1}{5}(d^\dagger \cdot d^\dagger)(\tilde{d} \cdot \tilde{d}) = -\tfrac{1}{5}\mathcal{C}_2[\mathrm{SO}(5)] + \tfrac{1}{5}\hat{n}_d(\hat{n}_d + 3) \tag{8.41}$$

while other interactions require some SO(3) recoupling manipulations [8]. For the application of the hamiltonian (8.40) to several nuclei, the $\hat{N}\hat{n}_d$ interaction is independent from the others in (8.40) and can be expressed in terms of the two-body interactions in (8.28) as

$$\hat{N}\hat{n}_d = \tfrac{7}{5}\sum_L \sqrt{2L+1}[[d^\dagger \times d^\dagger]^{(L)} \times [\tilde{d} \times \tilde{d}]^{(L)}]_0^{(0)} + \tfrac{3}{5}\hat{n}_d$$
$$+\sqrt{5}[[d^\dagger \times d^\dagger]^{(0)} \times [\tilde{s} \times \tilde{s}]^{(0)} + [s^\dagger \times s^\dagger]^{(0)} \times [\tilde{d} \times \tilde{d}]^{(0)}]_0^{(0)} \tag{8.42}$$

which is diagonal in the chain (8.20). From the form (8.40) we see that the general IBM hamiltonian is a mixture of the three dynamical-symmetry chains discussed in the previous section, to which we refer henceforth as the U(5), SO(6), and SU(3) limits. In the next section we study the first of them, the U(5) dynamical symmetry (8.20).

8.4 Vibrational nuclei: The U(5) limit

The U(5) limit of the IBM hamiltonian (8.40) corresponds to taking $\xi = \beta = 0$,

$$\boxed{\hat{H}_\mathrm{I} = E''_0 + \epsilon \hat{n}_d + \alpha \hat{n}_d^2 + \eta_3 \hat{N}\hat{n}_d + \zeta \mathcal{C}_2[\mathrm{SO}(5)] + \gamma \mathcal{C}_2[\mathrm{SO}(3)]} \tag{8.43}$$

The reduction rule associated with $U(6) \supset U(5)$ is given by the possible number of d bosons among the total number of bosons:

$$n_d = 0, 1, \ldots, N \tag{8.44}$$

The U(5) ⊃ SO(5) reduction rule takes the form

$$v = n_d, n_d - 2, \ldots, 1 \text{ or } 0 \qquad (8.45)$$

This reduction rule will be derived explicitly from the construction of the wave functions. Note that it follows the same pattern of decrease by two units observed in the U(3) ⊃ SO(3) and U(4) ⊃ SO(4) reduction rules (5.55) and (5.128). This rule is in fact valid for the general U(n) ⊃ SO(n) reduction from the fully symmetric U(n) representation [N].

It is illustrative to derive this result in a different way, which also supplies the eigenvalue of $C_2[SO(n)]$ for symmetric representations of SO(n). To this end we return to the general definition (8.34) for the SO(n) generators and proceed to derive the explicit form of the Casimir invariant:

$$\begin{aligned}
C_2[SO(n)] &\equiv \tfrac{1}{2} \sum_{ij=1}^{n} \Lambda_{ij}\Lambda_{ji} = \tfrac{1}{2} \sum_{ij=1}^{n} (\mathcal{G}_i^j - \mathcal{G}_j^i)(\mathcal{G}_j^i - \mathcal{G}_i^j) \\
&= \sum_{ij=1}^{n} (\mathcal{G}_i^j \mathcal{G}_j^i - \mathcal{G}_i^j \mathcal{G}_i^j) = C_2[U(n)] - \sum_{ij=1}^{n} b_i^\dagger b_j b_i^\dagger b_j \\
&= \hat{N}(\hat{N} + n - 1) - \sum_{ij=1}^{n} b_i^\dagger [b_j, b_i^\dagger] b_j - \sum_{i=1}^{n} b_i^\dagger b_i^\dagger \sum_{j=1}^{n} b_j b_j \\
&= \hat{N}(\hat{N} + n - 2) - \sum_{i=1}^{n} b_i^\dagger b_i^\dagger \sum_{j=1}^{n} b_j b_j \qquad (8.46)
\end{aligned}$$

which generalizes previous expressions for $n = 5$ and $n = 6$. Denoting the U(n) ⊃ SO(n) eigenvectors by $|[N]\rho\rangle$ (where ρ is associated with SO(n) and all other quantum numbers are omitted), we find, from (8.46),

$$C_2[SO(n)]|[N]\rho\rangle = N(N + n - 2)|[N]\rho\rangle - \sum_{i=1}^{n} b_i^\dagger b_i^\dagger \sum_{j=1}^{n} b_j b_j |[N]\rho\rangle \qquad (8.47)$$

The second term on the right-hand side of this relation should be diagonal in the basis $|[N]\rho\rangle$, by definition. Note that this operator contains the scalar

$$\mathcal{P} \equiv \sum_{i=1}^{n} b_i b_i \qquad (8.48)$$

8.4 Vibrational nuclei: The U(5) limit

which, acting on a state, annihilates all pairs of bosons coupled to zero angular momentum while

$$\mathcal{P}^\dagger \equiv \sum_{i=1}^{n} b_i^\dagger b_i^\dagger \tag{8.49}$$

just creates them back. Hence these operators are associated with the number of such zero angular momentum pairs, which should be a well-defined number in the $U(n) \supset SO(n)$ basis. Since the wave function may contain no pairs, one pair, etc., we may call this number n_0 and write the wave functions in the form

$$|[N]\rho\rangle = \left(\mathcal{P}^\dagger\right)^{n_0} \mathcal{Q}_\rho(b_k^\dagger)|0\rangle \tag{8.50}$$

where the ρ index denotes the remaining number of bosons, that is, $\mathcal{Q}_\rho(b_k^\dagger)$ contains no pairs coupled to zero angular momentum and thus satisfies

$$\mathcal{P}\mathcal{Q}_\rho(b_k^\dagger)|0\rangle = 0 \tag{8.51}$$

Also, since the total boson number is N, it is clear from (8.50) that

$$N = \rho + 2n_0 \tag{8.52}$$

Since n_0 is an integer number, we arrive at the rule

$$\rho = N, N-2, \ldots, 1 \text{ or } 0 \tag{8.53}$$

which is what we wanted to prove. The $SO(n)$ label can thus be identified with the number of bosons that are not coupled in pairs to zero angular momentum. It is now straightforward to derive the eigenvalue of $\mathcal{C}_2[SO(n)]$ using (8.47), (8.50), and (8.51). Applying $\mathcal{P}$ to $|[N]\rho\rangle$ and noting that the action of b_i is equivalent to that of $\partial/\partial b_i^\dagger$, we find

$$\sum_{i=1}^{n} b_i b_i \left(\sum_{j=1}^{n} b_j^\dagger b_j^\dagger\right)^{n_0} \mathcal{Q}_\rho(b_k^\dagger)|0\rangle$$

$$= \sum_{i=1}^{n} b_i \left[2n_0 \left(\sum_{j=1}^{n} b_j^\dagger b_j^\dagger\right)^{n_0-1} b_i^\dagger + \left(\sum_{j=1}^{n} b_j^\dagger b_j^\dagger\right)^{n_0} b_i\right] \mathcal{Q}_\rho(b_k^\dagger)|0\rangle$$

$$\begin{aligned}
&= \sum_{i=1}^{n} \left[4n_0(n_0-1) \left(\sum_{j=1}^{n} b_j^\dagger b_j^\dagger\right)^{n_0-2} (b_i^\dagger b_i^\dagger) + 2n_0 \left(\sum_{j=1}^{n} b_j^\dagger b_j^\dagger\right)^{n_0-1} \right.\\
&\qquad + 2n_0 \left(\sum_{j=1}^{n} b_j^\dagger b_j^\dagger\right)^{n_0-1} (b_i^\dagger b_i) + 2n_0 \left(\sum_{j=1}^{n} b_j^\dagger b_j^\dagger\right)^{n_0-1} (b_i^\dagger b_i) \\
&\qquad \left. + \left(\sum_{j=1}^{n} b_j^\dagger b_j^\dagger\right)^{n_0-1} (b_i b_i) \right] \mathcal{Q}_\rho(b_k^\dagger)|0\rangle \\
&= \left(\sum_{j=1}^{n} b_j^\dagger b_j^\dagger\right)^{n_0-1} \left(2n_0(2n_0-2+n) + 4n_0 \sum_{i=1}^{n} b_i^\dagger b_i\right) \mathcal{Q}_\rho(b_k^\dagger)|0\rangle \\
&= 2n_0(2n_0-2+n+2\rho) \left(\sum_{j=1}^{n} b_j^\dagger b_j^\dagger\right)^{n_0-1} \mathcal{Q}_\rho(b_k^\dagger)|0\rangle \qquad (8.54)
\end{aligned}$$

where use is made of (8.47) and the number operator equation

$$\sum_{i=1}^{n} b_i^\dagger b_i \mathcal{Q}_\rho(b_k^\dagger) = \rho \mathcal{Q}_\rho(b_k^\dagger) \qquad (8.55)$$

Substituting the result (8.54) into (8.47) and using the relation (8.52), we find

$$\begin{aligned}
\mathcal{C}_2[\mathrm{SO}(n)]|[N]\rho\rangle &= [N(N+n-2) - (N-\rho)(N+\rho-2+n)]|[N]\rho\rangle \\
&= \rho(\rho+n-2)|[N]\rho\rangle \qquad (8.56)
\end{aligned}$$

Note that for $n=3$, (8.56) gives (with $\rho = L$) the expected result for $\hat{L}^2 = \mathcal{C}_2[\mathrm{SO}(3)]$, while for $n=4$ it reproduces the $\mathcal{C}_2[\mathrm{SO}(4)]$ eigenvalue found in Chapter 5.

Using these results we may write the eigenvalues for the U(5) hamiltonian $\hat{H}_\mathrm{I}$ as

$$\begin{aligned}
&E_\mathrm{I}(N, n_\mathrm{d}, v, L) \\
&= E_0'' + \epsilon \hat{n}_\mathrm{d} + \alpha \hat{n}_\mathrm{d}^2 + \eta_3 \hat{N} \hat{n}_\mathrm{d} + \zeta v(v+3) + \gamma L(L+1) \quad (8.57)
\end{aligned}$$

where we now have at our disposal all necessary reduction rules except for SO(5) ⊃ SO(3), which is as follows. If we write the SO(5) quantum number as

$$v = 3n_\Delta + \mu \qquad (8.58)$$

8.4 Vibrational nuclei: The U(5) limit

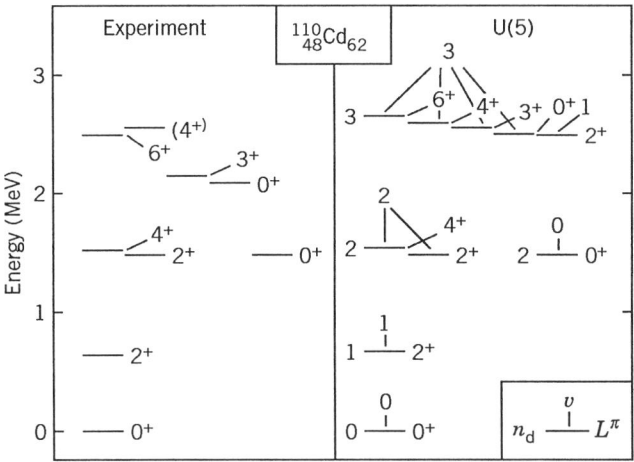

Figure 8.1: Partial U(5) eigenspectrum of $\hat{H}_\mathrm{I}$ for $N = 7$ bosons and its comparison to the observed levels of the nucleus ^{110}Cd. The calculated spectrum is obtained with the parameters (in keV) $\epsilon + \eta_3/N = 156$, $\alpha = 97$, $\zeta = -2.5$, and $\gamma = 5$, with the energy of the ground state normalized to zero. Levels are labeled by n_d, v, and the angular momentum and parity L^π.

where n_Δ and μ are non-negative integers, then L is given by [15,16]

$$L = \mu, \mu + 1, \ldots, 2\mu - 2, 2\mu \tag{8.59}$$

where we note the absence of $2\mu - 1$ in the possible L values. This rule is obtained by explicitly constructing the eigenstates associated with the U(5) limit, as is shown below. In Figure 8.1 we show a typical U(5) spectrum arising from (8.57) and we compare it with the observed levels of $^{110}_{48}\mathrm{Cd}_{62}$ [16].

The s-boson dependence of the U(5) wave functions is trivial, as both n_s and n_d are well defined in this basis. The s-boson part corresponds to a simple one-dimensional oscillator, and thus the states may be factorized in the form

$$|[N]n_\mathrm{d} v n_\Delta L M_L\rangle = \frac{(s^\dagger)^{N-n_\mathrm{d}}}{\sqrt{(N-n_\mathrm{d})!}}|n_\mathrm{d} v n_\Delta L M_L\rangle \tag{8.60}$$

The general form of the U(5) $\supset$ SO(5) $\supset$ SO(3) eigenstates $|n_\mathrm{d} v n_\Delta L M_L\rangle$ was first derived by Chacón et al. [24,25] in connection with the problem of quadrupole surface vibrations in nuclei [5]. The construction of a complete basis for the latter system is equivalent to the determination of the present d-boson basis, and in what follows we review the derivation of Chacón et al., following a method analogous to the one used in Chapter 5.

The states $|n_\mathrm{d} v n_\Delta L M_L\rangle$ in (8.60) are eigenstates of the following operators:

$$\begin{aligned} \hat{n}_\mathrm{d}|n_\mathrm{d} v n_\Delta L M_L\rangle &= n_\mathrm{d}|n_\mathrm{d} v n_\Delta L M_L\rangle \\ \mathcal{C}_2[\mathrm{SO}(5)]|n_\mathrm{d} v n_\Delta L M_L\rangle &= v(v+3)|n_\mathrm{d} v n_\Delta L M_L\rangle \\ \hat{L}^2|n_\mathrm{d} v n_\Delta L M_L\rangle &= L(L+1)|n_\mathrm{d} v n_\Delta L M_L\rangle \\ \hat{L}_0|n_\mathrm{d} v n_\Delta L M_L\rangle &= M_L|n_\mathrm{d} v n_\Delta L M_L\rangle \end{aligned} \qquad (8.61)$$

We first concentrate on the construction of states classified by both n_d and L, leaving a discussion of the SO(5) $\supset$ SO(3) labels v and n_Δ until later. To this end we consider the highest-weight states with $M_L = L$, since we may always use the Wigner–Eckart theorem to deal with $M_L \neq L$ (or apply the lowering operator $\hat{L}_{-1}$). We are thus interested in the polynomials $\mathcal{Q}(d_m^\dagger)$ that satisfy the equations

$$\hat{n}_\mathrm{d}\mathcal{Q}(d_m^\dagger) = n_\mathrm{d}\mathcal{Q}(d_m^\dagger), \quad \hat{L}_{+1}\mathcal{Q}(d_m^\dagger) = 0, \quad \hat{L}_0\mathcal{Q}(d_m^\dagger) = L\mathcal{Q}(d_m^\dagger) \quad (8.62)$$

with the operators $\hat{n}_\mathrm{d}$ and $\hat{L}_\mu$ given by (8.8) and (8.12). We recall that the annihilation operators d_m act on $\mathcal{Q}(d_m^\dagger)$ as differential operators, $d_m = \partial/\partial d_m^\dagger$. Given the explicit form (8.8) for $\hat{n}_\mathrm{d}$, the first equation in (8.62) then implies that $\mathcal{Q}(d_m^\dagger)$ is a homogeneous polynomial in the $d_m^\dagger$ operators, and thus we may write it as

$$\mathcal{Q}(d_m^\dagger) = (d_{+2}^\dagger)^{n_\mathrm{d}} \mathcal{Q}'\left(\frac{d_{+1}^\dagger}{d_{+2}^\dagger}, \frac{d_0^\dagger}{d_{+2}^\dagger}, \frac{d_{-1}^\dagger}{d_{+2}^\dagger}, \frac{d_{-2}^\dagger}{d_{+2}^\dagger}\right) \qquad (8.63)$$

where $\mathcal{Q}'$ is an arbitrary polynomial of the indicated ratios, of degree not exceeding n_d. To apply the other two equations in (8.62), we require that $\mathcal{Q}(d_m^\dagger)$ carries good angular momentum, so we introduce the following polynomial functions:

$$\begin{aligned} (2,0) &\equiv \sqrt{5}[d^\dagger \times d^\dagger]_0^{(0)} \\ (3,3) &\equiv -\sqrt{\tfrac{14}{3}}[[d^\dagger \times d^\dagger]^{(2)} \times d^\dagger]_3^{(3)} \\ (2,2) &\equiv \sqrt{7}[d^\dagger \times d^\dagger]_2^{(2)} \end{aligned} \qquad (8.64)$$

8.4 Vibrational nuclei: The U(5) limit

The notation (n_d, L) indicates the d-boson number and maximum angular momentum projection $M_L = L$ of the polynomials. The square-root factors are introduced to simplify their explicit form which, from the appropriate Clebsch–Gordan coefficients, becomes

$$\begin{aligned}
\frac{(2,0)}{(d^\dagger_{+2})^2} &= 2\frac{d^\dagger_{-2}}{d^\dagger_{+2}} - 2\frac{d^\dagger_{-1}d^\dagger_0}{d^\dagger_{+2}d^\dagger_{+2}} + \frac{(d^\dagger_0)^2}{(d^\dagger_{+2})^2} \\
\frac{(3,3)}{(d^\dagger_{+2})^3} &= 2\frac{d^\dagger_{-1}}{d^\dagger_{+2}} - \sqrt{6}\frac{d^\dagger_{+1}d^\dagger_0}{d^\dagger_{+2}d^\dagger_{+2}} + \frac{(d^\dagger_{+1})^3}{(d^\dagger_{+2})^3} \\
\frac{(2,2)}{(d^\dagger_{+2})^2} &= 2\sqrt{2}\frac{d^\dagger_0}{d^\dagger_{+2}} - \sqrt{3}\frac{(d^\dagger_{+1})^2}{(d^\dagger_{+2})^2}
\end{aligned} \qquad (8.65)$$

where we have divided by a power of $d^\dagger_{+2}$ to compare with (8.63). These equations imply that we could have written, instead of (8.63), the equivalent polynomials

$$\mathcal{Q}(d^\dagger_m) = (d^\dagger_{+2})^{n_d} \mathcal{Q}''\left(\frac{d^\dagger_{+1}}{d^\dagger_{+2}}, \frac{(2,0)}{(d^\dagger_{+2})^2}, \frac{(3,3)}{(d^\dagger_{+2})^3}, \frac{(2,2)}{(d^\dagger_{+2})^2}\right) \qquad (8.66)$$

where $\mathcal{Q}''$ is again an arbitrary polynomial of the new variables. Although it seems that the dependence on $d^\dagger_{+1}/d^\dagger_{+2}$ is still a problem, we note that $d^\dagger_{+2}$ and the polynomials in (8.64) have $M_L = L$, and thus the raising operator $\hat{L}_{+1}$ acting on them gives zero,

$$\hat{L}_{+1}d^\dagger_{+2} = \hat{L}_{+1}(2,0) = \hat{L}_{+1}(3,3) = \hat{L}_{+1}(2,2) = 0 \qquad (8.67)$$

Computing the $\hat{L}_\mu$ operators (8.12) in explicit form, we find

$$\hat{L}_{+1} = -\sqrt{2}d^\dagger_{-1}d_{-2} - \sqrt{3}d^\dagger_0 d_{-1} - \sqrt{3}d^\dagger_{+1}d_0 - \sqrt{2}d^\dagger_{+2}d_{+1} \qquad (8.68)$$

and

$$\hat{L}_0 = \sum_m m\, d^\dagger_m d_m \qquad (8.69)$$

Using these relations, the second equation in (8.62) may be written as

$$\begin{aligned}
\hat{L}_{+1}\mathcal{Q}(d^\dagger_m) &= (d^\dagger_{+2})^{n_d}(-\sqrt{2}d^\dagger_{+2}d_{+1})\mathcal{Q}'' \\
&= -\sqrt{2}(d^\dagger_{+2})^{n_d}\frac{\partial}{\partial(d^\dagger_{+1}/d^\dagger_{+2})}\mathcal{Q}'' = 0
\end{aligned} \qquad (8.70)$$

which shows that Q'' does not, in fact, depend on $d^\dagger_{+1}/d^\dagger_{+2}$. Since the polynomial $Q(d^\dagger_m)$ should have a total boson number of n_d, it may be expressed in the form

$$Q(d^\dagger_m) = \sum_{n_1 n_2 n_3} B_{n_1 n_2 n_3}(d^\dagger_{+2})^{n_d - 2n_1 - 3n_2 - 2n_3}(2,0)^{n_1}$$
$$\times (3,3)^{n_2}(2,2)^{n_3} \qquad (8.71)$$

where so far the $B_{n_1 n_2 n_3}$ are arbitrary coefficients. The third equation in (8.62) together with the explicit form for $\hat{L}_0$ imply, however, that

$$3n_2 + 2n_3 + 2(n_d - 2n_1 - 3n_2 - 2n_3) = 2n_d - 4n_1 - 3n_2 - 2n_3 = L \quad (8.72)$$

and we may eliminate n_3 from (8.71)

$$Q(d^\dagger_m) = \sum_{n_1 n_2} B'_{n_1 n_2}(d^\dagger_{+2})^{L - n_d + 2n_1}(2,0)^{n_1}$$
$$\times (3,3)^{n_2}(2,2)^{(2n_d - L - 3n_2)/2 - 2n_1} \qquad (8.73)$$

We note that for $Q(d^\dagger_m)$ to be a polynomial, n_2 must be even (odd) for L even (odd). It would seem that we find the appropriate polynomials by taking $B'_{n_1 n_2} = 1$ for a particular value of (n_1, n_2) and $B'_{n_1 n_2} = 0$ for all others, with the restriction that all exponents should be non-negative. This is not the case, however, since the exponent of $d^\dagger_{+2}$ does not depend on n_2. This means that

$$\sum_{n_2} B'_{n_1 n_2}(3,3)^{n_2}(2,2)^{(2n_d - L - 3n_2)/2 - 2n_1} \qquad (8.74)$$

could be divisible by $d^\dagger_{+2}$ for some coefficients $B'_{n_1 n_2}$. In this case the exponent of $d^\dagger_{+2}$ in (8.73) could be negative. To avoid this problem, we introduce the new polynomial

$$(3,0) \equiv \sqrt{35}[[d^\dagger \times d^\dagger]^{(2)} \times d^\dagger]^{(0)}_0 \qquad (8.75)$$

which is related to the previously defined ones (8.64) by

$$(d^\dagger_{+2})^3(3,0) = -\tfrac{3}{4}\sqrt{3}(3,3)^2 + \tfrac{3}{2}(2,0)(2,2)(d^\dagger_{+2})^2 - \tfrac{1}{4}(2,2)^3 \qquad (8.76)$$

8.4 Vibrational nuclei: The U(5) limit

Proving (8.76) involves simple but rather lengthy algebraic manipulations. Let us first consider the case of L even, which implies that n_2 is also even,

$$n_2 = 2n_\Delta \tag{8.77}$$

We may now use (8.76) to express $(3,3)^{2n_\Delta}$ in (8.73) in terms of $(3,0)^{n_\Delta}$ $(d^\dagger_{+2})^{3n_\Delta}$ and the polynomials already included there, leading to

$$\mathcal{Q}(d^\dagger_m) = \sum_{n_1 n_\Delta} B''_{n_1 n_\Delta} (d^\dagger_{+2})^{L-2n_\mathrm{d}+2n_1+3n_\Delta}(2,0)^{n_1}$$
$$\times (3,0)^{n_\Delta}(2,2)^{(2n_\mathrm{d}-L)/2-2n_1-3n_\Delta} \tag{8.78}$$

These polynomials cannot have negative coefficients [24]. The set of even-L polynomials

$$\mathcal{Q}^\mathrm{even}_{n_\mathrm{d} n_1 n_\Delta L}(d^\dagger_m) = (d^\dagger_{+2})^{L-2n_\mathrm{d}+2n_1+3n_\Delta}(2,0)^{n_1}$$
$$\times (3,0)^{n_\Delta}(2,2)^{(2n_\mathrm{d}-L)/2-2n_1-3n_\Delta} \tag{8.79}$$

then satisfies (8.62) with all exponents being non-negative. For L odd we take

$$n_2 = 2n_\Delta + 1 \tag{8.80}$$

and find in a similar way

$$\mathcal{Q}^\mathrm{odd}_{n_\mathrm{d} n_1 n_\Delta L}(d^\dagger_m) = (d^\dagger_{+2})^{L-2n_\mathrm{d}-2n_1+3n_\Delta}(3,3)(2,0)^{n_1}$$
$$\times (3,0)^{n_\Delta}(2,2)^{(2n_\mathrm{d}-L-3)/2-2n_1-3n_\Delta} \tag{8.81}$$

again with non-negative exponents. The polynomials (8.79) and (8.81) are a complete (but non-orthonormal) set of states classified by the representations n_d of U(5) and L of SO(3). We have found that there are two additional labels, the non-negative integers n_1 and n_Δ, associated with the powers of the polynomials $(2,0)$ and $(3,0)$, respectively.

Although the derivation of the U(5) $\supset$ SO(3) basis states and reduction rules (specified by the non-negative exponents) is somewhat lengthy, it illustrates the general procedure to determine eigenstates classified by algebras and associated reduction rules. The polynomials (8.79) and (8.81), however, do not carry the SO(5) label v. We now indicate a method to obtain the full U(5) $\supset$ SO(5) $\supset$ SO(3) states as linear combinations of these polynomials.

The defining equation for SO(5)-classified polynomials is

$$\mathcal{C}_2[SO(5)]\mathcal{Q}(d_m^\dagger) = v(v+3)\mathcal{Q}(d_m^\dagger) \quad (8.82)$$

Comparing with the expression for $\mathcal{C}_2[SO(5)]$,

$$\mathcal{C}_2[SO(5)] = \hat{n}_d(\hat{n}_d + 3) - (d^\dagger \cdot d^\dagger)(\tilde{d} \cdot \tilde{d}) \quad (8.83)$$

we see that the highest-weight polynomials (i.e., those with $v = n_d$, sometimes also referred to as having *maximum seniority*) satisfy the equation

$$(\tilde{d} \cdot \tilde{d})\mathcal{Q}(d_m^\dagger) = \sum_m (-)^m d_m d_{-m} \mathcal{Q}(d_m^\dagger) = 0 \quad (8.84)$$

Applying $\tilde{d} \cdot \tilde{d}$ to the U(5) $\supset$ SO(3) polynomials (8.79) and (8.81) with $v = n_d$, we find that, as they are, they do not satisfy (8.84). To impose this additional condition, we use a method first suggested in [26]. We introduce modified boson operators

$$h_m^\dagger = d_m^\dagger - (d^\dagger \cdot d^\dagger)(2\hat{n}_d + 5)^{-1}\tilde{d}_m \quad (8.85)$$

Note that the inverse operator $(2\hat{n}_d+5)^{-1}$ is well defined when acting on homogeneous polynomials in the $d_m^\dagger$ operators. By using the identities

$$\begin{aligned}(2\hat{n}_d+5)^{-1} d_m^\dagger &= d_m^\dagger (2\hat{n}_d+7)^{-1} \\ (2\hat{n}_d+5)^{-1} \tilde{d}_m &= \tilde{d}_m (2\hat{n}_d+3)^{-1}\end{aligned} \quad (8.86)$$

which again hold when applied to homogeneous polynomials $\mathcal{Q}(d_m^\dagger)$, and the commutator

$$[\tilde{d}_m, (2,0)] = \sum_{m'}(-)^{m+m'}[d_{-m}, d_{m'}^\dagger d_{-m'}^\dagger] = 2d_m^\dagger \quad (8.87)$$

one can show that

$$[h_m^\dagger, h_{m'}^\dagger] = 0 \quad (8.88)$$

This implies that the modified operators (8.85) satisfy the appropriate boson commutation relations. We also calculate the scalar product $h^\dagger \cdot h^\dagger$ and find

$$h^\dagger \cdot h^\dagger = \sum_m (-)^m h_m^\dagger h_{-m}^\dagger = (4\hat{n}_d^2 - 1)^{-1}(d^\dagger \cdot d^\dagger)^2 \sum_m (-)^m d_m d_{-m} \quad (8.89)$$

8.4 Vibrational nuclei: The U(5) limit

Replacing $d_m^\dagger$ by $h_m^\dagger$ in the polynomials $\mathcal{Q}_{n_d n_1 n_\Delta L}(d_m^\dagger)$ (8.79) and (8.81), we note from (8.89) that they vanish when acting on the vacuum state $|0\rangle$, unless $n_1 = 0$. Assuming $n_d = v$ and $n_1 = 0$, we define the new polynomials

$$\mathcal{Q}_{n_d = v\, n_\Delta L}(h_m^\dagger)|0\rangle \equiv |v n_\Delta L\rangle \tag{8.90}$$

which are linear combinations of terms of the form $d_{m_1}^\dagger d_{m_2}^\dagger \ldots d_{m_v}^\dagger |0\rangle$, that is, they are still homogeneous of degree v in the operators $d_m^\dagger$ and, by construction, have well-defined angular momentum. In contrast to the original polynomials (8.79) and (8.81), however, they also satisfy (8.84) since

$$\begin{aligned}
& (4\hat{n}_d^2 - 1)^{-1}(d^\dagger \cdot d^\dagger)^2 \sum_m (-)^m d_m d_{-m} |v n_\Delta L\rangle \\
&= (h^\dagger \cdot h^\dagger) \mathcal{Q}_{n_d = v\, n_\Delta L}(h_m^\dagger)|0\rangle \\
&= \mathcal{Q}_{n_d = v\, n_\Delta L}(h_m^\dagger)(h^\dagger \cdot h^\dagger)|0\rangle \\
&= \mathcal{Q}_{n_d = v\, n_\Delta L}(h_m^\dagger)(4\hat{n}_d^2 - 1)^{-1}(d^\dagger \cdot d^\dagger)^2 \sum_m (-)^m d_m d_{-m} |0\rangle \\
&= 0
\end{aligned} \tag{8.91}$$

where use is made of the commutation property (8.88) and the expression (8.89). Since $(4n_d^2 - 1)^{-1}(d^\dagger \cdot d^\dagger)^2$ does not vanish identically, it follows that the states (8.90) are classified by the U(5) $\supset$ SO(5) $\supset$ SO(3) chain of algebras. Denoting by $[n_d, L]$ the polynomials (8.64) and (8.75) with $d_m^\dagger$ replaced by $h_m^\dagger$, including $[1,2] \equiv h_{+2}^\dagger$, we find from (8.79) and (8.81) the explicit expression

$$|v n_\Delta L\rangle = [1,2]^{L-v+3n_\Delta}[3,3]^{\delta_L}[3,0]^{n_\Delta}[2,2]^{(2v-L-3\delta_L)/2-3n_\Delta} \tag{8.92}$$

where we have introduced the label δ_L which takes the values

$$\delta_L = \begin{cases} 0 & \text{even } L \\ 1 & \text{odd } L \end{cases} \tag{8.93}$$

to include both even and odd L's in the same formula. The missing label n_Δ is seen from (8.92) to be an integer corresponding to the number of *modified* boson triplets coupled to angular momentum zero. In addition, since all exponents must be non-negative, we find the inequalities

$$-v + 3n_\Delta + L \geq 0, \quad v - 3n_\Delta - \tfrac{1}{2}(L + 3\delta_L) \geq 0, \quad n_\Delta \geq 0 \tag{8.94}$$

Defining
$$\mu \equiv v - 3n_\Delta \tag{8.95}$$
we may rewrite these inequalities as
$$v = 3n_\Delta + \mu, \qquad 2\mu - 3\delta_L \geq L \geq \mu, \qquad n_\Delta \geq 0 \tag{8.96}$$

From the second inequality in (8.94) we also see that μ is a non-negative integer. These rules coincide with the $SO(5) \supset SO(3)$ reduction rule from v to L given at the beginning of this section.

Although the states (8.92) are indeed particular $U(5) \supset SO(5) \supset SO(3)$ states (i.e., those with $n_d = v$), the direct substitution of the modified bosons $h_m^\dagger$ is not an easy matter. We omit this last step here, since it involves rather lengthy calculations, referring the interested reader to the original publication [25]. The final result is

$$\boxed{\begin{aligned}|vn_\Delta L\rangle &= \sum_{rn} C_{rn}^{\sigma\tau n_\Delta}(d_{+2}^\dagger)^{\sigma+\tau-n}(3,3)^{\delta_L}(2,0)^{3r-\tau+n} \\ &\quad \times (3,0)^{n_\Delta+\tau-2r-n}(2,2)^n|0\rangle\end{aligned}} \tag{8.97}$$

where
$$\sigma = -v + 3n_\Delta + L, \qquad \tau = v - 3n_\Delta - \tfrac{1}{2}(L + 3\delta_L), \qquad \sigma, \tau \geq 0 \tag{8.98}$$

and

$$\begin{aligned}C_{rn}^{\sigma\tau n_\Delta} &= \frac{(-)^r 2^r 3^n \sigma!(3r)!(v+3\delta_L)!(2n_\Delta + 2\tau - 2r + \delta_L)!}{2^{n_\Delta+n} n!(2v+6\delta_L+1)!r!(n_\Delta+\tau-r)!(n_\Delta+\tau-n-2r)!} \\ &\quad \times \sum_s \frac{(-)^s 4^s (\tau+s)!(2v+6\delta_L+1-2s)!}{s!(\sigma-s)!(\tau-n+s)!(3r-\tau+n-s)!(v+3\delta_L-s)!}\end{aligned} \tag{8.99}$$

The combinations (n_d, L) in (8.97) are the $d_m^\dagger$ polynomials (8.64) and (8.75). Note that the number of d-boson triplets $(3,0)$ is *not* conserved for the $U(5) \supset SO(5) \supset SO(3)$ states and that the label n_Δ coincides with the maximum number of these triplets in the sum (8.97).

We may now complete our discussion of the $U(5)$ basis states by noting that the $n_d \neq v$ states are simply given by

$$\boxed{|n_d v n_\Delta L L\rangle = B_{n_d v}(d^\dagger \cdot d^\dagger)^{(n_d-v)/2}|vn_\Delta L\rangle} \tag{8.100}$$

8.4 Vibrational nuclei: The U(5) limit

where $B_{n_d v}$ is a normalization coefficient. The relation (8.100) follows directly from the properties

$$\hat{n}_d|n_d v n_\Delta LL\rangle = \left[2\left(\tfrac{1}{2}(n_d - v)\right) + v\right]|n_d v n_\Delta LL\rangle = n_d|n_d v n_\Delta LL\rangle \tag{8.101}$$

and

$$\begin{aligned}
\mathcal{C}_2[&\text{SO}(5)]|n_d v n_\Delta LL\rangle \\
&= \hat{n}_d(\hat{n}_d + 3)|n_d v n_\Delta LL\rangle - B_{n_d v}(d^\dagger \cdot d^\dagger)(\tilde{d} \cdot \tilde{d})(d^\dagger \cdot d^\dagger)^{(n_d - v)/2}|v n_\Delta L\rangle \\
&= n_d(n_d + 3)|n_d v n_\Delta LL\rangle - (n_d - v)(n_d + v + 3)|n_d v n_\Delta LL\rangle \\
&= v(v + 3)|n_d v n_\Delta LL\rangle
\end{aligned} \tag{8.102}$$

where use is made of (8.54).

Equations (8.60), (8.97), and (8.100) define the complete, though not orthonormal, basis states for chain (8.20). The non-orthogonality arises solely from the missing label n_Δ since, in contrast to the other quantum numbers, it is not associated to an invariant operator in the chain. The non-orthogonality is not a serious problem, though, since one may carry out a Schmidt orthogonalization procedure in the label n_Δ [24,25].

The U(5) spectrum corresponds to that of an anharmonic vibrator, as illustrated in the right-hand part of Figure 8.1. The spectra of many nuclei with a small number of valence neutrons and a larger number of valence protons (or *vice versa*) display such vibrational characteristics. The example of ^{110}Cd is shown in the left-hand part of Figure 8.1.

Clearly, to confirm the existence of a dynamical symmetry, we need more than predictions for the energies and we should also consider electromagnetic transitions and other spectroscopic properties in our analysis. We illustrate these calculations in the algebraic framework by evaluating some E2 transition rates using the explicit form of the U(5) eigenstates. In first approximation, the quadrupole operator of the U(5) limit is given by [16]

$$\hat{T}^{(E2)}_\mu = q[d^\dagger \times \tilde{s} + s^\dagger \times \tilde{d}]^{(2)}_\mu \tag{8.103}$$

where q is an effective boson charge. It has the units of [charge] $\times$ [length]^2 (usually eb, where b stands for barn = 10^4 fm^2 = 10^{-26} m^2)

since it includes the r^2 dependence of the electric quadrupole transition operator. The $B(\text{E2})$ values are defined by the relation

$$B(\text{E2}; L \to L') = \frac{2L'+1}{2L+1} |\langle L' \| \hat{T}^{(\text{E2})} \| L \rangle|^2 \qquad (8.104)$$

where the reduced matrix elements are evaluated by means of the Wigner–Eckart theorem (1.96). (The form (8.104) differs by the factor $(2L'+1)$ from the usual definition in nuclear physics (see, e.g.,[8]), but this comes about because of a different convention for the reduced matrix elements, producing the same definition of the $B(\text{E2})$ value.) As a first example we consider matrix elements among states with $v = n_\text{d}$ and $L = 2n_\text{d}$. From (8.60), (8.97), and (8.100) we note that the states $|[N] n_\text{d} v n_\Delta L M_L\rangle$ take the simple polynomial form

$$|[N] n_\text{d} n_\text{d} 0 \, 2n_\text{d} \, 2n_\text{d}\rangle = \frac{(s^\dagger)^{N-n_\text{d}} (d^\dagger_{+2})^{n_\text{d}}}{\sqrt{(N-n_\text{d})! n_\text{d}!}} |0\rangle \qquad (8.105)$$

where we include the factor $(n_\text{d}!)^{-1/2}$ to normalize the d-boson polynomial. Taking the $d^\dagger \tilde{s}$ part of (8.103) first, we find that

$$\langle [N] n_\text{d}+1 \, n_\text{d}+1 \, 0 \, 2n_\text{d}+2 \, 2n_\text{d}+2 | d^\dagger_{+2} \tilde{s} | [N] n_\text{d} n_\text{d} 0 \, 2n_\text{d} \, 2n_\text{d}\rangle$$

$$= \frac{\langle 0 | (d_{+2})^{n_\text{d}+1} (s)^{N-n_\text{d}-1} (d^\dagger_{+2} s)(s^\dagger)^{N-n_\text{d}} (d^\dagger_{+2})^{n_\text{d}} | 0\rangle}{\sqrt{(N-n_\text{d}-1)!(N-n_\text{d})! n_\text{d}! (n_\text{d}+1)!}}$$

$$= \frac{\langle 0 | (d_{+2})^{n_\text{d}+1} (s)^{N-n_\text{d}} (s^\dagger)^{N-n_\text{d}} (d^\dagger_{+2})^{n_\text{d}+1} | 0\rangle}{\sqrt{(N-n_\text{d}-1)!(N-n_\text{d})! n_\text{d}! (n_\text{d}+1)!}}$$

$$= \sqrt{(N-n_\text{d})(n_\text{d}+1)} \qquad (8.106)$$

where we make use of the simple result

$$\langle 0 | (b)^k (b^\dagger)^k | 0\rangle = k! \qquad (8.107)$$

for both the s and d bosons. The hermitian conjugate of (8.106) gives the corresponding result for $s^\dagger \tilde{d}_{+2}$,

$$\langle [N] n_\text{d} n_\text{d} 0 \, 2n_\text{d} \, 2n_\text{d} | s^\dagger \tilde{d}_{+2} | [N] n_\text{d}+1 \, n_\text{d}+1 \, 0 \, 2n_\text{d}+2 \, 2n_\text{d}+2\rangle$$

$$= \sqrt{(N-n_\text{d})(n_\text{d}+1)} \qquad (8.108)$$

8.4 Vibrational nuclei: The U(5) limit

From the Wigner–Eckart theorem and the definition (8.104) we arrive at the reduced transitions

$$B(\text{E2}; [N]n_\text{d}+1\ v=n_\text{d}+1\ n_\Delta=0\ L=2n_\text{d}+2$$
$$\to [N]n_\text{d}\ v=n_\text{d}\ n_\Delta=0\ L'=2n_\text{d})$$
$$= q^2(N-n_\text{d})(n_\text{d}+1)$$
$$= \tfrac{1}{4}q^2(2N-L')(L'+2) \tag{8.109}$$

A slightly more complicated example is provided by the E2 matrix elements connecting states with $v = n_d$ and $L = 2n_d$ to those with $v \neq n_d$ (see (8.100)),

$$|[N]n_\text{d}+1\ n_\text{d}-1\ 0\ 2n_\text{d}-2\ 2n_\text{d}-2\rangle = \mathcal{N}_{n_\text{d}} \frac{(d^\dagger \cdot d^\dagger)(s^\dagger)^{N-n_\text{d}-1}(d^\dagger_{+2})^{n_\text{d}-1}}{\sqrt{(N-n_\text{d}-1)!}}|0\rangle \tag{8.110}$$

where $\mathcal{N}_{n_\text{d}}$ is a normalization coefficient for the d-boson part that has to be determined. Using again (8.54) we may compute the matrix element

$$\langle 0|(d_{+2})^{n_\text{d}-1}(\tilde{d}\cdot\tilde{d})(d^\dagger\cdot d^\dagger)(d^\dagger_{+2})^{n_\text{d}-1}|0\rangle$$
$$= \langle 0|(d_{+2})^{n_\text{d}-1}2(2n_\text{d}+3)(d^\dagger_{+2})^{n_\text{d}-1}|0\rangle$$
$$= 2(2n_\text{d}+3)(n_\text{d}-1)! \tag{8.111}$$

and thus find

$$\mathcal{N}_{n_\text{d}} = \frac{1}{\sqrt{2(2n_\text{d}+3)(n_\text{d}-1)!}} \tag{8.112}$$

We can now evaluate the $s^\dagger \tilde{d}_{+2}$ matrix element

$$\langle[N]n_\text{d}n_\text{d}0\ 2n_\text{d}\ 2n_\text{d}|s^\dagger\tilde{d}_{+2}|[N]n_\text{d}+1\ n_\text{d}-1\ 0\ 2n_\text{d}-2\ 2n_\text{d}-2\rangle$$
$$= \mathcal{N}_{n_\text{d}} \frac{\langle 0|(d_{+2})^{n_\text{d}}(s)^{N-n_\text{d}}(s^\dagger d_{-2})(d^\dagger\cdot d^\dagger)(s^\dagger)^{N-n_\text{d}-1}(d^\dagger_{+2})^{n_\text{d}-1}|0\rangle}{\sqrt{(N-n_\text{d}-1)!(N-n_\text{d})!n_\text{d}!}}$$
$$= \left[\frac{2(N-n_\text{d})n_\text{d}}{2n_\text{d}+3}\right]^{1/2} \tag{8.113}$$

The corresponding reduced matrix element has the same value since the Clebsch–Gordan coefficient connecting the two equals 1. Using the definition (8.104), we arrive at the final result:

$$B(E2; [N]n_d+1 \; v=n_d-1 \; n_\Delta=0 \; L=2n_d-2$$
$$\to [N]n_d \; v=n_d \; n_\Delta=0 \; L'=2n_d)$$
$$= q^2 \frac{2(N-n_d)n_d(4n_d+1)}{(2n_d+3)(4n_d-3)}$$
$$= q^2 \frac{(2N-L')L'(2L'+1)}{2(L'+3)(2L'-3)} \quad (8.114)$$

Other matrix elements can be evaluated in a similar way, although given the general form (8.100) and the complexity of (8.97), they may require a significant amount of work. Other techniques are discussed in [16]. In Figure 8.2 we give the $B(E2)$ values of transitions between some low-energy states in the U(5) limit.

The basis states (8.60) form a complete set in which we can carry out the diagonalization of the general hamiltonian of the form (8.28) or (8.40). Note that the only operators not already diagonal in this basis are $\mathcal{C}_2[\mathrm{SO}(6)]$ and $\mathcal{C}_2[\mathrm{SU}(3)]$, so we only need to compute the corresponding matrix elements. The former may be evaluated analytically, using an SU(1,1) method analogous to the one presented in Chapter 5,

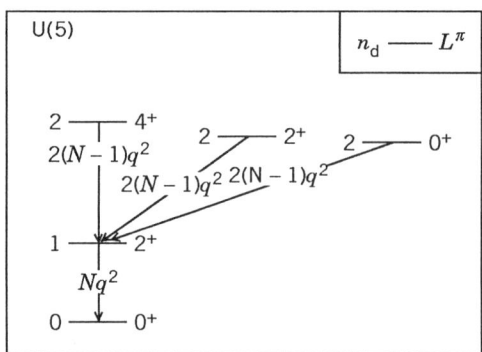

Figure 8.2: Allowed transitions in the U(5) limit induced by the quadrupole generator $q[d^\dagger \times \tilde{s} + s^\dagger \times \tilde{d}]^{(2)}_\mu$. The expressions alongside the arrows are the $B(E2)$ values. Labels are as in Figure 8.1.

8.4 Vibrational nuclei: The U(5) limit

as we proceed to show. We first define the pairing operators

$$\mathcal{P}^\dagger \equiv \tfrac{1}{2}(d^\dagger \cdot d^\dagger + s^\dagger s^\dagger), \qquad \tilde{\mathcal{P}} \equiv \tfrac{1}{2}(\tilde{d} \cdot \tilde{d} + \tilde{s}\tilde{s}) \tag{8.115}$$

which are similar to (5.94). Using the explicit expression of the SO(6) quadratic invariant, we find

$$4\mathcal{P}^\dagger \tilde{\mathcal{P}} = \hat{N}(\hat{N} + 4) - \mathcal{C}_2[\text{SO}(6)] \tag{8.116}$$

which shows that the diagonalization of $\mathcal{C}_2[\text{SO}(6)]$ is equivalent to that of $\mathcal{P}^\dagger \tilde{\mathcal{P}}$. In analogy with (5.98) we define the SU(1,1) generators

$$\hat{T}_+ \equiv \tfrac{1}{2}(d^\dagger \cdot d^\dagger), \quad \hat{T}_- \equiv \tfrac{1}{2}(\tilde{d} \cdot \tilde{d}), \quad \hat{T}_z \equiv -\tfrac{1}{4}(d^\dagger \cdot \tilde{d} + \tilde{d} \cdot d^\dagger) \tag{8.117}$$

which can be shown to satisfy again the commutation relations (5.99), while $\hat{T}_z$ can be rewritten as

$$\hat{T}_z = \tfrac{1}{2}\left(\hat{n}_d + \tfrac{5}{2}\right) \tag{8.118}$$

Using the explicit expression of the SO(5) quadratic invariant, we write

$$d^\dagger \cdot d^\dagger \tilde{d} \cdot \tilde{d} = \hat{n}_d\left(\hat{n}_d + 3\right) - \mathcal{C}_2[\text{SO}(5)] \tag{8.119}$$

which leads to the SU(1,1) Casimir invariant

$$\begin{aligned}
\hat{T}^2 &= -\hat{T}_+\hat{T}_- - \hat{T}_z + \hat{T}_z^2 \\
&= -\tfrac{1}{4}d^\dagger \cdot d^\dagger \tilde{d} \cdot \tilde{d} - \tfrac{1}{4}(2\hat{n}_d + 5) + \tfrac{1}{16}(2\hat{n}_d + 5)^2 \\
&= \tfrac{1}{4}\left(\mathcal{C}_2[\text{SO}(5)] + \tfrac{5}{4}\right)
\end{aligned} \tag{8.120}$$

Since the eigenvalues of $\hat{T}_z$ and $\hat{T}^2$ are m_t and $t(t-1)$, respectively, (8.118) implies

$$m_t = \tfrac{1}{2}\left(n_d + \tfrac{5}{2}\right) \tag{8.121}$$

while (8.56) and (8.120) give

$$t = \tfrac{1}{2}\left(v + \tfrac{5}{2}\right) \tag{8.122}$$

From the U(5) $\supset$ SO(5) reduction rule $v = n_d, n_d - 2, \ldots$ we thus find, for fixed t,

$$m_t = t, t+1, \ldots \tag{8.123}$$

which coincides with the result found in Chapter 5. The subsequent analysis follows the same lines as that of Chapter 5, except that now m_t and t are given by (8.121) and (8.122). From (5.104) we find in this case

$$\langle n_\mathrm{d}+2\,vn_\Delta LM_L|d^\dagger\cdot d^\dagger|n_\mathrm{d}vn_\Delta LM_L\rangle = \sqrt{(n_\mathrm{d}-v+2)(n_\mathrm{d}+v+5)}$$
$$\langle n_\mathrm{d}-2\,vn_\Delta LM_L|\tilde{d}\cdot\tilde{d}|n_\mathrm{d}vn_\Delta LM_L\rangle = \sqrt{(n_\mathrm{d}-v)(n_\mathrm{d}+v+3)}$$
(8.124)

in close correspondence with the U(3) result. This derivation based on SU(1,1) can be extended to arbitrary dimension to deal with the $U(n) \supset SO(n)$ reduction.

We note that matrix elements of $d^\dagger \cdot d^\dagger$ and $\tilde{d} \cdot \tilde{d}$ can be alternatively derived directly from formulas (8.54) and (8.100). Using these equations, we find

$$\langle vn_\Delta L|(\tilde{d}\cdot\tilde{d})^{n_0}(d^\dagger\cdot d^\dagger)^{n_0}|vn_\Delta L\rangle$$
$$= 2n_0(2n_0+2v+3)\langle vn_\Delta L|(\tilde{d}\cdot\tilde{d})^{n_0-1}(d^\dagger\cdot d^\dagger)^{n_0-1}|vn_\Delta L\rangle$$
$$\cdots$$
$$= \frac{(2n_0)!!(2n_0+2v+3)!!}{(2v+3)!!} \quad (8.125)$$

From this result we obtain the normalization coeffcient $B_{n_\mathrm{d}v}$ in (8.100),

$$B_{n_\mathrm{d}v} = (-)^{(n_\mathrm{d}-v)/2}\left[\frac{(2v+3)!!}{(n_\mathrm{d}-v)!!(n_\mathrm{d}+v+3)!!}\right]^{1/2} \quad (8.126)$$

and the matrix element

$$\langle n_\mathrm{d}+2\,vn_\Delta LM_L|d^\dagger\cdot d^\dagger|n_\mathrm{d}vn_\Delta LM_L\rangle$$
$$= B_{n_\mathrm{d}+2\,v}B_{n_\mathrm{d}v}\langle vn_\Delta L|(\tilde{d}\cdot\tilde{d})^{(n_\mathrm{d}-v)/2+1}(d^\dagger\cdot d^\dagger)^{(n_\mathrm{d}-v)/2+1}|vn_\Delta L\rangle$$
$$= B_{n_\mathrm{d}+2\,v}B_{n_\mathrm{d}v}\frac{(n_\mathrm{d}-v+2)!!(n_\mathrm{d}+v+5)!!}{(2v+3)!!}$$
$$= -\sqrt{(n_\mathrm{d}-v+2)(n_\mathrm{d}+v+5)} \quad (8.127)$$

The hermitian conjugate matrix element is found in the same way. The difference in sign compared to the result derived previously is a consequence of the phase $(-)^{(n_\mathrm{d}-v)/2}$ in the normalization coefficient $B_{n_\mathrm{d}v}$

8.4 Vibrational nuclei: The U(5) limit

which conforms with the U(3) ⊃ SO(3) and U(4) ⊃ SO(4) conventions of Chapter 5.

Returning to the matrix elements of the full pairing operator $\mathcal{P}^\dagger\mathcal{P}$ operator, we note that the action of $s^\dagger$ and $\tilde{s}$ on the U(6) ⊃ U(5) states (8.60) is trivially given by

$$s^\dagger|[N]n_d v n_\Delta L M_L\rangle = \sqrt{N - n_d + 1}|[N+1]n_d v n_\Delta L M_L\rangle$$
$$\tilde{s}|[N]n_d v n_\Delta L M_L\rangle = \sqrt{N - n_d}|[N-1]n_d v n_\Delta L M_L\rangle \quad (8.128)$$

The off-diagonal matrix of the pairing operator is then given by

$$\langle[N]n_d+2\,v n_\Delta L M_L|4\mathcal{P}^\dagger\tilde{\mathcal{P}}|[N]n_d v n_\Delta L M_L\rangle$$
$$= \langle[N]n_d+2\,v n_\Delta L M_L|d^\dagger \cdot d^\dagger \tilde{s}\tilde{s}|[N]n_d v n_\Delta L M_L\rangle$$
$$= \sqrt{(N-n_d-1)(N-n_d)}\langle n_d+2\,v n_\Delta L M_L|d^\dagger \cdot d^\dagger|n_d v n_\Delta L M_L\rangle$$
$$= \sqrt{(N-n_d-1)(N-n_d)(n_d-v+2)(n_d+v+5)} \quad (8.129)$$

while the diagonal ones are

$$\langle[N]n_d v n_\Delta L M_L|4\mathcal{P}^\dagger\tilde{\mathcal{P}}|[N]n_d v n_\Delta L M_L\rangle$$
$$= (N-n_d-1)(N-n_d) + (n_d-v)(n_d+v+3) \quad (8.130)$$

The resulting matrix is of tridiagonal form, with non-zero elements for $\Delta n_d = 0, \pm 2$. Its knowledge is sufficient to (numerically) determine the SO(6) states (8.21). In Section 8.5 we consider yet another method to obtain them.

As a final point in this section, we discuss the evaluation of the matrix elements of $\mathcal{C}_2[SU(3)]$ in the U(5) basis. Given its explicit expression it is clear that we need to determine the matrix elements of $\hat{Q}^2$, which, from (8.23), reads

$$\hat{Q}^2 = \sqrt{5}\left(\tfrac{7}{4}[[d^\dagger \times \tilde{d}]^{(2)} \times [d^\dagger \times \tilde{d}]^{(2)}]^{(0)}_0\right.$$
$$-\sqrt{7}[[d^\dagger \times \tilde{d}]^{(2)} \times [d^\dagger \times \tilde{s} + s^\dagger \times \tilde{d}]^{(2)}]^{(0)}_0$$
$$\left.+[[d^\dagger \times \tilde{s} + s^\dagger \times \tilde{d}]^{(2)} \times [d^\dagger \times \tilde{s} + s^\dagger \times \tilde{d}]^{(2)}]^{(0)}_0\right) \quad (8.131)$$

By means of angular momentum recoupling, this can be written in the alternative form [15]

$$\hat{Q}^2 = -\tfrac{11}{2}\hat{n}_d - \tfrac{7}{2}\hat{n}_d^2 + 2\hat{N}(2\hat{n}_d + 5) - \mathcal{C}_2[SO(6)]$$
$$+ \tfrac{1}{2}\mathcal{C}_2[SO(5)] + \tfrac{1}{8}\hat{L}^2 - \hat{T}_3 \quad (8.132)$$

where

$$\hat{T}_3 = \sqrt{35}[[d^\dagger \times d^\dagger]^{(2)} \times [\tilde{d} \times \tilde{s}]^{(2)}]^{(0)} + [[d^\dagger \times s^\dagger]^{(2)} \times [\tilde{d} \times \tilde{d}]^{(2)}]^{(0)}$$
$$= (3,0)\tilde{s} + s^\dagger(3,0)^\dagger \tag{8.133}$$

with $(3,0)$ defined in (8.75). Since only $C_2[SO(6)]$ and $\hat{T}_3$ in (8.132) are non-diagonal in the U(5) basis and the matrix elements of $C_2[SO(6)]$ are known from the previous discussion, only the matrix elements of $\hat{T}_3$ remain to be determined. While the s-boson part is trivial, the remaining part

$$\langle n_d+1 \, v'n'_\Delta LL|(3,0)|n_d v n_\Delta LL \rangle \tag{8.134}$$

involves a large number of sums. We do not write the result here but refer the reader to the paper by Castaños and Frank [27], where these matrix elements are evaluated in closed form.

These results in principle constitute a complete algebraic analysis of the IBM since the SO(6) and SU(3) wave functions may be determined by diagonalization in the U(5) basis. Whereas the latter cannot be analytically determined in general, the SO(6) wave functions are obtained in the next section in closed form.

8.5 γ-unstable nuclei: The SO(6) limit

The SO(6) limit of the IBM corresponds to taking $\epsilon = \alpha = \eta_3 = \xi = 0$ in the hamiltonian (8.40),

$$\boxed{\hat{H}_{\mathrm{II}} = E_0'' + \beta C_2[SO(6)] + \zeta C_2[SO(5)] + \gamma C_2[SO(3)]} \tag{8.135}$$

with eigenvalues

$$E_{\mathrm{II}}(\sigma, v, L) = E_0'' + \beta\sigma(\sigma+4) + \zeta v(v+3) + \gamma L(L+1) \tag{8.136}$$

where use is made of (8.56). The SO(5) $\supset$ SO(3) reduction rule is given in the previous section while the U(6) $\supset$ SO(6) reduction rule is a special case of (8.53),

$$\sigma = N, N-2, \ldots, 1 \text{ or } 0 \tag{8.137}$$

8.5 γ-unstable nuclei: The SO(6) limit

We still need one remaining branching rule in (8.21), namely the one corresponding to $SO(6) \supset SO(5)$.

In Chapter 5 the $SO(4) \supset SO(3)$ reduction (5.129) is derived using the $SO(4) \simeq SU_r(2) \times SU_{r'}(2)$ isomorphism. In the present case there is no simple isomorphism which can help us determine the σ to v branching rule in terms of known results [28]. Although in Appendix B we present the general $SO(n) \supset SO(n-1)$ reduction, from our previous analysis it is not difficult to deduce the result for this particular case (i.e., for fully symmetric representations). Since the SO(6) states only differ from the U(5) ones in the n_d label and the latter form a complete basis, it is possible to express the former as

$$|[N]\sigma v n_\Delta L M_L\rangle = \sum_{n_d} B_{n_d}^{N\sigma v} |[N] n_d v n_\Delta L M_L\rangle \tag{8.138}$$

where the abbreviation $B_{n_d}^{N\sigma v} = \langle [N] n_d v n_\Delta L | [N] \sigma v n_\Delta L \rangle$ is introduced as a short-hand notation for the transformation brackets between the two bases. Because of the $U(6) \supset U(5)$ and $U(5) \supset SO(5)$ reduction rules, the allowed values of n_d in the sum must satisfy $n_d \leq N$, with $n_d - v$ an even number. Consider now the highest-weight states with $N = \sigma$, for which (8.138) reads

$$|[\sigma]\sigma v n_\Delta L M_L\rangle = \sum_{n_d} B_{n_d}^{\sigma\sigma v} |[\sigma] n_d v n_\Delta L M_L\rangle \tag{8.139}$$

The analysis of the previous section indicates that v in the U(5) states $|[\sigma]n_d v n_\Delta L M_L\rangle$ can take the values

$$v = 0, 1, \ldots, N = \sigma \tag{8.140}$$

when we allow all permissible values for the intermediate label n_d. This gives the required $SO(6) \supset SO(5)$ branching rule. The explicit form of the transformation brackets $B_{n_d}^{N\sigma v}$ in (8.138) is determined at the end of this section.

In Figure 8.3 we show a typical SO(6) spectrum arising from (8.136) and compare it with the observed levels of $^{196}_{78}\text{Pt}_{118}$.

The SO(6) limit can be related to the geometric model of Bohr and Mottelson [5,6]. In this model collective excitations of the nucleus are describeded in terms of quadrupole oscillations of the nuclear surface.

328   8 The Interacting Boson Model

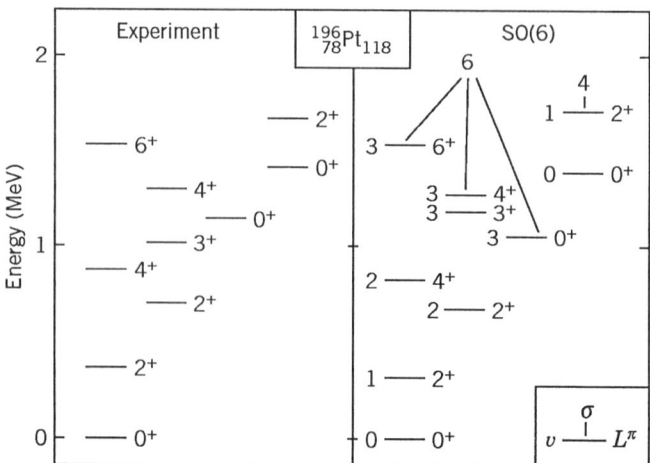

Figure 8.3: Partial SO(6) eigenspectrum of $\hat{H}_{\mathrm{II}}$ for $N = 6$ bosons and its comparison to the observed levels of the nucleus ^{196}Pt. The calculated spectrum is obtained with the parameters (in keV) $\beta = -50$, $\zeta = 58$, and $\gamma = 11$, with the energy of the ground state normalized to zero. Levels are labeled by σ, v, and the angular momentum and parity L^π.

If it is assumed that the total energy of the system is only weakly dependent on deviations from axial symmetry, the resulting γ-unstable model [29] has properties closely related to those of the SO(6) limit.

We now turn our attention to the determination of the SO(6) states. The defining equations for the U(6) ⊃ SO(6) ⊃ SO(5) ⊃ SO(3) basis are

$$\begin{aligned}
\hat{N}|[N]\sigma v n_\Delta L M_L\rangle &= N|[N]\sigma v n_\Delta L M_L\rangle \\
\mathcal{C}_2[\mathrm{SO}(6)]|[N]\sigma v n_\Delta L M_L\rangle &= \sigma(\sigma+4)|[N]\sigma v n_\Delta L M_L\rangle \\
\mathcal{C}_2[\mathrm{SO}(5)]|[N]\sigma v n_\Delta L M_L\rangle &= v(v+3)|[N]\sigma v n_\Delta L M_L\rangle \quad (8.141) \\
\hat{L}^2|[N]\sigma v n_\Delta L M_L\rangle &= L(L+1)|[N]\sigma v n_\Delta L M_L\rangle \\
\hat{L}_0|[N]\sigma v n_\Delta L M_L\rangle &= M_L|[N]\sigma v n_\Delta L M_L\rangle
\end{aligned}$$

which, except for the second equation, are identical to those satisfied by the U(5) states. In analogy with the procedure followed in Chapter 5 for the SO(4) states it is convenient to introduce a coordinate repre-

8.5 γ-unstable nuclei: The SO(6) limit

sentation and translate (8.141) into a set of differential equations. We first define coordinates appropriate for the U(5) states of the previous section. The substitutions

$$s^\dagger = \frac{1}{\sqrt{2}}\left(\bar{x} - \frac{\partial}{\partial \bar{x}}\right), \qquad \tilde{s} = \frac{1}{\sqrt{2}}\left(\bar{x} + \frac{\partial}{\partial \bar{x}}\right) \qquad (8.142)$$

and

$$d_m^\dagger = \frac{1}{\sqrt{2}}\left(\alpha_m - (-)^m \frac{\partial}{\partial \alpha_m}\right), \qquad \tilde{d}_m = \frac{1}{\sqrt{2}}\left(\alpha_m + (-)^m \frac{\partial}{\partial \alpha_m}\right) \qquad (8.143)$$

define a scalar $\bar{x}$ and a second rank tensor α_m (compare with (5.61) and (5.137)). Note that these definitions satisfy the correct boson commutation relations. For example,

$$\begin{aligned}[\tilde{d}_m, d_{m'}^\dagger] &= (-)^m [d_{-m}, d_{m'}^\dagger] = (-)^m \delta_{-mm'} \\ &= \tfrac{1}{2}\left(-[\alpha_m, (-)^{m'}\frac{\partial}{\partial \alpha_{-m'}}] + [(-)^m \frac{\partial}{\partial \alpha_{-m}}, \alpha_{m'}]\right) \\ &= \tfrac{1}{2}\left((-)^{m'} \delta_{m-m'} + (-)^m \delta_{-mm'}\right) = (-)^m \delta_{-mm'}\end{aligned} \qquad (8.144)$$

where use is made of the commutator $[\partial/\partial\alpha_m, \alpha_{m'}] = \delta_{mm'}$. The α_m variables can be identified with coordinates describing quadrupole oscillations of the nuclear surface [24]. One may define a set of hyperspherical coordinates in five-dimensional space through the transformation

$$\alpha_m = \sum_{m'} D^2_{mm'}(\theta_1, \theta_2, \theta_3) a_{m'} \qquad (8.145)$$

in terms of a Wigner D function for $j = 2$. The θ_i are the three Euler angles that define the direction of the principal axes of the nuclear surface [30]. It is customary to transform to the "intrinsic variables" β and γ defined as

$$a_0 = \beta \cos \gamma, \qquad a_{-1} = a_{+1} = 0, \qquad a_{-2} = a_{+2} = \sqrt{\tfrac{1}{2}}\beta \sin \gamma \qquad (8.146)$$

Using the orthogonality of the **D** matrices, we find

$$\alpha \cdot \alpha = \sum_m (-)^m \alpha_m \alpha_{-m} = \beta^2 \qquad (8.147)$$

which shows that β is the hyperradius in five dimensions. The β, γ, and θ_i variables were used by Chacón et al. [24] to find a coordinate realization of the U(5) $\supset$ SO(5) $\supset$ SO(3) states considered in the previous section. They can be written in the form

$$\Phi_{n_d v n_\Delta L M_L}(\beta,\gamma,\theta_i) = f_v^{n_d}(\beta)\chi^v_{n_\Delta L M_L}(\gamma,\theta_i) \qquad (8.148)$$

where

$$f_v^{n_d}(\beta) = \left[\frac{2((n_d - v)/2)!}{\Gamma((n_d + v + 5)/2)}\right]^{1/2} \beta^v L^{v+3/2}_{(n_d-v)/2}(\beta^2) e^{-\beta^2/2} \qquad (8.149)$$

with L_n^α being a Laguerre polynomial [31]. The functions $\chi^v_{n_\Delta L M_L}(\gamma,\theta_i)$ play the same role for the SO(5) $\supset$ SO(3) $\supset$ SO(2) reduction as the spherical harmonics $Y^L_{M_L}(\theta,\phi)$ for SO(3) $\supset$ SO(2). The derivation of (8.149) is straightforward since, in terms of the variables β, γ, and θ_i, the number operator $\hat{n}_d$ takes the form

$$\hat{n}_d = \frac{1}{2}\left[-\frac{1}{\beta^4}\frac{\partial}{\partial\beta}\left(\beta^4\frac{\partial}{\partial\beta}\right) + \beta^2 + \frac{1}{\beta^2}\mathcal{C}_2[\text{SO}(5)] - 5\right] \qquad (8.150)$$

which applied to (8.148) leads to a differential equation of the Laguerre type, in close analogy with the procedures used in Section 5.5. The $\chi^v_{n_\Delta L M_L}(\gamma,\theta_i)$ have a complex structure [24] which is closely related to the polynomials (8.97) with $d^\dagger_m$ substituted by the coordinates α_m, as we proceed to show.

The functions $\chi^v_{n_\Delta L M_L}(\gamma,\theta_i)$ satisfy the differential equations

$$\begin{aligned}
\mathcal{C}_2[\text{SO}(5)]\chi^v_{n_\Delta L M_L}(\gamma,\theta_i) &= v(v+3)\chi^v_{n_\Delta L M_L}(\gamma,\theta_i) \\
\hat{L}^2 \chi^v_{n_\Delta L M_L}(\gamma,\theta_i) &= L(L+1)\chi^v_{n_\Delta L M_L}(\gamma,\theta_i) \\
\hat{L}_0 \chi^v_{n_\Delta L M_L}(\gamma,\theta_i) &= M_L \chi^v_{n_\Delta L M_L}(\gamma,\theta_i)
\end{aligned} \qquad (8.151)$$

A complete U(6) $\supset$ U(5) $\supset$ SO(5) $\supset$ SO(3) state, including the s-boson dependence, is given by (8.148) multiplied with the one-dimensional oscillator result,

$$\Psi_{N n_d v n_\Delta L M_L}(\bar{x},\beta,\gamma,\theta_i) = g(\bar{x})\Phi_{n_d v n_\Delta L M_L}(\beta,\gamma,\theta_i) \qquad (8.152)$$

8.5 γ-unstable nuclei: The SO(6) limit

where (see Section 1.1)

$$g(\bar{x}) = \frac{1}{\sqrt{2^{N-n_d}\sqrt{\pi}(N-n_d)!}} H_{N-n_d}(\bar{x}) e^{-\bar{x}^2/2} \tag{8.153}$$

We point out that this coordinate realization for the U(5) states is useful for the determination of matrix elements [27]. In particular, the matrix elements of $C_2[SO(6)]$ can be simply derived from recursion relation properties of the Laguerre polynomials in (8.149) [15].

The variables defined in (8.142) and (8.143) and the subsequent transformations (8.145) and (8.146) are suitable for the description of wave functions in the U(5) limit where the s- and d-boson parts can be separated. In the SO(6) limit, however, the s- and d-boson dependence must be dealt with simultaneously. From our previous discussion concerning the U(5) ⊃ SO(5) ⊃ SO(3) coordinates, we conclude that we should now define *six-dimensional* hyperspherical coordinates in order to have well-defined properties under U(6) ⊃ SO(6). This is achieved by the transformations (8.142) to (8.146), followed by a new one,

$$\bar{x} = \rho\cos\varphi, \qquad \beta = \rho\sin\varphi \tag{8.154}$$

which defines the six-dimensional hyperradius

$$\rho^2 \equiv \bar{x}^2 + \beta^2 \tag{8.155}$$

and the new angle φ

$$\tan\varphi \equiv \beta/\bar{x} \tag{8.156}$$

instead of β and $\bar{x}$. The variables γ and θ_i remain the same as before, as the functions $\chi^v_{n_\Delta L M_L}(\gamma,\theta_i)$ still satisfy the relations (8.151) for the SO(6) states. In the new coordinates the total number operator takes the form

$$\hat{N} = \frac{1}{2}\left[-\frac{1}{\rho^5}\frac{\partial}{\partial\rho}\left(\rho^5\frac{\partial}{\partial\rho}\right) + \rho^2 + \frac{1}{\rho^2}C_2[SO(6)] - 6\right] \tag{8.157}$$

while the SO(6) Casimir invariant is given by

$$C_2[SO(6)] = -\frac{\partial^2}{\partial\varphi^2} - \frac{4}{\tan\varphi}\frac{\partial}{\partial\varphi} + \frac{1}{\sin^2\varphi}C_2[SO(5)] \tag{8.158}$$

Note the close analogy with the expressions of corresponding operators in the SO(4) limit of the vibron model ($\hat{N}$ and $\mathcal{C}_2[\mathrm{SO}(4)] = \hat{L}^2 + \hat{D}^2$) discussed in Section 5.5. The differential equations (8.141) are separable in the variables ρ, φ, γ, and θ_i, and consequently the SO(6) states can be written in the factored form

$$\Psi_{N\sigma v n_\Delta L M_L}(\rho,\varphi,\gamma,\theta_i) = A_{N\sigma v} f_\sigma^N(\rho) g_v^\sigma(\varphi) \chi_{n_\Delta L M_L}^v(\gamma,\theta_i) \qquad (8.159)$$

where $A_{N\sigma v}$ is a normalization coefficient. From the explicit expressions of the Casimir invariants and their eigenvalues we find the equations

$$\left(-\frac{d^2}{d\varphi^2} - \frac{4}{\tan\varphi}\frac{d}{d\varphi} + \frac{v(v+3)}{\sin^2\varphi}\right) g_v^\sigma(\varphi) = \sigma(\sigma+2) g_v^\sigma(\varphi)$$

$$\frac{1}{2}\left[-\frac{1}{\rho^5}\frac{d}{d\rho}\left(\rho^5 \frac{d}{d\rho}\right) + \rho^2 + \frac{\sigma(\sigma+4)}{\rho^2} - 6\right] f_\sigma^N(\rho) = N f_\sigma^N(\rho)$$

$$(8.160)$$

After a substitution $f_\sigma^N(\rho) = \rho^\sigma \exp(-\rho^2/2) h(\rho^2)$ the second of these is reduced to a Laguerre equation [31] (with $x = \rho^2$),

$$\left[x\frac{d^2}{dx^2} + (\sigma+3-x)\frac{d}{dx} + \frac{N-\sigma}{2}\right] h(x) = 0 \qquad (8.161)$$

while the first can be transformed to the Gegenbauer equation [31] through the transformation $g_v^\sigma(\varphi) = (\sin\varphi)^v p(\cos\varphi)$:

$$\left[(1-y^2)\frac{d^2}{dy^2} - (2v+5)y\frac{d}{dy} + (\sigma-v)(\sigma+v+4)\right] p(y) = 0 \qquad (8.162)$$

where $y = \cos\varphi$. The SO(6) wave functions can thus be written as

$$\begin{aligned}\Psi_{N\sigma v n_\Delta L M_L}&(\rho,\varphi,\gamma,\theta_i) \\ &= A_{N\sigma v}\, \rho^\sigma e^{-\rho^2/2} L_{(N-\sigma)/2}^{\sigma+2}(\rho^2)(\sin\varphi)^v \\ &\quad \times C_{\sigma-v}^{v+2}(\cos\varphi)\chi_{n_\Delta L M_L}^v(\gamma,\theta_i)\end{aligned} \qquad (8.163)$$

where we assume the functions $\chi_{n_\Delta L M_L}^v(\gamma,\theta_i)$ to be properly orthonormalized. We emphasize again the similarity between (8.163) and the SO(4) form (5.149), suggesting that the result can be generalized to arbitrary dimensions. The coefficients $A_{N\sigma v}$ are determined by the same

8.5 γ-unstable nuclei: The SO(6) limit

normalization condition as in Section 5.5 with the appropriate substitutions $L \to v+1$, $\omega \to \sigma+1$, and $N \to N+1$, from which we find the result

$$A_{N\sigma v} = 2^{v+2}(v+1)! \left[\frac{2^{\sigma+2}(\sigma+2)(N-\sigma)!!(\sigma-v)!}{\pi(N+\sigma+4)!!(\sigma+v+3)!}\right]^{1/2} \quad (8.164)$$

We can now return to the boson representation by applying Dragt's theorem (see Appendix C) to the highest-weight state

$$\Psi_{\sigma\sigma v n_\Delta L M_L}(\rho,\varphi,\gamma,\theta_i) = A_{\sigma\sigma v}\rho^\sigma e^{-\rho^2/2}(\sin\varphi)^v C_{\sigma-v}^{v+2}(\cos\varphi)\chi^v_{n_\Delta L M_L}(\gamma,\theta_i) \quad (8.165)$$

which in this case implies the replacements

$$\alpha_m \to \frac{d^\dagger_m}{\sqrt{2}}, \quad \bar{x} \to \frac{s^\dagger}{\sqrt{2}}, \quad \frac{e^{-\rho^2/2}}{\pi^{3/2}} \to |0\rangle \quad (8.166)$$

or in hyperspherical coordinates,

$$\begin{aligned}\rho &\to \left(\frac{d^\dagger \cdot d^\dagger + s^\dagger s^\dagger}{2}\right)^{1/2} \\ \cos\varphi &\to \frac{s^\dagger}{\sqrt{d^\dagger \cdot d^\dagger + s^\dagger s^\dagger}} \\ \sin\varphi &\to \left(\frac{d^\dagger \cdot d^\dagger}{d^\dagger \cdot d^\dagger + s^\dagger s^\dagger}\right)^{1/2}\end{aligned} \quad (8.167)$$

The transformation for γ and θ_i has a complicated form which we fortunately do not need, as we have already determined the boson form for $\chi^v_{n_\Delta LL}(\gamma,\theta_i)$ in the previous section. (The restriction to $M_L = L$ is of no consequence as explained before.) Using again Dragt's theorem, but now for the U(5) ⊃ SO(5) ⊃ SO(3) states and in the opposite direction, the maximum-seniority states $|v n_\Delta L\rangle$ of (8.97) may be translated into the wave functions

$$\Phi_{n_d = v\, v n_\Delta L M_L}(\alpha_m) \propto \sum_{rn} C^{\sigma\tau n_\Delta}_{rn}(\alpha_{+2})^{\sigma+\tau-n}\{3,3\}^{\delta_L}\{2,0\}^{3r-\tau+n}$$
$$\times \{3,0\}^{n_\Delta+\tau-2r-n}\{2,2\}^n e^{-\beta^2/2} \quad (8.168)$$

through the replacements

$$d_m^\dagger \to \sqrt{2}\alpha_m, \qquad |0\rangle \to \frac{e^{-\beta^2/2}}{\pi^{5/4}} \tag{8.169}$$

with β^2 given by (8.147). These states are not normalized. The $\{n_d, L\}$ are defined through the polynomials (8.64) and (8.75) with $d_m^\dagger$ substituted by α_m. On the other hand, from (8.148) and (8.149) for $n_d = v$, we find

$$\Phi_{n_d=v\,vn_\Delta LL}(\alpha_m) = \left[\frac{2}{\Gamma(v+5/2)}\right]^{1/2} \beta^v e^{-\beta^2/2} \chi^v_{n_\Delta LL}(\gamma,\theta_i) \tag{8.170}$$

Comparison of the two expressions for $\Phi_{n_d=v\,vn_\Delta LL}(\alpha_m)$ determines the functions $\chi^v_{n_\Delta LL}(\gamma,\theta_i)$ in terms of the α_m. In the application of Dragt's theorem to the SO(6) state $\Psi_{\sigma\sigma vn_\Delta LM_L}(\rho,\varphi,\gamma,\theta_i)$ we can thus make the direct replacement

$$\chi^v_{n_\Delta LL}(\gamma,\theta_i) \to \left[\frac{2}{\Gamma(v+5/2)}\right]^{1/2} \pi^{5/4} 2^{-v/2} (d^\dagger \cdot d^\dagger)^{-v/2} |vn_\Delta L\rangle \tag{8.171}$$

with $|vn_\Delta L\rangle$ given by (8.97). The boson realization of the highest-weight U(6) $\supset$ SO(6) $\supset$ SO(5) $\supset$ SO(3) state is then found to be

$$\boxed{|[\sigma]\sigma vn_\Delta LL\rangle = \frac{\pi^{3/2} A_{\sigma\sigma v}}{2^{v/2}}(d^\dagger \cdot d^\dagger + s^\dagger s^\dagger)^{(\sigma-v)/2} C^{v+2}_{\sigma-v}(t^\dagger)|vn_\Delta L\rangle} \tag{8.172}$$

where the notation

$$t^\dagger = \frac{s^\dagger}{\sqrt{d^\dagger \cdot d^\dagger + s^\dagger s^\dagger}} \tag{8.173}$$

is used. Note that the $|vn_\Delta L\rangle$ in (8.172) is the same boson polynomial as in (8.97) but in this case acting on the six-dimensional vacuum state $|0\rangle$.

The final step is to generalize the expression (8.172) for the highest-weight states to arbitrary SO(6) representations. This is accomplished by means of (8.50), leading to the final result

$$\boxed{|[N]\sigma vn_\Delta LL\rangle = B_{N\sigma}(d^\dagger \cdot d^\dagger + s^\dagger s^\dagger)^{(N-\sigma)/2}|[\sigma]\sigma vn_\Delta LL\rangle} \tag{8.174}$$

8.5 γ-unstable nuclei: The SO(6) limit

where $B_{N\sigma}$ is a normalization coefficient which is determined from repeated application of (8.54). We find the result

$$B_{N\sigma} = (-)^{(N-\sigma)/2} \left[\frac{(\sigma+2)!}{2^{N-\sigma}((N-\sigma)/2)!((N+\sigma)/2+2)!} \right]^{1/2} \quad (8.175)$$

The factor $(-)^{(N-\sigma)/2}$ is chosen so that the phases of the states (8.163) and (8.174) coincide [15].

One may now use the explicit expressions of the SO(6) states to derive matrix elements of the $s^\dagger$ operator in the same way as for the SO(4) states in Chapter 5, making use of recurrence relations in the Gegenbauer polynomials. For operators involving d bosons, however, closed expressions are only obtainable for particular states, given the complexity of the states $|vn_\Delta L\rangle$ of (8.97). Instead of proceeding further in this direction, as a last point in this section we derive the transformation brackets $B_{n_d}^{N\sigma v}$ between the SO(6) and U(5) bases.

Since $|vn_\Delta L\rangle$ is common to the two bases, we can select, for the purpose of determining $B_{n_d}^{N\sigma v}$, the states $|v0\, 2v\rangle \propto (d_{+2}^\dagger)^v |0\rangle$ in both the U(5) and SO(6) bases. From (8.60) and the application of Dragt's theorem we find, for U(5) states,

$$|[N]vv0\, 2v\, 2v\rangle = \left[\frac{2}{\Gamma(v+5/2)} \right]^{1/2} \pi^{5/4} 2^{-v/2} \frac{(s^\dagger)^{N-v}(d_{+2}^\dagger)^v}{\sqrt{(N-v)!}} \quad (8.176)$$

The more general U(5) states $|[N]n_d v0\, 2v\, 2v\rangle$ are determined through (8.50) to be

$$\begin{aligned}
&|[N]n_d v0\, 2v\, 2v\rangle \\
&= (-)^{(n_d-v)/2} \left[\frac{((n_d+v)/2+1)!(2v+3)!}{(N-n_d)!((n_d-v)/2)!(v+1)!(n_d+v+3)!} \right]^{1/2} \\
&\quad \times \left[\frac{2}{\Gamma(v+5/2)} \right]^{1/2} \pi^{5/4} 2^{-v/2} (s^\dagger)^{N-n_d} (d^\dagger \cdot d^\dagger)^{(n_d-v)/2} (d_{+2}^\dagger)^v |0\rangle
\end{aligned} \quad (8.177)$$

where the normalization coefficient is obtained by repeated application of (8.54). The factor $(-)^{(n_d-v)/2}$ is such that the phase convention coincides with that of (8.148). On the other hand, from (8.172) and (8.174), we find the SO(6) states

$$|[N]\sigma v 0\, 2v\, 2v\rangle$$
$$= B_{N\sigma} A_{\sigma\sigma v} \frac{\pi^{3/2} 2^{\sigma/2-v}}{\sqrt{v!}(v+1)!} (d^\dagger_{+2})^v$$
$$\times \sum_s \frac{(-)^s(\sigma+1-s)!}{2^{2s}(\sigma-v-2s)!s!} (s^\dagger)^{\sigma-v-2s} (d^\dagger \cdot d^\dagger + s^\dagger s^\dagger)^{(N-\sigma)/2+s} |0\rangle$$
(8.178)

where we have used the power series expansion (5.179) of the Gegenbauer polynomials [31]. We remark that the above expressions for $|[N]n_\mathrm{d} v 0\, 2v\, 2v\rangle$ and $|[N]\sigma v 0\, 2v\, 2v\rangle$ are not normalized. They lack a *common* normalization coefficient due to the non-orthonormality of the $\chi^v_{0\,2v\,2v}(\gamma,\theta_i)$ functions [15]. To find the transformation brackets between the two bases, it is not necessary to include it.

The last step in the analysis consists of a binomial expansion of $(d^\dagger \cdot d^\dagger + s^\dagger s^\dagger)^{(N-\sigma)/2+s}$ and an interchange of sums, in close analogy with the calculation of Section 5.6. We find in the end that the SO(6) states are given by

$$|[N]\sigma v 0\, 2v\, 2v\rangle = \sum_{n_\mathrm{d}} B^{N\sigma v}_{n_\mathrm{d}} |[N]n_\mathrm{d} v 0\, 2v\, 2v\rangle \qquad (8.179)$$

where summation is restricted to even values of $n_\mathrm{d} - v$ and with

$$B^{N\sigma v}_{n_\mathrm{d}} = (-)^{(N-\sigma+n_\mathrm{d}-v)/2} \left[\frac{2^{-N+2\sigma-v}(N-n_\mathrm{d})!}{((N-\sigma)/2)!((N+\sigma)/2+2)!} \right]^{1/2}$$
$$\times \left[\frac{(\sigma+2)!(\sigma-v)!(n_\mathrm{d}+v+3)!}{(\sigma+1)!(\sigma+v+3)!((n_\mathrm{d}-v)/2)!((n_\mathrm{d}+v)/2+1)!} \right]^{1/2}$$
$$\times \sum_s \frac{(-)^s((N-\sigma)/2+s)!(\sigma+1-s)!}{2^{2s} s!((N-\sigma)/2+s)!(\sigma-v-2s)!} \qquad (8.180)$$

Since the coefficients $B^{N\sigma v}_{n_\mathrm{d}}$ do not depend on the labels n_Δ, L, and M_L, this result gives the general expression for the transformation brackets. All matrix elements of interest in the SO(6) limit can now be obtained by transforming to the U(5) basis.

In the next section we use the results presented so far for the description of the collective states of the nucleus $^{196}_{78}\mathrm{Pt}_{118}$.

8.6 Example: The nucleus ^{196}Pt

In the previous section, Figure 8.3, we chose the spectrum of $^{196}_{78}$Pt$_{118}$ as an illustration of an SO(6)-like nucleus. Although a one-to-one correspondence is found between the experimental and theoretical low-energy levels of this nucleus and, furthermore, the states seem to follow the typical SO(6) pattern [8,12] of several of $L^\pi = 0^+, 2^+, 2^+$ sequences, it is clear that a more stringent test for such a symmetry must arise from a comparison of electromagnetic transition rates. We now illustrate how these calculations are carried out in the SO(6) limit, by considering the evaluation of E2 transition rates. One possible approach is to use the explicit form of the SO(6) eigenstates, either in coordinate space, (8.163), or in the boson representation (8.179). We shall, however, follow the alternative route provided by the knowledge of the SO(6)–U(5) transformation brackets discussed in the previous section, since the matrix elements of interest in the U(5) limit can be evaluated using techniques illustrated in Section 8.4. We consider the case of maximum-weight states $N = \sigma$ in (8.179). In this case the sum in $B^{\sigma\sigma v}_{n_d}$ can be explicitly evaluated,

$$\sum_s \frac{(-)^s(\sigma + 1 - s)!}{2^{2s}(s - ((n_d - v)/2)!(\sigma - v - 2s)!}$$
$$= (-)^{(n_d - v)/2} \frac{(\sigma + v + 3)!((n_d + v)/2 + 1)!}{2^{\sigma - v}(\sigma - n_d)!(n_d + v + 3)!} \quad (8.181)$$

which leads to the closed form

$$B^{\sigma\sigma v}_{n_d} = (-)^{(n_d - v)/2}$$
$$\times \left[\frac{(\sigma + v + 3)!(\sigma - v)!((n_d + v)/2 + 1)!}{2^{\sigma - v}(\sigma + 1)!((n_d - v)/2)!(\sigma - n_d)!(n_d + v + 3)!} \right]^{1/2}$$
$$(8.182)$$

Using the SO(6) generator as the quadrupole operator,

$$\hat{T}^{(\text{E2})}_\mu = q\, i[d^\dagger \times \tilde{s} - s^\dagger \times \tilde{d}]^{(2)}_\mu \quad (8.183)$$

we now compute some $B(\text{E2})$ values. From (8.138) we find

$$\langle [\sigma']\sigma' v' n'_\Delta L' \| \hat{T}^{(\text{E2})} \| [\sigma]\sigma v n_\Delta L \rangle$$

$$= \sum_{n_d n'_d} B^{\sigma'\sigma'v'}_{n'_d} B^{\sigma\sigma v}_{n_d} \langle [\sigma'] n'_d v' n'_\Delta L' \| \hat{T}^{(E2)} \| [\sigma] n_d v n_\Delta L \rangle \quad (8.184)$$

where we use the SO(3)-reduced matrix elements in order to compute the $B(E2)$ values from (8.104). We may further simplify this equation by noting that the form (8.184) implies a number of selection rules. First, since $\hat{T}^{(E2)}_\mu$ is an SO(6) generator, the SO(6) quantum number σ cannot change; thus σ' must equal σ to have a non-zero result. We next note that the form of the E2 transition operator ensures that $\Delta n_d = \pm 1$. From (8.179) and the tensor properties of the $d^\dagger_m$ and $\tilde{d}_m$ operators, we find the additional selection rule $\Delta v = \pm 1$. To see the last point, consider the case of $d^\dagger_m$ and the $U(5) \supset SO(5) \supset SO(3)$ states of Section 8.4. Clearly

$$d^\dagger_m |0\rangle = |n_d = 1\ v = 1\ n_\Delta = 0\ L = 2\ M_L = m\rangle \quad (8.185)$$

which implies that $d^\dagger_m$ transforms as a tensor $\hat{T}^{n_d}_{vn_\Delta L m} = \hat{T}^1_{102m}$ in the same basis. The selection rule $\Delta v = \pm 1$ follows directly from the SO(5) direct product [32]

$$(1) \otimes (v) = (v-1) \oplus (v+1) \quad (8.186)$$

We do not attempt to give a general discussion of products of SO(5) representations, but rather verify it for the case $v = 1$. From (8.100) we find the possible $n_d = 2$ states:

$$\begin{aligned}|20000\rangle &= \alpha(2,0)|0\rangle = \alpha(d^\dagger \cdot d^\dagger)|0\rangle \\ |2202m\rangle &= \beta[d^\dagger \times d^\dagger]^{(2)}_m |0\rangle \\ |2204m\rangle &= \gamma[d^\dagger \times d^\dagger]^{(4)}_m |0\rangle \end{aligned} \quad (8.187)$$

where α, β, and γ are normalization coefficients that we need not worry about. The action of $d^\dagger_m$ on the $n_d = 1$ state gives, on the other hand,

$$d^\dagger_m |1102m'\rangle = d^\dagger_m d^\dagger_{m'} |0\rangle = \sum_L \langle 22\ mm' | L M_L \rangle [d^\dagger \times d^\dagger]^{(L)}_{M_L} \quad (8.188)$$

which is a linear combination of the three states in (8.187). Thus

$$\hat{T}^1_{102m} |1102m'\rangle = \alpha' |20000\rangle + \beta' |2202\ m+m'\rangle + \gamma' |2204\ m+m'\rangle \quad (8.189)$$

8.6 Example: The nucleus ^{196}Pt

and the SO(5) product $(1)\oplus(1) = (0)\oplus(2)$ is indeed satisfied in (8.188).

Returning to (8.184) and using the selection rule (8.186), we find, for the case $v' = v+1$,

$$\langle [\sigma]\sigma\, v+1\, n'_\Delta L' \| i[d^\dagger \times \tilde{s} - s^\dagger \times \tilde{d}]^{(2)} \| [\sigma]\sigma v n_\Delta L\rangle$$
$$= i\sum_{n_d} B^{\sigma\sigma\, v+1}_{n_d+1} \left(B^{\sigma\sigma v}_{n_d} \langle [\sigma]n_d+1\, v+1\, n'_\Delta L' \| [d^\dagger \times \tilde{s}]^{(2)} \| [\sigma]n_d v n_\Delta L\rangle \right.$$
$$\left. - B^{\sigma\sigma v}_{n_d+2} \langle [\sigma]n_d+1\, v+1\, n'_\Delta L' \| [s^\dagger \times \tilde{d}]^{(2)} \| [\sigma]n_d+2\, v n_\Delta L\rangle \right)$$
(8.190)

The action of both $s^\dagger$ and $\tilde{s}$ on the right-hand side in this equation is given in Section 8.4,

$$s^\dagger |[\sigma]n_d+2\, v n_\Delta L M_L\rangle = \sqrt{\sigma - n_d - 1}|[\sigma+1]n_d+2\, v n_\Delta L M_L\rangle$$
$$\tilde{s}|[\sigma]n_d v n_\Delta L M_L\rangle = \sqrt{\sigma - n_d}|[\sigma-1]n_d v n_\Delta L M_L\rangle$$
(8.191)

while the d matrix elements may be reduced using the formulas

$$\langle n_d+1\, v+1\, n'_\Delta L'L'|d^\dagger_{+2}|n_d v n_\Delta LL\rangle$$
$$= \left[\frac{n_d+v+5}{2v+5}\right]^{1/2} \langle v+1\, v+1\, n'_\Delta L'L'|d^\dagger_{+2}|v v n_\Delta LL\rangle$$
$$\langle n_d+1\, v+1\, n'_\Delta L'L'|\tilde{d}_{+2}|n_d+2\, v n_\Delta LL\rangle$$
$$= \left[\frac{n_d-v+2}{2}\right]^{1/2} \langle v+1\, v+1\, n'_\Delta L'L'|\tilde{d}_{+2}|v+2\, v n_\Delta LL\rangle$$
(8.192)

Combining (8.191) and (8.192), we can rewrite (8.190) in the form

$$\langle [\sigma]\sigma\, v+1\, n'_\Delta L' \| i[d^\dagger \times \tilde{s} - s^\dagger \times \tilde{d}]^{(2)} \| [\sigma]\sigma v n_\Delta L\rangle$$
$$= i\sum_{n_d} B^{\sigma\sigma\, v+1}_{n_d+1} \left(\left[\frac{(\sigma-n_d)(n_d+v+5)}{2v+5}\right]^{1/2} a_{L'L}(v) B^{\sigma\sigma v}_{n_d} \right.$$
$$\left. - \left[\frac{(\sigma-n_d-1)(n_d-v+2)}{2}\right]^{1/2} b_{L'L}(v) B^{\sigma\sigma v}_{n_d+2} \right)$$
(8.193)

where

$$a_{L'L}(v) = \langle v+1\, v+1\, n'_\Delta L' \| d^\dagger \| v v n_\Delta L\rangle$$
$$b_{L'L}(v) = \langle v+1\, v+1\, n'_\Delta L' \| d^\dagger \| v+2\, v n_\Delta L\rangle$$
(8.194)

and the B coefficients are given in (8.182). Using the latter formula, (8.193) can be rewritten in the simpler form

$$\langle [\sigma]\sigma\, v+1\, n'_\Delta L' \| i[d^\dagger \times \tilde{s} - s^\dagger \times \tilde{d}]^{(2)} \| [\sigma]\sigma v n_\Delta L \rangle$$
$$= i\left[\frac{(\sigma+v+4)}{(\sigma-v)(2v+5)}\right]^{1/2} a_{L'L}(v) \sum_{n_d=\text{even}} (\sigma - n_d)\left(B^{\sigma\sigma v}_{n_d}\right)^2$$
$$+ i\left[\frac{\sigma-v}{2(\sigma+v+4)}\right]^{1/2} b_{L'L}(v) \sum_{n_d=\text{odd}} (\sigma - n_d)\left(B^{\sigma\sigma\, v+1}_{n_d}\right)^2 \quad (8.195)$$

Equation (8.195) can be used to compute these matrix elements once the U(5) $\supset$ SO(5) ones in (8.194) are known. These may be determined by the methods discussed in Section 8.4. Arima and Iachello [17] give explicit forms for these reduced matrix elements in a number of cases of interest. Furthermore, they are able to carry out the sums in (8.195) by substituting them by their integral representations. Using this technique, (8.195) reduces to

$$\langle [\sigma]\sigma\, v+1\, n'_\Delta L' \| i[d^\dagger \times \tilde{s} - s^\dagger \times \tilde{d}]^{(2)} \| [\sigma]\sigma v n_\Delta L \rangle$$
$$= i\left[\frac{(\sigma-v)(\sigma+v+4)}{4(\sigma+1)^2}\right]^{1/2}$$
$$\times \left[(\sigma+v+3)\frac{a_{L'L}(v)}{\sqrt{2v+5}} + (\sigma-v-1)\frac{b_{L'L}(v)}{\sqrt{2}}\right] \quad (8.196)$$

In Section 8.4 particular cases of the reduced matrix elements (8.194) are evaluated. Recalling the definition of the reduced matrix element given in Section 1.7 and canceling the s-boson matrix elements $\sqrt{N-v}$, we find, from (8.106) and (8.113),

$$a_{2v+2\,2v}(v) = \sqrt{v+1}, \qquad b_{2v+2\,2v}(v) = \left[\frac{2(v+1)}{2v+5}\right]^{1/2} \quad (8.197)$$

which lead to the $B(E2)$ values

$$B(E2; [\sigma]\sigma\, v+1\, 0\, 2v+2 \to [\sigma]\sigma v 0\, 2v)$$
$$= q^2 \frac{v+1}{2v+5}(\sigma-v)(\sigma+v+4) \quad (8.198)$$

8.6 Example: The nucleus ^{196}Pt

To arrive at this result, initial and final states in (8.196) have been interchanged, that is,

$$\langle L \| \hat{T}^{(E2)} \| L+2 \rangle^2 = \frac{2L+5}{2L+1} \langle L+2 \| \hat{T}^{(E2)} \| L \rangle^2 \qquad (8.199)$$

a relation obtained from the definition of the reduced matrix element and the Clebsch–Gordan coefficient [20]

$$\langle L+2\,L+2\,22|LL\rangle = \left[\frac{2L+1}{2L+5}\right]^{1/2} \qquad (8.200)$$

Replacing L by $2v$ we can rewrite (8.198) as

$$B(\text{E2}; [\sigma]\sigma\,v+1\,0\,2v+2 \to [\sigma]\sigma v 0\,2v)$$
$$= q^2 \frac{L+2}{8(L+5)}(2\sigma - L)(2\sigma + L + 8) \qquad (8.201)$$

Many other transition matrix elements in the SO(6) limit can be computed in closed form using the $a_{L'L}(v)$ and $b_{L'L}(v)$ coefficients of [17], and all of them may be obtained numerically from the U(5) matrix elements by means of (8.190). The $B(\text{E2})$ values of transitions between some low-energy levels in the SO(6) limit are shown in Figure 8.4. A characteristic feature of this limit is that the $4_1^+ \to 2_1^+$ and $2_2^+ \to 2_1^+$ transitions are predicted to have the same E2 strength. This prediction is independent of the value of the parameter q and hence constitutes a good test of the SO(6) symmetry.

In Table 8.4 we present a comparison between the measured and the calculated $B(\text{E2})$ values in the nucleus $^{196}_{78}\text{Pt}_{118}$. It is seen that all calculated transition rates fall within the experimental errors. Furthermore, the $2_2^+ \to 0_1^+$ transition, which is forbidden in SO(6), is observed to be very weak indeed. Nevertheless, the SO(6) description of $^{196}_{78}\text{Pt}_{118}$ has severe limitations: $\Delta v = 0$ E2 transitions are strictly forbidden in this limit, and as a result, all states are predicted to have vanishing quadrupole moments. This is at variance with the spectroscopic observations in $^{196}_{78}\text{Pt}_{118}$ and suggests that a more general E2 transition operator is needed in this nucleus [33] or—more likely—that SO(6)-breaking terms must be included in the hamiltonian.

In the last section of this chapter we analyze the third chain (8.26), the SU(3) limit of the IBM.

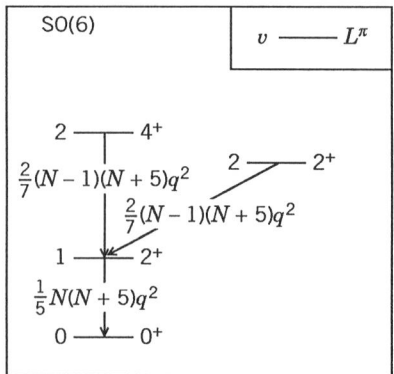

Figure 8.4: Allowed transitions in the SO(6) limit induced by the quadrupole generator $qi[d^\dagger \times \tilde{s} - s^\dagger \times \tilde{d}]^{(2)}_\mu$. The expressions alongside the arrows are the $B(E2)$ values. Labels are as in Figure 8.3.

8.7 Rotational nuclei: The SU(3) limit

The SU(3) limit of the IBM corresponds to taking $\epsilon = \alpha = \eta_3 = \beta = \zeta = 0$ in the hamiltonian (8.40),

$$\hat{H}_{\text{III}} = E'_0 + \xi \mathcal{C}_2[\text{SU}(3)] + \gamma \mathcal{C}_2[\text{SO}(3)] \tag{8.202}$$

or, using the explicit expression for the Casimir invariants,

$$\hat{H}_{\text{III}} = E'_0 + 2\xi \hat{Q}^2 + \left(\tfrac{3}{4}\xi + \gamma\right) \hat{L}^2 \tag{8.203}$$

To find the eigenvalue associated with $\mathcal{C}_2[\text{SU}(3)]$ (or $\hat{Q}^2$), we first consider the U(3) Casimir invariant $\mathcal{C}_2[\text{U}(3)]$ which in terms of the uncoupled generators $\mathcal{G}^j_i$ reads

$$\mathcal{C}_2[\text{U}(3)] = \sum_{ij=1}^{3} \mathcal{G}^j_i \mathcal{G}^i_j \tag{8.204}$$

By an analysis equivalent to the one carried out in Section 2.3 the U(3) irreducible representations are found to be characterized by the Young partition

$$h \geq h' \geq h'' \geq 0, \qquad h + h' + h'' = n \tag{8.205}$$

8.7 Rotational nuclei: The SU(3) limit

Table 8.4: $B(E2)$ values in ^{196}Pt

$E_i{}^a$	$(\sigma,v,L)_i$	$E_f{}^a$	$(\sigma,v,L)_f$	$B(E2; L_i \to L_f)^b$	
				Experiment	Theory^c
356	6,1,2	0	6,0,0	0.276(14)	0.276
689	6,2,2	0	6,0,0	$\leq 2 \times 10^{-6}$	0
689	6,2,2	356	6,1,2	0.340(30)	0.362
877	6,2,4	356	6,1,2	0.380(30)	0.362
1527	6,3,6	877	6,2,4	0.400(110)	0.368

^{a} In units of keV.
^{b} In units of e^{2}b^2.
^{c} With $q = 0.152$ eb.

of the number n. (See also Appendix B.) Recalling that this U(3) algebra should be a subalgebra of the IBM's U(6) dynamical algebra, we now relate n to the U(6) label N. Since n should be the eigenvalue of the number operator in U(3), we compute its value using the second-quantized form

$$\hat{n} = \sum_{lm}\sum_{l'm'} \langle 2lm|\hat{H}_0 - \tfrac{3}{2}|2l'm'\rangle b^\dagger_{lm} b_{l'm'} \qquad (8.206)$$

where $\hat{H}_0$ is the hamiltonian of a three-dimensional oscillator, $\hat{H}_0 = \hat{n} + 3/2$, where $\hat{n}$ is the corresponding number operator. Equation (8.206) is analogous to (8.23), used to find the expression for $\hat{Q}_\mu$. Since the matrix elements in (8.206) are diagonal and $\langle 2lm|\hat{H}_0 - 3/2|2l'm'\rangle = 2\delta_{ll'}\delta_{mm'}$, we find

$$\hat{n} = 2\sum_{lm} b^\dagger_{lm} b_{lm} = 2\hat{N} \qquad (8.207)$$

Equation (8.205) now reads

$$h \geq h' \geq h'' \geq 0, \qquad h + h' + h'' = 2N \qquad (8.208)$$

This shows that the fundamental representation $[1,0,0,0,0,0] \equiv [1]$ of U(6) contains the representation $[2,0,0]$ of U(3), that is,

$$\begin{array}{cc} \text{U}(6) \supset & \text{U}(3) \\ | & | \\ [1] & [2,0,0] \end{array} \qquad (8.209)$$

Although the partition $[1,1,0]$ of U(3) also satisfies (8.208) for $N = 1$, this possibility is eliminated by the fact that the $N = 1$ U(6) representation (generated by s and d bosons) should contain $L = 0$ and $L = 2$ SO(3) representations. Indeed, we have that

$$\begin{array}{ccc} \text{U}(3) & \supset & \text{SO}(3) \\ | & & | \\ [2,0,0] & & 0,2 \end{array} \qquad (8.210)$$

while the U(3) representation $[1,1,0]$ only contains the angular momentum $L = 1$ (see Appendix B). The reduction (8.210) is a general result, independent of the particular realization of the U(3) representation. We may thus generate a U(3) algebra in terms of p bosons as in the vibron model of Chapter 5, where $[2,0,0]$ corresponds to $n_p = 2$. (All states in the vibron model correspond to U(3) representations of the form $[n_p, 0, 0]$.) From (5.54) we find that (8.210) is indeed the correct reduction, as can be also verified by means of the general results of Appendix B. Equations (8.209) and (8.210) give the fundamental branching rules in the chain $\text{U}(6) \supset \text{U}(3) \supset \text{SO}(3)$.

The U(3) second-order Casimir invariant has the eigenvalue (see Appendix B)

$$h(h+2) + h'^2 + h''(h''-2) \qquad (8.211)$$

It is possible to eliminate the number operator from U(3) and thus arrive at its SU(3) subalgebra. This is achieved by defining the SU(3) generators

$$\tilde{\mathcal{G}}_i^j = \mathcal{G}_i^j - \tfrac{1}{3}\left(\sum_{i=1}^{3} \mathcal{G}_i^i\right)\delta_{ij} = \mathcal{G}_i^j - \tfrac{1}{3}\hat{n}\delta_{ij} = \mathcal{G}_i^j - \tfrac{2}{3}\hat{N}\delta_{ij} \qquad (8.212)$$

where use is made of (8.207). In this way we find $\sum_i \tilde{\mathcal{G}}_i^i = 0$ and hence

$$\begin{aligned} \mathcal{C}_2[\text{U}(3)] &= \sum_{ij=1}^{3} \left(\tilde{\mathcal{G}}_i^j + \tfrac{2}{3}\hat{N}\delta_{ij}\right)\left(\tilde{\mathcal{G}}_j^i + \tfrac{2}{3}\hat{N}\delta_{ij}\right) \\ &= \sum_{ij=1}^{3} \tilde{\mathcal{G}}_i^j \tilde{\mathcal{G}}_j^i + \tfrac{4}{3}\hat{N}^2 \equiv \tfrac{2}{3}\mathcal{C}_2[\text{SU}(3)] + \tfrac{4}{3}\hat{N}^2 \end{aligned} \qquad (8.213)$$

Introducing (8.208) and defining the SU(3) labels (following Elliott [21])

$$\lambda \equiv h - h', \qquad \mu \equiv h' - h'' \qquad (8.214)$$

8.7 Rotational nuclei: The SU(3) limit

we find from (8.211) and (8.213) the $\mathcal{C}_2[\mathrm{SU}(3)]$ eigenvalue

$$\lambda^2 + \mu^2 + \lambda\mu + 3(\lambda + \mu) \tag{8.215}$$

Note that in (8.213) a factor 2/3 is introduced for convenience, so that we have defined

$$\mathcal{C}_2[\mathrm{SU}(3)] \equiv \tfrac{3}{2} \sum_{ij=1}^{3} \tilde{\mathcal{G}}_i^j \tilde{\mathcal{G}}_j^i \tag{8.216}$$

Using the explicit expression for $\mathcal{C}_2[\mathrm{SU}(3)]$, we also obtain from (8.215) the $\hat{Q}^2$ eigenvalue

$$\tfrac{1}{2}[\lambda^2 + \mu^2 + \lambda\mu + 3(\lambda + \mu)] - \tfrac{3}{8}L(L+1) \tag{8.217}$$

From this analysis we can return to either (8.202) or (8.203) and obtain the eigenvalue of the SU(3) limit of the IBM:

$$E_{\mathrm{III}}(\lambda, \mu, L) = E_0' + \xi[\lambda^2 + \mu^2 + \lambda\mu + 3(\lambda + \mu)] + \gamma L(L+1) \tag{8.218}$$

To solve completely the eigenvalue problem in this limit, we need to generalize reductions (8.209) and (8.210) to arbitrary values of the boson number N and (λ, μ), respectively. The U(6) ⊃ U(3) (or U(6) ⊃ SU(3)) reduction can be found using the building-up principle [28] (see also Section 6.4), which starts from the fundamental reduction ((8.209) in this case) to derive the reduction rules for more general representations. In order to deduce the latter, we need to know the rules for evaluating the product of representations in both U(6) and U(3) and to use dimension formulas. The product of symmetric U(2) representations is analyzed in Chapter 2, and the general rules for U(n) products can be found in Appendix B. Dimension formulas for Lie algebra representations are well known and can be found, for example, in [34]. We illustrate the procedure by considering the U(6) ⊃ U(3) reduction for $N = 2$. We first consider the U(6) product

$$\square \otimes \square = \square\square \oplus \begin{array}{c}\square\\\square\end{array} \tag{8.219}$$

where we use the Young tableau notation introduced in Section 2.3, which can be rewritten as

$$[1] \otimes [1] = [2] \oplus [1,1]$$
$$6 \times 6 = 21 + 15 \tag{8.220}$$

Clearly, in this simple case the fundamental representations on the left-hand side can only be coupled to symmetric [2] or antisymmetric [1,1] representations, which have dimensions 21 and 15, respectively, as indicated in (8.220). Due to (8.209) these U(6) representations contain the U(3) representations arising from the U(3) product

$$[2] \otimes [2] = [4] \oplus [3,1] \oplus [2,2]$$
$$6 \times 6 = 15 + 15 + 6$$
(8.221)

where we have again indicated the dimension below each representation. Since the symmetric U(3) representation must be contained in the symmetric U(6) representation, we find that the [2] U(6) representation contains the U(3) representations [4] and [2,2], while the [1,1] U(6) representation reduces to [3,1] in U(3). In terms of Elliott's SU(3) labels we thus find that the symmetric U(6) representation [2] contains $(4,0)$ and $(0,2)$, and the antisymmetric representation [1,1] contains $(2,1)$. We can continue this process and evaluate the U(6) product $[2] \otimes [1]$ and the corresponding products in U(3), that is, $[4] \otimes [2]$ and $[2,2] \otimes [2]$. Using dimensional arguments we then find that [3] of U(6) contains $(6,0)$, $(2,2)$, and $(0,0)$ of SU(3). From these successive multiplications the general rule emerges [17] that the symmetric representation $[N]$ of U(6) contains the following SU(3) representations:

$$(2N,0), (2N-4,2), \ldots, \left\{ \begin{array}{c} (0,N) \\ (2,N-1) \end{array} \right\},$$
$$(2N-6,0), (2N-10,2), \ldots, \left\{ \begin{array}{c} (0,N-3) \\ (2,N-4) \end{array} \right\},$$
$$(2N-12,0), (2N-16,2), \ldots, \left\{ \begin{array}{c} (0,N-6) \\ (2,N-7) \end{array} \right\},$$
$$\ldots$$
(8.222)

where the upper (lower) representations in the curly brackets must be taken for even (odd) values of N.

The reduction from SU(3) to SO(3) is given in Section 6.4 and can be found in a similar way by application of the building-up principle. This method is often useful to study algebra reductions and fairly simple to apply, as illustrated by these examples.

8.7 Rotational nuclei: The SU(3) limit

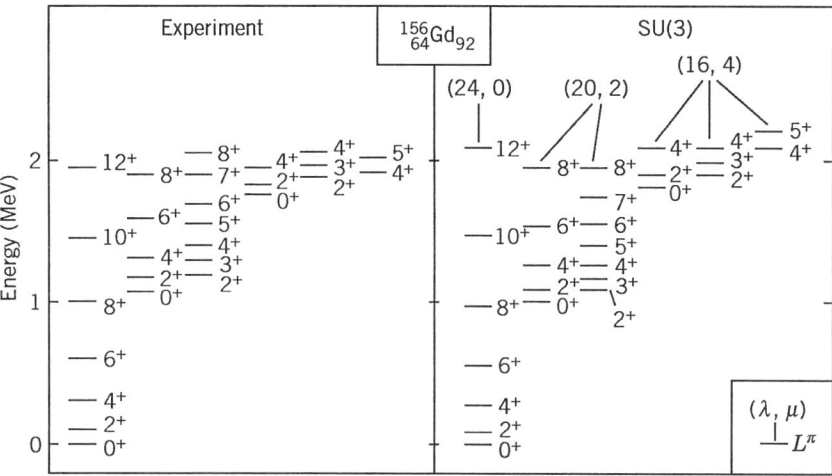

Figure 8.5: Partial SU(3) eigenspectrum of $\hat{H}_{\text{III}}$ for $N = 12$ bosons and its comparison to the observed levels of the nucleus ^{156}Gd. The calculated spectrum is obtained with the parameters (in keV) $\xi = -7$ and $\gamma = 13$, with the energy of the ground state normalized to zero. Levels are labeled by (λ, μ) and the angular momentum and parity L^π.

Armed with the branching rule (8.222) and the SU(3) $\supset$ SO(3) reduction, the energy formula (8.218) can be used to generate the spectrum associated with the SU(3) symmetry. An example is shown in Figure 8.5, where this spectrum is compared to the low-energy levels of $^{156}_{64}\text{Gd}_{92}$.

The SU(3) wave functions may be numerically obtained from the diagonalization of the $\hat{Q}^2$ operator, which requires the matrix elements of the (3,0) operator, as explained at the end of Section 8.4. The result of this diagonalization can be expressed in the form

$$|[N](\lambda,\mu)\tilde{K}LM_L\rangle = \sum_{n_\text{d} v \tilde{n}_\Delta} B^{N(\lambda,\mu)\tilde{K}L}_{n_\text{d} v \tilde{n}_\Delta} |[N]n_\text{d} v \tilde{n}_\Delta LM_L\rangle \qquad (8.223)$$

where the $B^{N(\lambda,\mu)\tilde{K}L}_{n_\text{d} v \tilde{n}_\Delta} = \langle [N]n_\text{d} v \tilde{n}_\Delta L | [N](\lambda,\mu)\tilde{K}L\rangle$ are transformation brackets from SU(3) to U(5). We place tildes on top of the indices n_Δ and K to indicate that prior to diagonalization one must construct appropriate linear combinations of the U(5) states in the n_Δ label to render the basis orthonormal, using, for example, a Gram–Schmidt procedure. The corresponding SU(3) states in (8.223) are likewise orthonormal linear combinations of the Elliott states in the K label.

The appearance of missing labels in both the U(5) and SU(3) wave functions explains the difficulty of deriving analytic expressions for the transformation brackets in this case. An exception is the $(2N, 0)$, $L = 0$ SU(3) ground state for which no multiplicity is present and which has the explicit form

$$\langle [N]n_\mathrm{d}v00|[N](2N,0)00\rangle$$
$$= (-)^{(n_\mathrm{d}+v)/2} \left[\frac{2^{n_\mathrm{d}+v+1}(2N+1)(2v+3)N!((n_\mathrm{d}+v)/2+1)!}{3^N(N-n_\mathrm{d})!((n_\mathrm{d}-v)/2)!(n_\mathrm{d}+v+3)!}\right]^{1/2}$$
(8.224)

derived from a numerical calculation by an induction procedure on N. Although analytic formulas are not known in general for the SU(3) states, B(E2) values and other spectroscopic properties can be computed explicitly [17], using previous results obtained by Elliott [21] and Vergados [35]. Matrix elements of the U(6) generators can also be evaluated in an SU(3) basis [36,37]. These calculations involve the use of SU(3) $\supset$ SO(3) isoscalar factors [28], which are defined in a similar way as for the U(2) and U(4) examples of the previous chapters.

This concludes our discussion of the group-theoretical structure of the basic version of the IBM. Refinements and extensions of the model can now be considered analogous to the analyses presented in Parts I and II. In the next two chapters we consider the two most important extensions that have been proposed: the explicit treatment of the neutron and proton degrees of freedom and the coupling of fermions to the system of interacting bosons.

Summary

In this chapter the algebraic structure of the interacting boson model of collective nuclear excitations is reviewed. The basic building blocks of this model are s and d bosons resulting in a U(6) algebra. Although alternative presentations of the model are possible, the one adopted in this chapter follows closely the expositions in Chapters 1 and 5. Apart from dimensionality, the equations (8.2) and (8.3), defining the U(6) algebra, are identical to those for U(2) and U(4). As it is the case in the vibron model, the hamiltonian of the interacting boson model is required to be rotationally invariant, and hence the coupled representation (8.7) of the generators is most useful. The algebraic structure of the model is now largely determined by the subalgebras of U(6) that contain the angular momentum algebra SO(3). Three different algebraic reductions are shown to exist, each corresponding to a symmetry limit: the U(5), SO(6), and SU(3) limits. The corresponding classifications are given in (8.20), (8.21), and (8.26). In Section 8.3 the most general rotationally invariant boson hamiltonian is written as (8.28) and rewritten in (8.40) in terms of Casimir invariants of the various subalgebras of U(6). The latter form is the appropriate one for studying the symmetries of the hamiltonian. The subsequent sections then deal with each of the limits in turn. In Section 8.4 the vibrational U(5) limit is analyzed. The hamiltonian is given in (8.43), and a lengthy analysis leads to its eigensolutions (8.97) and (8.100) which appear as complicated polynomial functions of the $s^\dagger$ and $d_m^\dagger$ operators. This result enables the derivation of the matrix elements of various operators in a U(5) basis. It is also possible to obtain the U(5) solutions in coordinate representation; they are given in (8.148) and (8.149). In Section 8.5 the γ-unstable SO(6) limit with hamiltonian (8.135) is investigated. Its solutions are first obtained in a coordinate representation (8.163) and subsequently transformed into the polynomial form (8.172) and (8.174) by means of Dragt's theorem. Closed expressions for various matrix elements can be derived as a result. This is illustrated in Section 8.6 with the calculation of $B(E2)$ values of transitions between levels in the SO(6) limit, which are subsequently compared to the data in $^{196}_{78}\text{Pt}_{118}$. The last section of the chapter summarizes some properties of the rotational SU(3) limit with hamiltonian (8.202).

References

1. A. Arima and F. Iachello, "Collective nuclear states as representations of a SU(6) group," Phys. Rev. Lett. **35** (1975) 1069.

2. H. Feshbach and F. Iachello, "The interacting boson model structure of ^{16}O," Phys. Lett. B **45** (1973) 7.

3. H. Feshbach and F. Iachello, "The interacting boson model," Ann. Phys. (NY) **84** (1974) 211.

4. D. Janssen, R.V. Jolos, and F. Dönau, "An algebraic treatment of the nuclear quadrupole degree of freedom," Nucl. Phys. A **224** (1974) 93.

5. A. Bohr and B.R. Mottelson, "Collective and individual-particle aspects of nuclear structure," Mat. Fys. Dan. Vid. Selsk. **27** (1953) No 16.

6. A. Bohr and B.R. Mottelson, *Nuclear Structure. II. Nuclear Deformations*, Benjamin, New York, 1975.

7. A. Arima, T. Otsuka, F. Iachello, and I. Talmi, "Collective nuclear states as symmetric couplings of proton and neutron excitations," Phys. Lett. B **66** (1977) 205.

8. F. Iachello and A. Arima, *The Interacting Boson Model*, Cambridge University Press, Cambridge, 1987.

9. F. Iachello and P. Van Isacker, *The Interacting Boson–Fermion Model*, Cambridge University Press, Cambridge, 1991.

10. O. Scholten, "Single particle degrees of freedom in the interacting boson model," Prog. Part. Nucl. Phys. **14** (1985) 189.

11. F. Iachello and I. Talmi, "Shell-model foundations of the interacting boson model," Rev. Mod. Phys. **59** (1987) 339.

12. R.F. Casten and D.D. Warner, "The interacting boson approximation," Rev. Mod. Phys. **60** (1988) 389.

13. D. Bonatsos, *Interacting Boson Models of Nuclear Structure*, Oxford University Press, New York, 1988.

14. P.O. Lipas, P. von Brentano, and A. Gelberg, "Proton–neutron symmetry in boson models of nuclear structure," Rep. Prog. Phys. **53** (1990) 1355.

15. O. Castaños, E. Chacón, A. Frank, and M. Moshinsky, "Group theory of the interacting boson model of the nucleus," J. Math. Phys. **20** (1979) 35.

16. A. Arima and F. Iachello, "Interacting boson model of collective states. I. The vibrational limit," Ann. Phys. (NY) **99** (1976) 253.

References

17. A. Arima and F. Iachello, "Interacting boson model of collective nuclear states. II. The rotational limit," Ann. Phys. (NY) **111** (1978) 201.
18. A. Arima and F. Iachello, "Interacting boson model of collective nuclear states. III. The O(6) limit," Ann. Phys. (NY) **123** (1979) 468.
19. A.R. Edmonds, *Angular Momentum in Quantum Mechanics*, Princeton University Press, Princeton, 1957.
20. M.E. Rose, *Elementary Theory of Angular Momentum*, Wiley, New York, 1957.
21. J.P. Elliott, "Collective motion in the nuclear shell model. I. Classification schemes for states of mixed configurations," Proc. Roy. Soc. A **245** (1958) 128.
22. M. Moshinsky, *The Harmonic Oscillator in Modern Physics: From Atoms to Quarks*, Gordon & Breach, New York, 1969.
23. O. Castaños, P. Federman, A. Frank, and S. Pittel, "Study of the effective hamiltonian interacting boson approximation," Nucl. Phys. A **379** (1982) 61.
24. E. Chacón, M. Moshinsky, and R.T. Sharp, "U(5) $\supset$ O(5) $\supset$ O(3) and the exact solution for the problem of quadrupole vibrations of the nucleus," J. Math. Phys. **17** (1976) 668.
25. E. Chacón and M. Moshinsky, "Group theory of the collective model of the nucleus," J. Math. Phys. **18** (1977) 870.
26. M.A. Lohe, *The Development of the Boson Calculus for the Orthogonal and Symplectic Groups*, doctoral dissertation, University of Adelaide, 1974.
27. O. Castaños and A. Frank, "Compact expressions for the matrix elements of operators in the IBA model," KINAM **4** (1982) 33.
28. B.G. Wybourne, *Classical Groups for Physicists*, Wiley-Interscience, New York, 1974.
29. L. Wilets and M. Jean, "Surface oscillations in even–even nuclei," Phys. Rev. **102** (1956) 788.
30. J.M. Eisenberg and W. Greiner, *Nuclear Theory. I. Nuclear Models*, Benjamin, New York, 1975.
31. I.S. Gradshteyn and I.M. Ryzhik, *Table of Integrals, Series, and Products*, Academic, New York, 1965.

32. B.G. Wybourne, *Symmetry Principles and Atomic Spectroscopy*, Wiley-Interscience, New York, 1970.
33. P. Van Isacker, "Quadrupole moments and E2 transitions with $\Delta\tau = 0, \pm 2$ in the γ-unstable O(6) limit of the interacting boson model," Nucl. Phys. A **465** (1987) 497.
34. M. Hamermesh, *Group Theory and Its Application to Physical Problems*, Addison-Wesley, Reading, MA, 1962.
35. J.D. Vergados, "SU(3) $\supset$ R(3) Wigner coefficients in the 2s–1d shell," Nucl. Phys. A **111** (1968) 681.
36. P. Van Isacker, "Boson number dependence of E2 transition rates in the rotational limit of the interacting boson model," Phys. Rev. C **27** (1983) 2447.
37. G. Rosensteel, "Analytic formulae for interacting boson model matrix elements in the SU(3) basis," Phys. Rev. C **41** (1990) 730.

Chapter 9

The Neutron–Proton Interacting Boson Model

9.1 Introduction

The interacting boson model was originally proposed as a phenomenological model, in which the precise relation between the bosons and the actual neutrons and protons in the nucleus was not clear. Soon, however, it was realized that the bosons can be identified with Cooper pairs with angular momentum $l = 0$ or $l = 2$, formed by the nucleons in the valence shell (the last, partially filled shell in the nuclear shell model). The number of bosons, N, was thus seen to correspond to half the number of valence nucleons. It also became apparent that a connection between the interacting boson model and the shell model could not be established without explicitly distinguishing between neutrons and protons since in medium-mass and heavy nuclei these occupy different valence orbits, giving rise to a difference in the shell-model structure of the bosons. Consequently, an extended version of the model was proposed by Arima *et al.* [1] in which a distinction was made between neutrons and protons, referred to as the neutron–proton model or IBM-2, as opposed to the original version of the model, the IBM-1, where this distinction is absent (see Chapter 8).

The essential idea of the IBM-2 is to describe collective excitations in a nucleus in terms of a system of N_ν neutron bosons and N_π proton bosons, where N_ν (N_π) is half the number of valence neutrons (protons). The main reason for introducing the IBM-2 was to establish a microscopic foundation of the boson model, that is, a justification of it in terms of the nuclear shell model. The method through which this can be achieved was outlined and subsequently applied by Otsuka *et al.* [2,3]. The immediate consequence of this work was a better understanding of the parameters of the IBM-2 hamiltonian and of their dependence on the number of valence nucleons. This, in turn, gave rise to a series of systematic IBM-2 calculations for even–even medium-mass and heavy nuclei, examples of which are given in [4,5].

In these early applications of the IBM-2 group-theoretical concepts were used to a limited extent only, primarily to provide a classification of wave functions according to their neutron–proton symmetry. This can be achieved via the introduction of a quantum number called F *spin* [1], which is to the bosons what isospin is to the nucleons. Other than that, little interest was shown at first in the algebraic structure of the IBM-2. This attitude changed when it was realized that the IBM-2 predicts states (outside IBM-1) that are non-symmetric in neutron and proton bosons with peculiar magnetic dipole properties. With the experimental observation of these states [6] the interest in the group-theoretical aspects of the IBM-2 increased and algebraic techniques were applied to derive simple, analytic predictions concerning non-symmetric states in the various limits of the IBM-2.

In this chapter we review some of the algebraic aspects of the IBM-2 and begin with a brief presentation of the Lie algebra appropriate for the IBM-2.

9.2 The $U_\nu(6) \otimes U_\pi(6)$ algebra

The bosons in the IBM-2 have angular momentum $l = 0$ or $l = 2$ and are of neutron or proton type. We have thus the following boson creation and annihilation operators in the IBM-2:

$$d^\dagger_{\nu m} \equiv b^\dagger_{\nu 2m}, \quad s^\dagger_\nu \equiv b^\dagger_{\nu 00}, \quad d^\dagger_{\pi m} \equiv b^\dagger_{\pi 2m}, \quad s^\dagger_\pi \equiv b^\dagger_{\pi 00}$$
$$d_{\nu m} \equiv b_{\nu 2m}, \quad s_\nu \equiv b_{\nu 00}, \quad d_{\pi m} \equiv b_{\pi 2m}, \quad s_\pi \equiv b_{\pi 00}$$
(9.1)

which we also collectively denote as $b^\dagger_{\rho l m}$ and $b_{\rho l m}$ where $\rho = \nu, \pi$ refers to the neutron or proton character of the boson, $l = 0, 2$ is the angular momentum of the boson, and m is its projection on the z axis. These creation and annihilation operators satisfy the boson commutation relations

$$\boxed{[b_{\rho l m}, b^\dagger_{\rho' l' m'}] = \delta_{\rho \rho'} \delta_{l l'} \delta_{m m'}, \qquad \rho, \rho' = \nu, \pi, \quad l, l' = 0, 2} \qquad (9.2)$$

with all others being zero. Consider the following set of bilinear products of boson creation and annihilation operators:

$$\begin{array}{ll} b^\dagger_{\nu l m} b_{\nu l' m'} & 36 \\ b^\dagger_{\pi l m} b_{\pi l' m'} & 36 \end{array} \qquad (9.3)$$

The first set of operators in (9.3) refers to the neutrons, and as in Chapter 8, they generate the algebra $U_\nu(6)$; the second set refers to the protons and they generate $U_\pi(6)$. The sets commute with each other and consequently the operators (9.3) together generate $U_\nu(6) \otimes U_\pi(6)$. The generators of $U_\nu(6) \otimes U_\pi(6)$ can be chosen differently by requiring them to transform as spherical tensors, leading to

$$\boxed{\begin{array}{ll} [b^\dagger_{\nu l m} \times \tilde{b}_{\nu l' m'}]^{(\lambda)}_\mu & 36 \\ [b^\dagger_{\pi l m} \times \tilde{b}_{\pi l' m'}]^{(\lambda)}_\mu & 36 \end{array}} \qquad (9.4)$$

with $\tilde{b}_{\rho l m} \equiv (-)^{l+m} b_{\rho l -m}$.

9.3 F Spin

In analogy with the isospin formalism for neutrons and protons one may consider the neutron boson $b^\dagger_\nu$ and the proton boson $b^\dagger_\pi$ as identical bosons in two different charge states. To distinguish these two states, one introduces a label M_F which (by convention) is $-1/2$ for a neutron boson and $+1/2$ for a proton boson [2]. A single boson thus carries an F spin of $1/2$. Although isospin and F spin are very similar, they are not identical. Since the bosons are made up of pairs of nucleons with each nucleon having isospin $T = 1/2$, the isospin of a boson can be $T = 1$

(with three possible charge states associated with a neutron–neutron, a neutron–proton, and a proton–proton pair) or $T = 0$ (neutron–proton). In the IBM-2 only two members of the $T = 1$ isospin triplet occur: the neutron and the proton boson corresponding to a neutron–neutron and a proton–proton pair, respectively. For this reason a general many-boson IBM-2 state cannot be expected to correspond to a shell-model state with good total isospin since it can only involve two components of the $T = 1$ triplet. A natural extension of the IBM-2 consists of the inclusion of a third (neutron–proton) boson with $T = 1$ and $M_T = 0$, enabling the construction of an isospin-invariant boson model [7]. The main advantage of this more elaborate version of the IBM is that isospin is now a quantum number that can be defined in the shell-model as well as in the boson-model space, which facilitates the study of the correspondence between both approaches. This can be viewed as a test of the boson model in terms of the more fundamental nuclear shell model.

One should realize that isospin symmetry is a consequence of the charge independence of the strong interaction between nucleons and that its breaking can be understood in terms of Coulomb effects. The situation is different for F spin. There is no *a priori* reason why the interaction between the bosons should not depend on M_F. In fact, shell-model calculations indicate that neutron–neutron and proton–proton boson interactions tend to be different from the corresponding neutron–proton boson interactions [3], which would lead to a breaking of the F-spin quantum number. Nevertheless, based on the success of IBM-1 (where such interactions are implicitly assumed to be equal), as a starting point one may consider classification schemes in the IBM-2 which have F spin as a good quantum number and at a later stage introduce F-spin mixing terms to construct more realistic neutron–proton boson hamiltonians.

In the F-spin formalism the creation operators are denoted as

$$b^\dagger_{\nu lm} \to b^\dagger_{1/2-1/2,lm}, \qquad b^\dagger_{\pi lm} \to b^\dagger_{1/2+1/2,lm} \qquad (9.5)$$

where the subscripts $1/2 \pm 1/2$ refer to the F spin of the boson and its projection on the z axis in F-spin space. The annihilation operators $b_{1/2\sigma,lm}$ are the hermitian conjugates of the $b^\dagger_{1/2\sigma,lm}$. To construct annihilation operators which have the usual transformation properties under

9.3 F Spin

rotations in physical space *and* F-spin space, the following definition is introduced:
$$\tilde{b}_{1/2\sigma,lm} \equiv (-)^{1/2+\sigma+l+m} b_{1/2-\sigma,l-m} \tag{9.6}$$

Thus $\tilde{b}_{1/2\sigma,lm}$ transforms in the same way as does $b^\dagger_{1/2\sigma,lm}$, both in physical and in F-spin space.

Just as it is possible to couple the bilinear products (9.3) to definite angular momentum, they can also be coupled to definite F spin,

$$[b^\dagger_{1/2,l} \times \tilde{b}_{1/2,l'}]^{(\kappa,\lambda)}_{\zeta,\mu} = \sum_{\sigma\sigma'}\sum_{mm'} \langle 1/2\sigma\ 1/2\sigma'|\kappa\zeta\rangle \langle lm\ l'm'|\lambda\mu\rangle$$
$$\times b^\dagger_{1/2\sigma,lm}(-)^{1/2+\sigma'+l'+m'} b_{1/2-\sigma',l'-m'} \tag{9.7}$$

Clearly, F can take the value 0, giving an F-spin scalar (or, in short, an F-scalar) operator, or 1, leading to an F-vector operator. Let us consider as an example $l = l' = \lambda = \mu = 0$ in the angular momentum part of the operator (9.7), which in that case reduces to

$$[s^\dagger \times \tilde{s}]^{(\kappa,0)}_{\zeta,0} \tag{9.8}$$

The following operators are obtained:

$$\begin{aligned}
[s^\dagger \times \tilde{s}]^{(0,0)}_{0,0} &= \sqrt{\tfrac{1}{2}}(s^\dagger_\nu \tilde{s}_\nu + s^\dagger_\pi \tilde{s}_\pi) \\
[s^\dagger \times \tilde{s}]^{(1,0)}_{-1,0} &= s^\dagger_\nu \tilde{s}_\pi \\
[s^\dagger \times \tilde{s}]^{(1,0)}_{0,0} &= \sqrt{\tfrac{1}{2}}(s^\dagger_\pi \tilde{s}_\pi - s^\dagger_\nu \tilde{s}_\nu) \\
[s^\dagger \times \tilde{s}]^{(1,0)}_{+1,0} &= -s^\dagger_\pi \tilde{s}_\nu
\end{aligned} \tag{9.9}$$

Note that the operators $s^\dagger_\nu \tilde{s}_\pi$ and $s^\dagger_\pi \tilde{s}_\nu$ transform a neutron boson into a proton boson or *vice versa* and that they are not contained in the set (9.4). In fact, it can be shown that all operators (9.7) can be written in terms of

$$\begin{array}{|ll|}
\hline
[b^\dagger_{\nu lm} \times \tilde{b}_{\nu l'm'}]^{(\lambda)}_\mu & 36 \\
[b^\dagger_{\nu lm} \times \tilde{b}_{\pi l'm'}]^{(\lambda)}_\mu & 36 \\
[b^\dagger_{\pi lm} \times \tilde{b}_{\nu l'm'}]^{(\lambda)}_\mu & 36 \\
[b^\dagger_{\pi lm} \times \tilde{b}_{\pi l'm'}]^{(\lambda)}_\mu & 36 \\
\hline
\end{array} \tag{9.10}$$

358   9 The Neutron–Proton Interacting Boson Model

of which (9.4) clearly is a subset. Since an arbitrary unitary transformation among 12 (6 neutron and 6 proton) boson states can be constructed from the operators (9.10), they generate the algebra U(12), which contains $U_\nu(6) \otimes U_\pi(6)$ as a subalgebra.

At this point we must ask why it is that, in Section 9.1, the algebra $U_\nu(6) \otimes U_\pi(6)$ is introduced instead of the larger algebra U(12). If our aim is to describe states of a single even–even nucleus, all states in the model space have a fixed number of neutron as well as proton bosons. Given this restriction, the dynamical algebra associated with this physical system (i.e., *one* nucleus) only contains operators that preserve the boson numbers for the neutrons and the protons separately. This justifies the choice of $U_\nu(6) \otimes U_\pi(6)$ as the dynamical algebra in the IBM-2 description of a single nucleus. For the discussion of the properties of one nucleus, the larger algebra U(12) is not essential but— as we shall see below—it greatly simplifies the understanding of F spin, and for this reason it is introduced in this section.

The F-scalar operator $[s^\dagger \times \tilde{s}]^{(0,0)}_{0,0}$ is the sum of corresponding neutron and proton generators $s^\dagger_\rho \tilde{s}_\rho$. This property is valid in general and thus the all F-scalar operators are of the form

$$[b^\dagger_{1/2,l} \times \tilde{b}_{1/2,l'}]^{(0,\lambda)}_{0,\mu} = [b^\dagger_{\nu lm} \times \tilde{b}_{\nu l'm'}]^{(\lambda)}_\mu + [b^\dagger_{\pi lm} \times \tilde{b}_{\pi l'm'}]^{(\lambda)}_\mu \qquad (9.11)$$

In addition, it can be shown that these operators close under commutation and thus form a U(6) subalgebra of U(12), which is denoted as $U_{sd}(6)$. The subscript sd indicates that only the orbital information is retained in the generators while their F-spin character is lost via summation over neutron and proton indices. Likewise, a subalgebra can be constructed by contracting (or summing) over the orbital indices of the generators of U(12). This can be done in a simple way using the uncoupled angular momentum form of the generators, arriving at the four operators

$$\sum_{lm} [b^\dagger_{1/2,lm} \times \tilde{b}_{1/2,lm}]^{(\kappa)}_\zeta \qquad 4 \qquad (9.12)$$

where the coupling is defined in F-spin space only. Explicit expressions

9.3 F Spin

for these operators are

$$
\begin{array}{ll}
\kappa=0, \zeta=0 : & \sqrt{\tfrac{1}{2}}(d_\nu^\dagger \cdot \tilde{d}_\nu + d_\pi^\dagger \cdot \tilde{d}_\pi + s_\nu^\dagger \tilde{s}_\nu + s_\pi^\dagger \tilde{s}_\pi) \equiv \sqrt{\tfrac{1}{2}}\hat{N} \\
\kappa=1, \zeta=-1 : & d_\nu^\dagger \cdot \tilde{d}_\pi + s_\nu^\dagger \tilde{s}_\pi \equiv \hat{F}_- \\
\kappa=1, \zeta=0 : & \sqrt{\tfrac{1}{2}}(d_\pi^\dagger \cdot \tilde{d}_\pi - d_\nu^\dagger \cdot \tilde{d}_\nu + s_\pi^\dagger \tilde{s}_\pi - s_\nu^\dagger \tilde{s}_\nu) \equiv \sqrt{2}\hat{F}_z \\
\kappa=1, \zeta=+1 : & -d_\pi^\dagger \cdot \tilde{d}_\nu - s_\pi^\dagger \tilde{s}_\nu = -(\hat{F}_-)^\dagger \equiv -\hat{F}_+
\end{array}
$$

(9.13)

Note that $\hat{F}_z = (\hat{N}_\pi - \hat{N}_\nu)/2$. The boson number operator $\hat{N}$ is F scalar and hence is contained in $U_{sd}(6)$. The operators $\hat{F}_z$ and $\hat{F}_\pm$ satisfy SU(2)-type commutation relations,

$$[\hat{F}_z, \hat{F}_\pm] = \pm \hat{F}_\pm, \qquad [\hat{F}_+, \hat{F}_-] = 2\hat{F}_z \qquad (9.14)$$

and the corresponding algebra will be denoted as $SU_F(2)$. The set of operators (9.13) generate $U_F(2)$. In summary, the existence of two subalgebras $U_{sd}(6)$ and $U_F(2)$ (or $SU_F(2)$) of U(12) is established,

$$U(12) \supset U_{sd}(6) \otimes (U_F(2) \supset SU_F(2)) \qquad (9.15)$$

arising from the separation of the degrees of freedom contained in U(12) into its orbital (or s–d) and its F-spin part.

So far we have used F spin to classify the *generators* of U(12). In a similar way F spin can be used to label the *basis states* in the IBM-2. Consider as an example the two-boson case. States can be coupled to good angular momentum as well as good F spin:

$$
\begin{aligned}
[b_{1/2,l}^\dagger \times b_{1/2,l'}^\dagger]_{M_F, M_L}^{(F,L)} &= \sum_{\sigma\sigma'}\sum_{mm'} \langle 1/2\sigma\, 1/2\sigma'|FM_F\rangle \langle lm\, l'm'|LM_L\rangle \\
&\quad \times b_{1/2\sigma, lm}^\dagger b_{1/2\sigma', l'm'}^\dagger
\end{aligned}
\qquad (9.16)
$$

Acting on the appropriate vacuum, this operator creates a two-boson state with definite angular momentum L and F spin F. In Table 9.1 two-boson states are listed obtained from (9.16) after normalization. Note that M_F, the F-spin projection on the z axis, is determined by the number of neutron and proton bosons, $M_F = (N_\pi - N_\nu)/2$. If $M_F = \pm 1$, the states consist of identical (either neutron or proton) bosons, and hence they are analogous to the IBM-1 states, also shown

Table 9.1: F-spin classification of two-boson states in IBM-2 and IBM-1

M_F	N_ν	N_π	F	IBM-2 state	IBM-1 state
-1	2	0	1	$\sqrt{\frac{1}{2}}[s_\nu^\dagger \times s_\nu^\dagger]_0^{(0)}$	$\sqrt{\frac{1}{2}}[s^\dagger \times s^\dagger]_0^{(0)}$
				$[d_\nu^\dagger \times s_\nu^\dagger]_{M_L}^{(2)}$	$[d^\dagger \times s^\dagger]_{M_L}^{(2)}$
				$\sqrt{\frac{1}{2}}[d_\nu^\dagger \times d_\nu^\dagger]_{M_L}^{(L)}, L = 0, 2, 4$	$\sqrt{\frac{1}{2}}[d^\dagger \times d^\dagger]_{M_L}^{(L)}, L = 0, 2, 4$
0	1	1	1	$[s_\nu^\dagger \times s_\pi^\dagger]_0^{(0)}$	$\sqrt{\frac{1}{2}}[s^\dagger \times s^\dagger]_0^{(0)}$
				$\sqrt{\frac{1}{2}}[d_\nu^\dagger \times s_\pi^\dagger + d_\pi^\dagger \times s_\nu^\dagger]_{M_L}^{(2)}$	$[d^\dagger \times s^\dagger]_{M_L}^{(2)}$
				$[d_\nu^\dagger \times d_\pi^\dagger]_{M_L}^{(L)}, L = 0, 2, 4$	$\sqrt{\frac{1}{2}}[d^\dagger \times d^\dagger]_{M_L}^{(L)}, L = 0, 2, 4$
			0	$\sqrt{\frac{1}{2}}[d_\nu^\dagger \times s_\pi^\dagger - d_\pi^\dagger \times s_\nu^\dagger]_{M_L}^{(2)}$	—
				$[d_\nu^\dagger \times d_\pi^\dagger]_{M_L}^{(L)}, L = 1, 3$	—
1	0	2	1	$\sqrt{\frac{1}{2}}[s_\pi^\dagger \times s_\pi^\dagger]_0^{(0)}$	$\sqrt{\frac{1}{2}}[s^\dagger \times s^\dagger]_0^{(0)}$
				$[d_\pi^\dagger \times s_\pi^\dagger]_{M_L}^{(2)}$	$[d^\dagger \times s^\dagger]_{M_L}^{(2)}$
				$\sqrt{\frac{1}{2}}[d_\pi^\dagger \times d_\pi^\dagger]_{M_L}^{(L)}, L = 0, 2, 4$	$\sqrt{\frac{1}{2}}[d^\dagger \times d^\dagger]_{M_L}^{(L)}, L = 0, 2, 4$

in Table 9.1. In this case $F = 1$ for all states and consequently F spin is not useful as a classification label. If $M_F = 0$ (or $N_\nu = N_\pi = 1$), the IBM-2 states can have $F = 0$ or $F = 1$. Table 9.1 shows that $F = 1$ states are symmetric under the exchange of neutron and proton indices and have IBM-1 counterparts. The $F = 0$ states are antisymmetric under neutron–proton exchange and are absent from IBM-1. This case provides an example where, in a given nucleus, several F-spin values occur which can be used to classify the states according to their neutron–proton symmetry.

In principle, this procedure for finding the allowed F-spin values for a given N_ν and N_π can be used to classify many-boson states, but it is clear that the method becomes increasingly cumbersome for higher boson numbers. A simpler approach exists which is based on group-theoretical considerations. First we note that a U(6) classification label can be defined through the following reduction:

$$U_\nu(6) \otimes U_\pi(6) \supset U_{\nu\pi}(6) \quad (9.17)$$

The algebra $U_{\nu\pi}(6)$ is obtained by adding the corresponding generators

9.3 F Spin

of $U_\nu(6)$ and $U_\pi(6)$. For a given N_ν and N_π the $U_{\nu\pi}(6)$ labels follow from the standard multiplication rule for unitary algebras, which is derived for U(2) and U(4) in Chapters 2 and 6, respectively, and is given in general in Appendix B,

$$[N_\nu] \otimes [N_\pi] = \oplus \sum_f [N_\nu + N_\pi - f, f] \tag{9.18}$$

where f runs from zero to $\min(N_\nu, N_\pi)$. The sum (9.18) contains the representation $[N]$ of $U_{\nu\pi}(6)$ (states that are symmetric in neutron and proton bosons), the non-symmetric representation $[N-1,1]$, and so on. The $f \neq 0$ states are sometimes referred to as having a *mixed symmetry* in the sense that the orbital part of the wave function is neither symmetric nor antisymmetric under the exchange of neutrons and protons. From these qualitative considerations it is clear that there must be a relation between F spin and f defined in (9.18) since both provide a classification of neutron–proton symmetry. To establish the exact relationship, we note that the $U_{sd}(6)$ algebra consisting of the F-scalar operators (9.11) is the same as the $U_{\nu\pi}(6)$ algebra, and consequently, for a given N_ν and N_π, both algebras have the same representations (9.18). On the other hand, we know that $U_{sd}(6)$ and $U_F(2)$ in (9.15) carry the same representations $[N-f, f]$ associated with two-row Young tableaux. This property, known as complementarity, was discussed in Section 7.4 and follows from the overall symmetry under U(12) for an arbitrary neutron–proton s–d-boson state. Finally, by observing that F spin, associated with $SU_F(2)$, is given by half the difference between the first and second rows of the $U_F(2)$ representation, the following relation is derived:

$$F = \tfrac{1}{2}(N - 2f) \tag{9.19}$$

or, with the help of (9.18),

$$F = \tfrac{1}{2}N, \tfrac{1}{2}N - 1, \ldots, \tfrac{1}{2}|N_\nu - N_\pi| \tag{9.20}$$

In deriving the F-spin values allowed in a neutron–proton boson system, we have established the equivalence between F spin and the $U_{\nu\pi}(6)$ representation label f. Thus F-spin symmetry can alternatively be viewed as a U(6) symmetry, and in principle, we do not need to make use of the U(12) algebra to discuss its properties. The advantage of the SU(2) interpretation of F spin is that, due to our familiarity with angular momentum theory, many of its properties can be easily understood. We will find it convenient in this chapter to make use of both interpretations and the connection between them.

To conclude this section we return to the problem of choosing an appropriate dynamical algebra for the IBM-2. As already mentioned, if one is only interested in the description of excitations of a single nucleus at a time, this algebra coincides with $U_\nu(6) \otimes U_\pi(6)$. In view of the general methodology at the basis of algebraic models (see Appendix A), however, it is possible to attempt a unified description of a larger physical system through the introduction of a bigger dynamical algebra, and an obvious candidate emerging from our previous discussion is $U(12)$. What physical systems are described by a $U(12)$ representation? From the explicit form (9.10) we conclude that the $U(12)$ generators conserve the *total* number of bosons $N_\nu + N_\pi$, though not necessarily the bosons numbers N_ν and N_π separately. As a consequence, a single representation of $U(12)$ corresponds to nuclei with constant $N_\nu + N_\pi$. Since we are dealing with a system of identical bosons (the difference between neutrons and protons is taken care of by the F-spin label), the appropriate $U(12)$ representation must be a symmetric one, characterized by a single row $[N]$, where N is the total boson number, $N = N_\nu + N_\pi$. Given the interpretation of a boson as two valence nucleons (either of particle or of hole character), we conclude that a single representation $[N]$ of $U(12)$ in general contains states in several *different* nuclei. Such collections of states have been named *F-spin multiplets*. (This is a somewhat unfortunate name since it seems to imply a multiplet of states which have the same F spin but differ in M_F. We adopt here a definition based on $U(12)$ as a dynamical algebra, a single representation of which contains states with different F-spin values, given by (9.20), but we nevertheless keep referring to it as an F-spin multiplet.) An example of a $U(12)$ F-spin multiplet with $N = N_\nu + N_\pi = 12$ is shown in Figure 9.1 [8]. In this example the valence neutrons are counted from the closed shell at 82, and consequently they have a particle character. The valence protons are counted from the nearest closed shell, which is also at 82, and which results in treating the protons as holes. As a consequence, in the example of Figure 9.1, the condition of constant $N_\nu + N_\pi$ implies a connection between nuclei that differ by multiples of an α particle

9.3 F Spin

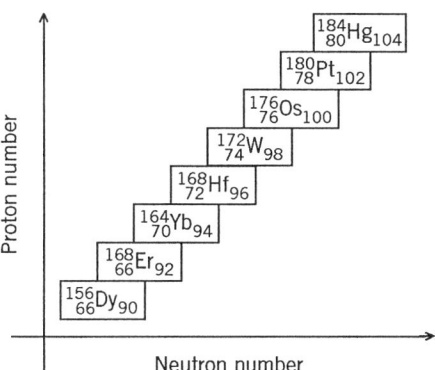

Figure 9.1: An example of a U(12) F-spin multiplet with total boson number $N = N_\nu + N_\pi = 12$.

(i.e., $2n$ neutrons and $2n$ protons with $n = 1, 2, \ldots$). Note also that the series stops at $^{156}_{66}\text{Dy}_{90}$ ($N_\nu = 4$ and $N_\pi = 8$); no member nucleus exists with $N_\pi > 8$ since the maximum number of allowed valence pairs, either particles or holes counted from the nearest closed shell, is 8 in the 50–82 shell. Other examples of F-spin multiplets are known, where neutrons and protons have both particle or both hole character. In these cases an F-spin multiplet consists of levels belonging to isobaric nuclei.

We emphasize that U(12) does not play the role of a symmetry algebra of the IBM-2. This would require the IBM-2 hamiltonian to commute with the generators of U(12),

$$[\hat{H}, b^\dagger_{\rho l m} b_{\rho' l' m'}] = 0, \qquad \rho, \rho' = \nu, \pi, \quad l, l' = 0, 2 \qquad (9.21)$$

a condition which is clearly not satisfied since nuclear states belonging to the same F-spin multiplet may have binding energies that differ by tens of MeV. The degeneracy associated with the U(12) symmetry can be lifted in a procedure of dynamical symmetry breaking, as explained in general terms in Section A.6. A possible algebraic reduction of U(12) is

$$\begin{array}{cccccc}
\text{U}(12) & \supset & (\text{U}_{\text{sd}}(6) & \supset \cdots) \otimes & (\text{U}_F(2) & \supset \text{SU}_F(2) \supset \text{SO}_F(2)) \\
| & & | & & | & | \qquad | \\
[N] & & [N-f, f] & & [N-f, f] & F \qquad M_F
\end{array} \qquad (9.22)$$

where N, f, and F are related through (9.19). The dots refer to further reductions associated with additional symmetry breaking terms in orbital space, as specified in Chapter 8. In other words, just as the U(12) symmetry must be broken with $\text{U}_{sd}(6)$ invariants in order to lift the degeneracy of states belonging to different nuclei, the $\text{U}_{sd}(6)$ symmetry must also be broken to lift the degeneracy of states belonging to a single nucleus. The only true symmetry remaining after this process of successive symmetry breakings in the orbital space is the one associated with $\text{SO}_{sd}(3)$ rotational invariance.

We can similarly investigate the symmetry character of the IBM-2 hamiltonian in F-spin space. Clearly, $\text{SO}_F(2)$ must be a symmetry algebra since the associated generator $\hat{F}_z$ commutes with any reasonable IBM-2 hamiltonian,

$$[\hat{H}, \hat{F}_z] = [\hat{H}, \tfrac{1}{2}(\hat{N}_\pi - \hat{N}_\nu)] = 0 \qquad (9.23)$$

This condition, together with $[\hat{H}, \hat{N}] = 0$, results in eigenstates with definite numbers of neutron and proton bosons, as it should. Of more interest is the symmetry character of the hamiltonian under $\text{SU}_F(2)$. A classification of the type (9.22) implicitly assumes an IBM-2 hamiltonian with eigenstates having good F spin. This is equivalent to the condition

$$[\hat{H}, \hat{F}^2] = 0 \qquad (9.24)$$

and hence establishes the role of $\text{SU}_F(2)$ as a dynamical-symmetry algebra. If, in addition, the condition

$$[\hat{H}, \hat{F}_\pm] = 0 \qquad (9.25)$$

is imposed, the hamiltonian $\hat{H}$ commutes with all generators of $\text{SU}_F(2)$, in which case the latter becomes a true symmetry algebra. One consequence of the condition (9.25) is that the same IBM-2 hamiltonian must be used for all nuclei in an F-spin multiplet and, in particular, that the same parameters are used throughout. This requirement, together with the condition of good F spin, results in the prediction that the low-energy spectra of nuclei in an F-spin multiplet, corresponding to states with maximal F spin $(N_\nu + N_\pi)/2$, should be identical.

This prediction can be tested easily, and several series of nuclei have been investigated in this way. An example is shown in Figure 9.2

9.3 F Spin

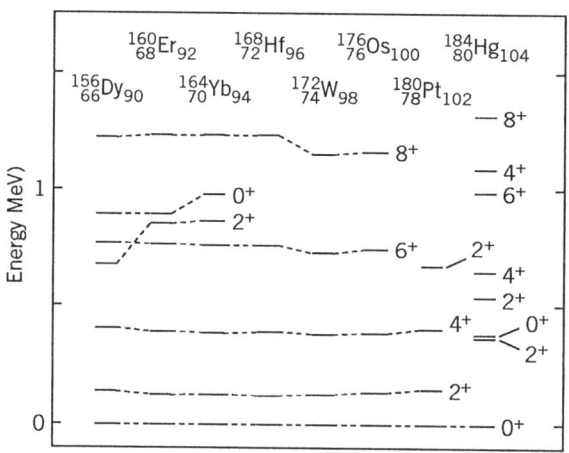

Figure 9.2: The observed low-energy excitation spectra corresponding to the U(12) F-spin multiplet with $N = N_\nu + N_\pi = 12$. Levels are labeled by the angular momentum and parity L^π. For each nucleus the members of the ground-state band are shown up to $L^\pi = 8^+$ as well as the bandheads of the γ and β bands, if known.

[8], which compares the low-energy spectra of nuclei belonging to the F-spin multiplet of Figure 9.1. The constancy of the energies of the ground-state and γ bands is quite remarkable, especially if one compares isotopes as far apart as, for example, $^{156}_{66}$Dy$_{90}$ and $^{176}_{76}$Os$_{100}$, differing in mass number by 20 mass units. The figure also reveals the violation of the SU$_F$(2) symmetry for certain levels and in certain nuclei: the excitation energy of the second $L^\pi = 0^+$ level (β bandhead) is not particularly constant and the spectrum of $^{184}_{80}$Hg$_{104}$ is markedly different from that of all other nuclei in the multiplet. These deviations would indicate either some violation of the conditions (9.24) and/or (9.25) or, in the case of $^{184}_{80}$Hg$_{104}$, perhaps a general breakdown of the assumption of quadrupole collectivity at the basis of the nuclear boson model.

In the remaining sections of this chapter we review some of the applications of the IBM-2 to single nuclei (rather than multiplets) and hence return to a description in terms of a $U_\nu(6) \otimes U_\pi(6)$ dynamical algebra.

9.4 Symmetry limits

The symmetries of the IBM-2 can be classified by studying the subalgebra structure of $U_\nu(6) \otimes U_\pi(6)$. In this section we give an overview of the symmetry limits that exist in the IBM-2, regardless of whether they have physical significance or not. In the next sections we study in more detail the cases which are relevant to nuclear structure applications.

In the classification of the symmetries of the IBM-2 much can be borrowed from the analogous problem in the IBM-1, where three subalgebra chains of U(6) exist: U(5), SO(6), and SU(3). The same algebraic structure exists for $U_\nu(6)$ and $U_\pi(6)$ separately:

$$U_\nu(6) \supset \left\{ \begin{array}{c} U_\nu(5) \supset SO_\nu(5) \\ SO_\nu(6) \supset SO_\nu(5) \\ SU_\nu(3) \end{array} \right\} \supset SO_\nu(3) \qquad (9.26)$$

and

$$U_\pi(6) \supset \left\{ \begin{array}{c} U_\pi(5) \supset SO_\pi(5) \\ SO_\pi(6) \supset SO_\pi(5) \\ SU_\pi(3) \end{array} \right\} \supset SO_\pi(3) \qquad (9.27)$$

The subalgebra structure of the direct product of $U_\nu(6)$ and $U_\pi(6)$ is obtained by combining the algebras appearing in (9.26) and (9.27) and is specified by the lattices shown in Figures 9.3a–c in the notation introduced in Chapter 5. The combined algebras are denoted by a subscript $\nu\pi$ and they are generated by the sum of generators of neutron and proton algebras. For instance, $U_{\nu\pi}(6)$ is generated by all operators of the form

$$[b_{\nu l m}^\dagger \times \tilde{b}_{\nu l' m'}]_\mu^{(\lambda)} + [b_{\pi l m}^\dagger \times \tilde{b}_{\pi l' m'}]_\mu^{(\lambda)}, \qquad l, l' = 0, 2 \qquad (9.28)$$

and similarly for the subalgebras of $U_{\nu\pi}(6)$.

The neutron and proton algebras can be combined in a different, non-equivalent way via the following particle–hole transformation on one of the bosons (say the proton bosons):

$$b_{\pi l m}^\dagger \to \tilde{b}_{\pi l m}, \qquad \tilde{b}_{\pi l m} \to -b_{\pi l m}^\dagger \qquad (9.29)$$

9.4 Symmetry limits

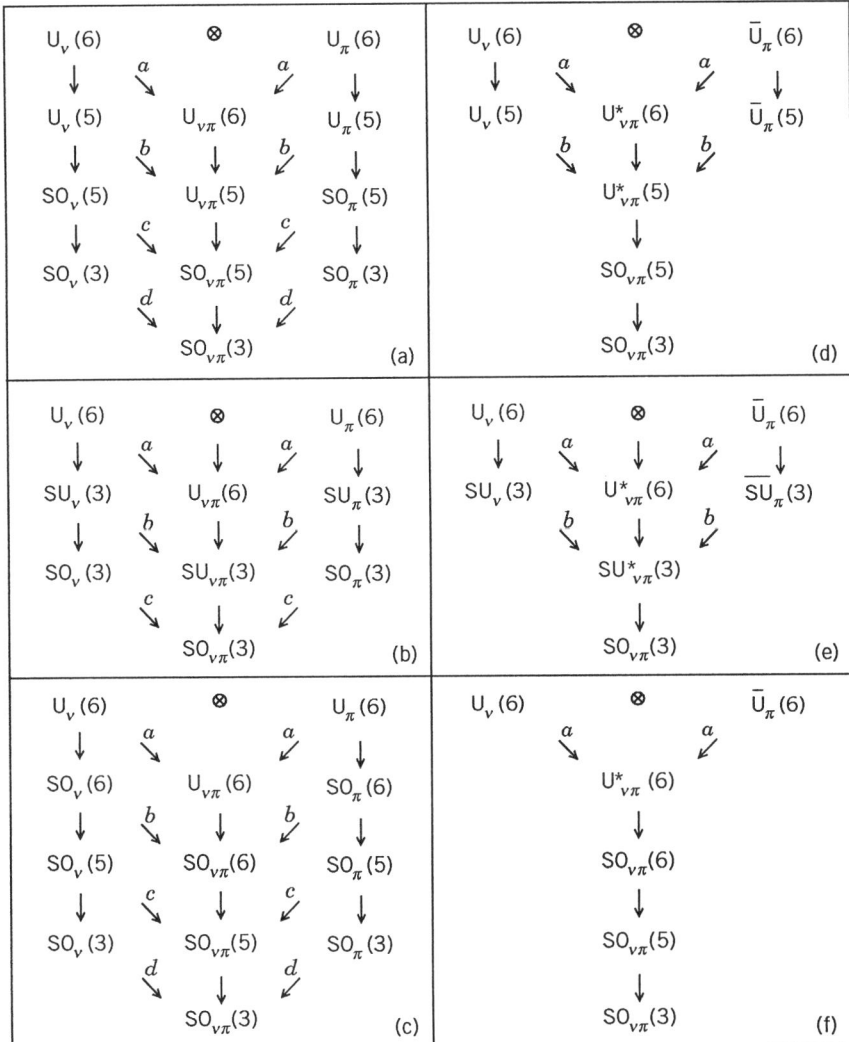

Figure 9.3: Lattices of algebras that exist in the neutron–proton interacting boson model.

Under this transformation the generators of $U_\pi(6)$ are changed into

$$[b^\dagger_{\pi l} \times \tilde{b}_{\pi l'}]^{(\lambda)}_\mu \to (-)^{\lambda+1}[b^\dagger_{\pi l'} \times \tilde{b}_{\pi l}]^{(\lambda)}_\mu - \sqrt{2l+1}\delta_{ll'}\delta_{\lambda 0}\delta_{\mu 0} \qquad (9.30)$$

and still generate a $U(6)$ algebra. Adding the generators of $U_\nu(6)$ and the modified $U_\pi(6)$ algebra leads to another $U(6)$, denoted as $U^*_{\nu\pi}(6)$ to distinguish it from $U_{\nu\pi}(6)$. As seen from (9.30), the particle–hole transformation leaves the generators of $SO(5)$ and $SO(3)$ unchanged since they have $\lambda = 1$ or $\lambda = 3$. As a consequence of the invariance of orthogonal algebras under a particle–hole transformation, only some of the classification schemes shown in Figures 9.3a–c are transformed into non-equivalent ones which are given in Figures 9.3d–f. In this respect the enumeration of possible lattices in the IBM-2 is similar to the corresponding problem in the $U(4)$ model discussed in Section 6.3.

The $U(6)$ tensor character of the s and d bosons depends on whether they are particles or holes. The six components $s^\dagger$ and $d^\dagger_m$ transform as a $[1]$ tensor under $U(6)$ if the bosons are particle-like; they transform as a $[1^5] \equiv [1,1,1,1,1,0]$ tensor for hole-like bosons. The usefulness of knowing the transformation properties of the bosons under various algebras can be illustrated with a simple example, namely finding the explicit form of the quadratic Casimir invariants of $U_{\nu\pi}(6)$ and $U^*_{\nu\pi}(6)$. In the preceding chapters we repeatedly derived expressions in coupled notation for the various quadratic Casimir invariants starting from their uncoupled forms $\sum_{ij} \mathcal{G}^j_i \mathcal{G}^i_j$ and $\sum_{ij} \Lambda_{ij}\Lambda_{ji}$. Here we present an alternative method for the calculation of invariants, which we illustrate with the examples of $\mathcal{C}_2[U_{\nu\pi}(6)]$ and $\mathcal{C}_2[U^*_{\nu\pi}(6)]$. A one-boson state always corresponds to a $[1]$ representation of $U(6)$ (see Table 9.2). The expectation value of $\mathcal{C}_2[U_{\nu\pi}(6)]$ in a one-boson state, $\langle[1]|\mathcal{C}_2[U_{\nu\pi}(6)]|[1]\rangle$, is equal to 6 (see Appendix B). A two-boson state can be either symmetric $[2]$ or antisymmetric $[1,1]$ under $U(6)$. The angular momentum content of the symmetric representation is known from the IBM-1 (i.e., $L = 0^2, 2^2, 4$), so what is left over from the product $[1] \otimes [1]$ must be contained in $[1,1]$ ($L = 1,2,3$). For a two-boson state we have $\langle[2]|\mathcal{C}_2[U_{\nu\pi}(6)]|[2]\rangle = 14$ or $\langle[1,1]|\mathcal{C}_2[U_{\nu\pi}(6)]|[1,1]\rangle = 10$. In this way we obtain the results shown in Table 9.2, where also the expectation values of the *two-body* part of $\mathcal{C}_2[U_{\nu\pi}(6)]$ are given. (The diagonal two-body part of an operator is found by calculating its expectation value in a two-boson state minus the expectation values of the operator in each

9.4 Symmetry limits

Table 9.2: Expectation value of $\mathcal{C}_2[U_{\nu\pi}(6)]$ and its two-body part for one- and two-boson states

$U_{\nu\pi}(6)$ label	Orbital content	L	$\langle\mathcal{C}_2[U_{\nu\pi}(6)]\rangle$	$\langle\mathcal{C}_2[U_{\nu\pi}(6)]\rangle_{\text{tb}}$
[1]	s	0	6	—
	d	2	6	—
[2]	s^2	0	14	2
	sd	2	14	2
	d^2	0,2,4	14	2
[1,1]	sd	2	10	-2
	d^2	1,3	10	-2

of the boson states separately.) The form of $\mathcal{C}_2[U_{\nu\pi}(6)]$ appropriate for an N-boson system can now be deduced: it consists of a term linear in N, a constant two-body interaction of strength 2, and a term which pushes down the antisymmetric states [1,1],

$$\mathcal{C}_2[U_{\nu\pi}(6)] = 6N + 2\left(\tfrac{1}{2}N(N-1)\right) - 4\hat{M}_{\nu\pi} \qquad (9.31)$$

where $\hat{M}_{\nu\pi}$ is the *Majorana operator* which gives 0 acting on a symmetric two-boson state and 1 acting on an antisymmetric one. The operator $\hat{M}_{\nu\pi}$ can be written explicitly in terms of s and d creation and annihilation operators to give

$$\begin{aligned}\mathcal{C}_2[U_{\nu\pi}(6)] &= N(N+5) - 4\sum_{L=1,3}[d_\nu^\dagger\times d_\pi^\dagger]^{(L)}\cdot[\tilde{d}_\pi\times\tilde{d}_\nu]^{(L)}\\&\quad -2[d_\nu^\dagger\times s_\pi^\dagger - d_\pi^\dagger\times s_\nu^\dagger]^{(2)}\cdot[\tilde{d}_\nu\times\tilde{s}_\pi - \tilde{d}_\pi\times\tilde{s}_\nu]^{(2)}\end{aligned}$$
$$(9.32)$$

The operator $\mathcal{C}_2[U_{\nu\pi}^*(6)]$ can be dealt with likewise. In this case the one-boson state can be either a particle [1] or a hole $[1^5]$ with expectation values $\langle[1]|\mathcal{C}_2[U_{\nu\pi}^*(6)]|[1]\rangle = 6$ or $\langle[1^5]|\mathcal{C}_2[U_{\nu\pi}^*(6)]|[1^5]\rangle = 10$, respectively. Two-boson states can be particle–particle, hole–hole, or particle–hole and are classified in Table 9.3. The form of $\mathcal{C}_2[U_{\nu\pi}^*(6)]$ appropriate for a system with N_ν neutron bosons and N_π proton bosons

9 The Neutron–Proton Interacting Boson Model

Table 9.3: Expectation value of $C_2^*[U_{\nu\pi}(6)]$ and its two-body part for one- and two-boson states

$U_{\nu\pi}^*(6)$ label	Orbital content	L	$\langle C_2[U_{\nu\pi}^*(6)]\rangle$	$\langle C_2[U_{\nu\pi}^*(6)]\rangle_{\text{tb}}$
[1]	s	0	6	—
	d	2	6	—
$[1^5]$	s	0	10	—
	d	2	10	—
[2]	s^2	0	14	2
	sd	2	14	2
	d^2	0,2,4	14	2
$[2^5]$	s^2	0	30	10
	sd	2	30	10
	d^2	0,2,4	30	10
$[1^6]$	s^2+d^2	0	6	−10
$[2,1^4]$	s^2-d^2	0	18	2
	sd	2^2	18	2
	d^2	1,2,3,4	18	2

is derived from Table 9.3 as

$$C_2[U_{\nu\pi}^*(6)] = 6N_\nu + 2\left(\tfrac{1}{2}N_\nu(N_\nu-1)\right) + 10N_\pi + 10\left(\tfrac{1}{2}N_\pi(N_\pi-1)\right)$$
$$+2N_\nu N_\pi - 12\hat{M}_{\nu\pi}^* \qquad (9.33)$$

where it is assumed that the neutrons are particles and the protons are holes. The Majorana operator $\hat{M}_{\nu\pi}^*$ for the particle–hole case always gives zero except when acting on the two-boson state with symmetry $[1^6]$. To find $\hat{M}_{\nu\pi}^*$ in terms of s and d creation and annihilation operators, we note that $[1^6]$ is equivalent to the scalar representation $[0]$ of U(6), and thus $\hat{M}_{\nu\pi}^*$ is of the form $\hat{T}_{\nu\pi}^\dagger \hat{T}_{\nu\pi}$ where $\hat{T}_{\nu\pi}^\dagger$ ($\hat{T}_{\nu\pi}$) creates (annihilates) a one-particle–one-hole boson state that is scalar under U(6). Including the appropriate normalization (such that $\hat{M}_{\nu\pi}^*$ gives zero on a $[1^6]$ state), we find

$$\hat{M}_{\nu\pi}^* = \tfrac{1}{6}(d_\nu^\dagger \cdot d_\pi^\dagger + s_\nu^\dagger s_\pi^\dagger)(\tilde{d}_\nu \cdot \tilde{d}_\pi + \tilde{s}_\nu \tilde{s}_\pi) \qquad (9.34)$$

9.5 Symmetry limits with good F spin

This leads to the final result:

$$\begin{aligned} \mathcal{C}_2[\mathrm{U}^*_{\nu\pi}(6)] &= N_\nu(N_\nu + 5) + 5N_\pi(N_\pi + 1) + 2N_\nu N_\pi \\ &\quad - 2(d^\dagger_\nu \cdot d^\dagger_\pi + s^\dagger_\nu s^\dagger_\pi)(\tilde{d}_\nu \cdot \tilde{d}_\pi + \tilde{s}_\nu \tilde{s}_\pi) \end{aligned} \quad (9.35)$$

valid for neutron particles and proton holes.

The operators $\mathcal{C}_2[\mathrm{U}_{\nu\pi}(6)]$ and $\mathcal{C}_2[\mathrm{U}^*_{\nu\pi}(6)]$ have a completely different form: the latter is (up to a constant) a pairing interaction between neutron and proton bosons whereas $\mathcal{C}_2[\mathrm{U}_{\nu\pi}(6)]$ contains interaction terms with angular momentum $L = 1, 2, 3$. In principle, the connection between the two operators can be established via the particle–hole transformation (9.29), but the Racah algebra involved in such a derivation is lengthy and rather tedious. The method discussed here is certainly much simpler as it only requires the classification of one- and two-boson states and their transformation properties under U(6). The technique can be used to derive the explicit form of the Casimir invariants of other algebras appearing in the lattices of Figure 9.3 and hence to determine the hamiltonian in each of the associated symmetry limits. Rather than giving a detailed description of all the limits of the IBM-2, in the next two sections, we confine our attention to those cases which are relevant to applications in nuclear structure physics.

9.5 Symmetry limits with good F spin

In Section 9.3 we introduced the notion of F spin to provide a classification of neutron–proton symmetry of states in the IBM-2 and showed how this label is related to $\mathrm{U}_{\nu\pi}(6)$, the algebra formed by adding the generators of $\mathrm{U}_\nu(6)$ and $\mathrm{U}_\pi(6)$. To find the symmetry limits of the IBM-2 which have F spin as a good quantum number, we must look among the lattices in Figure 9.3 for those reductions containing $\mathrm{U}_{\nu\pi}(6)$. The reduction schemes satisfying this condition are

$$\boxed{\mathrm{U}_\nu(6) \otimes \mathrm{U}_\pi(6) \supset \mathrm{U}_{\nu\pi}(6) \supset \left\{ \begin{array}{c} \mathrm{U}_{\nu\pi}(5) \supset \mathrm{SO}_{\nu\pi}(5) \\ \mathrm{SO}_{\nu\pi}(6) \supset \mathrm{SO}_{\nu\pi}(5) \\ \mathrm{SU}_{\nu\pi}(3) \end{array} \right\} \supset \mathrm{SO}_{\nu\pi}(3)}$$

$$(9.36)$$

Thus, imposing F-spin symmetry as a restriction, three different symmetry limits occur in the IBM-2, and not surprisingly, they are the analogues of the IBM-1 limits.

Each of the reductions (9.36) can be used to classify IBM-2 states. The analysis of each of the reduction schemes proceeds as in Chapters 2 and 6 but is now extended to the U(6) algebra with its subalgebra structure as discussed in Chapter 8. To avoid unnecessary repetitions, we illustrate the procedure for U(6) in one particular case, the SO(6) limit, which is of relevance in the example we discuss in the next section. We introduce the following notation to label states in this limit:

$$
\begin{array}{cccccc}
U_\nu(6) \otimes U_\pi(6) & \supset & U_{\nu\pi}(6) & \supset SO_{\nu\pi}(6) \supset SO_{\nu\pi}(5) \supset SO_{\nu\pi}(3) \\
| & & | & | & | & | \\
[N_\nu] \quad [N_\pi] & & [N-f,f] & (\sigma,\sigma',0) & (v,v') & \nu_\Delta L
\end{array}
$$
(9.37)

The label f and its relation to F spin is discussed in Section 9.3. Since $U_{\nu\pi}(6)$ is characterized by a two-row representation $[N-f,f]$, the representations of $SO_{\nu\pi}(6)$ and $SO_{\nu\pi}(5)$ in general will also have two rows, and hence each is characterized by *two* indices (σ,σ') and (v,v'), respectively. The index ν_Δ in (9.37) is a missing label, not associated with any algebra but necessary to completely specify the reduction $SO(5) \supset SO(3)$. Its role is discussed in detail in Chapter 8, where it appears in the $SO(5) \supset SO(3)$ reduction of one-row $SO(5)$ representations. A similar situation arises for two-row $SO(5)$ representations; since in most cases of interest in the IBM-2 no ν_Δ is needed, we henceforth omit it. For the symmetric representation of $U_{\nu\pi}(6)$, $f=0$, the entire state classification becomes identical to the one in IBM-1. Symmetric IBM-2 states are thus in one-to-one correspondence with the IBM-1 states.

The states (9.37) imply the coupling of various algebras and can formally be written as

$$
\begin{aligned}
&|[N_\nu],[N_\pi];[N-f,f](\sigma,\sigma',0)(v,v')LM_L\rangle \\
&= \sum_{\substack{\sigma_\nu v_\nu L_\nu \\ \sigma_\pi v_\pi L_\pi}} \xi^{\sigma_\nu v_\nu L_\nu, \sigma_\pi v_\pi L_\pi}_{[N-f,f](\sigma,\sigma',0)(v,v')L} |[N_\nu]\sigma_\nu v_\nu L_\nu, \sigma_\pi v_\pi L_\pi; LM_L\rangle
\end{aligned} \quad (9.38)
$$

This expression provides an expansion of the state (9.37) in terms of

9.5 Symmetry limits with good F spin

the IBM-1 SO(6) basis discussed in Chapter 8 and generalized coupling coefficients associated with the $U(6) \supset SO(6) \supset SO(5) \supset SO(3)$. These ξ coefficients can be written, according to Racah's factorization lemma, as a product of several factors, much in the same way as for $U_1(4) \otimes U_2(4)$:

$$\xi^{\sigma_\nu v_\nu L_\nu, \sigma_\pi v_\pi L_\pi}_{[N-f,f](\sigma,\sigma',0)(v,v')L}$$

$$= \left\langle \begin{array}{cc} [N_\nu] & [N_\pi] \\ (\sigma_\nu,0,0) & (\sigma_\pi,0,0) \end{array} \middle| \begin{array}{c} [N-f,f] \\ (\sigma,\sigma',0) \end{array} \right\rangle$$

$$\times \left\langle \begin{array}{cc} (\sigma_\nu,0,0) & (\sigma_\pi,0,0) \\ (v_\nu,0) & (v_\pi,0) \end{array} \middle| \begin{array}{c} (\sigma,\sigma',0) \\ (v,v') \end{array} \right\rangle \left\langle \begin{array}{cc} (v_\nu,0) & (v_\pi,0) \\ L_\nu & L_\pi \end{array} \middle| \begin{array}{c} (v,v') \\ L \end{array} \right\rangle$$

(9.39)

Thus the $U(6) \supset SO(6) \supset SO(5) \supset SO(3)$ coupling coefficient is written as the product of three isoscalar factors, associated with $U(6) \supset SO(6)$, $SO(6) \supset SO(5)$, and $SO(5) \supset SO(3)$.

So far the discussion of neutron–proton coupled systems proceeded along the same lines as in Chapters 2 and 6, leading to an expansion (9.38) of the wave functions which has a structure similar to (2.58) or (2.70) and similar expressions in Chapter 6. There is, however, one important difference: in contrast to the $U(2)$ or $U(4)$ case, the coupling coefficients in $U(6)$ are not known in general. Only some of the coupling coefficients in (9.38) are known in closed, analytic form; others are only numerically available. The principal reason for these technical problems is the occurrence of both missing and multiplicity labels in (9.37). In spite of the lack of an explicit form for the various coupling coefficients, it is still possible to derive closed formulas for many of the matrix elements of physical operators between states (9.38), as illustrated in the next section by means of some selected examples.

In the remainder of this section we discuss some additional consequences of F-spin symmetry. Its purpose is to find out what can be learned in general about matrix elements between good F-spin states. Such general rules (e.g., selection rules) are useful in the calculation of specific matrix elements considered in the next section.

We assume for the sake of concreteness that the initial and final states (i.e., bra and ket) of the matrix element belong to the same nucleus, appropriate when, for instance, the physical process under consideration is an electromagnetic transition. Other processes, such as two-nucleon or alpha-particle transfer, have initial and final states in different nuclei but can be studied in a similar fashion. In general, an electromagnetic transition operator of the IBM-2 can be written as

$$\hat{T} = t_\nu \hat{T}_\nu + t_\pi \hat{T}_\pi \qquad (9.40)$$

where t_ν (t_π) is the effective charge or gyromagnetic ratio associated with the neutron (proton) bosons, depending on whether $\hat{T}$ is an electric or a magnetic transition operator. For the subsequent discussion the multipolarity of the transition need not be specified, but it is assumed that $\hat{T}_\nu$ and $\hat{T}_\pi$ are one-body operators. Examples of the operator (9.40) are:

1. the magnetic dipole (M1) operator [9]

$$\hat{T}_\mu^{(\mathrm{M1})} = \left(\frac{3}{4\pi}\right)^{1/2} \left(g_\nu \hat{L}_{\nu\mu} + g_\pi \hat{L}_{\pi\mu}\right) \qquad (9.41)$$

with $\hat{L}_{\rho\mu} = \sqrt{10}[d_\rho^\dagger \times \tilde{d}_\rho]_\mu^{(1)}$, $\rho = \nu, \pi$;

2. the electric quadrupole (E2) operator [9]

$$\begin{aligned}\hat{T}_\mu^{(\mathrm{E2})} &= q_\nu \left([d_\nu^\dagger \times \tilde{s}_\nu + s_\nu^\dagger \times \tilde{d}_\nu]_\mu^{(2)} + \chi_\nu [d_\nu^\dagger \times \tilde{d}_\nu]_\mu^{(2)}\right) \\ &\quad + q_\pi \left([d_\pi^\dagger \times \tilde{s}_\pi + s_\pi^\dagger \times \tilde{d}_\pi]_\mu^{(2)} + \chi_\pi [d_\pi^\dagger \times \tilde{d}_\pi]_\mu^{(2)}\right) \end{aligned} \quad (9.42)$$

We make the additional assumption that the operators $\hat{T}_\nu$ and $\hat{T}_\pi$ have the same structure. This is satisfied automatically for the M1 operator; in case of the E2 operator it requires $\chi_\nu = \chi_\pi$. Given these two restrictions (i.e., $\hat{T}_\nu$ and $\hat{T}_\pi$ are one-body operators and they have the same structure), it follows that $\hat{T}_\nu + \hat{T}_\pi$ is a generator of $U_{\nu\pi}(6)$ and hence

$$\langle [N_\nu], [N_\pi]; [N-f, f]\alpha | \hat{T}_\nu + \hat{T}_\pi | [N_\nu], [N_\pi]; [N-f', f']\alpha' \rangle = 0 \quad (9.43)$$

if $f \neq f'$ and where α and α' stand for all additional labels (besides F spin) necessary to completely specify the states in bra and ket. (They

9.5 Symmetry limits with good F spin

could coincide with the labels in any of the three bases (9.36) or, for that matter, with any transitional basis as long as it has F spin as a good quantum number.) Equation (9.43) thus implies the following relation between neutron and proton matrix elements for a transition between initial and final states with different F spin:

$$\boxed{\frac{\langle[N_\nu],[N_\pi];[N-f,f]\alpha|\hat{T}_\nu|[N_\nu],[N_\pi];[N-f',f']\alpha'\rangle}{\langle[N_\nu],[N_\pi];[N-f,f]\alpha|\hat{T}_\pi|[N_\nu],[N_\pi];[N-f',f']\alpha'\rangle}=-1} \quad (9.44)$$

if $f \neq f'$.

A different relation can be derived for the matrix elements between states with the *same* F spin. We rewrite the state $|[N_\nu],[N_\pi];[N-f,f]\alpha\rangle$ as $|N\alpha;FM_F\rangle$ to more clearly indicate its transformation properties in F-spin space. The connection between the two notations is easily established through the relations $F=N/2-f$ and $M_F=(N_\pi-N_\nu)/2$. Likewise, we rewrite $\hat{T}_\nu$ and $\hat{T}_\pi$ as a linear combination of F-scalar and F-vector operators $\hat{T}_0^{(0)}$ and $\hat{T}_0^{(1)}$ (with $M_F=0$),

$$\hat{T}_\nu = \tfrac{1}{2}\left(\hat{T}_0^{(0)}-\hat{T}_0^{(1)}\right), \qquad \hat{T}_\pi = \tfrac{1}{2}\left(\hat{T}_0^{(0)}+\hat{T}_0^{(1)}\right) \quad (9.45)$$

Having established the transformation properties of states and operators, we may now apply the Wigner–Eckart theorem in F-spin space to obtain

$$\begin{aligned}
\langle N\alpha;FM_F|\hat{T}_\nu|N\alpha';FM_F\rangle \\
= \tfrac{1}{2}\Big(\langle N\alpha;F\|\hat{T}^{(0)}\|N\alpha';F\rangle \\
-\langle FM_F\,10|FM_F\rangle\langle N\alpha;F\|\hat{T}^{(1)}\|N\alpha';F\rangle\Big) \\
\langle N\alpha;FM_F|\hat{T}_\nu|N\alpha';FM_F\rangle \\
= \tfrac{1}{2}\Big(\langle N\alpha;F\|\hat{T}^{(0)}\|N\alpha';F\rangle \\
+\langle FM_F\,10|FM_F\rangle\langle N\alpha;F\|\hat{T}^{(1)}\|N\alpha';F\rangle\Big)
\end{aligned} \quad (9.46)$$

Inserting the appropriate expressions for the Clebsch–Gordan coefficients $\langle FM_F\,10|FM_F\rangle$, the following ratio of matrix elements is obtained:

$$\frac{\langle N\alpha;FM_F|\hat{T}_\nu|N\alpha';FM_F\rangle}{\langle N\alpha;FM_F|\hat{T}_\pi|N\alpha';FM_F\rangle} = \frac{\sqrt{F(F+1)}-r(N,F)M_F}{\sqrt{F(F+1)}+r(N,F)M_F} \quad (9.47)$$

where $r(N, F)$ is the F-vector–F-scalar ratio of $SU_F(2)$-reduced matrix elements,

$$r(N, F) = \frac{\langle N\alpha; F \| \hat{T}^{(1)} \| N\alpha'; F \rangle}{\langle N\alpha; F \| \hat{T}^{(0)} \| N\alpha'; F \rangle} \qquad (9.48)$$

Relation (9.47) becomes particularly simple for symmetric states, $F = N/2$. For $M_F = F = N/2$, which corresponds to an all-proton case, the ratio (9.47) must be zero and hence

$$r(N, F = \tfrac{1}{2}N) = \left[\frac{N+2}{2N}\right]^{1/2} \qquad (9.49)$$

Using again (9.47) with $F = N/2$ but arbitrary M_F, we finally obtain

$$\frac{\langle N\alpha; F = \tfrac{1}{2}N \ M_F | \hat{T}_\nu | N\alpha'; F = \tfrac{1}{2}N \ M_F \rangle}{\langle N\alpha; F = \tfrac{1}{2}N \ M_F | \hat{T}_\pi | N\alpha'; F = \tfrac{1}{2}N \ M_F \rangle} = \frac{N_\nu}{N_\pi} \qquad (9.50)$$

or in the notation of (9.44),

$$\boxed{\frac{\langle [N_\nu], [N_\pi]; [N]\alpha | \hat{T}_\nu | [N_\nu], [N_\pi]; [N]\alpha' \rangle}{\langle [N_\nu], [N_\pi]; [N]\alpha | \hat{T}_\pi | [N_\nu], [N_\pi]; [N]\alpha' \rangle} = \frac{N_\nu}{N_\pi}} \qquad (9.51)$$

The intuitive interpretation of (9.51) is clear: in a symmetric representation $[N]$, neutron and proton contributions to the total matrix element are proportional to the respective boson numbers. With the help of relation (9.51) all IBM-2 matrix elements between symmetric states can be derived from the corresponding IBM-1 result:

$$\langle [N_\nu], [N_\pi]; [N]\alpha | \hat{T}_\rho | [N_\nu], [N_\pi]; [N]\alpha' \rangle = \frac{N_\rho}{N} \langle [N]\alpha | \hat{T} | [N]\alpha' \rangle_{\text{IBM1}} \qquad (9.52)$$

The derivation of (9.51) relies entirely on the interpretation of F spin as an $SU(2)$ label, in contrast to the arguments leading to (9.44), which are based on the interpretation of $\hat{T}_\nu + \hat{T}_\pi$ as a generator of $U_{\nu\pi}(6)$ coupled with the result that a generator of an algebra cannot connect different representations of that algebra. In the latter derivation F spin is thus viewed as a $U(6)$ label. The two examples illustrate the convenience of both interpretations and how to make use of the equivalence between them.

The results (9.44) and (9.51) have a general character and do not presuppose one of the three classifications (9.36). In fact, the only condition for these results to be valid is that F spin is a good quantum number or, equivalently, that the IBM-2 hamiltonian has a U(6) symmetry. In the next section we consider a specific example where the usefuless of these general results is illustrated.

9.6 Example: Non-symmetric states in the nucleus ^{196}Pt

In the previous sections of this chapter we studied the general characteristics of the algebraic structure of the IBM-2. In this section we turn our attention to some of its more specific properties. To illustrate these ideas, we take the nucleus $^{196}_{78}\text{Pt}_{118}$ as an example.

In Section 8.6 we argued that many of the properties of low-energy levels of $^{196}_{78}\text{Pt}_{118}$ support an SO(6) interpretation of this nucleus. We use this as a starting point and choose a hamiltonian and transition operators of the IBM-2 that reproduce the IBM-1 results obtained in Section 8.6. The projection from IBM-1 into IBM-2 is not unique since many combinations of hamiltonian and transition operators can be chosen in IBM-2 that lead to approximately the same results after projection onto IBM-1. The simplest approach—which we will follow here—is to require the IBM-2 hamiltonian to have eigenstates with good F spin (i.e., to require that $\hat{H}[\text{IBM1}]$ commutes with $\hat{F}^2$). Under this restriction the correspondence between IBM-1 and IBM-2 becomes simple and exact: the complete energy spectrum of the general IBM-1 hamiltonian

$$\hat{H}[\text{IBM1}] = \epsilon \mathcal{C}_1[\text{U}(5)] + \alpha \mathcal{C}_2[\text{U}(5)] + \xi \mathcal{C}_2[\text{SU}(3)] + \beta \mathcal{C}_2[\text{SO}(6)] \\ + \zeta \mathcal{C}_2[\text{SO}(5)] + \gamma \mathcal{C}_2[\text{SO}(3)] \quad (9.53)$$

is reproduced by

$$\hat{H}[\text{IBM2}] = \eta \mathcal{C}_2[\text{U}_{\nu\pi}(6)] + \epsilon \mathcal{C}_1[\text{U}_{\nu\pi}(5)] + \alpha \mathcal{C}_2[\text{U}_{\nu\pi}(5)] + \xi \mathcal{C}_2[\text{SU}_{\nu\pi}(3)] \\ + \beta \mathcal{C}_2[\text{SO}_{\nu\pi}(6)] + \zeta \mathcal{C}_2[\text{SO}_{\nu\pi}(5)] + \gamma \mathcal{C}_2[\text{SO}_{\nu\pi}(3)] \quad (9.54)$$

in the usual notation of $\mathcal{C}_n[G]$ representing the nth-order Casimir invariant of the algebra G. For an arbitrary value of η and equal parameters

ϵ, α, ξ, β, ζ, and γ in $\hat{H}$ and $\hat{H}'$ a one-to-one correspondence is established between all IBM-1 states and the IBM-2 states with $F = N/2$. The relation between the IBM-1 matrix element

$$\langle [N]\alpha | t\hat{T} | [N]\alpha' \rangle_{\text{IBM1}} \qquad (9.55)$$

and the corresponding matrix element in IBM-2,

$$\langle [N_\nu], [N_\pi]; [N]\alpha | t_\nu \hat{T}_\nu + t_\pi \hat{T}_\pi | [N_\nu], [N_\pi]; [N]\alpha' \rangle_{\text{IBM2}} \qquad (9.56)$$

is readily established via the result (9.51): they coincide for $tN = t_\nu N_\nu + t_\pi N_\pi$. From this we conclude that the IBM-1 results quoted in Section 8.6 are reproduced *exactly* by the IBM-2 hamiltonian

$$\hat{H}_{\text{II}}[\text{IBM2}] = \eta \mathcal{C}_2[\text{U}_{\nu\pi}(6)] + \beta \mathcal{C}_2[\text{SO}_{\nu\pi}(6)] + \zeta \mathcal{C}_2[\text{SO}_{\nu\pi}(5)] + \gamma \mathcal{C}_2[\text{SO}_{\nu\pi}(3)] \qquad (9.57)$$

with $\beta = -50$, $\zeta = 58$, and $\gamma = 11$ in units of keV and an E2 transition operator of the form (9.42) with $(q_\nu N_\nu + q_\pi N_\pi)/N = q = 0.152$ in units of $e^2 b^2$ and $\chi_\nu = \chi_\pi = 0$.

The new feature of the IBM-2 calculation, as compared with the IBM-1, is the occurrence of non-symmetric states with $F \neq N/2$, so we now turn our attention to them. The excitation energies of these states are easily obtained from the eigenvalue expression corresponding to the hamiltonian $\hat{H}_{\text{II}}[\text{IBM2}]$:

$$\begin{aligned}
E_{\text{II}}&(f, \sigma, \sigma', v, v', L) \\
&= \eta[(N-f)(N-f+5) + f(f+3)] \\
&\quad + \beta[\sigma(\sigma+4) + \sigma'(\sigma'+2)] \\
&\quad + \zeta[v(v+3) + v'(v'+1)] + \gamma L(L+1) \qquad (9.58)
\end{aligned}$$

The parameters β, ζ, and γ appropriate for $^{196}_{78}\text{Pt}_{118}$ are given above; η should be determined from the empirical excitation energies of non-symmetric states, if some of them are observed and identified. Note that η should be negative to have the symmetric $F = N/2$ states lowest in energy. No non-symmetric state has been firmly identified in $^{196}_{78}\text{Pt}_{118}$, but the value $\eta \approx -100$ keV may be taken on the basis of systematics established in other nuclei [10]. A partial energy spectrum of $^{196}_{78}\text{Pt}_{118}$, calculated with (9.58), is shown in Figure 9.4.

9.6 Example: Non-symmetric states in the nucleus ^{196}Pt

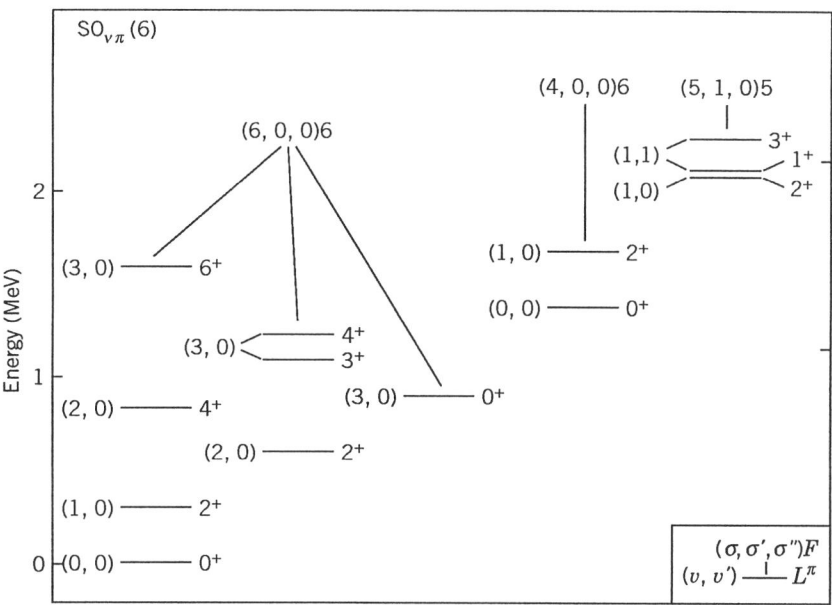

Figure 9.4: Partial SO$_{\nu\pi}$(6) eigenspectrum of $\hat{H}_{\text{II}}$[IBM2] for $N_\nu = 4$ and $N_\pi = 2$ bosons corresponding to the nucleus ^{196}Pt. The spectrum is obtained with the parameters (in keV) $\eta = 100$, $\beta = -50$, $\zeta = 58$, and $\gamma = 11$, with the energy of the ground state normalized to zero. Levels are labeled by $(\sigma, \sigma', \sigma'')$, (v, v'), angular momentum and parity L^π, and the value of F spin.

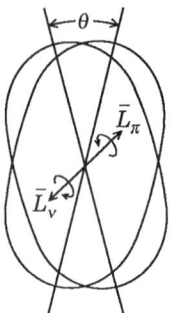

Figure 9.5: The geometric interpretation of non-symmetric states in deformed nuclei. The neutrons and protons are assumed to have density distributions with axially symmetric ellipsoidal shapes and to exhibit rotational oscillations in the angle θ between the axes of symmetry. Also shown are the collective rotational angular momenta $\mathbf{L}_\nu$ and $\mathbf{L}_\pi$.

Another property which is important for the identification of non-symmetric states is their electromagnetic decay to symmetric $F = N/2$ states and, in particular, the magnetic dipole strength from $L^\pi = 1^+$ states to the $L^\pi = 0^+$ ground state of the nucleus. Why M1 transitions are a characteristic signature of non-symmetric states can be understood intuitively from their geometric interpretation. Conventional collective states in nuclei can be pictured in terms of quadrupole oscillations around an equilibrium "liquid-drop" surface of the nucleus, which is either spherical or ellipsoidal in shape [11]. In the case of deformed nuclei (ellipsoidal shape) these vibrations are further combined with rotations. This approach can be extended to a two-fluid model, where the two fluids refer to the neutrons and protons, respectively. It can be shown [9] that symmetric IBM-2 states correspond, in the classical (or large-N) limit, to in-phase vibrations and rotations of the neutron and proton fluids whereas these are out of phase for non-symmetric states. A typical example is shown in Figure 9.5 in the case of axially symmetric ellipsoidal neutron and proton shapes. From this picture it is intuitively clear that the operator inducing a transition from the symmetric ground state to a non-symmetric excited configuration is of the form $\hat{L}_{\nu\mu} - \hat{L}_{\pi\mu}$. Given the form (9.41) of the M1 transition

9.6 Example: Non-symmetric states in the nucleus ^{196}Pt

operator, we may thus expect M1 matrix elements from symmetric to non-symmetric states to be proportional to $g_\nu - g_\pi$.

These intuitive arguments indicate that symmetric and non-symmetric levels are connected by non-zero M1 matrix elements, which could be crucial in the experimental detection of non-symmetric states. In the example of $^{196}_{78}$Pt$_{118}$ we need to evaluate reduced matrix elements between the SO(6) basis states,

$$\langle [N_\nu], [N_\pi]; [N-f, f](\sigma, \sigma', 0)(v, v')L \| \hat{T} \\ \| [N_\nu], [N_\pi]; [N](\sigma_0, 0, 0)(v_0, 0)L_0 \rangle \quad (9.59)$$

where the reduction is defined with respect to SO$_{\nu\pi}$(3). Unfortunately, the expansion (9.38) is not useful in the calculation of this matrix element since the coupling coefficients are unknown for general N_ν and N_π. Nevertheless, the matrix element can be computed in general, and we illustrate the procedure with the example of M1 transitions to the ground state, related to the matrix element

$$\langle [N_\nu], [N_\pi]; [N-f, f](\sigma, \sigma')(v, v')\gamma L \| \hat{T}^{(\text{M1})} \| [N_\nu], [N_\pi]; [N](N, 0)(0, 0)0 \rangle \quad (9.60)$$

with the M1 transition operator defined in (9.43). As $\hat{T}^{(\text{M1})}_\mu$ contains the angular momentum operators $\hat{L}_{\nu\mu}$ and $\hat{L}_{\pi\mu}$ which, according to (9.44), have matrix elements equal in magnitude but opposite in sign, we only require to evaluate the matrix element of one of them, say $\hat{L}_{\pi\mu}$.

To do so, we first determine the tensor character of the $\hat{L}_{\rho\mu}$ operators under the algebras U$_{\nu\pi}$(6), SO$_{\nu\pi}$(6), SO$_{\nu\pi}$(5), and SO$_{\nu\pi}$(3). Since $\hat{L}_{\rho\mu}$ is an operator of the form $[b^\dagger_{\rho l} \times \tilde{b}_{\rho l'}]^{(\lambda)}_\mu$, this involves the classification of the product $[1] \otimes [1^5] = [1^6] \oplus [2, 1^4]$ under U(6) $\supset$ SO(6) $\supset$ SO(5) $\supset$ SO(3) $\supset$ SO(2). This problem is similar to the classification of generators under U(4) $\supset$ SO(4) $\supset$ SO(3) $\supset$ SO(2) considered in Section 5.8 (see Table 5.5) and leads to the results shown in Table 9.4. It follows that $\hat{L}_{\rho\mu}$ has tensor character $[2, 1^4](1, 1, 0)(1, 1)1\mu$. Acting on the ground state,

$$|N_\nu, N_\pi; 0^+_g\rangle \equiv |[N_\nu], [N_\pi]; [N](N, 0, 0)(0, 0)0\rangle \quad (9.61)$$

this operator can excite only one state, namely

$$|N_\nu, N_\pi; 1^+_{\text{ns}}\rangle \equiv |[N_\nu], [N_\pi]; [N-1, 1](N-1, 1, 0)(1, 1)1\rangle \quad (9.62)$$

Table 9.4: Tensor character of U(6) generators under U(6) ⊃ SO(6) ⊃ SO(5) ⊃ SO(3) ⊃ SO(2)

	Tensor character				Operator
U(6)	SO(6)	SO(5)	SO(3)	SO(2)	
$[\vec{h}]$	$(\sigma, \sigma', \sigma'')$	(v, v')	λ	μ	
$[1^6]$	$(0,0,0)$	$(0,0)$	0	0	$\frac{1}{2}(\hat{n}_d + \hat{n}_s)$
$[2,1^4]$	$(1,1,0)$	$(1,0)$	2	μ	$i[d^\dagger \times \tilde{s} - s^\dagger \times \tilde{d}]^{(2)}_\mu$
		$(1,1)$	1,3	μ	$[d^\dagger \times \tilde{d}]^{(\lambda)}_\mu$
	$(2,0,0)$	$(0,0)$	0	0	$\frac{1}{2}\left(-\sqrt{\frac{1}{5}}\hat{n}_d + \sqrt{5}\hat{n}_s\right)$
		$(1,0)$	1	μ	$[d^\dagger \times \tilde{s} + s^\dagger \times \tilde{d}]^{(2)}_\mu$
		$(2,0)$	2,4	μ	$[d^\dagger \times \tilde{d}]^{(\lambda)}_\mu$

Consequently, our task is reduced to the calculation of a single matrix element

$$M(N_\nu, N_\pi) \equiv \langle N_\nu, N_\pi; 1^+_{\text{ns}} \| \hat{L}_\pi \| N_\nu, N_\pi; 0^+_g \rangle \qquad (9.63)$$

since all others are zero by virtue of U(6), SO(6), SO(5), and/or SO(3) selection rules.

Next, we apply the Wigner–Eckart theorem to (9.63) and obtain

$$M(N_\nu, N_\pi)$$
$$= \left\langle \begin{array}{cc} [N] & [2,1^4] \\ (N,0,0) & (1,1,0) \end{array} \middle| \begin{array}{c} [N-1,1] \\ (N-1,1,0) \end{array} \right\rangle$$
$$\times \left\langle \begin{array}{cc} (N,0,0) & (1,1,0) \\ (0,0) & (1,1) \end{array} \middle| \begin{array}{c} (N-1,1,0) \\ (1,1) \end{array} \right\rangle$$
$$\times \langle [N_\nu], [N_\pi]; [N-1,1] \| \hat{L}_\pi \| [N_\nu], [N_\pi]; [N] \rangle \qquad (9.64)$$

where use is made of the tensor character of $\hat{L}_{\pi\mu}$. The first two factors on the right-hand side are isoscalar factors associated with U(6) ⊃ SO(6) and SO(6) ⊃ SO(5). These depend only on the *total* boson number. The second factor is the U(6)-reduced matrix element of $\hat{L}_{\pi\mu}$ and has a dependence on both N_ν and N_π. The advantage of the application of the Wigner–Eckart theorem in this case is that the reduced matrix element—and hence the (N_ν, N_π) dependence—is a U(6) feature, common to all classification schemes with good F spin. To find

9.6 Example: Non-symmetric states in the nucleus ^{196}Pt

the (N_ν, N_π) dependence, we consider the simpler matrix element between U(5) states,

$$M'(N_\nu, N_\pi) \equiv \langle N_\nu, N_\pi; 2^+_{\text{ns}} \| i[d^\dagger_\pi \times \tilde{s}_\pi - s^\dagger_\pi \times \tilde{d}_\pi]^{(2)} \| N_\nu, N_\pi; 0^+_{\text{g}'} \rangle \quad (9.65)$$

where the bra and ket are the following *vibrational* (or U(5)) states:

$$\begin{aligned} |N_\nu, N_\pi; 0^+_{\text{g}'}\rangle &\equiv |[N_\nu], [N_\pi]; [N] n_{\text{d}} = 0 \ v = 0 \ L = 0\rangle \\ |N_\nu, N_\pi; 2^+_{\text{ns}}\rangle &\equiv |[N_\nu], [N_\pi]; [N-1, 1] n_{\text{d}} = 1 \ v = 1 \ L = 2\rangle \end{aligned} \quad (9.66)$$

Note that although the operators in (9.63) and (9.65) are different (i.e., $\hat{L}_{\pi\mu}$ and $[d^\dagger_\pi \times \tilde{s}_\pi - s^\dagger_\pi \times \tilde{d}_\pi]^{(2)}_\mu$), they both have $[2, 1^4]$ character under U(6), and consequently they have the same U(6)-reduced matrix element. The vibrational states appearing in (9.65) can be constructed explicitly with the techniques leading to Table 9.1. We find

$$|N_\nu, N_\pi; 0^+_{\text{g}'}\rangle = \frac{(s^\dagger_\nu)^{N_\nu}(s^\dagger_\pi)^{N_\pi}}{\sqrt{N_\nu! N_\pi!}} |0\rangle \quad (9.67)$$

and

$$|N_\nu, N_\pi; 2^+_{\text{ns}}\rangle = \frac{(s^\dagger_\nu)^{N_\nu-1}(s^\dagger_\pi)^{N_\pi} d^\dagger_\nu - (s^\dagger_\nu)^{N_\nu}(s^\dagger_\pi)^{N_\pi-1} d^\dagger_\pi}{\sqrt{N(N_\nu - 1)!(N_\pi - 1)!}} |0\rangle \quad (9.68)$$

The calculation of $M'(N_\nu, N_\pi)$ is now a matter of elementary algebra, which gives the result

$$M'(N_\nu, N_\pi) = i \left(\frac{N_\nu N_\pi}{N}\right)^{1/2} \quad (9.69)$$

This establishes the (N_ν, N_π) dependence of the U(6)-reduced matrix element, and as a result, we can write

$$M(N_\nu, N_\pi) = f(N)\sqrt{N_\nu N_\pi} \quad (9.70)$$

with $f(N)$ a function yet to be determined.

Since $f(N)$ is a function of the sum $N_\nu + N_\pi$ only and not of N_ν and N_π separately, it can be evaluated by a convenient choice of N_ν and N_π, namely $N_\nu = N - 1$ and $N_\pi = 1$. In that case the expansion

9 The Neutron–Proton Interacting Boson Model

(9.38) becomes tractable since all of the isoscalar factors involved are known (most of them reduce to unity) and the matrix element of $\hat{L}_{\pi\mu}$ can be computed directly:

$$M(N_\nu = N-1, N_\pi = 1) = \left\langle \begin{array}{cc} (N-1,0,0) & (1,0,0) \\ (1,0) & (1,0) \end{array} \middle| \begin{array}{c} (N,0,0) \\ (0,0) \end{array} \right\rangle$$
$$\times \langle d_\pi \| \hat{L}_\pi \| d_\pi \rangle \qquad (9.71)$$

The SO(3)-reduced matrix element of $\hat{L}_\pi$ can be calculated via standard angular momentum coupling. The isoscalar factor is associated with SO(6) $\supset$ SO(5) and is known in general for the multiplication $(\sigma,0,0) \otimes (1,0,0)$. Techniques for its calculation are discussed in the next chapter. Inserting the appropriate expression, we find the result

$$M(N_\nu = N-1, N_\pi = 1) = -\left[\frac{N-1}{N+1}\right]^{1/2} \qquad (9.72)$$

which implies

$$f(N) = -\frac{1}{\sqrt{N+1}} \qquad (9.73)$$

and hence

$$M(N_\nu, N_\pi) = -\left[\frac{N_\nu N_\pi}{N+1}\right]^{1/2} \qquad (9.74)$$

The above derivation of $M(N_\nu, N_\pi)$ is a typical example of the use of group theory in solving problems related to complex coupled systems, and it is perhaps useful to summarize the various steps leading to the final result (9.74):

1. Determine the tensor character of the operator describing the physical process under consideration.

2. Find the selection rules associated with this tensor character and determine which matrix elements are non-zero.

3. Apply the Wigner–Eckart theorem enabling one to write matrix elements as products of generalized coupling coefficients and reduced matrix elements.

4. Calculate matrix elements in specific, simple cases.

9.6 Example: Non-symmetric states in the nucleus ^{196}Pt

5. If the generalized coupling coefficients appearing in the Wigner–Eckart theorem are known, an expression for the reduced matrix element can be derived from 4. This completely determines all matrix elements covered by the Wigner–Eckart theorem.

6. In general expressions for the coupling coefficients are not available. Nevertheless, the Wigner–Eckart theorem can be useful because it provides information on the structure and functional dependence of the matrix elements on various quantum numbers. Knowledge of this functional dependence can lead to simplifications in the calculation.

Summarizing the result obtained in the SO(6) limit, we have shown that only one non-symmetric 1^+_{ns} state is M1-excited from the ground state and that the associated magnetic dipole transition strength is

$$B(\mathrm{M1}; 0^+_{\mathrm{g}} \to 1^+_{\mathrm{ns}}) = \frac{3}{4\pi}(g_\nu - g_\pi)^2 \frac{3}{N+1} N_\nu N_\pi \qquad (9.75)$$

To arrive at a prediction for the $0^+_{\mathrm{g}} \to 1^+_{\mathrm{ns}}$ transition strength, we still need an estimate for the boson gyromagnetic ratios g_ν and g_π, which can be achieved in two different ways. The quantities g_ν and g_π may be obtained via a microscopic calculation which relates matrix elements in IBM-2 to corresponding ones in the nuclear shell model. This leads to boson g factors of the order $g_\nu \approx 0$ and $g_\pi \approx 1$ in units of μ_{N} (nuclear magneton), the precise values depending somewhat on the mass and charge of the nucleus under consideration [12]. Alternatively, they can be obtained from a phenomenological analysis in which the expression $(g_\nu N_\nu + g_\pi N_\pi)/N$ is fitted to the observed g factors of the first-excited 2^+ state in even–even nuclei. In this way one finds for the Pt isotopes: $g_\nu = 0.12\ \mu_{\mathrm{N}}$ and $g_\pi = 0.68\ \mu_{\mathrm{N}}$ [13]. Using (9.75) with $N_\nu = 4$ and $N_\pi = 2$, appropriate for $^{196}_{78}\mathrm{Pt}_{118}$, the predicted M1 strength from the ground state to the non-symmetric 1^+_{ns} state ranges from 0.26 to 0.82 μ^2_{N}, the first value corresponding with the fitted boson g factors and the second with the microscopically calculated g_ν and g_π.

To better appreciate the magnitude of this $B(\mathrm{M1})$ value, it should be compared with the M1 strength typically observed between low-lying, collective states in even–even nuclei in this mass region, which is of the order 10^{-3} to 10^{-2} μ_N^2. The weakness of these low-lying M1 transitions can be understood on the basis of an F-spin selection rule. Assuming that both initial and final states in the transition have $F = N/2$, we find, according to (9.51),

$$\langle [N_\nu], [N_\pi]; [N]\alpha | \hat{T}^{(M1)} | [N_\nu], [N_\pi]; [N]\alpha' \rangle$$
$$= \left(\frac{3}{4\pi}\right)^{1/2} \frac{g_\nu N_\nu + g_\pi N_\pi}{N} \langle [N]\alpha | \hat{L} | [N]\alpha' \rangle_{\text{IBM1}} = 0$$
(9.76)

for $\alpha \neq \alpha'$. As a consequence M1 matrix elements between symmetric IBM-2 states vanish *exactly*. To the extent that low-energy states have $F = N/2$, M1 transitions beween them are forbidden, in approximate agreement with experimental observations.

We thus conclude that $F = N/2 \to F = N/2 - 1$ transitions have $B(M1)$ values which are large in comparison with $F = N/2 \to F = N/2$ transitions. This characteristic feature can be used to identify non-symmetric states experimentally. The first example of a strong $0_g^+ \to 1_{\text{ns}}^+$ transition was observed in ^{156}Gd [6]. Subsequently, additional non-symmetric states were found in many other nuclei [10,14] and their existence is by now a well-documented fact. Nevertheless, their experimental detection is by no means straightforward, due to the high level density in the energy region where non-symmetric states are expected to occur, and in the specific example of $^{196}_{78}$Pt$_{118}$, no 1_{ns}^+ states have been firmly identified as yet.

Additional information is needed about non-symmetric states other than the single $B(M1; 0_g^+ \to 1_{\text{ns}}^+)$ value: E2, M3, ... strengths from the ground state to $2_{\text{ns}}^+, 3_{\text{ns}}^+, \ldots$ states, the electromagnetic decay of these states, their magnetic dipole and electric quadrupole moments, etc. The prediction of these properties involves the calculation of various matrix elements and can be achieved with techniques very similar to the ones leading to the result (9.75). An overview of the three symmetry limits of IBM-2 which have good F spin is given in [15].

This concludes our discussion of the neutron–proton version of the IBM. We wish to emphasize again that we have been primarily concerned in this chapter with analyzing the algebraic structure of the IBM-2 rather than giving an overview of the applications of the model.

The reader interested in the latter is referred to the book by Iachello and Arima [9], where many of these applications are summarized and numerous relevant references are given. In the last chapter of this book we consider the introduction of fermion degrees of freedom in the U(6) model, again with emphasis on its algebraic ramifications.

Summary

The neutron–proton interacting boson model discussed in this chapter takes account of the simple observation that nuclei consist of neutrons and protons. Although many collective states can be described with a single set of s and d bosons, some of them require the explicit distinction between the neutron and proton degrees of freedom. This means that for the description of these states the U(6) algebra of Chapter 8 must be replaced by a direct product $U_\nu(6) \otimes U_\pi(6)$, where the first algebra is associated with the neutrons and the second with the protons. Two sets of s and d bosons are defined, satisfying the commutation relations (9.2). The 72 bilinear combinations (9.4) of these operators generate $U_\nu(6) \otimes U_\pi(6)$. A larger set (9.10), which consists of 144 generators and includes operators that transform neutron into proton bosons and *vice versa*, can also be defined and generates the algebra U(12). A single symmetric representation of U(12) contains states that do not have fixed numbers of neutron and proton bosons; only their sum is conserved. As a consequence, U(12) can be used as a dynamical algebra for the simultaneous description of several nuclei which have different neutron and proton boson numbers N_ν and N_π but the same sum $N_\nu + N_\pi$. Within U(12) is contained a U(2) algebra with generators (9.13), which has an associated quantum number called F spin. This quantum number arises by assigning each boson an F spin of $F = 1/2$ and a projection $\sigma = -1/2$ or $\sigma = +1/2$ for neutron or proton bosons, respectively. The F spin provides a convenient classification of the symmetry character under the exchange of neutron and proton bosons. In Section 9.4 the symmetry limits of $U_\nu(6) \otimes U_\pi(6)$ are enumerated and some of their properties, such as the invariant operators of the various subalgebras, are discussed. In Section 9.5 the limits which have F spin as a good quantum number are more closely examined. The three possible algebraic reductions satisfying this condition are given in (9.36). The classification in all three cases is similar and is illustrated with an example in the SO(6) limit: the labels are given in (9.37) and the wave function in (9.38). The two properties (9.44) and (9.51) are derived as consequences of the goodness of F spin; they amount to general statements about relations that should be satisfied by matrix elements of neutron and proton operators between states with good F spin. The final section of the chapter examines some of the predictions that arise from the neutron–proton version of the interacting boson model. A characteristic feature that emerges is the prediction of strong magnetic dipole transitions from symmetric to non-symmetric states.

References

1. A. Arima, T. Otsuka, F. Iachello, and I. Talmi, "Collective nuclear states as symmetric couplings of proton and neutron excitations," Phys. Lett. B **66** (1977) 205.
2. T. Otsuka, A. Arima , F. Iachello, and I. Talmi, "Shell model description of interacting bosons," Phys. Lett. B **76** (1978) 139.
3. T. Otsuka, A. Arima, and F. Iachello, "Nuclear shell model and interacting bosons," Nucl. Phys. A **309** (1978) 1.
4. F. Iachello, *Interacting Bosons in Nuclear Physics*, Plenum, New York, 1979.
5. F. Iachello, *Interacting Bose–Fermi Systems in Nuclei*, Plenum, New York, 1981.
6. B. Bohle, A. Richter, W. Steffen, A.E.L. Dieperink, N. Lo Iudice, F. Palumbo, and O. Scholten, "New magnetic dipole excitation mode studied in the heavy deformed nucleus ^{156}Gd by inelastic electron scattering," Phys. Lett. B **137** (1984) 27.
7. J.P. Elliott and A.P. White, "An isospin invariant form of the interacting boson model," Phys. Lett. B **97** (1980) 169.
8. P. von Brentano, A. Gelberg, H. Harter, and P. Sala, "F-spin multiplets in rare-earth nuclei," J. Phys. G **11** (1985) L85.
9. F. Iachello and A. Arima, *The Interacting Boson Model*, Cambridge University Press, Cambridge, 1987.

References

10. U. Hartmann, D. Bohle, T. Guhr, K.-D. Hummel, G. Kilgus, U. Milkau, and A. Richter, "Excitation energy and transition strength systematics of mixed symmetry $J^\pi = 1^+$ states from inelastic electron scattering," Nucl. Phys. A **465** (1987) 25.
11. A. Bohr and B.R. Mottelson, *Nuclear Structure. II. Nuclear Deformations*, Benjamin, New York, 1975.
12. M. Sambataro, O. Scholten, A.E.L. Dieperink, and G. Piccitto, "On magnetic properties in the neutron–proton IBA model," Nucl. Phys. A **423** (1984) 333.
13. A. Wolf, R.F. Casten, and D.D. Warner, "g Factors in heavy nuclei and the proton–neutron interaction," Phys. Lett. B **190** (1987) 19.
14. W.D. Hamilton, A. Irbäck, and J.P. Elliott, "Mixed-symmetry interacting-boson-model states in the nuclei ^{140}Ba, ^{142}Ce, and ^{144}Nd with $N = 84$," Phys. Rev. Lett. **53** (1984) 2469.
15. P. Van Isacker, K. Heyde, J. Jolie, and A. Sevrin, "The F-spin symmetric limits of the neutron–proton interacting boson model," Ann. Phys. (NY) **171** (1986) 253.

Chapter 10

The Interacting Boson–Fermion Model

10.1 Introduction

In the previous two chapters we argued that collective excitations in nuclei—traditionally [1] approximated as quadrupole surface oscillations of a nuclear liquid drop—can alternatively be described in terms of a set of interacting s and d bosons. The latter approach not only has the advantage of a simple, analytic treatment based on powerful group-theoretical techniques but, in addition, has a more direct connection to the nuclear shell model since the bosons can be identified with pairs of nucleons coupled to angular momentum $l = 0$ or $l = 2$. This interpretation of the microscopic structure of the bosons has the immediate consequence that the interacting boson model in its simplest forms (i.e., the IBM-1 and IBM-2 discussed in Chapters 8 and 9) only deals with nuclei with an even number of neutrons and protons. However, more than half of the nuclear species have an odd number of neutrons and/or protons, and for this reason alone an extension of the interacting boson model to cover these situations is of interest. In addition, low-energy states in odd-mass nuclei display a characteristic mixture of collective and single-particle properties, which makes their description a theoretical challenge and an important test for the algebraic approach to nuclear structure.

10 The Interacting Boson–Fermion Model

A natural way to extend the interacting boson model to odd-mass nuclei is to assume that all valence nucleons except one occur in pairs coupled to angular momentum $l = 0$ or $l = 2$, which are approximated as s or d bosons. The last, unpaired nucleon is treated explicitly as a fermion. If the bosons have a hole character, the fermion is also treated as a hole. The resulting model has been named the interacting boson–fermion model, or IBFM, and was proposed by Iachello and Scholten [2]. As with the IBM-2, predictions of the IBFM were at first obtained through numerical calculations in which sets of odd-mass isotopes are described with a hamiltonian with smoothly varying parameters. Examples of this approach and relevant references are given in [3]. An important contribution to the algebraic development of the model was made by Iachello [4], who proposed an analytical treatment of the IBFM through the use of boson–fermion symmetries. This work opened the road to a systematic classification of all dynamical boson–fermion symmetries that are of relevance to odd-mass nuclei. In addition, Iachello realized that the similarity between the treatments of even–even nuclei in the IBM-1 and odd-mass nuclei in the IBFM permitted the simultaneous description of both types of nuclei within a single framework. This approach is commonly referred to as supersymmetry since it is based upon the embedding into a superalgebra as introduced in Chapter 4.

The algebraic structure of the IBFM depends on the orbits that are available to the last, unpaired nucleon. The choice of these single-particle orbits, and in particular their angular momenta or j values, is governed by the nuclear shell model and varies from nucleus to nucleus. In some nuclei a single orbit suffices to give a reasonable description of the low-energy spectrum; in many others, however, a specific combination of several single-particle orbits is needed.

It is not our intention to give here a comprehensive overview of all analytically solvable IBFM hamiltonians since this is already available elsewhere [5]. Rather, after a brief description of the general algebraic structure of the IBFM, we illustrate the idea with two examples that are treated in some detail. The examples are chosen in such a way that extensions of the IBFM (e.g., to odd–odd nuclei or to the supersymmetric description of pairs or quartets of nuclei) can be illustrated in terms of them.

10.2 The $U^B(6) \otimes U^F(\Omega)$ algebra

The boson–fermion algebra of the IBFM is defined in terms of the boson operators $b_{lm}^\dagger$ and their hermitian conjugates b_{lm} with $l = 0, 2$, and the creation and annihilation fermion operators $a_{jm}^\dagger$ and a_{jm}, with the allowed j values depending on the nucleus under consideration. We sometimes use the short-hand notations $b_i^\dagger$ and b_i ($i = 1, 2, \ldots, 6$), and $a_{i'}^\dagger$ and $a_{i'}$ ($i' = 1, 2, \ldots, \Omega$), where Ω is the dimensionality of the fermion space, that is, $\Omega = \sum(2j + 1)$, with the summation going over all available single-particle orbits. The boson operators satisfy commutation relations

$$\boxed{[b_i, b_j^\dagger] = \delta_{ij}, \qquad [b_i, b_j] = [b_i^\dagger, b_j^\dagger] = 0, \qquad i, j = 1, 2, \ldots, 6} \quad (10.1)$$

while the fermion operators obey the anticommutators

$$\boxed{\{a_{i'}, a_{j'}^\dagger\} = \delta_{i'j'}, \quad \{a_{i'}, a_{j'}\} = \{a_{i'}^\dagger, a_{j'}^\dagger\} = 0, \quad i', j' = 1, 2, \ldots, \Omega} \quad (10.2)$$

Bosons and fermions commute:

$$\boxed{[b_i, a_{i'}] = [b_i, a_{i'}^\dagger] = [b_i^\dagger, a_{i'}] = [b_i^\dagger, a_{i'}^\dagger] = 0} \quad (10.3)$$

The corresponding algebra is defined by the bilinear operators

$$\boxed{\mathcal{G}_{Bi}^{Bj} \equiv b_i^\dagger b_j, \qquad \mathcal{G}_{Fi'}^{Fj'} \equiv a_{i'}^\dagger a_{j'}} \quad (10.4)$$

which satisfy the commutation relations

$$\boxed{\begin{aligned} [\mathcal{G}_{Bi}^{Bj}, \mathcal{G}_{Bk}^{Bl}] &= \mathcal{G}_{Bi}^{Bl}\delta_{jk} - \mathcal{G}_{Bk}^{Bj}\delta_{il}, & i, j, k, l &= 1, 2, \ldots, 6 \\ [\mathcal{G}_{Fi'}^{Fj'}, \mathcal{G}_{Fk'}^{Fl'}] &= \mathcal{G}_{Fi'}^{Fl'}\delta_{j'k'} - \mathcal{G}_{Fk'}^{Fj'}\delta_{i'l'}, & i', j', k', l' &= 1, 2, \ldots, \Omega \\ [\mathcal{G}_{Bi}^{Bj}, \mathcal{G}_{Fi'}^{Fj'}] &= 0, & i, j = 1, 2, \ldots, 6, \quad i', j' &= 1, 2, \ldots, \Omega \end{aligned}} \quad (10.5)$$

Since the boson and fermion operators commute with each other, the algebraic structure generated by $(\mathcal{G}_{Bi}^{Bj}, \mathcal{G}_{Fi'}^{Fj'})$ is the direct product $U^B(6) \otimes U^F(\Omega)$. We may, as usual, introduce the algebra in terms of the coupled generators

$$\boxed{\begin{aligned} \mathcal{B}_\mu^{(\lambda)}(l, l') &\equiv [b_l^\dagger \times \tilde{b}_{l'}]_\mu^{(\lambda)} & 36 \\ \mathcal{A}_\mu^{(\lambda)}(j, j') &\equiv [a_j^\dagger \times \tilde{a}_{j'}]_\mu^{(\lambda)} & \Omega^2 \end{aligned}} \quad (10.6)$$

10 The Interacting Boson–Fermion Model

with modified annihilation operators defined as $\tilde{b}_{lm} \equiv (-)^{l+m} b_{l-m}$ and $\tilde{a}_{jm} \equiv (-)^{j+m} a_{j-m}$. For the subsequent discussion it is also useful to know the commutators of the generators in their coupled form (10.6). The boson commutators are given in Chapter 5 (see (5.16)); to derive those for the fermion generators, one may follow the method of Chapter 5 and obtain

$$[\mathcal{A}_\mu^{(\lambda)}(j,j'), \mathcal{A}_{\mu'}^{(\lambda')}(j'',j''')]$$
$$= \sum_{\lambda''\mu''} \sqrt{(2\lambda+1)(2\lambda'+1)} \langle \lambda\mu\,\lambda'\mu' | \lambda''\mu''\rangle$$
$$\times \left[(-)^{\lambda''+j+j'''} \begin{Bmatrix} \lambda & \lambda' & \lambda'' \\ j''' & j & j' \end{Bmatrix} \delta_{j'j''} \mathcal{A}_{\mu''}^{(\lambda'')}(j,j''') \right.$$
$$\left. - (-)^{\lambda+\lambda'+j'+j''} \begin{Bmatrix} \lambda & \lambda' & \lambda'' \\ j'' & j' & j \end{Bmatrix} \delta_{jj'''} \mathcal{A}_{\mu''}^{(\lambda'')}(j'',j') \right] \quad (10.7)$$

Note that (5.16) and (10.7) have exactly the same structure but that in the former equation the arguments $l, l', \ldots$ are integer while in the latter the arguments $j, j', \ldots$ are fermion angular momenta and hence half-integer.

The substructure of the boson algebra $U^B(6)$ is known from the discussion in Chapter 8 and leads to three types of symmetry limits named after the largest subalgebra of $U^B(6)$ in each of the reductions: $U^B(5)$, $SO^B(6)$, and $SU^B(3)$. The substructure of the fermion algebra $U^F(\Omega)$, on the other hand, depends on the single-particle orbits available to the last, unpaired nucleon. As mentioned in the introduction, we do not examine here all possible cases but illustrate the procedure with two examples.

In the first example we consider a nucleon in a single orbit with $j = 3/2$. The generators of the corresponding fermion algebra $U^F(4)$ are

$$\begin{aligned}
\hat{n}_{3/2} &\equiv \sqrt{4}[a_{3/2}^\dagger \times \tilde{a}_{3/2}]_0^{(0)} & 1 \\
\hat{L}_{f\mu} &\equiv \sqrt{5}[a_{3/2}^\dagger \times \tilde{a}_{3/2}]_\mu^{(1)} & 3 \\
\hat{Q}_{f\mu} &\equiv [a_{3/2}^\dagger \times \tilde{a}_{3/2}]_\mu^{(2)} & 5 \\
\hat{O}_{f\mu} &\equiv [a_{3/2}^\dagger \times \tilde{a}_{3/2}]_\mu^{(3)} & 7
\end{aligned} \quad (10.8)$$

where the factors $\sqrt{4}$ and $\sqrt{5}$ in $\hat{n}_{3/2}$ and $\hat{L}_{f\mu}$ are needed for these operators to have the interpretation of fermion number and angular

10.2 The $U^B(6) \otimes U^F(\Omega)$ algebra

momentum operators, respectively. If we discard the number operator $\hat{n}_{3/2}$, the remaining generators still form a closed algebra, analogous to the decomposition $U(2) \simeq SU(2) \otimes U(1)$ discussed in Section 1.1. The operators (10.8) without $\hat{n}_{3/2}$ generate the algebra $SU^F(4)$.

To determine the substructure of a single-orbit $SU^F(2j+1)$ algebra, we must look for algebras G that satisfy

$$SU^F(2j+1) \supset G \supset SU^F(2) \qquad (10.9)$$

where $SU^F(2)$ is generated by the fermion angular momentum operators $\hat{L}_{f\mu}$. To find the possible algebras G, we use (10.7) in the specific case $j = j' = j'' = j'''$ when it reduces to

$$[A_\mu^{(\lambda)}(j,j), A_{\mu'}^{(\lambda')}(j,j)]$$
$$= \sum_{\lambda''\mu''} \sqrt{(2\lambda+1)(2\lambda'+1)} \langle \lambda\mu\, \lambda'\mu' | \lambda''\mu'' \rangle$$
$$\times \left[(-)^{\lambda+\lambda'} - (-)^{\lambda''}\right] \begin{Bmatrix} \lambda & \lambda' & \lambda'' \\ j & j & j \end{Bmatrix} A_{\mu''}^{(\lambda'')}(j,j) \quad (10.10)$$

This commutation relation shows that the generators with odd multipolarity (i.e., with odd λ) form a subalgebra of $SU^F(2j+1)$ which contains $SU^F(2)$. The algebra formed by these generators is called *symplectic* [6,7], and we denote it as $Sp(2j+1)$. In the specific case $j = 3/2$ we conclude that a subalgebra $G = Sp(4)$ exists generated by

$$\begin{aligned} \hat{L}_{f\mu} &\equiv \sqrt{5}[a_{3/2}^\dagger \times \tilde{a}_{3/2}]_\mu^{(1)} & 3 \\ \hat{O}_{f\mu} &\equiv [a_{3/2}^\dagger \times \tilde{a}_{3/2}]_\mu^{(3)} & 7 \end{aligned} \qquad (10.11)$$

and giving rise to the following chain of algebras:

$$\boxed{U^F(4) \supset SU^F(4) \supset Sp^F(4) \supset SU^F(2) \supset SO^F(2)} \qquad (10.12)$$

This completely determines the subalgebra structure of $U^F(4)$. How to combine the fermion algebras appearing in (10.12) with some of the boson algebras of Chapter 8, leading to boson–fermion symmetries of the IBFM, is a question that will be addressed in the next section.

Our second example concerns the case when the available single-particle orbits have angular momenta $j = 1/2, 3/2, 5/2$. This is a combination of single-particle orbits frequently occurring in the nuclear shell model. An example is the set of orbits $p_{1/2}$, $p_{3/2}$, and $f_{5/2}$ in the $N = 3$ and $N = 5$ major oscillator shells [8], which in many nuclei are isolated from other negative-parity levels because of the presence of an unnatural-parity orbit (either $g_{9/2}$ for $N = 3$ or $i_{13/2}$ for $N = 5$). The fermion creation operators in this case are $a^\dagger_{jm}$ with $j = 1/2, 3/2, 5/2$. Alternatively, single-particle levels can be characterized by quantum numbers in an l-s-type coupling, and the corresponding creation operators $a^\dagger_{\tilde{l}m,1/2\sigma}$ are defined in the following way in terms of the original $a^\dagger_{jm}$:

$$a^\dagger_{\tilde{l}m,1/2\sigma} = \sum_j \langle \tilde{l}m\, 1/2\sigma | jm + \sigma \rangle a^\dagger_{j\, m+\sigma} \tag{10.13}$$

It should be emphasized that $\tilde{l}$ is *not* the physical orbital momentum of the particle. In the example of the $p_{1/2}$, $p_{3/2}$, and $f_{5/2}$ orbits the physical orbital momenta are $l = 1$ and $l = 3$ whereas the $j = 1/2, 3/2, 5/2$ values above arise from the coupling of $\tilde{l} = 0$ and $\tilde{l} = 2$ to $\tilde{s} = 1/2$. For this reason $\tilde{l}$ is often referred to as the *pseudo*-orbital angular momentum and $\tilde{s}$ as the *pseudo*-spin [9,10]. Because of the presence of a strong $l \cdot s$ term in the nuclear shell-model potential, the *real* spin–orbit partners are far apart in energy—in particular in medium- and heavy-mass nuclei—and as a consequence the *real* l-s coupling scheme is not of much use in such nuclei. Empirically it is found, however, that *pseudo* spin–orbit partners (e.g., $p_{3/2}$ and $f_{5/2}$ in the example above) are much less split in energy and hence the pseudo quantum numbers $\tilde{l}$ and $\tilde{s}$ offer a more relevant classification scheme.

Although the difference between real and pseudo quantum numbers is of importance for physical considerations, and in particular, for the description of magnetic dipole properties, mathematically they can be handled in the same way. Thus, as in Section 7.2, we define the generators

$$[a^\dagger_{\tilde{l},1/2} \times \tilde{a}_{\tilde{l}',1/2}]^{(\lambda,\kappa)}_{\mu,\zeta} \qquad 144 \tag{10.14}$$

where $\tilde{a}_{\tilde{l}m,1/2\sigma} = (-)^{\tilde{l}+m+1/2+\sigma} a_{\tilde{l}-m,1/2-\sigma}$. The notation $[a^\dagger_{\tilde{l},1/2} \times \tilde{a}_{\tilde{l}',1/2}]^{(\lambda,\kappa)}_{\mu,\zeta}$ implies that the pseudo-orbital angular momenta are coupled to λ with projection μ while the pseudo-spins are coupled to κ with projection ζ. Since they are linear combinations of the original generators $\mathcal{A}^{(\lambda)}_\mu(j,j')$, the 144 operators (10.14) generate the same algebra $U^F(12)$.

10.2 The $U^B(6) \otimes U^F(\Omega)$ algebra

Two subalgebras of $U^F(12)$ can be constructed by considering the generators with $\kappa = 0$ and those with $\lambda = 0$, respectively. The first set consists of the operators

$$\mathcal{A}_\mu^{(\lambda)}(\tilde{l}, \tilde{l}') \equiv \sqrt{2}[a^\dagger_{\tilde{l},1/2} \times \tilde{a}_{\tilde{l}',1/2}]^{(\lambda,0)}_{\mu,0} \qquad 36 \qquad (10.15)$$

which together generate a $U^F(6)$ algebra; the second set constitutes the $SU^F(2)$ algebra formed by

$$\hat{S}_\zeta \equiv \sum_{\tilde{l}} \sqrt{\tfrac{1}{2}(2\tilde{l}+1)} [a^\dagger_{\tilde{l},1/2} \times \tilde{a}_{\tilde{l},1/2}]^{(0,1)}_{0,\zeta} \qquad 4 \qquad (10.16)$$

The factors $\sqrt{2}$ and $\sqrt{\tfrac{1}{2}}$ in the definition of $\mathcal{A}_\mu^{(\lambda)}(\tilde{l},\tilde{l}')$ and $\hat{S}_\zeta$ are included here for later convenience. The generators of $U^F(6)$ and $SU^F(2)$ can be rewritten in terms of the original generators $\mathcal{A}_\mu^{(\lambda)}(j,j')$. This amounts to a transformation from $\tilde{l}$-$\tilde{s}$ to j-j coupling:

$$\mathcal{A}_\mu^{(\lambda)}(\tilde{l},\tilde{l}')$$
$$= \sqrt{2} \sum_{jj'} \sqrt{(2\lambda+1)(2j+1)(2j'+1)} \begin{Bmatrix} \tilde{l} & 1/2 & j \\ \tilde{l}' & 1/2 & j' \\ \lambda & 0 & \lambda \end{Bmatrix} \mathcal{A}_\mu^{(\lambda)}(j,j')$$
$$= \sum_{jj'}(-)^{j'+\tilde{l}+1/2+\lambda}\sqrt{(2j+1)(2j'+1)} \begin{Bmatrix} j & j' & \lambda \\ \tilde{l}' & \tilde{l} & 1/2 \end{Bmatrix} \mathcal{A}_\mu^{(\lambda)}(j,j')$$
$$(10.17)$$

This expression, together with the commutation relation (10.7) for $\mathcal{A}_\mu^{(\lambda)}(j,j')$, can be used to derive the commutator properties of the $\mathcal{A}_\mu^{(\lambda)}(\tilde{l},\tilde{l}')$. This derivation is rather awkward as it involves tedious summations over $3j$ and $6j$ coefficients, and we quote here only the end result,

$$[\mathcal{A}_\mu^{(\lambda)}(\tilde{l},\tilde{l}'), \mathcal{A}_{\mu'}^{(\lambda')}(\tilde{l}'',\tilde{l}''')]$$
$$= \sum_{\lambda''\mu''} \sqrt{(2\lambda+1)(2\lambda'+1)} \langle \lambda\mu\,\lambda'\mu' | \lambda''\mu'' \rangle$$
$$\times \Bigg[(-)^{\lambda''+\tilde{l}+\tilde{l}'''} \begin{Bmatrix} \lambda & \lambda' & \lambda'' \\ \tilde{l}''' & \tilde{l} & \tilde{l}' \end{Bmatrix} \delta_{\tilde{l}'\tilde{l}''} \mathcal{A}_{\mu''}^{(\lambda'')}(\tilde{l},\tilde{l}''')$$
$$- (-)^{\lambda+\lambda'+\tilde{l}'+\tilde{l}''} \begin{Bmatrix} \lambda & \lambda' & \lambda'' \\ \tilde{l}'' & \tilde{l}' & \tilde{l} \end{Bmatrix} \delta_{\tilde{l}\tilde{l}'''} \mathcal{A}_{\mu''}^{(\lambda'')}(\tilde{l}'',\tilde{l}') \Bigg] \quad (10.18)$$

Since $\tilde{l}, \tilde{l}', \ldots$ take integer values, this commutator property is *exactly* the same as the one for the boson generators $\mathcal{B}_\mu^{(\lambda)}(l,l')$ (see (5.16)). The relation (10.18), together with the fact that in our example we consider $j = 1/2, 3/2, 5/2$ corresponding to pseudo-orbital angular momenta $\tilde{l} = 0, 2$, proves that the $\mathcal{A}_\mu^{(\lambda)}(\tilde{l}, \tilde{l}')$ generate a U(6) algebra. In the same way we can derive the expression for the $\mathrm{SU^F}(2)$ generators,

$$\hat{S}_\zeta = \sum_{\tilde{l}} (-)^{j+\tilde{l}+3/2} \sqrt{(2j+1)(2j'+1)} \left\{ \begin{array}{ccc} j & j' & 1 \\ 1/2 & 1/2 & \tilde{l} \end{array} \right\} \mathcal{A}_\zeta^{(1)}(j,j') \quad (10.19)$$

which can be used to prove the closure property of the $\hat{S}_\zeta$ generators as well as to show that they satisfy SU(2) commutation relations. In addition, it can be shown on the basis of these expressions that $\mathcal{A}_\mu^{(\lambda)}(\tilde{l}, \tilde{l}')$ and $\hat{S}_\zeta$ commute,

$$[\mathcal{A}_\mu^{(\lambda)}(\tilde{l}, \tilde{l}'), \hat{S}_\zeta] = 0 \quad (10.20)$$

We have thus established the decomposition

$$\boxed{\mathrm{U^F}(12) \supset \mathrm{U^F}(6) \otimes \mathrm{SU^F}(2)} \quad (10.21)$$

which corresponds to a separation of the fermion degrees of freedom into their pseudo-orbital and pseudo-spin parts. The subalgebra structure of $\mathrm{U^F}(6) \otimes \mathrm{SU^F}(2)$ is completely determined from our earlier analyses in Chapters 1 and 8. Hence we can now turn to the question of how to combine the fermion algebras considered in this section with the boson algebras introduced in Chapter 8.

10.3 Symmetry limits

Let us begin with a brief outline of the general structure of the IBFM hamiltonian. Including up to two-body interactions between the bosons and the fermions, it can be written as

$$\hat{H} = \hat{H}^\mathrm{B} + \hat{H}^\mathrm{F} + \hat{V}^\mathrm{BF} \quad (10.22)$$

where $\hat{H}^\mathrm{B}$ is the s–d-boson hamiltonian defined in Chapter 8 and $\hat{H}^\mathrm{F}$ is the fermion hamiltonian

$$\hat{H}^\mathrm{F} = E_0 + \sum_j \epsilon_j \hat{n}_j + \sum_{j_i L} e^L_{j_1 j_2 j_3 j_4} [[a^\dagger_{j_1} \times a^\dagger_{j_2}]^{(L)} \times [\tilde{a}_{j_3} \times \tilde{a}_{j_4}]^{(L)}]^{(0)}_0 \quad (10.23)$$

10.3 Symmetry limits

For odd-mass nuclei the IBFM considers a *single* fermion interacting with a system of bosons; in this case the last term in $\hat{H}^{\mathrm{F}}$ is of no importance. Odd–odd nuclei, on the other hand, are described as two fermions (one neutron and one proton) interacting with bosons and, in that case, a fermion–fermion interaction should be included in the hamiltonian. The hamiltonian (10.22) also contains a boson–fermion interaction of the form

$$\hat{V}^{\mathrm{BF}} = \sum_{l_i j_i L} v^{L}_{l_1 j_1 l_2 j_2} [[b^{\dagger}_{l_1} \times a^{\dagger}_{j_1}]^{(L)} \times [\tilde{b}_{l_2} \times \tilde{a}_{j_2}]^{(L)}]^{(0)}_0 \qquad (10.24)$$

This is the standard form of the IBFM hamiltonian, which has been used extensively to describe series of odd-mass isotopes in a numerical approach based upon the diagonalization of (10.22) in an appropriate basis. Examples are presented in [5], where also an overview of available calculations is given.

The question that concerns us here is how, for particular j values of the single-particle orbits and for particular values of the parameters in the IBFM hamiltonian, the associated eigenvalue problem can be solved analytically. We note that there is a trivial solution to this problem which consists of neglecting the interactions between the bosons and the fermion, except for the angular momentum coupling, leading to eigenstates that are products of boson and fermion wave functions. However, these so-called *weak-coupling* bases have a very limited use in the analysis of odd-mass spectra. Hence our main aim is to propose analytically solvable hamiltonians that include additional boson–fermion interaction. This we proceed to illustrate in the two examples of the previous section.

For a fermion in a $j = 3/2$ orbit we note the similarity between the $\mathrm{SU}^{\mathrm{F}}(4)$ generators $\hat{L}_{\mathrm{f}\mu}$, $\hat{Q}_{\mathrm{f}\mu}$, and $\hat{O}_{\mathrm{f}\mu}$ defined in (10.8) and the $\mathrm{SO}^{\mathrm{B}}(6)$ generators defined in Section 8.2. In fact, on the basis of the relations (5.16) and (10.7), we can establish the correspondence

$$\begin{aligned}
\sqrt{10}[d^{\dagger} \times \tilde{d}]^{(1)}_{\mu} &\leftrightarrow \sqrt{5}[a^{\dagger}_{3/2} \times \tilde{a}_{3/2}]^{(1)}_{\mu} \\
i[d^{\dagger} \times \tilde{s} - s^{\dagger} \times \tilde{d}]^{(2)}_{\mu} &\leftrightarrow [a^{\dagger}_{3/2} \times \tilde{a}_{3/2}]^{(2)}_{\mu} \\
[d^{\dagger} \times \tilde{d}]^{(3)}_{\mu} &\leftrightarrow -\sqrt{\tfrac{1}{2}}[a^{\dagger}_{3/2} \times \tilde{a}_{3/2}]^{(3)}_{\mu}
\end{aligned} \qquad (10.25)$$

where the operators on both sides of ↔ satisfy *exactly* the same commutation relations. As noted before, the corresponding algebras are thus isomorphic. We have thus established the isomorphisms $SO^B(6) \simeq SU^F(4)$, $SO^B(5) \simeq Sp^F(4)$, and $SO^B(3) \simeq SU^F(2)$. Since boson and fermion operators commute, we may define the three combined operators

$$\begin{aligned}\hat{J}_\mu &\equiv \sqrt{10}[d^\dagger \times \tilde{d}]^{(1)}_\mu + \sqrt{5}[a^\dagger_{3/2} \times \tilde{a}_{3/2}]^{(1)}_\mu & 3 \\ \hat{Q}_\mu &\equiv i[d^\dagger \times \tilde{s} - s^\dagger \times \tilde{d}]^{(2)}_\mu + [a^\dagger_{3/2} \times \tilde{a}_{3/2}]^{(2)}_\mu & 5 \\ \hat{O}_\mu &\equiv [d^\dagger \times \tilde{d}]^{(3)}_\mu - \sqrt{\tfrac{1}{2}}[a^\dagger_{3/2} \times \tilde{a}_{3/2}]^{(3)}_\mu & 7\end{aligned} \quad (10.26)$$

which together generate the boson–fermion algebra $SU^{BF}(4)$. The subalgebras of $SU^{BF}(4)$ are $Sp^{BF}(4)$ and $SU^{BF}(2)$, with generators $(\hat{J}_\mu, \hat{O}_\mu)$ and $(\hat{J}_\mu)$, respectively. We remark that these boson–fermion algebras are often denoted as $Spin^{BF}(6)$, $Spin^{BF}(5)$, and $Spin^{BF}(3)$ and referred to as *spinor algebras* [11]. This nomenclature can be clarified with the example of the angular momentum algebra which is denoted in three different ways: $SU(2)$, $SO(3)$, and $Spin(3)$. These notations do *not* distinguish between the algebras; rather, they refer to the representations associated with them. Denoting the algebra by $SU(2)$ implies integer as well as half-integer representations; using $SO(3)$ one restricts these to only integer representations, while with the $Spin(3)$ notation one is confined to half-integer ones. A similar disctinction is made for higher dimensions. For rotations in five dimensions one uses $SO(5)$ (integer), $Spin(5)$ (half-integer), or $Sp(4)$ (both), and similarly for six-dimensional rotations, $SO(6)$ (integer), $Spin(6)$ (half-integer), or $SU(4)$ (both). As we intend to use the boson–fermion algebras introduced earlier for the classification of states in even–even as well as odd-mass nuclei, leading to integer and half-integer representations (see Section 10.6), the unitary–symplectic notation seems the most appropriate and is used here.

We can summarize the possible symmetry limits for a $j = 3/2$ fermion coupled to a system of bosons by way of a lattice of algebras as introduced in Section 5.8. Analytically solvable IBFM hamiltonians only occur when the $j = 3/2$ fermion is coupled to an $SO(6)$ boson core. If that is the case, three different coupling schemes are possible,

10.3 Symmetry limits

corresponding to routes a, b, and c in the lattice

$$
\begin{array}{ccc}
U^B(6) & \otimes & U^F(4) \\
\downarrow & & \downarrow \\
SO^B(6) & & SU^F(4) \\
\downarrow \;\searrow^{\!a} & \;\swarrow^{\!a} & \downarrow \\
SO^B(5) & SU^{BF}(4) & Sp^F(4) \\
\downarrow \;\searrow^{\!b} & \downarrow & \;\swarrow^{\!b}\;\downarrow \\
SO^B(3) & Sp^{BF}(4) & SU^F(2) \\
\;\searrow^{\!c} & \downarrow & \;\swarrow^{\!c} \\
& SU^{BF}(2) &
\end{array}
\qquad (10.27)
$$

Route c gives rise to a weak-coupling basis; it includes the weakest possible core–particle (i.e., boson–fermion) interaction which is compatible with a rotational invariant hamiltonian. The other two routes are more interesting, in particular route a since it contains a quadrupole core–particle interaction which is known to play a crucial role in odd-mass nuclei [1].

In a similar way we can construct the symmetry limits for a fermion in three orbits with $j = 1/2, 3/2, 5/2$. In this case the correspondence between boson and fermion generators is

$$\mathcal{B}^{(\lambda)}_\mu(l,l') \leftrightarrow \mathcal{A}^{(\lambda)}_\mu(l,l') \qquad (10.28)$$

since we have shown in the previous section that they satisfy identical commutation relations. The summed operators

$$\mathcal{G}^{(\lambda)}_\mu(l,l') \equiv \mathcal{B}^{(\lambda)}_\mu(l,l') + \mathcal{A}^{(\lambda)}_\mu(l,l') \qquad 36 \qquad (10.29)$$

are the generators of the boson–fermion algebra $U^{BF}(6)$. It is clear that in this case analytical solutions are available for the three limits U(5), SO(6), and SU(3), and all three have been studied in great detail and have been applied to odd-mass nuclei [12–15]. For the purpose of our example in Section 10.5 we are interested here in the SO(6) symmetry limits associated with the lattice

$$
\begin{array}{c}
\mathrm{U^B(6)} \quad\quad\quad\quad\quad \otimes \quad\quad\quad \mathrm{U^F(12)} \\
\downarrow \quad\quad\quad\quad\quad\quad\quad\quad\quad\quad\quad \downarrow \\
\quad\quad {\scriptstyle a}\searrow \quad\quad\quad\quad \mathrm{U^F(6)} \quad \otimes \quad \mathrm{SU^F(2)} \\
\downarrow \quad\quad\quad {\scriptstyle a}\nearrow \quad \downarrow \quad\quad\quad\quad\quad \\
\mathrm{SO^B(6)} \quad \mathrm{U^{BF}(6)} \quad\quad \mathrm{SO^F(6)} \\
\downarrow \quad {\scriptstyle b}\searrow \quad \downarrow \quad {\scriptstyle b}\nearrow \quad \downarrow \\
\mathrm{SO^B(5)} \quad \mathrm{SO^{BF}(6)} \quad \mathrm{SO^F(5)} \\
\downarrow \quad {\scriptstyle c}\searrow \quad \downarrow \quad {\scriptstyle c}\nearrow \quad \downarrow \\
\mathrm{SO^B(3)} \quad \mathrm{SO^{BF}(5)} \quad \mathrm{SO^F(3)} \\
\quad {\scriptstyle d}\searrow \quad \downarrow \quad {\scriptstyle d}\nearrow \quad\quad\quad\quad \downarrow \\
\quad\quad \mathrm{SO^{BF}(3)} \quad\quad\quad\quad\quad \mathrm{SU^F(2)} \\
\quad\quad\quad \searrow \quad\quad\quad \swarrow \\
\quad\quad\quad\quad \mathrm{SU^{BF}(2)}
\end{array}
\quad (10.30)
$$

We note that four routes exist, each incorporating different types of core–particle interactions. The $\mathrm{SU^{BF}(2)}$ algebra in (10.30) consists of the total (i.e., boson plus fermion) angular momentum operators $\hat{J}_\mu$. The $\hat{J}_\mu$ are obtained by taking an appropriate linear combination of $\mathcal{G}_\mu^{(1)}(2,2)$ and $\hat{S}_\mu$,

$$\hat{J}_\mu \equiv \sqrt{10}\,\mathcal{G}_\mu^{(1)}(2,2) + \hat{S}_\mu \qquad 3 \qquad (10.31)$$

Inserting the expressions for $\mathcal{G}_\mu^{(1)}(2,2)$ and $\hat{S}_\mu$ in terms of the boson and fermion generators $\mathcal{B}_\mu^{(1)}(2,2)$ and $\mathcal{A}_\mu^{(1)}(2,2)$, we find

$$\hat{J}_\mu = \sqrt{10}\,\mathcal{B}_\mu^{(1)}(2,2) + \sum_j \sqrt{\tfrac{1}{3}j(j+1)(2j+1)}\,\mathcal{A}_\mu^{(1)}(j,j) \qquad (10.32)$$

which confirms its character as the total angular momentum operator.

In the next two sections we investigate the properties of the symmetry limits associated with routes a of the lattices (10.27) and (10.30) and show how these can be used in the interpretation of the spectroscopic properties of odd-mass nuclei.

10.4 Example 1: The gold nuclei

In this section we take a closer look at the dynamical symmetry corresponding to route a of lattice (10.27),

$$\mathrm{U^B(6) \otimes U^F(4) \supset SO^B(6) \otimes SU^F(4) \supset SU^{BF}(4) \supset Sp^{BF}(4) \supset SU^{BF}(2)} \tag{10.33}$$

which we refer to as the $\mathrm{SU^{BF}(4)}$ limit. This symmetry of the IBFM was the first to be investigated [4]. Its properties are analyzed in detail in [11] and compared to many odd-mass nuclei in several regions of the nuclear chart. A detailed overview of such comparisons is given in [16].

Since we are interested in the coupling of a system of N bosons to a single fermion, the appropriate representation of $\mathrm{U^B(6) \otimes U^F(4)}$ is $[N] \otimes [1]$. From Chapter 8 we know the branching rule for $\mathrm{U^B(6) \supset SO^B(6)}$,

$$\sigma_\mathrm{b} = N, N-2, \ldots, 1 \text{ or } 0 \tag{10.34}$$

which supplies the allowed representations of $\mathrm{SO^B(6)}$. These must subsequently be multiplied with the single-fermion representation of $\mathrm{SU^F(4)}$. Here we face the problem that although we know the algebras $\mathrm{SO^B(6)}$ and $\mathrm{SU^F(4)}$ to be isomorphic, we have so far been using a different notation for labeling their representations. A simple way to find the relation between the notations is to compare the dimension formulas in the two cases. A general $\mathrm{SO}(6)$ representation $(\sigma, \sigma', \sigma'')$ has the dimension [17]

$$\tfrac{1}{12}(\sigma+\sigma'+3)(\sigma+\sigma''+2)(\sigma'+\sigma''+1)(\sigma-\sigma'+1)(\sigma-\sigma''+2)(\sigma'-\sigma''+1) \tag{10.35}$$

while a general $\mathrm{SU}(4)$ representation is characterized by three labels $[h, h', h'', 0]$ and dimensionality

$$\tfrac{1}{12}(h+3)(h'+2)(h''+1)(h-h'+1)(h-h''+2)(h'-h''+1) \tag{10.36}$$

Comparison implies the correspondence

$$h = \sigma + \sigma', \qquad h' = \sigma \pm \sigma'', \qquad h'' = \sigma' \pm \sigma'' \tag{10.37}$$

or inversely,

$$\sigma = \tfrac{1}{2}(h+h'-h''), \qquad \sigma' = \tfrac{1}{2}(h-h'+h''), \qquad \sigma'' = \pm\tfrac{1}{2}(-h+h'+h'') \tag{10.38}$$

One can choose either of the signs in these relations. To be consistent with [17], we take the lower sign, in which case the single-fermion representation [1] of SU(4) corresponds to the SO(6) representation $(1/2, 1/2, 1/2)$. In the same way we find the connection between Sp(4) and SO(5) representations. In this case the respective dimension formulas are

$$\tfrac{1}{6}(2v+3)(2v'+1)(v+v'+2)(v-v'+1) \tag{10.39}$$

for an SO(5) representation (v, v') and

$$\tfrac{1}{6}(q+2)(q'+1)(q-q'+1)(q+q'+3) \tag{10.40}$$

for an Sp(4) representation $\{q, q'\}$, leading to the correspondence

$$q = v + v', \qquad q' = v - v' \tag{10.41}$$

or

$$v = \tfrac{1}{2}(q+q'), \qquad v' = \tfrac{1}{2}(q-q') \tag{10.42}$$

Thus the single-fermion representation $\{1\}$ of Sp(4) corresponds to the SO(5) representation $(1/2, 1/2)$. These relations between the unitary-symplectic labels $[h, h', h'']$ or $\{q, q'\}$ and the orthogonal ones $(\sigma, \sigma', \sigma'')$ or (v, v') are a generalization of the connection (1.24) in the U(2) model. It is a matter of convention which one of the two notations is used. In line with the accepted usage in the interacting boson model, we employ in the following the orthogonal labels.

We can now characterize the algebras by an appropriate set of labels

$$
\begin{array}{c}
\mathrm{U^B(6) \otimes U^F(4) \supset SO^B(6) \otimes\ SU^F(4)\ \supset SU^{BF}(4)} \\
|\qquad\ |\qquad\ \ |\qquad\qquad |\qquad\qquad\ | \\
[N]\quad\ [1]\quad (\sigma_b,0,0)\ \ (1/2,1/2,1/2)\ \ (\sigma,\sigma',\sigma'') \\
\\
\supset \mathrm{Sp^{BF}(4) \supset SU^{BF}(2)} \\
|\qquad\ \ | \\
(v,v')\quad \nu_\Delta J
\end{array} \tag{10.43}
$$

The allowed values of $(\sigma, \sigma', \sigma'')$ are obtained by multiplying the $\mathrm{SO^B(6)}$ representation $(\sigma_b, 0, 0)$ with $(1/2, 1/2, 1/2)$. This multiplication is found

10.4 Example 1: The gold nuclei

by reverting to the SU(4) notation and working it out according to the usual Young tableaux rules (see Appendix B):

$$\begin{aligned}(\sigma_{\rm b},0,0)\otimes(1/2,1/2,1/2) \\ \simeq\ & [\sigma_{\rm b},\sigma_{\rm b},0]\otimes[1,0,0] \\ =\ & [\sigma_{\rm b}+1,\sigma_{\rm b},0]\oplus[\sigma_{\rm b},\sigma_{\rm b},1] \\ \simeq\ & (\sigma_{\rm b}+1/2,1/2,1/2)\oplus(\sigma_{\rm b}-1/2,1/2,-1/2)\end{aligned} \quad (10.44)$$

The $SO^{BF}(6)$ representations with $\sigma'' = +1/2$ and $\sigma'' = -1/2$ can, for our purpose, be considered as equivalent. For example, the energy of eigenstates depends quadratically on σ'' and hence is independent of its sign. Henceforth we follow the convention to characterize the states with the absolute value $|\sigma''|$. Combination of the branching rule (10.34) and the multiplication (10.44) then shows that the allowed $SO^{BF}(6)$ representations are

$$(\sigma,\sigma',\sigma'') = (N+1/2,1/2,1/2),(N-1/2,1/2,1/2),\ldots,(1/2,1/2,1/2) \quad (10.45)$$

Next we need the reduction from $SU^{BF}(4)$ to $Sp^{BF}(4)$. The corresponding rules can be deduced from the "building-up" technique introduced in Section 6.4. We start from the observation that for a given boson number N the allowed $SU^{BF}(4)$ representations are known from (10.45). Likewise, the allowed $Sp^{BF}(4)$ representations can be determined in route b of the lattice (10.27) from the $SO(6) \supset SO(5)$ branching rule (known from Chapter 8) and the $SO(5)$ multiplication of $(v,0)$ with $(1/2,1/2)$ (see Appendix B). To find out precisely which $Sp^{BF}(4)$ representations are contained in a given $SU^{BF}(4)$ one, we begin by considering the simplest possible non-trivial case and from there build up more complicated reductions. For $N = 1$ the allowed $SU^{BF}(4)$ representations are, from (10.45), $(3/2,1/2,1/2)$ and $(1/2,1/2,1/2)$. On the other hand, the allowed $Sp^{BF}(4)$ representations are

$$\begin{aligned}[(1,0)\oplus(0,0)]\otimes(1/2,1/2) &\simeq [\{1,1\}\oplus\{0,0\}]\otimes\{1,0\} \\ &= \{2,1\}\oplus\{1,0\}^2 \\ &\simeq (3/2,1/2)\oplus(1/2,1/2)^2\end{aligned} \quad (10.46)$$

Since we know $(1/2,1/2,1/2)$ of $SU^{BF}(4)$ to contain $(1/2,1/2)$ of $Sp^{BF}(4)$, we deduce that $(3/2,1/2,1/2)$ reduces to $(3/2,1/2)$ and $(1/2,1/2)$. Continuing in this way for $N = 2, 3, \ldots$, we find that $(\sigma, 1/2, 1/2)$ of $\mathrm{SU}^{\mathrm{BF}}(4)$ contains the $\mathrm{Sp}^{\mathrm{BF}}(4)$ representations

$$(v, v') = (\sigma, 1/2), (\sigma - 1, 1/2), \ldots, (1/2, 1/2) \qquad (10.47)$$

The reduction from $\mathrm{Sp}^{\mathrm{BF}}(4)$ to $\mathrm{SU}^{\mathrm{BF}}(2)$ can be found with the same building-up technique, resulting in the following branching rule. The $\mathrm{Sp}^{\mathrm{BF}}(4)$ representation $(v, 1/2)$ contains the angular momenta

$$J = 2v - 6\nu_\Delta + \tfrac{1}{2}, 2v - 6\nu_\Delta - \tfrac{1}{2}, \ldots, v - 3\nu_\Delta + 1 - \tfrac{1}{4}\left(1 - (-)^{2\nu_\Delta}\right) \qquad (10.48)$$

with

$$\nu_\Delta = 0, \tfrac{1}{2}, 1, \tfrac{3}{2}, \ldots \qquad (10.49)$$

Note that in this case several angular momenta with the same J value may occur in one $\mathrm{Sp}^{\mathrm{BF}}(4)$ representation, distinguished by a different ν_Δ. This shows that the reduction $\mathrm{Sp}^{\mathrm{BF}}(4) \supset \mathrm{SU}^{\mathrm{BF}}(2)$ is not simply reducible and that ν_Δ plays the role of a missing label.

The hamiltonian diagonal in the basis (10.43) can be written in the usual way as a sum of Casimir invariants. Omitting terms that contribute to the binding energy only, it is of the form

$$\begin{aligned}\hat{H}_{\mathrm{II}}[\mathrm{SU}^{\mathrm{BF}}(4)] &= \beta' \mathcal{C}_2[\mathrm{SO}^{\mathrm{B}}(6)] + \beta'' \mathcal{C}_2[\mathrm{SU}^{\mathrm{BF}}(4)] + \zeta \mathcal{C}_2[\mathrm{Sp}^{\mathrm{BF}}(4)] \\ &\quad + \gamma \mathcal{C}_2[\mathrm{SU}^{\mathrm{BF}}(2)]\end{aligned} \qquad (10.50)$$

The Casimir invariant of the boson algebra $\mathrm{SO}^{\mathrm{B}}(6)$ is defined in Section 8.2; the Casimir invariants of the boson–fermion algebras can be written as

$$\begin{aligned}\mathcal{C}_2[\mathrm{SU}^{\mathrm{BF}}(4)] &= 2\hat{J} \cdot \hat{J} + \hat{Q} \cdot \hat{Q} + 2\hat{O} \cdot \hat{O} \\ \mathcal{C}_2[\mathrm{Sp}^{\mathrm{BF}}(4)] &= 2\hat{J} \cdot \hat{J} + 2\hat{O} \cdot \hat{O} \\ \mathcal{C}_2[\mathrm{SU}^{\mathrm{BF}}(2)] &= \hat{J} \cdot \hat{J}\end{aligned} \qquad (10.51)$$

in terms of the generators (10.26). The hamiltonian $\hat{H}_{\mathrm{II}}[\mathrm{SU}^{\mathrm{BF}}(4)]$ has the eigenvalues

$$\begin{aligned}&E_{\mathrm{II}}(\sigma_{\mathrm{b}}, \sigma, v, J) \\ &= \beta' \sigma_{\mathrm{b}}(\sigma_{\mathrm{b}} + 4) + \beta''[\sigma(\sigma + 4) + \sigma'(\sigma' + 2) + \sigma''^2] \\ &\quad + \zeta[v(v+3) + v'(v'+1)] + \gamma J(J+1) \\ &= \beta' \sigma_{\mathrm{b}}(\sigma_{\mathrm{b}} + 4) + \beta''\left[\sigma(\sigma+4) + \tfrac{3}{2}\right] + \zeta\left[v(v+3) + \tfrac{3}{4}\right] \\ &\quad + \gamma J(J+1)\end{aligned} \qquad (10.52)$$

10.4 Example 1: The gold nuclei

The energy spectrum of the hamiltonian is now completely determined from the expression (10.52) together with the branching rules given above.

For an odd-mass nucleus to exhibit the properties of the $SU^{BF}(4)$ limit, two conditions must be satisfied. First, the dominant orbit for the odd nucleon must have angular momentum $j = 3/2$ and, second, the core to which this nucleon is coupled must have an SO(6) symmetry. If both these conditions are met, we must then investigate whether the hamiltonian $\hat{H}_{II}[SU^{BF}(4)]$ contains the correct core–particle interaction terms. Many nuclei in several regions of the nuclear chart satisfy to some degree of approximation the above conditions [16]. We choose here a comparison with one such nucleus, $^{197}_{79}Au_{118}$, not because it is the best example (it is not), but because we use it in our subsequent discussion of supersymmetry and odd–odd nuclei.

In Figure 10.1 we show the energy spectrum of $^{197}_{79}Au_{118}$ and compare it with the $SU^{BF}(4)$ limit. We see that the basic structure of the odd-mass spectrum is correctly reproduced. Refinements can be introduced by considering the admixtures of additional single-particle orbits which are neglected in the present approach. The nuclear shell model predicts in $^{197}_{79}Au_{118}$ a dominant $d_{3/2}$ orbit for the odd proton—in line with the $SU^{BF}(4)$ assumption—but also an $s_{1/2}$ orbit at a slightly higher energy. The effect of this orbit can be properly considered in the context of the IBFM, but only in a numerical approach using the full IBFM hamiltonian (10.22) for two orbits with $j = 1/2, 3/2$.

One problem frequently encountered in the interpretation of nuclear spectra—and in particular those of odd-mass and odd–odd nuclei—is the difficulty in assigning the correct quantum numbers to the observed levels. This is especially so for those quantum numbers for which no simple physical interpretation is known. Standard spectroscopic techniques exist for the determination of the angular momentum J and parity π of a nuclear level. But how can we determine the σ or v quantum number of a level? The energy of a level is certainly an important property, but on its own it is often unreliable for a full characterization of that level, and additional experimental information (such as electromagnetic transition rates and moments or intensities of nucleon-transfer reactions) must be collected to that effect. Electromagnetic or nuclear processes obey certain selection rules in the angular momentum and

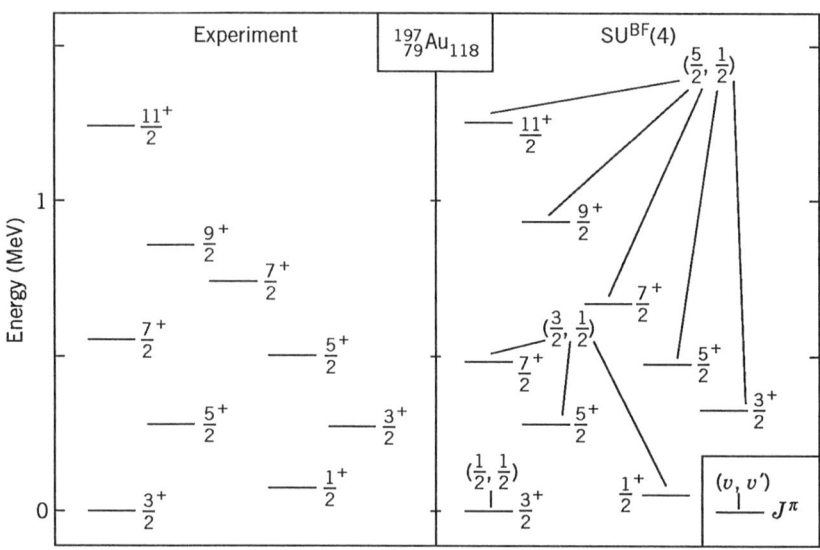

Figure 10.1: Partial eigenspectrum of $\hat{H}_{\text{II}}[\text{SU}^{\text{BF}}(4)]$ for $N = 5$ bosons and $M = 1$ fermion and its comparison to the observed levels of the nucleus ^{197}Au. The calculated spectrum is obtained with the parameters (in keV) $\zeta = 27$ and $\gamma = 28$, with the energy of the ground state normalized to zero. Levels are labeled by (v, v'), and the angular momentum and parity J^π. All levels have $\sigma_{\text{b}} = N$ and $\sigma = N + 1/2$, and hence the values of the parameters β' and β'' are undetermined.

parity, which permits the experimental identification of these quantum numbers; in the same way the σ and v labels can be determined from similar associated selection rules, as discussed in Chapter 8.

We illustrate these rules with the example of electric quadrupole transitions. The E2 transition operator in the $\text{SU}^{\text{BF}}(4)$ limit is of the form

$$\hat{T}^{(\text{E2})}_\mu = q_{\text{b}} i [d^\dagger \times \tilde{s} - s^\dagger \times \tilde{d}]^{(2)}_\mu + q_{\text{f}} [a^\dagger_{3/2} \times \tilde{a}_{3/2}]^{(2)}_\mu \quad (10.53)$$

In many nuclei it is found that $q_{\text{b}} \approx q_{\text{f}}$, and hence as a first-order approximation we may take $q_{\text{b}} = q_{\text{f}} \equiv q$, in which case the E2 transition

10.4 Example 1: The gold nuclei

operator reduces to

$$\hat{T}_\mu^{(\mathrm{E2})} = q\left(i[d^\dagger \times \tilde{s} - s^\dagger \times \tilde{d}]_\mu^{(2)} + [a_{3/2}^\dagger \times \tilde{a}_{3/2}]_\mu^{(2)}\right) = q\hat{Q}_\mu \qquad (10.54)$$

that is, it becomes proportional to the $\mathrm{SU}^{\mathrm{BF}}(4)$ generator. For the operator (10.54) we immediately deduce the selection rule $\Delta\sigma = 0$ since a generator of an algebra cannot connect its different representations. Because $\hat{Q}_\mu$ transforms as a $(1,0)$ tensor under $\mathrm{Sp}^{\mathrm{BF}}(4)$, we can also readily establish the v selection rule by working out the multiplication

$$\begin{aligned}
(v, 1/2) &\otimes (1, 0) \\
&\simeq \{v + 1/2, v - 1/2\} \otimes \{1, 1\} \\
&= \{v - 1/2, v - 3/2\} \oplus \{v + 1/2, v - 1/2\} \\
&\quad \oplus \{v + 3/2, v + 1/2\} \oplus \{v + 3/2, v - 3/2\} \\
&\simeq (v - 1, 1/2) \oplus (v, 1/2) \oplus (v + 1, 1/2) \oplus (v, 3/2) \qquad (10.55)
\end{aligned}$$

where we shuttle back and forth between the symplectic and orthogonal notations. It is always advisable to check in expressions like (10.55) whether left- and right-hand sides have the same dimension, as is indeed confirmed in this case by using the $\mathrm{Sp}(4)$ dimension formula. The result (10.55) implies that E2 transitions obey the selection rule $\Delta v = 0, \pm 1$. Note, however, that there is an important difference between the σ or v selection rules and the one for angular momentum ($\Delta J = 0, \pm 1, \pm 2$ for quadrupole transitions). The latter is rigorously valid since the rotational invariance of the hamiltonian is fundamental and exact; the σ and v selection rules are, at best, only approximate since the goodness of the associated quantum numbers depends on the specific form chosen for the hamiltonian, which is not a fundamental question but rather a phenomenological one.

In addition to the knowledge of selection rules for various processes, predictions for the actual strengths or intensities of the allowed transitions would be useful. These are related to matrix elements of the associated transition operator and can, in a symmetry limit, be obtained in closed form, as we now proceed to illustrate with the example of E2 transitions in the $\mathrm{SU}^{\mathrm{BF}}(4)$ limit.

The first task at hand is to derive an expression for the wave function of an $\mathrm{SU}^{\mathrm{BF}}(4)$ state. Formally, it can be expanded in terms of a weak-coupling basis,

$$
\begin{aligned}
&|[N](\sigma_{\rm b},0,0),[1](1/2,1/2,1/2);(\sigma,1/2,1/2)(v,1/2)\nu_\Delta J M_J\rangle \\
&= \sum_{v_{\rm b} n_\Delta L_{\rm b}} \left\langle \begin{array}{cc} (\sigma_{\rm b},0,0) & (1/2,1/2,1/2) \\ (v_{\rm b},0) & (1/2,1/2) \end{array} \middle| \begin{array}{c} (\sigma,1/2,1/2) \\ (v,1/2) \end{array} \right\rangle \\
&\quad \times \left\langle \begin{array}{cc} (v_{\rm b},0) & (1/2,1/2) \\ n_\Delta L_{\rm b} & 3/2 \end{array} \middle| \begin{array}{c} (v,1/2) \\ \nu_\Delta J \end{array} \right\rangle \\
&\quad \times |[N](\sigma_{\rm b},0,0)(v_{\rm b},0)n_\Delta L_{\rm b},[1](1/2,1/2,1/2)(1/2,1/2)3/2; JM_J\rangle
\end{aligned}
$$
(10.56)

where the coefficients are isoscalar factors, the first associated with SU(4) ⊃ Sp(4) and the second with Sp(4) ⊃ SU(2). We are thus faced with the calculation of these isoscalar factors, and in contrast to similar cases considered in Parts I and II, this problem cannot be reduced to one of standard angular momentum coupling and recoupling. Several methods exist for finding analytic expressions for isoscalar factors, and we present here one [11] which is conceptually straightforward but tedious in its execution.

We begin by introducing a simplified notation in which only the σ, v, and J quantum numbers of a state are shown. The expansion in a weak-coupling basis can then be written as

$$|\sigma,v,J\rangle = \sum_{v_{\rm b} L_{\rm b}} \xi^{\sigma_{\rm b} v_{\rm b} L_{\rm b}}_{\sigma v L} |(\sigma_{\rm b},v_{\rm b},L_{\rm b}),(1/2,1/2,3/2);J\rangle \qquad (10.57)$$

where we also introduce the short-hand notation $\xi^{\sigma_{\rm b} v_{\rm b} L_{\rm b}}_{\sigma v L}$ for the SU(4) ⊃ Sp(4) ⊃ SU(2) coupling coefficients. Furthermore, we assume that no missing labels (either n_Δ or ν_Δ) are necessary. The idea of the technique is to calculate matrix elements of an appropriate operator between an SU$^{\rm BF}$(4) and a weak-coupling state. Using two different forms of the operator—one in terms of Casimir invariants of boson–fermion algebras and another one in terms of Casimir invariants of the separate algebras—one derives relations among the ξ coefficients. In the present example, the appropriate operator is $\hat{Q}\cdot\hat{Q}$, where $\hat{Q}_\mu$ is the quadrupole generator of SU$^{\rm BF}$(4),

$$\hat{Q}_\mu = i[d^\dagger \times \tilde{s} - s^\dagger \times \tilde{d}]^{(2)}_\mu + [a^\dagger_{3/2} \times \tilde{a}_{3/2}]^{(2)}_\mu \equiv \hat{Q}_{{\rm b}\mu} + \hat{Q}_{{\rm f}\mu} \qquad (10.58)$$

This operator can be written as

$$\hat{Q}\cdot\hat{Q} = \mathcal{C}_2[{\rm SU}^{\rm BF}(4)] - \mathcal{C}_2[{\rm Sp}^{\rm BF}(4)] \qquad (10.59)$$

10.4 Example 1: The gold nuclei

or, alternatively, as

$$\begin{aligned}\hat{Q}\cdot\hat{Q} &= \hat{Q}_\mathrm{b}\cdot\hat{Q}_\mathrm{b}+\hat{Q}_\mathrm{f}\cdot\hat{Q}_\mathrm{f}+2\hat{Q}_\mathrm{b}\cdot\hat{Q}_\mathrm{f} \\ &= \mathcal{C}_2[\mathrm{SO^B}(6)]-\mathcal{C}_2[\mathrm{SO^B}(5)]+\mathcal{C}_2[\mathrm{SU^F}(4)]-\mathcal{C}_2[\mathrm{Sp^F}(4)] \\ &\quad +2\hat{Q}_\mathrm{b}\cdot\hat{Q}_\mathrm{f}\end{aligned} \quad (10.60)$$

Consider now the matrix element of $\hat{Q}\cdot\hat{Q}$ between the $\mathrm{SU^{BF}}(4)$ state

$$|\sigma_\mathrm{b}+1/2,v_\mathrm{b}+1/2,J\rangle \qquad (10.61)$$

and the weak-coupling state

$$|(\sigma_\mathrm{b},v_\mathrm{b},L_\mathrm{b}),(1/2,1/2,3/2);J\rangle \qquad (10.62)$$

Using the form (10.59), we find

$$\begin{aligned}&\langle(\sigma_\mathrm{b},v_\mathrm{b},L_\mathrm{b}),(1/2,1/2,3/2);J|\hat{Q}\cdot\hat{Q}|\sigma_\mathrm{b}+1/2,v_\mathrm{b}+1/2,J\rangle \\ &= \left[\left(\sigma_\mathrm{b}+\tfrac{1}{2}\right)\left(\sigma_\mathrm{b}+\tfrac{9}{2}\right)+\tfrac{3}{2}-\left(v_\mathrm{b}+\tfrac{1}{2}\right)\left(v_\mathrm{b}+\tfrac{7}{2}\right)-\tfrac{3}{4}\right]\xi^{\sigma_\mathrm{b}v_\mathrm{b}L_\mathrm{b}}_{\sigma_\mathrm{b}+1/2\,v_\mathrm{b}+1/2\,J}\end{aligned} \quad (10.63)$$

In turn, the form (10.60) of $\hat{Q}\cdot\hat{Q}$ gives rise to

$$\begin{aligned}&\langle(\sigma_\mathrm{b},v_\mathrm{b},L_\mathrm{b}),(1/2,1/2,3/2);J|\hat{Q}\cdot\hat{Q}|\sigma_\mathrm{b}+1/2,v_\mathrm{b}+1/2,J\rangle \\ &= \left[\sigma_\mathrm{b}(\sigma_\mathrm{b}+4)-v_\mathrm{b}(v_\mathrm{b}+3)+\tfrac{15}{4}-\tfrac{5}{2}\right]\xi^{\sigma_\mathrm{b}v_\mathrm{b}L_\mathrm{b}}_{\sigma_\mathrm{b}+1/2\,v_\mathrm{b}+1/2\,J} \\ &\quad +2\sum_{v'_\mathrm{b}L'_\mathrm{b}}(-)^{J+L'_\mathrm{b}+3/2}\sqrt{4(2L_\mathrm{b}+1)}\begin{Bmatrix}L_\mathrm{b}&3/2&J\\3/2&L'_\mathrm{b}&2\end{Bmatrix}\xi^{\sigma_\mathrm{b}v'_\mathrm{b}L'_\mathrm{b}}_{\sigma_\mathrm{b}+1/2\,v_\mathrm{b}+1/2\,J} \\ &\quad \times\langle\sigma_\mathrm{b},v_\mathrm{b},L_\mathrm{b}\|\hat{Q}_\mathrm{b}\|\sigma_\mathrm{b},v'_\mathrm{b},L'_\mathrm{b}\rangle\langle 1/2,1/2,3/2\|\hat{Q}_\mathrm{f}\|1/2,1/2,3/2\rangle\end{aligned} \quad (10.64)$$

From Section 8.5 we know that the reduced matrix element of $\hat{Q}_\mathrm{b}$ vanishes unless $v'_\mathrm{b}=v_\mathrm{b}\pm 1$. Since $v'_\mathrm{b}=(v_\mathrm{b}+1/2)\pm 1/2$, we conclude that only one v'_b value is allowed, $v'_\mathrm{b}=v_\mathrm{b}+1$. The fermion reduced matrix element is simple to evaluate:

$$\langle 1/2,1/2,3/2\|\hat{Q}_\mathrm{f}\|1/2,1/2,3/2\rangle \equiv \langle 3/2\|[a^\dagger_{3/2}\times\tilde{a}_{3/2}]^{(2)}\|3/2\rangle = \sqrt{\tfrac{5}{4}} \quad (10.65)$$

Equating the two forms of the $\hat{Q}\cdot\hat{Q}$ matrix elements leads to the relation

$$\xi^{\sigma_b v_b L_b}_{\sigma_b+1/2\, v_b+1/2\, J}$$
$$= \frac{2\sqrt{5(2L_b+1)}}{\sigma_b - v_b} \sum_{L'_b} (-)^{J+L'_b+3/2} \begin{Bmatrix} L_b & 3/2 & J \\ 3/2 & L'_b & 2 \end{Bmatrix}$$
$$\times \langle \sigma_b, v_b, L_b \| \hat{Q}_b \| \sigma_b, v_b+1, L'_b \rangle \xi^{\sigma_b v_b+1\, L'_b}_{\sigma_b+1/2\, v_b+1/2\, J} \qquad (10.66)$$

The reduced matrix elements of the boson generator $\hat{Q}_b$ are known from Chapter 8. A second relation is found by calculating the matrix element of $\hat{Q}\cdot\hat{Q}$ between

$$|\sigma_b + 1/2, v_b - 1/2, J\rangle \qquad (10.67)$$

and a weak-coupling state, resulting in

$$\xi^{\sigma_b v_b L_b}_{\sigma_b+1/2\, v_b-1/2\, J}$$
$$= \frac{2\sqrt{5(2L_b+1)}}{\sigma_b + v_b + 3} \sum_{L'_b} (-)^{J+L'_b+3/2} \begin{Bmatrix} L_b & 3/2 & J \\ 3/2 & L'_b & 2 \end{Bmatrix}$$
$$\times \langle \sigma_b, v_b, L_b \| \hat{Q}_b \| \sigma_b, v_b-1, L'_b \rangle \xi^{\sigma_b v_b-1\, L'_b}_{\sigma_b+1/2\, v_b-1/2\, J} \qquad (10.68)$$

These two relations in the ξ coefficients, together with the normalization condition

$$\sum_{v_b L_b} \left(\xi^{\sigma_b v_b L_b}_{\sigma_b+1/2\, vJ} \right)^2 = 1 \qquad (10.69)$$

completely determine the isoscalar factors in all multiplicity-free cases. For example, consider $v = v_b - 1/2$ and $J = 2v_b - 1/2$, in which case we find

$$\xi^{\sigma_b v_b L_b}_{\sigma_b+1/2\, v_b-1/2\, 2v_b-1/2}$$
$$= \frac{2\sqrt{5(2L_b+1)}}{\sigma_b + v_b + 3} \begin{Bmatrix} L_b & 3/2 & 2v_b-1/2 \\ 3/2 & 2v_b-2 & 2 \end{Bmatrix} \xi^{\sigma_b v_b-1\, 2v_b-2}_{\sigma_b+1/2\, v_b-1/2\, 2v_b-1/2}$$
$$\times \langle \sigma_b, v_b, L_b \| \hat{Q}_b \| \sigma_b, v_b-1, 2v_b-2 \rangle \qquad (10.70)$$

10.4 Example 1: The gold nuclei

Equation (10.70) can be applied for $L_{\rm b}' = 2v_{\rm b}$ and $L_{\rm b}' = 2v_{\rm b} - 2$. Taking the ratio of the two, we find

$$\frac{\xi^{\sigma_{\rm b} v_{\rm b} \, 2v_{\rm b}}_{\sigma_{\rm b}+1/2 \, v_{\rm b}-1/2 \, 2v_{\rm b}-1/2}}{\xi^{\sigma_{\rm b} v_{\rm b} \, 2v_{\rm b}-2}_{\sigma_{\rm b}+1/2 \, v_{\rm b}-1/2 \, 2v_{\rm b}-1/2}} = \frac{\sqrt{4v_{\rm b}+1} \left\{ \begin{array}{ccc} 2v_{\rm b} & 3/2 & 2v_{\rm b}-1/2 \\ 3/2 & 2v_{\rm b}-2 & 2 \end{array} \right\}}{\sqrt{4v_{\rm b}-3} \left\{ \begin{array}{ccc} 2v_{\rm b}-2 & 3/2 & 2v_{\rm b}-1/2 \\ 3/2 & 2v_{\rm b}-2 & 2 \end{array} \right\}}$$

$$\times \frac{\langle \sigma_{\rm b}, v_{\rm b}, 2v_{\rm b} \| \hat{Q}_{\rm b} \| \sigma_{\rm b}, v_{\rm b}-1, 2v_{\rm b}-2 \rangle}{\langle \sigma_{\rm b}, v_{\rm b}, 2v_{\rm b}-2 \| \hat{Q}_{\rm b} \| \sigma_{\rm b}, v_{\rm b}-1, 2v_{\rm b}-2 \rangle} \qquad (10.71)$$

Since the $SU(4) \supset Sp(4)$ isoscalar factor is common to the ξ coefficient in numerator and denominator, this represents a ratio of two $Sp(4) \supset SU(2)$ isoscalar factors. Inserting the expressions

$$\langle \sigma_{\rm b}, v_{\rm b}, 2v_{\rm b} \| \hat{Q}_{\rm b} \| \sigma_{\rm b}, v_{\rm b}-1, 2v_{\rm b}-2 \rangle$$
$$= \left[\frac{(\sigma_{\rm b} - v_{\rm b} + 1)(\sigma_{\rm b} + v_{\rm b} + 3)v_{\rm b}}{2v_{\rm b} + 3} \right]^{1/2}$$
$$\langle \sigma_{\rm b}, v_{\rm b}, 2v_{\rm b}-2 \| \hat{Q}_{\rm b} \| \sigma_{\rm b}, v_{\rm b}-1, 2v_{\rm b}-2 \rangle \qquad (10.72)$$
$$= \left[\frac{2(\sigma_{\rm b} - v_{\rm b} + 1)(\sigma_{\rm b} + v_{\rm b} + 3)(2v_{\rm b} - 1)}{(2v_{\rm b} + 3)(4v_{\rm b} - 5)} \right]^{1/2}$$

known from Chapter 8, and

$$\left\{ \begin{array}{ccc} 2v_{\rm b} & 3/2 & 2v_{\rm b}-1/2 \\ 3/2 & 2v_{\rm b}-2 & 2 \end{array} \right\} = -\left[\frac{2v_{\rm b}+1}{20v_{\rm b}(4v_{\rm b}-1)} \right]^{1/2}$$
$$\left\{ \begin{array}{ccc} 2v_{\rm b}-2 & 3/2 & 2v_{\rm b}-1/2 \\ 3/2 & 2v_{\rm b}-2 & 2 \end{array} \right\} \qquad (10.73)$$
$$= -\left[\frac{(v_{\rm b}-1)(4v_{\rm b}-5)}{10(2v_{\rm b}-1)(4v_{\rm b}-3)(4v_{\rm b}-1)} \right]^{1/2}$$

taken from [20], we obtain

$$\frac{\left\langle \begin{array}{cc} (v_{\rm b},0) & (1/2,1/2) \\ 2v_{\rm b} & 3/2 \end{array} \middle| \begin{array}{c} (v_{\rm b}-1/2,1/2) \\ 2v_{\rm b}-1/2 \end{array} \right\rangle}{\left\langle \begin{array}{cc} (v_{\rm b},0) & (1/2,1/2) \\ 2v_{\rm b}-2 & 3/2 \end{array} \middle| \begin{array}{c} (v_{\rm b}-1/2,1/2) \\ 2v_{\rm b}-1/2 \end{array} \right\rangle} = \left[\frac{(2v_{\rm b}+1)(4v_{\rm b}+1)}{4(v_{\rm b}-1)} \right]^{1/2}$$

$$(10.74)$$

or, using the orthonormality of the Sp(4) ⊃ SU(2) isoscalar factors,

$$\left\langle \begin{matrix} (v_b,0) & (1/2,1/2) \\ 2v_b & 3/2 \end{matrix} \middle| \begin{matrix} (v_b - 1/2, 1/2) \\ 2v_b - 1/2 \end{matrix} \right\rangle = \left[\frac{(2v_b + 1)(4v_b + 1)}{(2v_b + 3)(4v_b - 1)} \right]^{1/2}$$

$$\left\langle \begin{matrix} (v_b,0) & (1/2,1/2) \\ 2v_b - 2 & 3/2 \end{matrix} \middle| \begin{matrix} (v_b - 1/2, 1/2) \\ 2v_b - 1/2 \end{matrix} \right\rangle = \left[\frac{4(v_b - 1)}{(2v_b + 3)(4v_b - 1)} \right]^{1/2}$$

(10.75)

We have thus found analytic expressions for the Sp(4) ⊃ SU(2) isoscalar factor $\langle (v_b,0)L_b \otimes (1/2,1/2)3/2 | (v,1/2)J \rangle$ in the specific case $J = 2v_b - 1/2$ and $v = v_b - 1/2$. The isoscalar factors with $v = v_b + 1/2$ are easily obtained from orthogonality. Analogous derivations become increasingly difficult for lower values of J. Due to the missing label in the Sp(4) ⊃ SU(2) reduction, no general, closed formula exists for the associated isoscalar factor, but many cases of interest are available (see, e.g., [5]).

We may now introduce the expressions for the Sp(4) ⊃ SU(2) isoscalar factors into equation (10.70) (either for $L_b = 2v_b$ or $L_b = 2v_b - 2$, since both lead to the same result) to find

$$\frac{\left\langle \begin{matrix} (\sigma_b,0,0) & (1/2,1/2,1/2) \\ (v_b,0) & (1/2,1/2) \end{matrix} \middle| \begin{matrix} (\sigma_b + 1/2, 1/2, 1/2) \\ (v_b - 1/2, 1/2) \end{matrix} \right\rangle}{\left\langle \begin{matrix} (\sigma_b,0,0) & (1/2,1/2,1/2) \\ (v_b - 1,0) & (1/2,1/2) \end{matrix} \middle| \begin{matrix} (\sigma_b + 1/2, 1/2, 1/2) \\ (v_b - 1/2, 1/2) \end{matrix} \right\rangle}$$

$$= \left[\frac{\sigma_b - v_b + 1}{\sigma_b + v_b + 3} \right]^{1/2} \left\langle \begin{matrix} (v_b - 1, 0) & (1/2, 1/2) \\ 2v_b - 2 & 3/2 \end{matrix} \middle| \begin{matrix} (v_b - 1/2, 1/2) \\ 2v_b - 1/2 \end{matrix} \right\rangle$$

(10.76)

The Sp(4) ⊃ SU(2) isoscalar factor on the right-hand side of this equation reduces to unity since it is *stretched*, that is, given that $J = 2v_b - 1/2$, only one L_b value is possible, namely $L_b = 2v_b - 2$. Taking account of the orthonormality of the SU(4) ⊃ Sp(4) isoscalar factors, we then obtain

$$\left\langle \begin{matrix} (\sigma_b,0,0) & (1/2,1/2,1/2) \\ (v_b,0) & (1/2,1/2) \end{matrix} \middle| \begin{matrix} (\sigma_b + 1/2, 1/2, 1/2) \\ (v_b - 1/2, 1/2) \end{matrix} \right\rangle = \left[\frac{\sigma_b - v_b + 1}{2(\sigma_b + 2)} \right]^{1/2}$$

10.4 Example 1: The gold nuclei

$$\left\langle \begin{matrix} (\sigma_{\rm b},0,0) & (^{1}/_{2},^{1}/_{2},^{1}/_{2}) \\ (v_{\rm b}-1,0) & (^{1}/_{2},^{1}/_{2}) \end{matrix} \middle| \begin{matrix} (\sigma_{\rm b}+^{1}/_{2},^{1}/_{2},^{1}/_{2}) \\ (v_{\rm b}-^{1}/_{2},^{1}/_{2}) \end{matrix} \right\rangle = \left[\frac{\sigma_{\rm b}+v_{\rm b}+3}{2(\sigma_{\rm b}+2)} \right]^{1/2} \tag{10.77}$$

Note that these are general expressions for the SU(4) $\supset$ Sp(4) isoscalar factors covering all cases that occur in the SU$^{\rm BF}$(4) limit.

We have given here a rather detailed derivation with the purpose to illustrate how isoscalar factors can be calculated from basic principles using a method that is relatively simple in its idea. We note, however, that more powerful techniques have been developed to deal with this problem [18,19], but they involve rather refined group-theoretical concepts which are outside the scope of this book.

Calculations of a similar type (but for different algebra reductions) are repeatedly required in the study of the various symmetry limits of the IBFM. They are an essential part of the analysis since without them no quantitative predictions can be made as regards the intensities of electromagnetic transitions and nucleon-transfer processes. Henceforth we do not carry out calculations of isoscalar factors—since they are invariably tedious but all follow the same idea—but content ourselves with citing the appropriate reference.

Returning to the example of E2 transitions in the SU$^{\rm BF}$(4) limit, we can now readily derive the matrix elements of the E2 transition operator (10.54) using the expansion (10.57), resulting in

$$\langle \sigma, v, J \| \hat{Q} \| \sigma, v', J' \rangle$$
$$= \sqrt{(2J'+1)} \sum_{\sigma_{\rm b} v_{\rm b} L_{\rm b}} \sum_{\sigma'_{\rm b} v'_{\rm b} L'_{\rm b}} \xi^{\sigma_{\rm b} v_{\rm b} L_{\rm b}}_{\sigma v J} \xi^{\sigma_{\rm b} v'_{\rm b} L'_{\rm b}}_{\sigma v' J'} (-)^{J'+L_{\rm b}+3/2}$$
$$\times \left[\sqrt{2L_{\rm b}+1} \begin{Bmatrix} L & J & 3/2 \\ J' & L' & 2 \end{Bmatrix} \langle \sigma_{\rm b}, v_{\rm b}, L_{\rm b} \| \hat{Q}_{\rm b} \| \sigma_{\rm b}, v'_{\rm b}, L'_{\rm b} \rangle \delta_{\sigma_{\rm b}\sigma'_{\rm b}} \right.$$
$$\left. +(-)^{J-J'}\sqrt{5} \begin{Bmatrix} 3/2 & J & L \\ J' & 3/2 & 2 \end{Bmatrix} \delta_{v_{\rm b}v'_{\rm b}} \delta_{L_{\rm b}L'_{\rm b}} \right] \tag{10.78}$$

Insofar as the ξ coefficients are known—which is the case for all states at low energies—this expression determines the strength of the E2 transition, with the corresponding B(E2) value related to the reduced matrix element in the usual way (see (8.104)). The B(E2) values of transitions

10 The Interacting Boson–Fermion Model

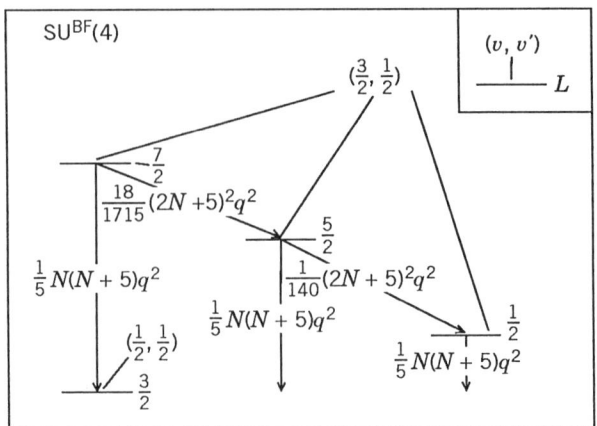

Figure 10.2: Allowed transitions in the $\mathrm{SU^{BF}}(4)$ limit induced by the quadrupole generator $q\hat{Q}_\mu$ of $\mathrm{SU^{BF}}(4)$. The expressions alongside the arrows are the $B(E2)$ values. Labels are as in Figure 10.1.

between states belonging to the $v = 1/2$ and $v = 3/2$ multiplets are shown in Figure 10.2. Note that the $\Delta v = 0$ E2 transition strengths are predicted to be much smaller than those with $\Delta v = 1$, a result which persists for higher v values. Another quantitative prediction is that all three transitions from the $v = 3/2$ multiplet to the ground state have the same $B(E2)$ values. This represents a parameter-free test of the symmetry.

In Table 10.1 we present a comparison between the measured and the calculated $B(E2)$ values in the nucleus $^{197}_{79}\mathrm{Au}_{118}$. The three $v = 3/2$ to $v = 1/2$ transitions are indeed observed to have equal $B(E2)$ values to an approximation of about 10%, and $\Delta v > 1$ transitions are seen to be significantly weaker.

The $\mathrm{SU^{BF}}(4)$ limit has been applied to other gold and iridium isotopes, and in general, a reasonable agreement with the data is obtained for E2 transition rates. The limit is much less successful, however, in reproducing the observed M1 rates, presumably due to the incompleteness of the single-particle space.

10.5 Example 2: The platinum nuclei

Table 10.1: $B(\text{E2})$ values in ^{197}Au

$E_\text{i}{}^a$	$(\sigma, v, J)_\text{i}$	$E_\text{f}{}^a$	$(\sigma, v, J)_\text{f}$	$B(\text{E2}; J_\text{i} \to J_\text{f})^b$	
				Expt	Theory^c
77	$11/2, 3/2, 1/2$	0	$11/2, 1/2, 3/2$	0.260(14)	0.231
279	$11/2, 3/2, 5/2$	0	$11/2, 1/2, 3/2$	0.209(5)	0.231
547	$11/2, 3/2, 7/2$	0	$11/2, 1/2, 3/2$	0.226(9)	0.231
269	$11/2, 5/2, 3/2$	0	$11/2, 1/2, 3/2$	0.083(6)	0
503	$11/2, 5/2, 5/2$	0	$11/2, 1/2, 3/2$	≤ 0.003	0
737	$11/2, 5/2, 7/2$	0	$11/2, 1/2, 3/2$	≤ 0.004	0
855	$11/2, 5/2, 9/2$	279	$11/2, 3/2, 5/2$	0.258(38)	0.228
1231	$11/2, 5/2, 11/2$	547	$11/2, 3/2, 7/2$	0.269(60)	0.290

^{a} In units of keV.
^{b} In units of $e^2 b^2$.
^{c} With $q_\text{b} - q_\text{f} = 0.152$ eb.

10.5 Example 2: The platinum nuclei

In this section we analyze in some detail the properties of the dynamical symmetry corresponding to route a of the lattice (10.30),

$$\begin{array}{ccccccccc}
\text{U}^\text{B}(6) \otimes \text{U}^\text{F}(12) & \supset & \text{U}^\text{B}(6) & \otimes & \text{U}^\text{F}(6) & \otimes & \text{SU}^\text{F}(2) & & \\
\downarrow & & \downarrow & & \downarrow & & \downarrow & & \\
[N] \quad [1] & & [N] & & [1] & & 1/2 & & \\
\\
\supset \text{U}^\text{BF}(6) \otimes \text{SU}^\text{F}(2) & \supset & \text{SO}^\text{BF}(6) & \otimes & \text{SU}^\text{F}(2) & & & & \\
\downarrow \qquad\qquad \downarrow & & \downarrow & & \downarrow & & & & \\
[h, h'] \qquad 1/2 & & (\sigma, \sigma', \sigma'') & & 1/2 & & & & \\
\\
\supset \text{SO}^\text{BF}(5) \otimes \text{SU}^\text{F}(2) & \supset & \text{SO}^\text{BF}(3) & \otimes & \text{SU}^\text{F}(2) & \supset & \text{SU}^\text{BF}(2) & & \\
\downarrow \qquad\qquad \downarrow & & \downarrow & & \downarrow & & \downarrow & & \\
(v, v') \qquad 1/2 & & \nu_\Delta L & & 1/2 & & J & &
\end{array} \quad (10.79)$$

which we refer to as the $\text{SO}^\text{BF}(6) \otimes \text{SU}^\text{F}(2)$ limit. This type of symmetry has been investigated in detail with extensive applications to odd-mass platinum nuclei. The first study of this kind [21] followed route b of the lattice (10.30) and compared with the energy spectrum of $^{195}_{78}\text{Pt}_{117}$.

Subsequent work indicated that route a is more appropriate and extended the comparison to more levels at higher energies as well as to E2 transition rates and neutron-transfer strengths. An up-to-date assignment to levels in $^{195}_{78}\text{Pt}_{117}$ is discussed in [22] while a general review of the applicabiltiy of the $\text{SO}^{\text{BF}}(6) \otimes \text{SU}^{\text{F}}(2)$ limit is given in [16].

The labels introduced in (10.79) can be found in the usual way through the application of branching and multiplication rules. The allowed values of h and h' are obtained from the U(6) multiplication $[N] \otimes [1]$,

$$[N] \otimes [1] = [N+1,0] \oplus [N,1] \qquad (10.80)$$

For the symmetric representation $[N+1]$ of $\text{U}^{\text{BF}}(6)$ all subsequent branching rules are known from Chapter 8 and are not repeated here. Those for the $\text{U}^{\text{BF}}(6)$ representation $[N,1]$ can be derived with the building-up technique explained in the previous section. The derivation here is entirely analogous and we only quote the results. The $\text{SO}^{\text{BF}}(6)$ representations contained in $[N,1]$ are

$$\begin{aligned}(\sigma,\sigma',\sigma'') &= (N-1,0,0),(N-3,0,0),\ldots,\left\{\begin{array}{c}(2,0,0)\\(1,0,0)\end{array}\right\},\\ &\quad (N,1,0),(N-2,1,0),\ldots,\left\{\begin{array}{c}(1,1,0)\\(2,1,0)\end{array}\right\}\end{aligned} \qquad (10.81)$$

where the upper (lower) labels in the curly brackets should be taken for odd (even) N. The $\text{SO}^{\text{BF}}(5)$ representations contained in $(\sigma,0,0)$ are again known from Chapter 8 while those contained in $(\sigma,1,0)$ are

$$\begin{aligned}(v,v') &= (\sigma,0),(\sigma-1,0),\ldots,(1,0),\\ &\quad (\sigma,1),(\sigma-1,1),\ldots,(1,1)\end{aligned} \qquad (10.82)$$

The allowed L values in the $\text{SO}^{\text{BF}}(5)$ representation $(v,1)$ are

$$L = 2v - 6\nu_\Delta + 1, 2v - 6\nu_\Delta, \ldots, v - 3\nu_\Delta + 1 + \delta_{\nu_\Delta 0} \qquad (10.83)$$

with

$$\nu_\Delta = 0, \tfrac{1}{3}, \tfrac{2}{3}, 1, \ldots \qquad (10.84)$$

Clearly, the total angular momentum can be $J = L - 1/2$ or $J = L + 1/2$.

10.5 Example 2: The platinum nuclei

The $SO^{BF}(6) \otimes SU^F(2)$ hamiltonian is an expansion in terms of Casimir invariants of the algebras in (10.79),

$$\begin{aligned}\hat{H}_{\text{II}}[SO^{BF}(6) \otimes SU^F(2)] \\ = \eta \mathcal{C}_2[U^{BF}(6)] + \beta \mathcal{C}_2[SO^{BF}(6)] + \zeta \mathcal{C}_2[SO^{BF}(5)] \\ + \gamma' \mathcal{C}_2[SO^{BF}(3)] + \gamma'' \mathcal{C}_2[SU^{BF}(2)]\end{aligned} \quad (10.85)$$

where terms that only contribute to the binding energy are omitted. The explicit expressions of the Casimir invariants are given in Chapter 8 but now involve the boson–fermion generators $\mathcal{G}_\mu^{(\lambda)}(l,l')$ defined in (10.29) instead of the boson generators $\mathcal{B}_\mu^{(\lambda)}(l,l')$. The hamiltonian $\hat{H}_{\text{II}}[SO^{BF}(6) \otimes SU^F(2)]$ has the eigenvalues

$$\begin{aligned}E_{\text{II}}(h, h', \sigma, \sigma', v, v', L, J) \\ = \eta[h(h+5) + h'(h'+3)] + \beta[\sigma(\sigma+4) + \sigma'(\sigma'+2)] \\ + \zeta[v(v+3) + v'(v'+1)] \\ + \gamma' L(L+1) + \gamma'' J(J+1)\end{aligned} \quad (10.86)$$

The expression (10.86), together with the branching rules, fully determines the energy spectrum of $\hat{H}_{\text{II}}[SO^{BF}(6) \otimes SU^F(2)]$.

In the application of the $SO^{BF}(6) \otimes SU^F(2)$ limit one is, as in the example of the previous section, restricted to odd-mass nuclei with an SO(6) symmetry for the core. In addition, the odd nucleon must occupy orbits with $j = 1/2, 3/2, 5/2$. Both these conditions are satisfied in several odd-mass platinum isotopes when the neutron occupies the $p_{1/2}$, $p_{3/2}$, and $f_{5/2}$ orbits of the $N = 5$ major shell. In Figure 10.3 we show the energy spectrum of $^{197}_{78}Pt_{119}$ and compare it with the $SO^{BF}(6) \otimes SU^F(2)$ limit. This nucleus displays a complex type of excitation spectrum, but nevertheless a one-to-one correspondence can be established between observed and calculated levels. This correspondence is all the more remarkable since the measured spectrum is known to be complete up to about 600 keV, that is, it includes *all* levels up to that energy. Such complete spectroscopy has proven to be a stringent test of nuclear models in general and has led to a much more reliable assignment of the various quantum numbers in this particular example.

In complex spectra of the type shown in Figure 10.3 it is important to verify the assignment of quantum numbers, proposed on the basis of

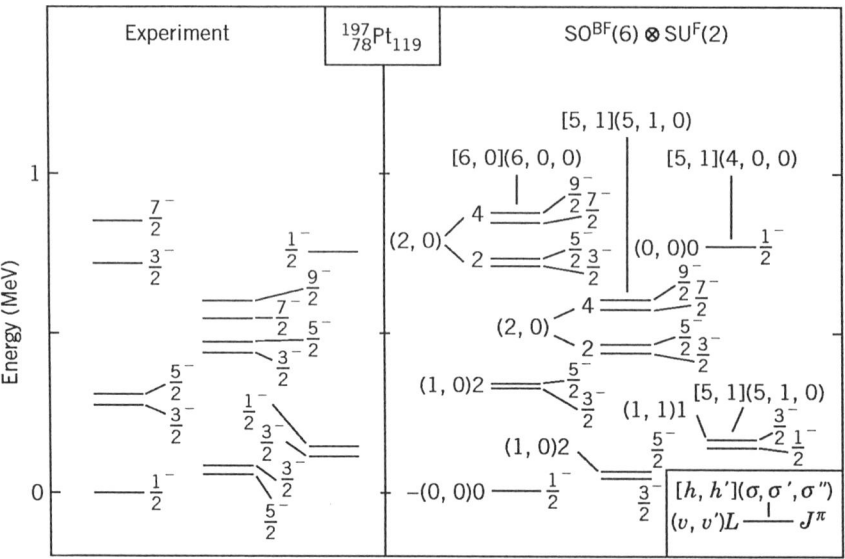

Figure 10.3: Partial eigenspectrum of $\hat{H}_{\mathrm{II}}[\mathrm{SO}^{\mathrm{BF}}(6) \otimes \mathrm{SU}^{\mathrm{F}}(2)]$ for $N = 5$ bosons and $M = 1$ fermion and its comparison to the observed levels of the nucleus ^{197}Pt. The calculated spectrum is obtained with the parameters (in keV) $\eta = 86$, $\beta = -64$, $\zeta = 65$, $\gamma' = 7$, and $\gamma'' = 3$, with the energy of the ground state normalized to zero. Levels are labeled by $[h, h']$, $(\sigma, \sigma', \sigma'')$, (v, v'), L, and the angular momentum and parity J^π.

energies, against other experimental data. Not much is known about the electromagnetic decay properties of $^{197}_{78}\mathrm{Pt}_{119}$, but in the neighboring isotope, $^{195}_{78}\mathrm{Pt}_{117}$, a wealth of information about electric quadrupole transitions is available and hence it seems natural to study the predictions of the $\mathrm{SO}^{\mathrm{BF}}(6) \otimes \mathrm{SU}^{\mathrm{F}}(2)$ limit with respect to E2 transition rates in this nucleus. The E2 transition operator in the $\mathrm{SO}^{\mathrm{BF}}(6) \otimes \mathrm{SU}^{\mathrm{F}}(2)$ limit is of the form

$$\hat{T}^{(\mathrm{E2})}_\mu = q_{\mathrm{b}} i [d^\dagger \times \tilde{s} - s^\dagger \times \tilde{d}]^{(2)}_\mu + q_{\mathrm{f}} i [\mathcal{A}^{(2)}_\mu(2,0) - \mathcal{A}^{(2)}_\mu(0,2)]$$
$$\equiv q_{\mathrm{b}} \hat{Q}_{\mathrm{b}\mu} + q_{\mathrm{f}} \hat{Q}_{\mathrm{f}\mu} \qquad (10.87)$$

Contrary to the case discussed in the previous section, $q_{\mathrm{b}} \approx q_{\mathrm{f}}$ is not a good approximation. In fact, microscopic calculations suggest that in

10.5 Example 2: The platinum nuclei

the platinum isotopes one should take $q_b \approx -q_f$, and in any case the E2 transition operator cannot be taken as proportional to the $SO^{BF}(6)$ generator. We may thus expect some non-zero E2 transitions between levels belonging to different $U^{BF}(6)$ or $SO^{BF}(6)$ representations, which would have been forbidden if $q_b = q_f$.

The calculation of the transition strengths proceeds by expanding the $SO^{BF}(6) \otimes SU^{F}(2)$ states (10.79) in terms of a weak-coupling basis,

$$|[N],[1];[h,h'](\sigma,\sigma',0)(v,v')\nu_\Delta L, 1/2; JM_J\rangle$$
$$= \sum_{\sigma_b v_b \nu_\Delta L_b} \sum_{v_f L_f} \xi^{\sigma_b v_b n_\Delta L_b, v_f L_f}_{[h,h'](\sigma,\sigma',0)(v,v')\nu_\Delta L}$$
$$\times |[N](\sigma_b,0,0)(v_b,0)n_\Delta L_b, [1](1,0,0)(v_f,0)L_f; L, 1/2; JM_J\rangle$$
(10.88)

where the transformation brackets are coupling coefficients associated with the reduction $U(6) \supset SO(6) \supset SO(5) \supset SO(3)$. According to Racah's factorization lemma, the ξ coefficients can be written as the product of the three isoscalar factors associated with the separate reductions,

$$\xi^{\sigma_b v_b L_b, v_f L_f}_{[h,h'](\sigma,\sigma',0)(v,v')L}$$
$$= \left\langle \begin{array}{cc} [N] & [1] \\ (\sigma_b,0,0) & (1,0,0) \end{array} \middle| \begin{array}{c} [h,h'] \\ (\sigma,\sigma',0) \end{array} \right\rangle \left\langle \begin{array}{cc} (\sigma_b,0,0) & (1,0,0) \\ (v_b,0) & (v_f,0) \end{array} \middle| \begin{array}{c} (\sigma,\sigma',0) \\ (v,v') \end{array} \right\rangle$$
$$\times \left\langle \begin{array}{cc} (v_b,0) & (v_f,0) \\ L_b & L_f \end{array} \middle| \begin{array}{c} (v,v') \\ L \end{array} \right\rangle$$
(10.89)

assuming that no missing labels n_Δ or ν_Δ are needed. Again we face the problem of the calculation of the isoscalar factors, which can be done with the technique explained in the previous section. We do not repeat the calculation here but mention that the relevant expressions for low-energy states are derived in [12,14]. To compute the matrix elements of the E2 transition operator (10.87), we note that $\hat{T}^{(E2)}_\mu$ is scalar under the pseudo-spin algebra $SU^{BF}(2)$. We can thus use the reduction formula (see (6.133))

$$\langle \alpha L, 1/2; J \| \hat{T}^{(E2)} \| \alpha' L, 1/2; J' \rangle$$
$$= \begin{bmatrix} L' & 1/2 & J' \\ 2 & 0 & 2 \\ L & 1/2 & J \end{bmatrix} \langle \alpha L \| \hat{T}^{(E2)} \| \alpha' L \rangle$$

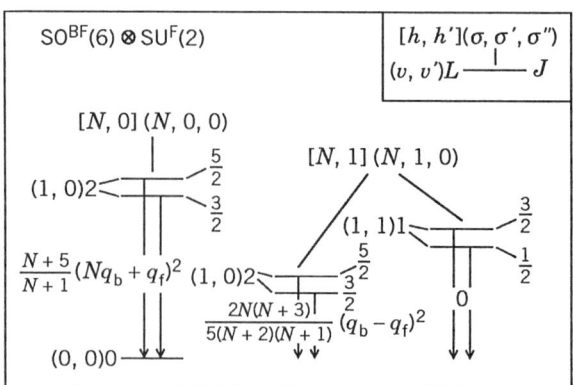

Figure 10.4: Allowed transitions to the ground state in the $SO^{BF}(6) \otimes SU^F(2)$ limit induced by the quadrupole operator $q_b \hat{Q}_{b\mu} + q_f \hat{Q}_{f\mu}$. The expressions alongside the arrows are the $B(E2)$ values. Labels are as in Figure 10.3.

$$= (-)^{J'+L+1/2}\sqrt{(2L+1)(2J'+1)} \begin{Bmatrix} J & L & 1/2 \\ L' & J' & 2 \end{Bmatrix} \langle \alpha L \| \hat{T}^{(E2)} \| \alpha' L \rangle$$

(10.90)

where α and α' are short-hand notations for the various labels needed in bra and ket. The remaining reduced matrix element is found by using the expansion in a weak-coupling basis to obtain an expression which is analogous to (10.78) but now involving the coupling coefficients (10.89).

The $B(E2)$ values of transitions from some low-energy levels to the ground state of the $SO^{BF}(6) \otimes SU^F(2)$ limit are shown in Figure 10.4. We confirm our earlier statement that, for $q_b \neq q_f$, E2 transitions between different $U^{BF}(6)$ or $SO^{BF}(6)$ representations are allowed. We note, however, that $\Delta \sigma = \Delta \sigma' = 0$ transitions have $B(E2)$ values proportional to $(Nq_b + q_f)^2$ while other $B(E2)$ values are proportional to $(q_b - q_f)^2$ and hence are expected to be much smaller. Furthermore, even with $q_b \neq q_f$, the selection rule $\Delta(v + v') = 0$ is predicted to hold rigorously. We also note the existence of parameter-free relations between various transition rates and in particular the equality of the $B(E2)$ values for transitions from the two members of a pseudo-L doublet to the ground state. We emphasize that these are *entirely* parameter-free predictions,

10.6 Supersymmetry

in the sense that they do not depend on the parameters in the hamiltonian or those in the transition operator.

In Table 10.2 the E2 transition rates predicted in the SO$^{\rm BF}$(6) ⊗ SU$^{\rm F}$(2) limit are compared with the available data in $^{195}_{78}$Pt$_{117}$. In view of the simplicity of the approach and the fact that all calculated transition rates depend on a single parameter $q_{\rm b} = -q_{\rm f}$, the agreement between the calculated and observed $B({\rm E2})$ values can be considered as remarkable.

Results of similar quality are obtained for intensities of one-neutron transfer reactions starting from or leading to $^{195}_{78}$Pt$_{117}$. The SO$^{\rm BF}$(6) ⊗ SU$^{\rm F}$(2) limit has also been applied to other odd-mass platinum isotopes, showing a good overall agreement for $A = 193$ to $A = 199$, with $^{195}_{78}$Pt$_{117}$ being the best example. In addition, the applicability of symmetries of the IBFM is increased further by considerations involving supersymmetry, as discussed in the next section.

10.6 Supersymmetry

In the previous two sections we presented calculations for the odd-mass nuclei $^{197}_{79}$Au$_{118}$ and $^{197}_{78}$Pt$_{119}$ in which a fermion is coupled to a system of bosons with SO(6) symmetry. In each case the calculation proceeded along the same lines, the basic ingredient being a hamiltonian expanded in terms of linear and quadratic Casimir invariants of coupled boson–fermion algebras with coefficients that are determined from an adjustment to experimental energies. An almost identical procedure was followed in Chapter 8 in the treatment of even–even nuclei, with the only difference that in that case the expansion involves invariants of boson algebras. Furthermore, a comparison of the hamiltonians used to fit neighboring even–even and odd-mass isotopes reveals that the coefficients in front of corresponding Casimir invariants often have similar values. For example, the parameter ζ, appearing in front of $\mathcal{C}_2[{\rm SO}(6)]$ in the SO(6) limit and of $\mathcal{C}_2[{\rm SO}^{\rm BF}(6)]$ in the SO$^{\rm BF}$(6) ⊗ SU$^{\rm BF}$(2) limit, is found to have nearly equal values ($\zeta \approx 60$ keV) in $^{196}_{78}$Pt$_{118}$ and $^{197}_{78}$Pt$_{119}$ (see Figures 8.3 and 10.3). This suggests that a reasonable agreement between calculated and observed energies could be obtained with a single value for ζ. Similarly, the parameter $q_{\rm b}$ multiplying the

Table 10.2: $B(E2)$ values in ^{195}Pt

E_i^a	$((\sigma,\sigma'),(v,v'),L,J)_i$	E_f^a	$((\sigma,\sigma'),(v,v'),L,J)_f$	$B(E2; J_i \to J_f)^b$ Expt	Theory^c
211	$(7,0),(1,0),2,3/2$	0	$(7,0),(0,0),0,1/2$	0.190(10)	0.179
239	$(7,0),(1,0),2,5/2$	0	$(7,0),(0,0),0,1/2$	0.170(10)	0.179
525	$(7,0),(2,0),2,3/2$	0	$(7,0),(0,0),0,1/2$	0.017(1)	0
544	$(7,0),(2,0),2,5/2$	0	$(7,0),(0,0),0,1/2$	0.008(4)	0
99	$(6,1),(1,0),2,3/2$	0	$(7,0),(0,0),0,1/2$	0.038(6)	0.035
130	$(6,1),(1,0),2,5/2$	0	$(7,0),(0,0),0,1/2$	0.066(4)	0.035
420	$(6,1),(2,0),2,3/2$	0	$(7,0),(0,0),0,1/2$	0.015(1)	0
455	$(6,1),(2,0),2,5/2$	0	$(7,0),(0,0),0,1/2$	≤ 0.00004	0
199	$(6,1),(1,1),1,3/2$	0	$(7,0),(0,0),0,1/2$	0.025(2)	0
389	$(6,1),(1,1),3,5/2$	0	$(7,0),(0,0),0,1/2$	0.007(1)	0
613	$(7,0),(2,0),4,7/2$	211	$(7,0),(1,0),2,3/2$	0.170(70)	0.215
508	$(6,1),(2,0),4,7/2$	211	$(7,0),(1,0),2,3/2$	0.055(17)	0.020
525	$(7,0),(2,0),2,3/2$	239	$(7,0),(1,0),2,5/2$	≤ 0.019	0.072
667	$(7,0),(2,0),4,9/2$	239	$(7,0),(1,0),2,5/2$	0.200(40)	0.239
563	$(6,1),(2,0),4,9/2$	239	$(7,0),(1,0),2,5/2$	0.091(22)	0.022
239	$(7,0),(1,0),2,5/2$	99	$(6,1),(1,0),2,3/2$	0.060(20)	0
525	$(7,0),(2,0),2,3/2$	99	$(6,1),(1,0),2,3/2$	≤ 0.033	0.007
613	$(7,0),(2,0),4,7/2$	99	$(6,1),(1,0),2,3/2$	0.005(3)	0.009
420	$(6,1),(2,0),2,3/2$	99	$(6,1),(1,0),2,3/2$	0.005(4)	0.177
508	$(6,1),(2,0),4,7/2$	99	$(6,1),(1,0),2,3/2$	0.240(50)	0.228
389	$(6,1),(1,1),3,5/2$	99	$(6,1),(1,0),2,3/2$	0.200(70)	0.219
525	$(7,0),(2,0),2,3/2$	130	$(6,1),(1,0),2,5/2$	0.009(5)	0.003
667	$(7,0),(2,0),4,9/2$	130	$(6,1),(1,0),2,5/2$	0.012(3)	0.010
563	$(6,1),(2,0),4,9/2$	130	$(6,1),(1,0),2,5/2$	0.240(40)	0.253
389	$(6,1),(1,1),3,5/2$	130	$(6,1),(1,0),2,5/2$	≤ 0.014	0.055

^{a} In units of keV.
^{b} In units of e^{2}b^2.
^{c} With $q_b = -q_f = 0.151$ eb.

10.6 Supersymmetry

boson part $\hat{Q}_{b\mu}$ of the E2 transition operator has approximately the same value in $^{196}_{78}\text{Pt}_{118}$, $^{197}_{79}\text{Au}_{118}$, and $^{197}_{78}\text{Pt}_{119}$ ($q_b \approx 0.15$ eb). Note that these parameters have been obtained for each of the nuclei *separately* and that, in principle, they need not be equal; the fact that they are suggests the existence of some relation between the different isotopes.

It is our aim here to cast these observed equalities of parameters into a formal framework and, specifically, to establish an algebraic connection between neighboring even–even and odd-mass nuclei. In a first step, we attempt to link properties of a doublet of nuclei—one even–even and one odd-mass. We subsequently show how this idea can be extended in a natural way to encompass a quartet of nuclei with an even–even, an even–odd, an odd–even, and odd–odd member.

For the sake of concreteness we explain the procedure with the example of the doublet $^{196}_{78}\text{Pt}_{118}$–$^{197}_{79}\text{Au}_{118}$. Let us begin by summarizing the *separate* descriptions of these nuclei in the IBM and IBFM. The quadrupole collective excitations of even–even nuclei can be approximated, in the IBM, in terms of a system of N s and d bosons. The algebra $\text{U}^\text{B}(6)$ can be considered as the dynamical algebra of the problem since one of its representations, namely the symmetric one $[N]$, contains all the states we wish to describe. In the specific case of $^{196}_{78}\text{Pt}_{118}$ we have $N = 6$, arising from four neutron bosons and two proton bosons, all with hole character in their respective valence shells.

Quadrupole collective states of odd-mass nuclei are described, in the IBFM, in terms of N bosons coupled to a single fermion. The dynamical algebra is now $\text{U}^\text{B}(6) \otimes \text{U}^\text{F}(\Omega)$, where Ω is the degeneracy of orbits available to the fermion, and the appropriate representation is $[N] \otimes [1]$. In the specific case of $^{197}_{79}\text{Au}_{118}$ we have $N = 5$ bosons and $M = 1$ fermion. Note that we may consider—although somewhat artificially—$\text{U}^\text{B}(6) \otimes \text{U}^\text{F}(\Omega)$ as the dynamical algebra for the collective states of an even–even nucleus since these are contained in its $[N] \otimes [0]$ representation. This argument is equivalent to the statement that the IBFM can also be used to describe even–even nuclei by diagonalizing an appropriate IBFM hamiltonian in a space of N-boson states. In other words, the general IBFM hamiltonian contains all the boson–boson interactions that are included in the IBM.

With the preceding argument we have established a relation between the descriptions of even–even and odd-mass nuclei, but note that these are still *separate* descriptions since they involve *different* representations of the dynamical algebra $U^B(6) \otimes U^F(\Omega)$. This situation invites the following question. Can we add some operators to $U^B(6) \otimes U^F(\Omega)$ such that one representation of this enlarged dynamical algebra contains even–even as well as odd-mass nuclei? This idea is similar to the one proposed in Chapter 9, where the dynamical algebra of the IBM-2 is enlarged from $U_\nu(6) \otimes U_\pi(6)$ to $U(12)$ to arrive at a simultaneous description of several even–even nuclei (so-called F-spin multiplets). In the present case, however, we need a unified treatment of a boson system and a boson–fermion system, which cannot be done in terms of classical Lie algebras but necessarily requires the use of superalgebras as introduced in Chapter 4. According to the discussion of Section 4.2, the appropriate superalgebra is obtained by adding the generators $b_i^\dagger a_{i'}$ and $a_{i'}^\dagger b_i$ to the set contained in $U^B(6) \otimes U^F(\Omega)$, giving rise to the superalgebra $U(6/\Omega)$. In the case of $^{196}_{78}Pt_{118}$ or $^{197}_{79}Au_{118}$ the appropriate superalgebra is $U(6/4)$.

The fact that $^{196}_{78}Pt_{118}$ and $^{197}_{79}Au_{118}$ are contained in the same representation of $U(6/4)$ can be understood in several ways. We must choose a representation of the superalgebra which classifies the bosons symmetrically and the fermions antisymmetrically. As shown in Section 4.2 the appropriate $U(6/4)$ representation is the supersymmetric one $[\mathcal{N}\}$, containing the following $U^B(6) \otimes U^F(4)$ products:

$$[\mathcal{N}] \otimes [0], [\mathcal{N}-1] \otimes [1], [\mathcal{N}-2] \otimes [1^2], \ldots$$
$$\ldots, \begin{cases} [\mathcal{N}-4] \otimes [1^4] & \text{for } \mathcal{N} \geq 4 \\ [0] \otimes [1^\mathcal{N}] & \text{for } \mathcal{N} \leq 4 \end{cases} \qquad (10.91)$$

The representations corresponding to $^{196}_{78}Pt_{118}$ and $^{197}_{79}Au_{118}$, $[6] \otimes [0]$ and $[5] \otimes [1]$, are indeed obtained for $\mathcal{N} = 6$. Note that (10.91) contains additional products $[4] \otimes [1^2], \ldots$; some of these correspond to more complicated nuclear excitations, as is shown below.

The occurrence of $^{196}_{78}Pt_{118}$ and $^{197}_{79}Au_{118}$ in the same supermultiplet (i.e., in the same representation of the superalgebra $U(6/4)$) can be understood in an alternative way by investigating the action of the fermion-sector generators of the superalgebra. Acting with the genera-

10.6 Supersymmetry

tor $a_{i'}^\dagger b_i$ on the even–even nucleus, we find, schematically,

$$a_{i'}^\dagger b_i \; ^{196}\text{Pt} \to a_{i'}^\dagger \; ^{198}\text{Hg} \to \; ^{197}\text{Au} \qquad (10.92)$$

since the bosons represent fermion pairs, and in addition, in this example both bosons and fermions have a hole character. All states that can be connected by the generators of a dynamical algebra belong to a single representation of that algebra, and hence (10.92) indeed suggests that $^{196}_{78}\text{Pt}_{118}$ and $^{197}_{79}\text{Au}_{118}$ are in the same supermultiplet. We now act with $a_{i'}^\dagger b_i$ on $^{197}_{79}\text{Au}_{118}$ to find the next member of the supermultiplet, the nucleus $^{198}_{80}\text{Hg}^*_{118}$ associated with the $U^B(6) \otimes U^F(4)$ representation $[4] \otimes [1^2]$. This is again an even–even nucleus, but note that it is now described as a system of $N = 4$ bosons coupled to $M = 2$ fermions. The corresponding excitations are *not* the low-energy, collective levels in $^{198}_{80}\text{Hg}_{118}$ since they involve a pair of uncorrelated fermions, in contrast to the two fermions making up a boson, which have a correlated or collective structure. These non-collective states are sometimes referred to as quasi-particle excitations [8] and are denoted here with an asterisk—or asterisks, if several pairs of uncorrelated fermions are involved.

In this way we find that in the example of $^{196}_{78}\text{Pt}_{118}$ the $U(6/4)$ supermultiplet contains the three nuclei $^{196}_{78}\text{Pt}_{118}$, $^{197}_{79}\text{Au}_{118}$, and $^{198}_{80}\text{Hg}^*_{118}$, shown in Figure 10.5. There seems to be an inconsistency between this result and the reduction rule (10.91), which predicts, for $\mathcal{N} = 6$, a su-

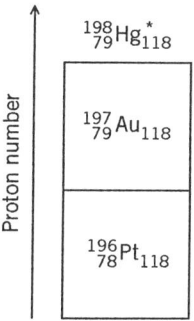

Figure 10.5: An example of a $U(6/4)$ supermultiplet with $\mathcal{N} = 6$. The nuclear systems studied experimentally are boxed.

permultiplet consisting of five members. However, in relations such as (10.92), we assume all operators to have a proton character; as there are only two proton bosons available in $^{196}_{78}\text{Pt}_{118}$, triple action of $a^\dagger_i b_i$ gives zero and the series stops after $^{198}_{80}\text{Hg}^*_{118}$. The assumption that only proton bosons are involved is a natural one, and it would be difficult to justify a relation between nuclei where a neutron boson is transformed into a proton fermion. Nevertheless, the formalism presented so far does not distinguish between neutron and proton bosons and (10.91) is a reflection of that fact: it allows *all* bosons to be transformed into fermions, including the neutron bosons. We show below how this situation can be remedied by simply extending the description of the core from IBM-1 to IBM-2. We should, nevertheless, mention that the problem is not a serious one and that, from a practical point of view, the present formalism is adequate enough. Owing to the severe experimental difficulties in detecting quasi-particle excitations, one is confined to comparing properties of the first two members of supermultiplets, which are identical in either (10.91) or Figure 10.5.

A single hamiltonian can now simultaneously describe the states in the nuclei belonging to a given representation of the superalgebra. Confining ourselves to the low-energy collective nuclear excitations, this hamiltonian becomes

$$\hat{H}_{\text{II}}[\text{U}(6/4)] = \beta' \mathcal{C}_2[\text{SO}^{\text{B}}(6)] + \beta'' \mathcal{C}_2[\text{SU}^{\text{BF}}(4)] + \zeta \mathcal{C}_2[\text{Sp}^{\text{BF}}(4)] \\ + \gamma \mathcal{C}_2[\text{SU}^{\text{BF}}(2)] \quad (10.93)$$

Since (10.93) is identical to the one used in the $\text{SU}^{\text{BF}}(4)$ limit, it is evidently appropriate for $^{197}_{79}\text{Au}_{118}$. For $^{196}_{78}\text{Pt}_{118}$ it effectively reduces to

$$\hat{H}_{\text{II}}[\text{SO}(6)] = (\beta'+\beta'')\mathcal{C}_2[\text{SO}^{\text{B}}(6)] + \zeta \mathcal{C}_2[\text{SO}^{\text{BF}}(5)] + \gamma \mathcal{C}_2[\text{SO}^{\text{B}}(3)] \quad (10.94)$$

It should be emphasized that the calculations for the two nuclei are performed with a *single* hamiltonian of the form (10.93) with *constant* parameters and that (10.94) is the form this hamiltonian acquires when acting on the $[N]\otimes[0]$ representation. Note also that terms contributing to binding energies alone are omitted from (10.93). This form of the hamiltonian is thus only appropriate for calculating relative energies within a single nucleus. The hamiltonian can be easily extended to include additional terms so that it can be used to investigate relations

10.6 Supersymmetry

between binding energies of members of a single supermultiplet. This is a useful application of supersymmetry, studied in some detail in [23].

The experimental spectra of $^{196}_{78}$Pt$_{118}$ and $^{197}_{79}$Au$_{118}$, shown in Figure 10.6, can be compared with the ones calculated with the hamiltonian (10.93), shown in Figure 10.7. On average, the observed and calculated energies deviate more than in the case of the separate fits shown in Figures 8.3 and 10.1. This is a consequence of fitting both nuclei with the same hamiltonian, which results in a compromise. This is also illustrated in Table 10.3 where the parameters of the various calculations are summarized. The separate fits to the nuclei $^{196}_{78}$Pt$_{118}$ and $^{197}_{79}$Au$_{118}$ give rather different values for the parameters ζ and $\gamma' + \gamma''$ while the supersymmetric fit ^{196}Pt–^{197}Au results in intermediate values. Comparison of Figures 10.6 and 10.7 shows, however, that with these modified parameters a good agreement is also obtained for the even–

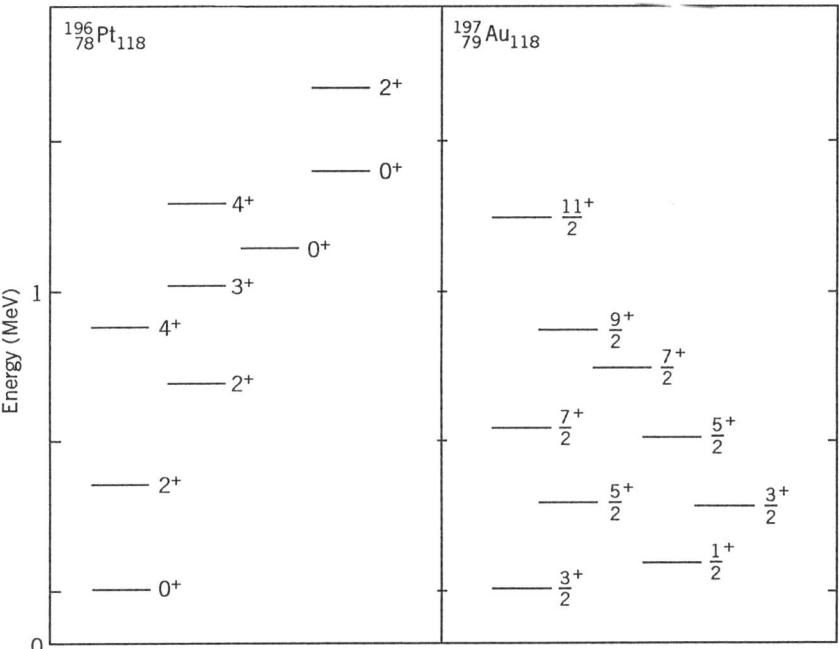

Figure 10.6: The observed excitation spectra of the pair of nuclei ^{196}Pt and ^{197}Au belonging to the U(6/4) supermultiplet with $\mathcal{N} = 6$. Levels are labeled by the angular momentum and parity.

430 10 The Interacting Boson–Fermion Model

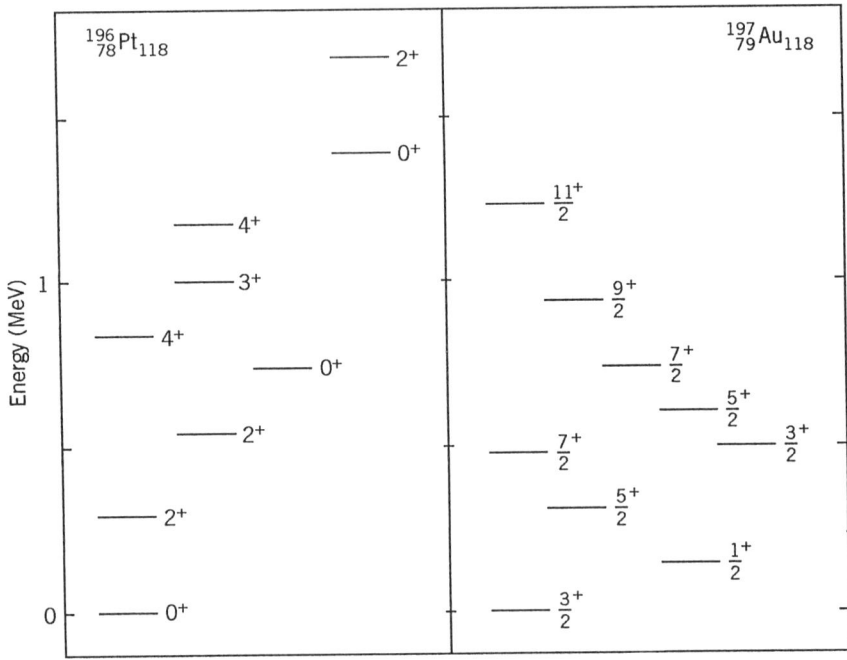

Figure 10.7: Partial eigenspectrum of $\hat{H}_{\mathrm{II}}[\mathrm{U}(6/4)]$ in the supersymmetric representation $[\mathcal{N}=6\}$ corresponding to the pair of nuclei ^{196}Pt and ^{197}Au. The spectra are obtained with the parameters (in keV) $\beta' + \beta'' = -50$, $\zeta = 42$, and $\gamma' + \gamma'' = 21$, with the energy of the ground state in each nucleus normalized to zero. Levels are labeled by the angular momentum and parity; all other quantum numbers are given in Figures 8.3 and 10.1.

Table 10.3: Parameters (in keV) used in the calculation of energy levels

	^{196}Pt	^{197}Au	^{197}Pt	^{196}Pt–^{197}Au	^{196}Pt–^{197}Pt	Triplet
η	—	—	86	—	71	68
$\beta' + \beta''$	−50	—	−64	−50	−52	−52
ζ	58	27	65	42	59	45
$\gamma' + \gamma''$	11	28	10	21	11	20
γ'	—	—	7	—	11	21
γ''	—	—	3	—	0	−1

10.6 Supersymmetry

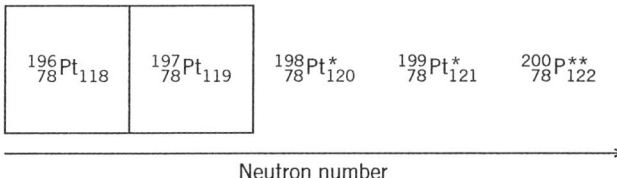

Figure 10.8: An example of a U(6/12) supermultiplet with $\mathcal{N} = 6$. The nuclear systems studied experimentally are boxed.

even and odd-mass spectra. We conclude that for the pair $^{196}_{78}\text{Pt}_{118}$ and $^{197}_{79}\text{Au}_{118}$ supersymmetry succeeds in reproducing the density of levels as well as the basic structure of the even–even and odd-mass spectra.

The analysis for the second pair $^{196}_{78}\text{Pt}_{118}$ and $^{197}_{78}\text{Pt}_{119}$ proceeds along the same lines. In this case the relevant superalgebra is U(6/12). The supersymmetric representation $[\mathcal{N}\}$ can be decomposed in the same way as in the previous example,

$$\boxed{\begin{array}{l} [\mathcal{N}] \otimes [0], [\mathcal{N}-1] \otimes [1], [\mathcal{N}-2] \otimes [1^2], \ldots \\ \qquad \ldots, \begin{cases} [\mathcal{N}-12] \otimes [1^{12}] & \text{for } \mathcal{N} \geq 12 \\ [0] \otimes [1^{\mathcal{N}}] & \text{for } \mathcal{N} \leq 12 \end{cases} \end{array}} \quad (10.95)$$

where the product representations now refer to $U^B(6) \otimes U^F(12)$. A more physical picture is obtained by acting with the fermion-sector generator $a^\dagger_{i'} b_i$ on the even–even nucleus, resulting in

$$a^\dagger_{i'} b_i \,{}^{196}\text{Pt} \to a^\dagger_{i'} \,{}^{198}\text{Pt} \to {}^{197}\text{Pt} \quad (10.96)$$

The difference between (10.92) and (10.96) is that in the former all operators refer to protons whereas in the latter they pertain to neutrons. As mentioned before, this is not explicitly recognized by the formalism presented so far but rather implicitly assumed. Continued action with the operator $a^\dagger_{i'} b_i$ gives rise to the supermultiplet shown in Figure 10.8. As in the previous example only the first two members of the supermultiplet can be studied experimentally.

For the pair of nuclei $^{196}_{78}\text{Pt}_{118}$ and $^{197}_{78}\text{Pt}_{119}$ the appropriate hamiltonian is of the form

$$\begin{aligned}\hat{H}_{\text{II}}[\text{U}(6/12)] &= \eta \mathcal{C}_2[\text{U}^{\text{BF}}(6)] + \beta \mathcal{C}_2[\text{SO}^{\text{BF}}(6)] + \zeta \mathcal{C}_2[\text{SO}^{\text{BF}}(5)] \\ &\quad + \gamma' \mathcal{C}_2[\text{SO}^{\text{BF}}(3)] + \gamma'' \mathcal{C}_2[\text{SU}^{\text{BF}}(2)] \end{aligned} \quad (10.97)$$

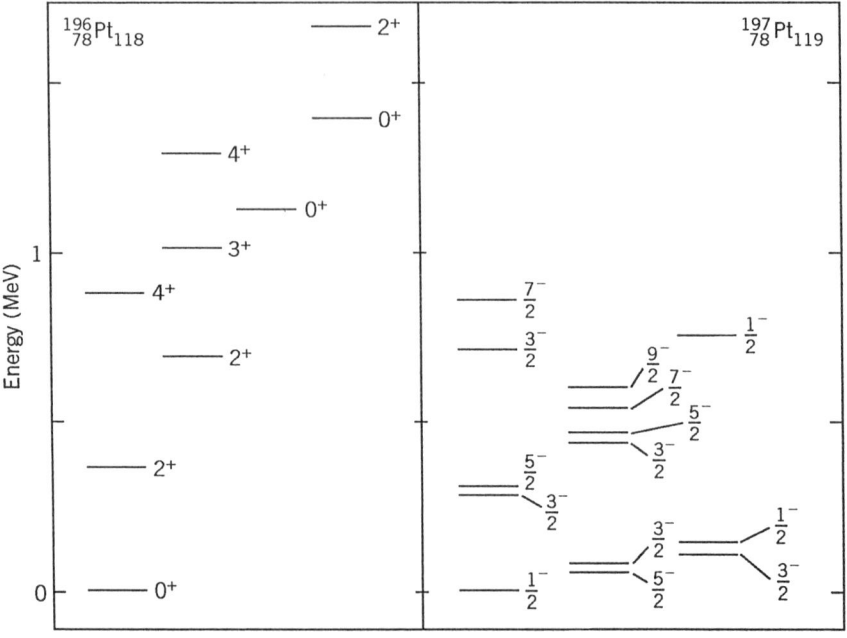

Figure 10.9: The observed excitation spectra of the pair of nuclei ^{196}Pt and ^{197}Pt belonging to the U(6/12) supermultiplet with $\mathcal{N} = 6$. Levels are labeled by the angular momentum and parity.

This hamiltonian is used in the $SO^{BF}(6) \otimes SU^F(2)$ limit, appropriate for $^{197}_{78}\text{Pt}_{119}$. For $^{196}_{78}\text{Pt}_{118}$ it effectively reduces to

$$\hat{H}_{\text{II}}[SO(6)] = \beta\mathcal{C}_2[SO^B(6)] + \zeta\mathcal{C}_2[SO^{BF}(5)] + (\gamma+\gamma'')\mathcal{C}_2[SO^B(3)] \quad (10.98)$$

where, as usual, we neglect constant contributions to the energy. The observed and calculated spectra for the pair of nuclei $^{196}_{78}\text{Pt}_{118}$ and $^{197}_{78}\text{Pt}_{119}$ are compared in Figures 10.9 and 10.10. Comparing the fits to the separate nuclei (see Figures 8.3 and 10.3) with the supersymmetric fit of Figure 10.10, there is, in this case, almost no deterioration in quality. This is also illustrated by the parameter values listed in Table 10.3. The separate fits to $^{196}_{78}\text{Pt}_{118}$ and $^{197}_{78}\text{Pt}_{119}$ give rise to nearly

10.6 Supersymmetry

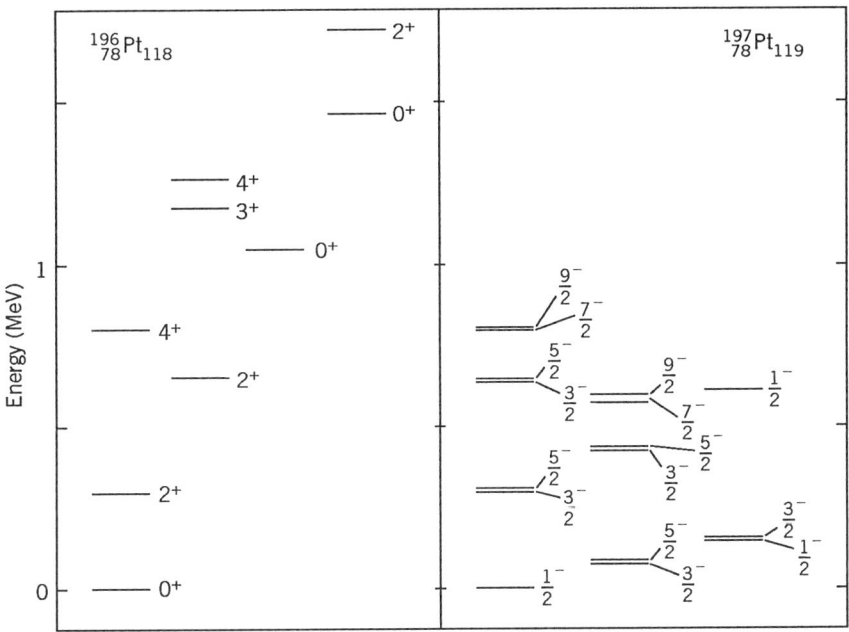

Figure 10.10: Partial eigenspectrum of $\hat{H}_{\mathrm{II}}[\mathrm{U}(6/12)]$ in the supersymmetric representation $[\mathcal{N}=6]$ corresponding to the pair of nuclei ^{196}Pt and ^{197}Pt. The spectra are obtained with the parameters (in keV) $\eta=71$, $\beta=-52$, $\zeta=59$, $\gamma'=11$, and $\gamma''=0$, with the energy of the ground state in each nucleus normalized to zero. Levels are labeled by the angular momentum and parity; all other quantum numbers are given in Figures 8.3 and 10.3.

identical parameter values, and as a consequence, the supersymmetric compromise listed in the column ^{196}Pt–^{197}Pt is rather good. Thus we conclude that for excitation spectra of the platinum isotopes the supersymmetric description is an excellent approximation.

10 The Interacting Boson–Fermion Model

Since the nucleus $^{196}_{78}\text{Pt}_{118}$ is at the same time a member of the U(6/4) and U(6/12) supermultiplets, it is natural to investigate whether a dynamical algebra can be constructed containing all three nuclei (and possibly more) in a single representation. This can indeed be achieved by explicitly introducing the neutron and proton degrees of freedom for the bosons. The appropriate dynamical algebra is now the direct product of superalgebras

$$U_\nu(6/12) \otimes U_\pi(6/4) \quad (10.99)$$

which consists of the generators

$$\mathbf{G}_\nu \equiv \begin{bmatrix} \mathcal{G}^{B\nu j}_{B\nu i} & \vdots & \mathcal{G}^{F\nu l'}_{B\nu k} \\ \cdots & \cdot & \cdots \\ \mathcal{G}^{B\nu l}_{F\nu k'} & \vdots & \mathcal{G}^{F\nu j'}_{F\nu i'} \end{bmatrix}, \quad \mathbf{G}_\pi \equiv \begin{bmatrix} \mathcal{G}^{B\pi j}_{B\pi i} & \vdots & \mathcal{G}^{F\pi l'}_{B\pi k} \\ \cdots & \cdot & \cdots \\ \mathcal{G}^{B\pi l}_{F\pi k'} & \vdots & \mathcal{G}^{F\pi j'}_{F\pi i'} \end{bmatrix} \quad (10.100)$$

for the neutrons and protons, respectively. To determine what nuclei are contained in a product of supersymmetric representations $\{\mathcal{N}_\nu\} \otimes \{\mathcal{N}_\pi\}$ of $U_\nu(6/12) \otimes U_\pi(6/4)$, we may proceed as before and study the action of the generators (10.100) on the even–even member of the supermultiplet, in our example $^{196}_{78}\text{Pt}_{118}$. We find, for a neutron generator,

$$a^\dagger_{\nu i'} b_{\nu i} \; ^{196}\text{Pt} \to \; ^{197}\text{Pt} \quad (10.101)$$

and, for a proton generator,

$$a^\dagger_{\pi i'} b_{\pi i} \; ^{196}\text{Pt} \to \; ^{197}\text{Au} \quad (10.102)$$

which shows that the triplet of nuclei $^{196}_{78}\text{Pt}_{118}$–$^{197}_{78}\text{Pt}_{119}$–$^{197}_{79}\text{Au}_{118}$ belongs to the representation $[4\} \otimes [2\}$ of the dynamical algebra $U_\nu(6/12) \otimes U_\pi(6/4)$. We may continue in this way with the action of various generators to find that the complete supermultiplet consists of the nuclei shown in Figure 10.11.

It is instructive to derive this result in a formal way by studying the reduction of $\{\mathcal{N}_\nu\} \otimes \{\mathcal{N}_\pi\}$ into representations of $U^B_\nu(6) \otimes U^F_\nu(12) \otimes U^B_\pi(6) \otimes U^F_\pi(4)$. Application of the reduction rules (10.91) and (10.95) from $U(n/m)$ to $U(n) \otimes U(m)$ shows that $\{\mathcal{N}_\nu\} \otimes \{\mathcal{N}_\pi\}$ contains the

10.6 Supersymmetry

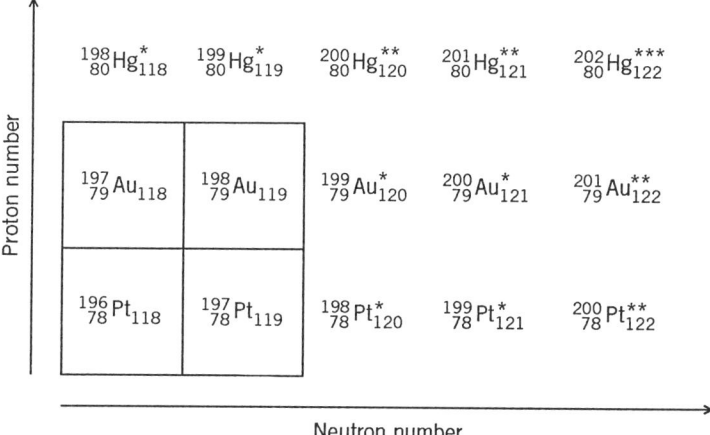

Figure 10.11: An example of a $U_\nu(6/12) \otimes U_\pi(6/4)$ supermultiplet with $\mathcal{N}_\nu = 4$ and $\mathcal{N}_\pi = 2$. The nuclear systems studied experimentally are boxed.

representations

$$[\mathcal{N}_\nu] \otimes [0] \otimes [\mathcal{N}_\pi] \otimes [0], [\mathcal{N}_\nu - 1] \otimes [1] \otimes [\mathcal{N}_\pi] \otimes [0], \ldots$$
$$\ldots, [0] \otimes [1^{\mathcal{N}_\nu}] \otimes [\mathcal{N}_\pi] \otimes [0],$$
$$[\mathcal{N}_\nu] \otimes [0] \otimes [\mathcal{N}_\pi - 1] \otimes [1], [\mathcal{N}_\nu - 1] \otimes [1] \otimes [\mathcal{N}_\pi - 1] \otimes [1], \ldots$$
$$\ldots, [0] \otimes [1^{\mathcal{N}_\nu}] \otimes [\mathcal{N}_\pi - 1] \otimes [1],$$
$$\ldots$$
$$[\mathcal{N}_\nu] \otimes [0] \otimes [0] \otimes [1^{\mathcal{N}_\pi}], [\mathcal{N}_\nu - 1] \otimes [1] \otimes [0] \otimes [1^{\mathcal{N}_\pi}], \ldots$$
$$\ldots, [0] \otimes [1^{\mathcal{N}_\nu}] \otimes [0] \otimes [1^{\mathcal{N}_\pi}]$$

(10.103)

where we assume $\mathcal{N}_\nu \leq 12$ and $\mathcal{N}_\pi \leq 4$. The supermultiplet in our example, shown in Figure 10.11 for $\mathcal{N}_\nu = 4$ and $\mathcal{N}_\pi = 2$ and obtained from the action of the fermion-sector generators of $U_\nu(6/12) \otimes U_\pi(6/4)$, is now consistent with this general result. This is a consequence of the explicit introduction of neutron and proton bosons.

There is one nucleus of particular interest in this supermultiplet, namely the one associated with the representation $[\mathcal{N}_\nu - 1] \otimes [1] \otimes [\mathcal{N}_\pi - 1] \otimes [1]$ of $U_\nu^B(6) \otimes U_\nu^F(12) \otimes U_\pi^B(6) \otimes U_\pi^F(4)$. In our example it corresponds to $^{198}_{79}Au_{119}$. This nucleus has an odd number of neutrons and an odd

10 The Interacting Boson–Fermion Model

number of protons; given that neutrons and protons occupy different valence shells, its low-energy excitations involve the coupling of two uncorrelated fermions (one neutron and one proton) to an even–even core. We thus find that, besides a triplet consisting of an even–even, an even–odd, and an odd–even nucleus, the supermultiplet contains a fourth, odd–odd member whose excitations are, in principle, accessible to experimental detection.

To predict the spectroscopic properties of this quartet of nuclei, a reduction of the dynamical algebra $U_\nu(6/12) \otimes U_\pi(6/4)$ is needed. Several different bases are possible, and we choose here one which has been studied in some detail [24]. The algebras appearing in the reduction and their associated quantum numbers are

$$
\begin{array}{c}
U_\nu(6/12) \otimes U_\pi(6/4) \supset U_\nu^B(6) \otimes U_\nu^F(12) \otimes U_\pi^B(6) \otimes U_\pi^F(4) \\
\qquad |\qquad\qquad |\qquad\qquad |\qquad\qquad |\qquad\qquad |\qquad\qquad | \\
\{\mathcal{N}_\nu\} \quad\;\; \{\mathcal{N}_\pi\} \quad\;\; [N_\nu] \quad\;\; [1^{M_\nu}] \quad\;\; [N_\pi] \quad\;\; [1^{M_\pi}] \\[4pt]
\supset U_{\nu\pi}^B(6) \otimes U_\nu^F(6) \otimes SU_\nu^F(2) \otimes \quad SU_\pi^F(4) \\
\qquad |\qquad\qquad |\qquad\qquad |\qquad\qquad\qquad | \\
[\bar{h}, \bar{h}'] \quad\;\; [1^{M_\nu}] \quad\;\; \tilde{s} \quad\;\; (\sigma_f, \sigma_f', \sigma_f'') \\[4pt]
\supset U_{\nu\pi}^{BF}(6) \otimes SU_\nu^F(2) \otimes \quad SU_\pi^F(4) \\
\qquad |\qquad\qquad\qquad |\qquad\qquad\qquad | \\
[h, h', h''] \quad\;\; \tilde{s} \quad\;\; (\sigma_f, \sigma_f', \sigma_f'') \\[4pt]
\supset SO_{\nu\pi}^{BF}(6) \otimes SU_\nu^F(2) \otimes \quad SU_\pi^F(4) \\
\qquad |\qquad\qquad\qquad |\qquad\qquad\qquad | \\
(\bar{\sigma}, \bar{\sigma}', \bar{\sigma}'') \quad\;\; \tilde{s} \quad\;\; (\sigma_f, \sigma_f', \sigma_f'') \\[4pt]
\supset SU_{\nu\pi}^{BF}(4) \otimes SU_\nu^F(2) \supset Sp_{\nu\pi}^{BF}(4) \otimes SU_\nu^F(2) \\
\qquad |\qquad\qquad\qquad |\qquad\qquad\qquad |\qquad\qquad\qquad | \\
(\sigma, \sigma', \sigma'') \quad\;\; \tilde{s} \quad\;\; (v, v') \quad\;\; \tilde{s} \\[4pt]
\supset \overline{SU}_{\nu\pi}^{BF}(2) \otimes SU_\nu^F(2) \supset SU_{\nu\pi}^{BF}(2) \\
\qquad |\qquad\qquad\qquad |\qquad\qquad\qquad | \\
\nu_\Delta L \quad\;\; \tilde{s} \quad\;\; J
\end{array}
\qquad (10.104)
$$

10.6 Supersymmetry

while the corresponding hamiltonian reads

$$\begin{aligned}\hat{H}_{\mathrm{II}}&[\mathrm{U}_\nu(6/12)\otimes\mathrm{U}_\pi(6/4)]\\&= \eta_\mathrm{b}\mathcal{C}_2[\mathrm{U}^\mathrm{B}_{\nu\pi}(6)] + \eta\mathcal{C}_2[\mathrm{U}^{\mathrm{BF}}_{\nu\pi}(6)] + \beta'\mathcal{C}_2[\mathrm{SO}^{\mathrm{BF}}_{\nu\pi}(6)]\\&+ \beta''\mathcal{C}_2[\mathrm{SU}^{\mathrm{BF}}_{\nu\pi}(4)] + \zeta\mathcal{C}_2[\mathrm{Sp}^{\mathrm{BF}}_{\nu\pi}(4)] + \gamma'\mathcal{C}_2[\overline{\mathrm{SU}}^{\mathrm{BF}}_{\nu\pi}(2)]\\&+\gamma''\mathcal{C}_2[\mathrm{SU}^{\mathrm{BF}}_{\nu\pi}(2)]\end{aligned} \quad (10.105)$$

where terms that only contribute to binding energies are again neglected. Although this example is complex in its details, the underlying principles are simple extensions of the cases discussed previously. Note that because of the embedding into the superalgebras $\mathrm{U}_\nu(6/12)$ and $\mathrm{U}_\pi(6/4)$, the neutron and proton boson and fermion numbers of the nuclei in the supermultiplet are not constant but vary according to $N_\nu + M_\nu = \mathcal{N}_\nu$ and $N_\pi + M_\pi = \mathcal{N}_\pi$, as in (10.91) and (10.95). The quantum numbers of the subalgebras in the reduction then depend on these numbers N_ν, M_ν, N_π, and M_π. For example, in the absence of any neutron fermions ($M_\nu = 0$) the neutron pseudo-spin is $\tilde{s} = 0$, while for $M_\nu = 1$ it is $\tilde{s} = 1/2$, as in (10.79). Likewise, the $\mathrm{SU}^\mathrm{F}_\pi(4)$ labels $(\sigma_\mathrm{f}, \sigma'_\mathrm{f}, \sigma''_\mathrm{f})$ are $(0,0,0)$ for $M_\pi = 0$ and $(1/2, 1/2, 1/2)$ for $M_\pi = 1$. The basis (10.104) requires the coupling of three U(6) algebras: first the neutron and proton boson U(6) algebras are combined to $\mathrm{U}^\mathrm{B}_{\nu\pi}(6)$, which is then coupled with $\mathrm{U}^\mathrm{F}_\nu(6)$ to form $\mathrm{U}^{\mathrm{BF}}_{\nu\pi}(6)$. All these couplings involve the addition of the relevant generators. In general, the representations of $\mathrm{U}^\mathrm{B}_{\nu\pi}(6)$ are characterized by two labels $[\bar{h}, \bar{h}']$, but as the non-symmetric states occur at much higher energies ($E_\mathrm{x} > 2$ MeV) than we are interested in the present application, we may assume a symmetric $\mathrm{U}^\mathrm{B}_{\nu\pi}(6)$ representation, $\bar{h}' = 0$. Subsequent reduction and multiplication rules that are necessary to find the values of labels in (10.104) are either identical to those given in Sections 10.4 and 10.5 or can be derived with similar techniques.

Under the assumption $\bar{h}' = 0$ the first term in the hamiltonian (10.105) reduces to a constant and its eigenvalue becomes

$$\begin{aligned}E_{\mathrm{II}}&(h, h', \bar{\sigma}, \bar{\sigma}', \sigma, \sigma', \sigma'', v, v', L, J)\\&= \eta[h(h+5) + h'(h'+3)] + \beta'[\bar{\sigma}(\bar{\sigma}+4) + \bar{\sigma}'(\bar{\sigma}'+2)]\\&+ \beta''[\sigma(\sigma+4) + \sigma'(\sigma'+2) + \sigma''^2] + \zeta[v(v+3) + v'(v'+1)]\\&+ \gamma'L(L+1) + \gamma''J(J+1)\end{aligned} \quad (10.106)$$

The hamiltonian (10.105) can be used to calculate the spectra of all the nuclei belonging to a given supermultiplet. For the odd–odd nucleus the hamiltonian is associated with the space $[N_\nu] \otimes [1] \otimes [N_\pi] \otimes [1]$ of $U_\nu^B(6) \otimes U_\nu^F(12) \otimes U_\pi^B(6) \otimes U_\pi^F(4)$ and all terms in (10.105) or (10.106) are independent. For the odd-mass nuclei the number of independent terms is reduced by 1 while for the even–even nucleus it is reduced by 2. For example, for the pair of nuclei $^{196}_{78}\text{Pt}_{118}$ and $^{197}_{79}\text{Au}_{118}$ the hamiltonian reduces to

$$\hat{H}_{II}[U_\pi(6/4)] = \beta' \mathcal{C}_2[\text{SO}^{BF}_{\nu\pi}(6)] + \beta'' \mathcal{C}_2[\text{SU}^{BF}_{\nu\pi}(4)] + \zeta \mathcal{C}_2[\text{Sp}^{BF}_{\nu\pi}(4)]$$
$$+ (\gamma' + \gamma'') \mathcal{C}_2[\text{SU}^{BF}_{\nu\pi}(2)] \qquad (10.107)$$

which is of the type (10.93). In turn, for the pair $^{196}_{78}\text{Pt}_{118}$ and $^{197}_{78}\text{Pt}_{119}$ it reduces to

$$\hat{H}_{II}[U_\nu(6/12)] = \eta \mathcal{C}_2[\text{U}^{BF}_{\nu\pi}(6)] + (\beta' + \beta'') \mathcal{C}_2[\text{SO}^{BF}_{\nu\pi}(6)]$$
$$+ \zeta \mathcal{C}_2[\text{SO}^{BF}_{\nu\pi}(5)] + \gamma' \mathcal{C}_2[\text{SO}^{BF}_{\nu\pi}(3)] + \gamma'' \mathcal{C}_2[\text{SU}^{BF}_{\nu\pi}(2)]$$
$$(10.108)$$

which is of the type (10.97).

An appealing aspect of this extension of supersymmetry is that the spectra of the even–even, even–odd, and odd–even members of the supermultiplet can be fitted to provide a *prediction* of the fourth, odd–odd nucleus. This prediction can then be used in the interpretation of the odd–odd spectrum, which in most cases is poorly and incompletely known experimentally. As an example of this approach we show in Figure 10.12 the calculated spectra of $^{196}_{78}\text{Pt}_{118}$, $^{197}_{78}\text{Pt}_{119}$, $^{197}_{79}\text{Au}_{118}$, and $^{198}_{79}\text{Au}_{119}$. The fit with the hamiltonian (10.105) to the levels in the first three nuclei results in the parameter values shown in Table 10.3 under "triplet." As before, they represent a compromise between the values obtained for the separate fits. The parameters resulting from this triplet fit are then used to predict the spectrum of $^{198}_{79}\text{Au}_{119}$. This procedure gives the correct density of levels in $^{198}_{79}\text{Au}_{119}$ [24], but owing to the experimental uncertainties in the spin–parity assignments, it is difficult to establish a one-to-one correspondence between calculated and observed levels solely on the basis of excitation energies. Some attempts in this direction have been made [25], but they must be treated

10.6 Supersymmetry

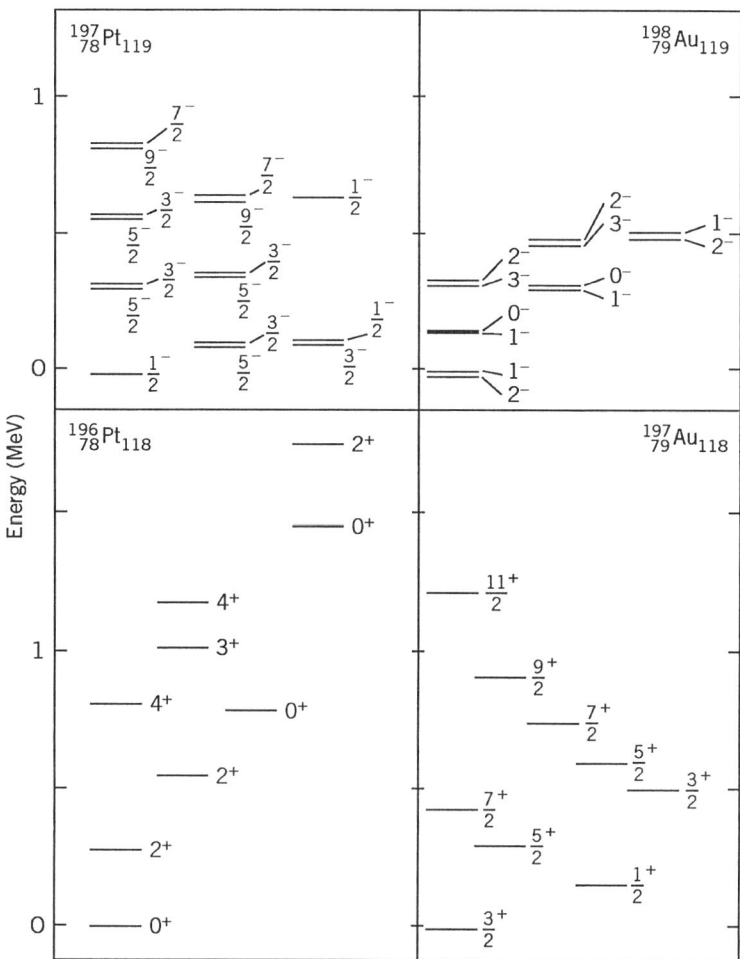

Figure 10.12: Partial eigenspectrum of $\hat{H}_{II}[U_\nu(6/12) \otimes U_\pi(6/4)]$ in the supersymmetric representation $[\mathcal{N}_\nu = 4\} \otimes [\mathcal{N}_\pi = 2\}$ corresponding to the quartet of nuclei ^{196}Pt, ^{197}Pt, ^{197}Au, and ^{198}Au. The spectra are obtained with the parameters (in keV) $\eta = 68$, $\beta' + \beta'' = -52$, $\zeta = 45$, $\gamma' = 21$, and $\gamma'' = -1$, with the energy of the ground state in each nucleus normalized to zero. Levels are labeled by the angular momentum and parity; other quantum numbers are given in Figures 8.3, 10.1, and 10.3.

as tentative since they have not been corroborated by additional spectroscopic properties.

Similar supersymmetric arguments can be used to relate other properties of nuclei belonging to a supermultiplet. A good example of this approach is provided by E2 transitions. One seeks to evaluate $B(E2)$ values in supersymmetric partners using a single E2 transition operator. In $U_\nu(6/12) \otimes U_\pi(6/4)$ it is of the form

$$\hat{T}^{(E2)}_\mu = q_{b\nu}\hat{Q}_{b\nu\mu} + q_{b\pi}\hat{Q}_{b\pi\mu} + q_{f\nu}\hat{Q}_{f\nu\mu} + q_{f\pi}\hat{Q}_{f\pi\mu} \qquad (10.109)$$

in terms of previously defined quadrupole operators, but now with explicit neutron and proton indices, that is,

$$\begin{aligned}\hat{Q}_{b\rho\mu} &= i[d^\dagger_\rho \times \tilde{s}_\rho - s^\dagger_\rho \times \tilde{d}_\rho]^{(2)}_\mu, \qquad \rho = \nu, \pi \\ \hat{Q}_{f\nu\mu} &= i[\mathcal{A}^{(2)}_{\nu\mu}(2,0) - \mathcal{A}^{(2)}_{\nu\mu}(0,2)] \qquad (10.110) \\ \hat{Q}_{f\pi\mu} &= [a^\dagger_{\pi 3/2} \times \tilde{a}_{\pi 3/2}]^{(2)}_\mu \end{aligned}$$

With reference to the analyses presented in Sections 8.6, 10.4, and 10.5 it is clear that a single E2 transition operator with effective boson charges $(N_\nu q_{b\nu} + N_\pi q_{b\pi})/N = -q_{f\nu} = q_{f\pi} = 0.152$ eb leads to a satisfactory agreement in the three nuclei $^{196}_{78}\text{Pt}_{118}$, $^{197}_{79}\text{Au}_{118}$, and $^{197}_{78}\text{Pt}_{119}$.

To conclude, we would like to stress again that the algebraic methods presented in this book can be useful in many areas other than the molecular and nuclear structure calculations that we have described. The evaluation of physically meaningful matrix elements in any conceivable situation where a symmetry is present, be it of geometric or dynamical origin, requires the same basic techniques. Even systems which do not display obvious symmetries can often be analyzed in terms of a hypothetical symmetry which has been dynamically broken. We hope that in this book we have transmitted to the reader a small amount of knowledge and interest in the beautiful subject of symmetry.

Summary

This final chapter deals with the coupling of fermions to the system of bosons analyzed in Chapters 8 and 9. The ensuing interacting boson–fermion model is aimed at describing nuclei with an odd number of

Summary

neutrons and/or protons. The relevant algebra is the direct product $U^B(6) \otimes U^F(\Omega)$, where Ω is the dimensionality of the fermion space and its defining equations (10.1) to (10.6) are given for general Ω. The subalgebra structure of $U^B(6) \otimes U^F(\Omega)$ cannot be analyzed in general terms but depends on the specific orbits available to the fermion. Two examples are studied in this chapter: a proton in a $j = 3/2$ orbit and a neutron in orbits with $j = 1/2, 3/2, 5/2$. The corresponding fermion classifications are given in (10.12) and (10.21), respectively. In the subsequent sections the coupling of the fermion to an SO(6) boson core is investigated for both cases. The possible symmetry limits in the $j = 3/2$ case are specified by the lattice of algebras (10.27). The labels in the most useful limit, which includes a quadrupole boson–fermion interaction, are given in (10.43). The corresponding wave functions (10.56) involve isoscalar factors associated with the nested algebras $SU(4) \supset Sp(4)$ and $Sp(4) \supset SU(2)$ and their derivation is explained in detail. The explicit expression for the wave function can then be used to derive various matrix elements, such as those of the E2 transition operator. A comparison is made with the observed electric quadrupole transitions in $^{197}_{79}Au_{118}$. The symmetry limits of the $j = 1/2, 3/2, 5/2$ case are those of the lattice (10.30). Again, states can be appropriately labeled as in (10.79) and the expression (10.88) for the wave functions can be derived. Section 10.5 concludes with the example of $^{195}_{78}Pt_{117}$. The final section deals with the extension to larger dynamical algebras. In the $j = 3/2$ case the dynamical algebra can be extended to the superalgebra $U(6/4)$. A single supersymmetric representation of $U(6/4)$ contains the $U^B(6) \otimes U^F(4)$ representations given in (10.91), the first of which corresponds to a nucleus with an even number of protons while for the second this number is odd. The choice of a dynamical superalgebra thus leads to a simultaneous description of even–even and odd-mass nuclei. Similarly, in the $j = 1/2, 3/2, 5/2$ case the dynamical algebra can be chosen as $U(6/12)$, leading to the $U^B(6) \otimes U^F(12)$ representations given in (10.95) and a simultaneous description of nuclei with different neutron numbers. Finally, this scheme can be further extended to the product of superalgebras $U_\nu(6/12) \otimes U_\pi(6/4)$, a single representation of which contains an even–even, an even–odd, an odd–even, and an odd–odd nucleus. Throughout this last section these ideas are illustrated with several examples of platinum and gold spectra.

References

1. A. Bohr and B.R. Mottelson, *Nuclear Structure. II. Nuclear Deformations*, Benjamin, New York, 1975.
2. F. Iachello and O. Scholten, "Interacting boson–fermion model of collective states in odd-A nuclei," Phys. Rev. Lett. **43** (1979) 679.
3. F. Iachello, *Interacting Bose–Fermi Systems in Nuclei*, Plenum, New York, 1981.
4. F. Iachello, "Dynamical supersymmetries in nuclei," Phys. Rev. Lett. **44** (1980) 772.
5. F. Iachello and P. Van Isacker, *The Interacting Boson–Fermion Model*, Cambridge University Press, Cambridge, 1991.
6. R. Gilmore, *Lie Groups, Lie Algebras, and Some of Their Applications*, Wiley-Interscience, New York, 1974.
7. B.G. Wybourne, *Classical Groups for Physicists*, Wiley-Interscience, New York, 1974.
8. A. Bohr and B.R. Mottelson, *Nuclear Structure. I. Single-Particle Motion*, Benjamin, New York, 1969.
9. A. Arima, M. Harvey, and K. Shimizu, "Pseudo LS coupling and pseudo SU_3 coupling schemes," Phys. Lett. B **30** (1969) 517.
10. K.T. Hecht and A. Adler, "Generalized seniority for favored $J \neq 0$ pairs in mixed configurations," Nucl. Phys. A **137** (1969) 129.
11. F. Iachello and S. Kuyucak, "Interacting boson–fermion model of collective states. I. The Spin(6) limit," Ann. Phys. (NY) **136** (1981) 19.
12. P. Van Isacker, A. Frank, and H.Z. Sun, "The U(6/12) supersymmetric limit of the interacting boson–fermion model," Ann. Phys. (NY) **157** (1984) 183.
13. R. Bijker and V.K.B. Kota, "Interacting boson–fermion model of collective states. II. Boson–fermion symmetries connected with the U(5) limit," Ann. Phys. (NY) **156** (1984) 110.
14. R. Bijker and F. Iachello, "Interacting boson–fermion model of collective states. III. The $SO(6) \otimes SU(2)$ limit," Ann. Phys. (NY) **161** (1985) 360.

15. R. Bijker and V.K.B. Kota, "Interacting boson–fermion model of collective states. IV. The $SU(3) \otimes U(2)$ limit," Ann. Phys. (NY) **187** (1988) 148.
16. J. Vervier, "Boson–fermion symmetries and supersymmetries in nuclear physics," Rivista Nuovo Cimento **10** (1987) No. 9.
17. M. Hamermesh, *Group Theory and Its Application to Physical Problems*, Addison-Wesley, Reading, MA, 1962.
18. K.T. Hecht and J.P. Elliott, "Coherent-state theory for the proton neutron quasispin group," Nucl. Phys. A **438** (1985) 29.
19. K.T. Hecht, "Wigner coefficients for the proton–neutron quasispin group: An application of the vector coherent state technique," Nucl. Phys. A **493** (1989) 29.
20. A.R. Edmonds, *Angular Momentum in Quantum Mechanics*, Princeton University Press, Princeton, 1957.
21. A.B. Balentekin, I. Bars, R. Bijker, and F. Iachello, "New class of supersymmetry in nuclei," Phys. Rev. C **27** (1983) 1761.
22. A. Mauthofer, K. Stelzer, J. Gerl, Th.W. Elze, Th. Happ, G. Eckert, T. Faestermann, A. Frank, and P. Van Isacker, "New supersymmetry classification of nuclear levels in ^{195}Pt," Phys. Rev. C **34** (1986) 1958.
23. A.B. Balentekin, I. Bars, and F. Iachello, "U(6/4) supersymmetry in nuclei," Nucl. Phys. A **370** (1981) 284.
24. P. Van Isacker, J. Jolie, K. Heyde, and A. Frank, "Extension of supersymmetry in nuclear structure," Phys. Rev. Lett. **54** (1985) 653.
25. D.D. Warner, R.F. Casten, and A. Frank, "Average resonance capture studies of ^{198}Au: A test of symmetry schemes for odd-odd nuclei," Phys. Lett. B **180** (1986) 207.

Appendix A

Group Theory and the Algebraic Approach

A.1 Introduction

Symmetry and its mathematical framework—group theory—play an increasingly important role in physics. Both classical and quantum systems usually display great complexity, but the analysis of their symmetry properties often gives rise to simplifications and new insights which can lead to a deeper understanding. In addition, symmetries themselves can point the way toward the formulation of a correct physical theory by providing constraints and guidelines in an otherwise intractable situation. It is remarkable that, in spite of the wide variety of systems one may consider, all the way from classical ones to molecules, nuclei, and elementary particles, group theory applies the same basic principles and extracts the same kind of useful information from all of them. This universality in the applicability of symmetry considerations is one of the most attractive features of group theory.

Most people have an intuitive understanding of symmetry, particularly in its most obvious manifestation in terms of geometric transformations that leave a body or system invariant. This interpretation, however, is not enough to readily grasp its deep connections with physics, and it thus becomes necessary to generalize the notion of symmetry transformations to encompass more abstract ideas. The mathematical theory of these transformations is the subject matter of group theory. When these operations are of a continuous nature, one can always consider the case of infinitesimal transformations and study the behavior of the systems subject to the latter. The mathematical theory of such transformations was first considered by Marius Sophus Lie, who introduced the basic concepts and operations of what are now called Lie algebras [1].

A.2 Some definitions

An abstract group G is defined[1] by a set of elements $(\hat{G}_1, \hat{G}_2, \ldots, \hat{G}_n)$ for which a "multiplication" rule combining these elements exists and which satisfies the following conditions:

1. *Closure.* If $\hat{G}_i$ and $\hat{G}_j$ are elements of the set, so is their product $\hat{G}_i \hat{G}_j$.

2. *Associativity.* The following property is always valid:
$$\hat{G}_i(\hat{G}_j \hat{G}_k) = (\hat{G}_i \hat{G}_j)\hat{G}_k$$

3. *Identity.* There exists an element $\hat{E}$ of G satisfying
$$\hat{E}\hat{G}_i = \hat{G}_i \hat{E} = \hat{G}_i$$

4. *Inverse.* For every $\hat{G}_i$ there exists an element $\hat{G}_i^{-1}$ such that
$$\hat{G}_i \hat{G}_i^{-1} = \hat{G}_i^{-1} \hat{G}_i = \hat{E}$$

The number n of elements is called the *order* of the group.

For continuous (or Lie) groups all elements may be obtained by exponentiation in terms of a basic set of elements $\hat{g}_i, i = 1, 2, \ldots, s$,

---

[1] In this appendix groups and algebras (to be defined below) are distinguished by upper- and lowercase roman letters, G and g, while the corresponding elements are denoted as $\hat{G}_i$ and $\hat{g}_i$, respectively. We adhere to this convention in this appendix only; in the main text of the book, as well as in the examples given in Section A.6, Lie algebras and their elements are denoted with uppercase letters.

A.2 Some definitions

called *generators*, which together form the *Lie algebra* associated with the Lie group. A simple example is provided by the SO(2) group of rotations in two-dimensional space, with elements that may be realized as

$$\hat{G}(\alpha) = e^{-i\alpha \hat{l}_z} \tag{A.1}$$

where α is the angle of rotation and

$$\hat{l}_z = -i\left(x\frac{\partial}{\partial y} - y\frac{\partial}{\partial x}\right) \tag{A.2}$$

is the generator of these transformations in the x–y plane. Three-dimensional rotations require the introduction of two additional generators, associated with rotations in the z–x and y–z planes,

$$\hat{l}_y = -i\left(z\frac{\partial}{\partial x} - x\frac{\partial}{\partial z}\right), \qquad \hat{l}_x = -i\left(y\frac{\partial}{\partial z} - z\frac{\partial}{\partial y}\right) \tag{A.3}$$

Finite rotations can then be parametrized by three angles (which may be chosen to be the Euler angles) and expressed as a product of exponentials of the so(3) generators (A.2) and (A.3) [2]. Evaluating the commutators of these operators, we find

$$[\hat{l}_x, \hat{l}_y] = i\hat{l}_z, \qquad [\hat{l}_y, \hat{l}_z] = i\hat{l}_x, \qquad [\hat{l}_z, \hat{l}_x] = i\hat{l}_y \tag{A.4}$$

which illustrates the closure property of the group generators. In general, the s operators $\hat{g}_i, i = 1, 2, \ldots, s$, define a *Lie algebra* if they close under commutation,

$$[\hat{g}_i, \hat{g}_j] = \sum_k c_{ij}^k \hat{g}_k \tag{A.5}$$

and satisfy the Jacobi identity [3]

$$[\hat{g}_i, [\hat{g}_j, \hat{g}_k]] + [\hat{g}_k, [\hat{g}_i, \hat{g}_j]] + [\hat{g}_j, [\hat{g}_k, \hat{g}_i]] = 0 \tag{A.6}$$

The set of constants c_{ij}^k are called *structure constants*, and their values determine the properties of both the Lie algebra and its associated (Lie) group. All Lie groups have been classified by Cartan [3,4], and many of their properties have been established.

A.3 Symmetry transformations

From a general point of view symmetry transformations of a physical system may be defined in terms of the equations of motion for the system [5]. Suppose we consider the system of equations

$$\mathcal{O}_i \psi_i = 0, \quad i = 1, 2, \ldots \quad (A.7)$$

where the functions $\psi_i(\mathbf{x})$ denote a vector column with a finite or infinite number of components, or a more general structure such as a matrix depending on the variables x_i. The operators $\mathcal{O}_i$ are quite arbitrary, and (A.7) may correspond, for example, to Maxwell, Schrödinger, or Dirac equations. The operators $\hat{g}_{ij}$ such that

$$\sum_j \mathcal{O}_i(\hat{g}_{ij}\psi_j) = 0, \quad i = 1, 2, \ldots \quad (A.8)$$

are called symmetry transformations, since they transform the solutions $\vec{\psi}$ to other solutions $g\vec{\psi}$ of the equations (A.7). As a particular example we consider the time-dependent Schrödinger equation (with $\hbar = 1$)

$$\left(\hat{H}(\mathbf{x}, \mathbf{p}) - i\frac{\partial}{\partial t}\right)\psi(\mathbf{x}, t) = 0 \quad (A.9)$$

One can verify that $\hat{k}_j(\mathbf{x}, \mathbf{p}, t)\psi(\mathbf{x}, t)$ is also a solution of (A.9) as long as $\hat{k}_j$ satisfies the equation

$$[\hat{H}, \hat{k}_j] - i\frac{\partial \hat{k}_j}{\partial t} = 0 \quad (A.10)$$

which means that $\hat{k}_j$ is an operator associated with a conserved quantity. The last statement follows from the definition of the total derivative of an operator $\hat{A}_j$

$$\frac{d\hat{A}_j}{dt} = \frac{\partial \hat{A}_j}{\partial t} + i[\hat{H}, \hat{A}_j] \quad (A.11)$$

where $\hat{H}$ is the quantum-mechanical hamiltonian [6]. If $\hat{k}_1$ and $\hat{k}_2$ satisfy (A.10), their commutator is again a constant of the motion since

$$\begin{aligned}
\frac{d}{dt}[\hat{k}_1, \hat{k}_2] &= \frac{\partial}{\partial t}[\hat{k}_1, \hat{k}_2] + i[\hat{H}, [\hat{k}_1, \hat{k}_2]] \\
&= \frac{\partial}{\partial t}[\hat{k}_1, \hat{k}_2] - [\frac{\partial \hat{k}_1}{\partial t}, \hat{k}_2] - [\hat{k}_1, \frac{\partial \hat{k}_2}{\partial t}] = 0 \quad (A.12)
\end{aligned}$$

A.4 Constants of the motion and state labeling

where use is made of (A.10) and the Jacobi identity (A.6). A particularly interesting situation arises when the set $(\hat{k}_i)$ is such that $[\hat{k}_i, \hat{k}_j]$ closes under commutation to form a Lie algebra as in (A.5). In this case we refer to $(\hat{k}_i)$ as the generators of the *symmetry* (Lie) algebra of the time-dependent quantum system (A.9) [7]. Note that in general these operators do not commute with the hamiltonian but rather satisfy (A.10),

$$[\hat{H} - i\frac{\partial}{\partial t}, \hat{k}_j] = 0 \qquad (A.13)$$

What about the time-independent Schrödinger equation? This case corresponds to substituting $\psi(\mathbf{x},t) = \psi_n(x)e^{-iE_n t}$ in (A.9), leading to

$$(\hat{H}(\mathbf{x},\mathbf{p}) - E_n)\psi_n(\mathbf{x}) = 0 \qquad (A.14)$$

The set $\hat{k}_j(\mathbf{x},\mathbf{p},t=0)$ still satisfies the same commutation relations as before but due to (A.10) are not in general integrals of the motion anymore. These operators constitute the *dynamical algebra* for the time-independent Schrödinger equation (A.14) and connect all solutions $\psi_n(x)$ with each other, including states at different energies. Due again to (A.10), only those $\hat{k}_j$ generators that are time independent satisfy

$$[\hat{H}, \hat{k}_j] = 0 \qquad (A.15)$$

which implies that they are constants of the motion for the system (A.14). Equation (A.15) (together with the closure of the $\hat{k}_j$s) constitutes the familiar definition of the symmetry algebra for a time-independent system. The connection between the dynamical algebra $(\hat{k}_j(0))$ and the symmetry algebra of the corresponding time-dependent system $(\hat{k}_j(t))$ allows a unique definition of the dynamical algebra [7].

A.4 Constants of the motion and state labeling

From the previous discussion we see that the symmetry Lie algebras associated with both the time-dependent and time-independent Schrödinger equations supply integrals of the motion for physical systems. In addition, the dynamical algebra of the latter is such that all solutions $\psi_n(\mathbf{x})$ are connected by means of its generators. This means that the dynamical algebra implicitly defines the appropriate Hilbert space for the description of the physical system. For any Lie algebra one may construct one or more operators $\mathcal{C}_l$ which commute with all the generators $\hat{k}_j$,

$$[\mathcal{C}_l, \hat{k}_j] = 0, \quad l = 1, 2, \ldots, r, \quad j = 1, 2, \ldots, s \quad (A.16)$$

These operators are called *Casimir operators* or *Casimir invariants*, and examples are given throughout the book for the u(n) and so(n) algebras. They may be linear, quadratic, or of higher order in the generators. The number r of linearly independent Casimir operators is called the rank of the algebra [3]. This number coincides with the maximum subset of generators which commute among themselves (called *weight generators*)

$$[\hat{k}_\alpha, \hat{k}_\beta] = 0, \quad \alpha, \beta = 1, 2, \ldots, r \quad (A.17)$$

where we use greek labels to indicate that they belong to the subset satisfying (A.17). The operators $(\mathcal{C}_i, \hat{k}_\alpha)$ may be simultaneously diagonalized and their eigenvalues used to label the corresponding eigenstates.

To illustrate these definitions, we consider the su(2) algebra $(\hat{j}_x, \hat{j}_y, \hat{j}_z)$ with commutation relations

$$[\hat{j}_x, \hat{j}_y] = i\hat{j}_z, \quad [\hat{j}_z, \hat{j}_x] = i\hat{j}_y, \quad [\hat{j}_y, \hat{j}_z] = i\hat{j}_x \quad (A.18)$$

isomorphic to the so(3) commutators given in (A.4). From (A.18) we conclude that $r = 1$ and we may choose $\hat{j}_z$ as the generator to diagonalize together with the Casimir invariant

$$\hat{j}^2 = \hat{j}_x^2 + \hat{j}_y^2 + \hat{j}_z^2 \quad (A.19)$$

The eigenvalues and branching rules for the commuting set $(\mathcal{C}_l, \hat{k}_\alpha)$ can be determined solely from the commutation relations (A.5). In the case of su(2) the eigenvalue equations are

$$\hat{j}^2 |jm\rangle = n_j |jm\rangle, \quad \hat{j}_z |jm\rangle = m |jm\rangle \quad (A.20)$$

where j is an index to distinguish the different $\hat{j}^2$ eigenvalues. Defining the raising and lowering operators

$$\hat{j}_\pm = \hat{j}_x \pm i\hat{j}_y \quad (A.21)$$

A.5 Eigenfunctions and representations

and using (A.18), one finds the well-known results [2]
$$n_j = j(j+1), \quad j = 0, 1/2, 1, \ldots, \quad m = -j, -j+1, \ldots, j \quad (A.22)$$
As a bonus, the action of $\hat{j}_\pm$ on the $|jm\rangle$ eigenstates is also determined to be
$$\hat{j}_\pm |jm\rangle = \sqrt{(j \mp m)(j \pm m + 1)} |j\, m\pm 1\rangle \quad (A.23)$$
In the case of a general Lie algebra (A.5) the procedure can be quite complicated but requires the same basic steps. The analysis leads to the algebraic determination of eigenvalues, branching rules, and matrix elements of raising and lowering operators [3].

Returning to the time-independent Schrödinger equation, it follows from our discussion that the *symmetry* algebra provides constants of the motion, which in turn lead to quantum numbers that label the states associated with a given energy eigenvalue. The raising and lowering operators in this algebra only connect degenerate states. The dynamical algebra, however, defines the whole set of eigenstates associated with a given system. The generators are no longer constants of the motion as not all commute with the hamiltonian. The raising and lowering operators may now connect all states with each other.

A.5 Eigenfunctions and representations

For a given group G of physical operations ($\hat{R}$) one may introduce a set of operators $\hat{P}_R$ which are defined by their action on an arbitrary scalar function $f(\mathbf{x})$:
$$\hat{P}_R f(\mathbf{x}) = f(\hat{R}\mathbf{x}) \quad (A.24)$$
The correspondence $\hat{R} \to \hat{P}_R$ is an isomorphism, as $\hat{S}\hat{R} \to \hat{P}_S\hat{P}_R = \hat{P}_{SR}$, as can be shown from (A.24). A simple example is provided by the two-dimensional rotations (A.1). To deduce their explicit form we apply (A.24), using polar coordinates
$$\hat{P}_\alpha f(r, \phi) = f(r, \phi - \alpha) \quad (A.25)$$
and expand in a Taylor series,
$$f(r, \phi - \alpha) = \sum_{n=0}^{\infty} (-\alpha)^n \frac{1}{n!} \frac{\partial^n f(r, \phi)}{\partial \phi^n}$$

452   A   Group Theory and the Algebraic Approach

$$= \sum_{n=0}^{\infty} \frac{1}{n!}\left(-\alpha\frac{\partial}{\partial\phi}\right)^n f(r,\phi)$$
$$= e^{-\alpha\partial/\partial\phi} f(r,\phi) \qquad (A.26)$$

leading to

$$\hat{P}_\alpha = e^{-i\alpha \hat{l}_z}, \qquad \hat{l}_z = -i\frac{\partial}{\partial\phi} = -i\left(x\frac{\partial}{\partial y} - y\frac{\partial}{\partial x}\right) \qquad (A.27)$$

which coincides with (A.1) and (A.2).

Now consider the defining equation

$$\hat{H}(\mathbf{x})f(\mathbf{x}) = g(\mathbf{x}) \qquad (A.28)$$

where $\hat{H}(\mathbf{x})$ is an operator. Using this definition and the property (A.24), we find the following two relations:

$$\begin{aligned}\hat{P}_R \hat{H}(\mathbf{x}) \hat{P}_R^{-1} \hat{P}_R f(\mathbf{x}) &= \hat{P}_R g(\mathbf{x}) = g(\hat{R}\mathbf{x}) = \hat{H}(\hat{R}\mathbf{x}) f(\hat{R}\mathbf{x}) \\ \hat{P}_R \hat{H}(\mathbf{x}) \hat{P}_R^{-1} \hat{P}_R f(\mathbf{x}) &= \hat{P}_R \hat{H}(\mathbf{x}) \hat{P}_R^{-1} f(\hat{R}\mathbf{x})\end{aligned} \qquad (A.29)$$

Since $f(\mathbf{x})$ is an arbitrary function, comparison of the right-hand sides of these equations shows that operators transform as

$$\hat{P}_R \hat{H}(\mathbf{x}) \hat{P}_R^{-1} = \hat{H}(\hat{R}\mathbf{x}) \qquad (A.30)$$

If for all $\hat{R}$ we have

$$\hat{P}_R \hat{H}(\mathbf{x}) \hat{P}_R^{-1} = \hat{H}(\mathbf{x}) \qquad (A.31)$$

then $\hat{H}(\mathbf{x})$ is said to be invariant under the action of the group $G = (\hat{R})$ or that G is a symmetry group for $H(\mathbf{x})$. This definition coincides with our general discussion leading to (A.15), as (A.31) implies

$$[\hat{P}_R, \hat{H}(\mathbf{x})] = 0 \qquad (A.32)$$

Let us return to the time-independent Schrödinger equation

$$\hat{H}\psi = E\psi \qquad (A.33)$$

and use (A.32). We find

$$\hat{H}(\hat{P}_R \psi) = E(\hat{P}_R \psi) \qquad (A.34)$$

A.5 Eigenfunctions and representations 453

Suppose that the eigenvalue E is degenerate and that l independent eigenfunctions $\psi_1, \psi_2, \ldots, \psi_l$ are associated with it. Since (A.34) implies that $\hat{P}_R \psi$ is also an eigenfunction of $\hat{H}$ associated with E, it must be a linear combination of the ψ_is,

$$\hat{P}_R \psi_i(\mathbf{x}) = \sum_{j=1}^{l} D_{ji}(\hat{R}) \psi_j(\mathbf{x}), \qquad i = 1, 2, \ldots, l \qquad (A.35)$$

The matrices $D_{ji}(\hat{R})$ are called a *representation* of the group G, and it is easy to prove that they satisfy the matrix product

$$D(\hat{S}) D(\hat{R}) = D(\hat{S}\hat{R}) \qquad (A.36)$$

The l independent eigenfunctions $\psi_1, \psi_2, \ldots, \psi_l$ are said to constitute a basis for this representation. In addition, if the ψ_i's are such that no change of basis transformation

$$\phi_i = \sum_j U_{ij} \psi_j \qquad (A.37)$$

can take all the **D** matrices to block-diagonal form, that is, to the form

$$\mathbf{U}^{-1} \mathbf{D} \mathbf{U} \to \begin{bmatrix} \mathbf{D}_1 & \vdots & \mathbf{0} \\ \cdots & \cdot & \cdots \\ \mathbf{0} & \vdots & \mathbf{D}_2 \end{bmatrix} \qquad (A.38)$$

we then say that the representation is *irreducible* and that the ψ_i's are a basis for an *irreducible representation* of G. The form (A.38) would imply that two subsets of the l ψ_i's transform only among themselves under the action of $G = (\hat{R})$.

As an example we return to the SO(3) group where the appropriate basis for the irreducible representations is given by the spherical harmonics [2] $Y_m^l(\theta, \phi)$. The action of the rotation-group elements gives

$$\hat{P}_R(\theta_1, \theta_2, \theta_3) Y_m^l(\theta, \phi) = \sum_{m'} D_{m'm}^l(\theta_1, \theta_2, \theta_3) Y_{m'}^l(\theta, \phi) \qquad (A.39)$$

where Wigner's **D** matrices are introduced [2], which play the role of SO(3) irreducible representations. We further note that the $Y_{lm}(\theta, \phi)$ satisfy the eigenvalue equations

$$\hat{l}^2 Y_m^l(\theta, \phi) = l(l+1) Y_m^l(\theta, \phi), \qquad \hat{l}_z Y_m^l(\theta, \phi) = m Y_m^l(\theta, \phi) \qquad (A.40)$$

where $\hat{l}^2$ is the SO(3) Casimir invariant

$$\hat{l}^2 = \hat{l}_x^2 + \hat{l}_y^2 + \hat{l}_z^2 \qquad (A.41)$$

This symmetry group (and its algebra) applies for all hamiltonians invariant under physical rotations. For arbitrary Lie groups relation (A.39) is generalized to

$$\hat{P}_R f_\mu^\lambda(\mathbf{x}) = \sum_{\mu'} D_{\mu'\mu}^\lambda(\hat{R}) f_{\mu'}^\lambda(\mathbf{x}) \qquad (A.42)$$

where λ denotes in general a set of quantum numbers that label the irreducible representations of the group $G = (\hat{R})$ and μ (and μ') label the different functions in the representation. They are often chosen to correspond to sets of quantum numbers that label the irreducible representations of subgroups of G. Likewise, (A.40) is generalized to

$$\mathcal{C}_l f_\mu^\lambda(\mathbf{x}) = h_l(\lambda) f_\mu^\lambda(\mathbf{x}), \qquad \hat{k}_\alpha f_\mu^\lambda(\mathbf{x}) = h_\alpha(\mu) f_\mu^\lambda(\mathbf{x}) \qquad (A.43)$$

where $\mathcal{C}_l$ and $\hat{k}_\alpha$ are the Casimir invariants and weight generators defined in Section A.4. The eigenvalues $h_l(\lambda)$ and $h_\alpha(\mu)$ may be determined from the commutation relations that define the Lie algebra associated with G, as explained in the previous section.

A.6 The algebraic approach

In this section we show how the concepts presented in the previous sections of this appendix lead to an algebraic approach to the study of diverse physical systems.

We start by considering again (A.15), which describes the invariance of a hamiltonian under the algebra $g = (\hat{k}_j)$,

$$[\hat{H}, \hat{k}_j] = 0 \qquad (A.44)$$

implying that g plays the role of symmetry algebra for the system. Equation (A.34), on the other hand, implies that an eigenstate of $\hat{H}$ with energy E may be written as $|\lambda\mu\rangle$, where λ labels the irreducible representations of the group G corresponding to g and μ distinguishes

A.6 The algebraic approach

between the different eigenstates with energy E (and may be chosen to correspond to irreducible representations of subgroups of G). The energy eigenvalues of the hamiltonian in (A.44) thus depend only on λ,

$$\hat{H}|\lambda\mu\rangle = E(\lambda)|\lambda\mu\rangle \tag{A.45}$$

and furthermore, (A.42) implies that the generators $\hat{k}_i$ (and their corresponding group operators $\hat{P}_R$) do not admix states with different λ's.

The use of the mutually commuting set of Casimir invariants and generators described in the previous section then leads to the full specification of the states $|\lambda\mu\rangle$ through (A.43).

We now consider the chain of algebras

$$g_1 \supset g_2 \tag{A.46}$$

which will lead us to introduce the concept of *dynamical symmetry*. If g_1 is a symmetry algebra for $\hat{H}$, we may label its eigenstates as $|\lambda_1\mu_1\rangle$. Since $g_2 \subset g_1$, g_2 must also be a symmetry algebra for $\hat{H}$ and, consequently, its eigenvalues labeled as $|\lambda_2\mu_2\rangle$. Combination of the two properties leads to the eigenequation

$$\hat{H}|\lambda_1\lambda_2\mu_2\rangle = E(\lambda_1)|\lambda_1\lambda_2\mu_2\rangle \tag{A.47}$$

where the role of μ_1 is played by $\lambda_2\mu_2$ and hence the eigenvalues depend only on λ_1. This process may be continued when there are further subalgebras, that is, $g_1 \supset g_2 \supset g_3 \supset \cdots$, in which case μ_2 is substituted by $\lambda_3\mu_3$, and so on.

In many physical applications the original assumption that g_1 is a symmetry algebra of the hamiltonian is found to be too strong and must be relaxed, that is, one is led to consider the breaking of this symmetry. An elegant way to do so is by considering a hamiltonian of the form

$$\hat{H}' = a\mathcal{C}_{l_1}(g_1) + b\mathcal{C}_{l_2}(g_2) \tag{A.48}$$

where $\mathcal{C}_{l_i}(g_i)$ is a Casimir invariant of g_i. Since $[\hat{H}', \hat{k}_i] = 0$ for $\hat{k}_i \in g_2$, $\hat{H}'$ is invariant under g_2, but not anymore under g_1 because $[\mathcal{C}_{l_2}(g_2), \hat{k}_i] \neq 0$ for $\hat{k}_i \notin g_2$. The new *symmetry algebra* is thus g_2 while g_1 now plays the role of *dynamical algebra* for the system, as long as all states we wish to describe are those originally associated with $E(\lambda_1)$. The extent of the symmetry breaking depends on the ratio b/a. Furthermore, since $\hat{H}'$ is given as a combination of Casimir operators, its eigenvalues can be obtained in closed form using (A.43):

$$\hat{H}'|\lambda_1\lambda_2\mu_2\rangle = (aE_{l_1}(\lambda_1) + bE_{l_2}(\lambda_2))|\lambda_1\lambda_2\mu_2\rangle \qquad (A.49)$$

The kind of symmetry breaking caused by interactions of the form (A.48) is known as *dynamical-symmetry breaking* and the remaining symmetry is called a *dynamical symmetry* of the hamiltonian $\hat{H}'$. From (A.49) we conclude that even if $\hat{H}'$ is not invariant under g_1, its eigenstates are the same as those of $\hat{H}$ in (A.47). The dynamical-symmetry breaking thus splits but does not admix the eigenstates.

The algebraic approach often makes use of dynamical symmetries to compute energy eigenvalues, but it goes further in order to describe all relevant aspects of a system in purely algebraic terms. To do so, it follows a number of steps:

1. A given system is described in terms of a dynamical algebra g_1 which spans all possible states in the system within a fixed irreducible representation. The choice of this algebra is often dictated by physical considerations (such as the quadrupole nature of collective nuclear excitations or the dipole character of diatomic molecular vibrations).

2. The hamiltonian and all other operators in the system, such as electromagnetic multipole operators, should be expressed entirely in terms of the generators of the dynamical algebra. Since the matrix elements of the generators can be evaluated from the commutation properties of the dynamical algebra, this implies that all observables of the system can be calculated algebraically.

3. The appropriate bases for the computation of matrix elements are supplied by the different dynamical symmetries associated with the hamiltonian. Physically meaningful chains are those where the symmetry algebra of the hamiltonian is a subalgebra of the dynamical algebra in the chain $g_1 \supset g_2 \supset \cdots$ chosen to label these bases.

A.6 The algebraic approach

4. Branching rules for the different algebra chains as well as eigenvalues of their Casimir operators need to be evaluated to fully determine the dynamical symmetry bases and their associated energy eigenvalues.

5. When several dynamical symmetry chains containing the symmetry algebra are present in the system, the hamiltonian will in general not be diagonal in any given chain but rather include invariant operators of all possible subalgebras. In that case the hamiltonian should be diagonalized in one of these bases. Dynamical symmetries are still useful as limiting cases where all observables can be analytically determined.

We remark that the condition 1, namely that all states of the system should be spanned by a single irreducible representation of the dynamical algebra g_1, assures that all states of the system can be reached by means of the generators of g_1. If this condition is not satisfied (e.g., if two or more irreducible representations would span the states), step 2 indicates that the physical operators would not connect the states in different irreducible representations and would constitute independent sets.

Some of these ideas can be illustrated with well-known examples. In 1932 Heisenberg considered the occurrence of isospin multiplets in nuclei [8]. To a first approximation neutrons and protons in nuclei interact through isospin-invariant forces, that is, to this approximation the electromagnetic effects are neglected compared with the strong interaction. In the notation used above (without making the distinction between algebras and groups), G_1 is in this case the isospin algebra $SU_T(2)$, consisting of the operators $\hat{T}_x$, $\hat{T}_y$, and $\hat{T}_z$ which satisfy commutation relations (A.18), and G_2 can be identified with $SO_T(2) = (\hat{T}_z)$. An isospin-invariant hamiltonian commutes with $\hat{T}_x$, $\hat{T}_y$, and $\hat{T}_z$, and hence the eigenstates $|TM_T\rangle$ with fixed T and $M_T = -T, -T+1, \ldots, T$ are degenerate in energy. The next approximation is to take into account the electromagnetic interaction which breaks isospin invariance and lifts the degeneracy of the states $|TM_T\rangle$. It is assumed that this symmetry breaking occurs dynamically, and since the Coulomb force has a two-body character, the breaking terms are at most quadratic in

A Group Theory and the Algebraic Approach

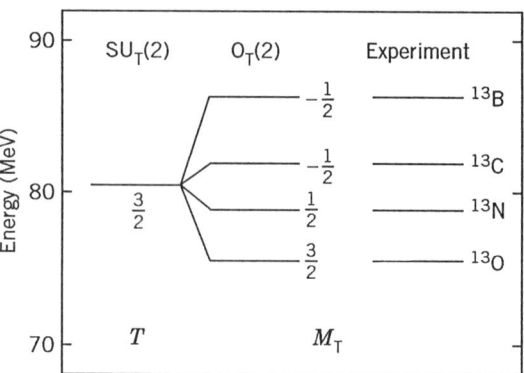

Figure A.1: Binding energies of the $T = 3/2$ isobaric analog states with angular momentum and parity $J^\pi = 1/2^-$ in ^{13}B, ^{13}C, ^{13}N, and ^{13}O. The column on the left is obtained for an exact $SU_T(2)$ *symmetry*, which predicts states with different M_T to be degenerate. The middle column is obtained in the case of an $SU_T(2)$ *dynamical symmetry*, equation (A.50) with parameters (in MeV) $a = 80.59$, $b = -2.96$, and $c = -0.26$.

$\hat{T}_z$ [9]. The energies of the corresponding nuclear states with the same T are then given by

$$E(M_T) = a + bM_T + cM_T^2 \qquad (A.50)$$

and $SU_T(2)$ becomes the dynamical symmetry for the system while $SO_T(2)$ is the symmetry algebra. The dynamical symmetry breaking thus implies that the eigenstates of the nuclear hamiltonian have well-defined values of T and M_T. Extensive tests have shown that indeed this is the case to a good approximation, at least at low excitation energies and in light nuclei [10]. Formula (A.50) can be tested in a number of cases. In Figure A.1 a $T = 3/2$ multiplet consisting of states in ^{13}B, ^{13}C, ^{13}N, and ^{13}O is compared with the theoretical prediction (A.50).

A less trivial example of dynamical-symmetry breaking is provided by the Gell-Mann–Okubo mass-splitting formula for elementary particles [11,12]. The SU(3) model of Gell-Mann and Ne'eman [13] classifies hadrons as SU(3) multiplets, that is, a given irreducible representation

A.6 The algebraic approach

(λ,μ) of SU(3) of dimension d contains d particles. For example, the neutron and proton are placed in the eight dimensional representation (1,1), the so-called *octet* representation. Besides isospin T a new quantum number is needed to fully classify the SU(3) states. This turns out to be an additive number Y, called *hypercharge* [9], associated with the chain of algebras

$$\begin{array}{cccc} \mathrm{U}(3) & \supset & \mathrm{U}_Y(1) \otimes (\mathrm{SU}_T(2) & \supset & \mathrm{SO}_T(2)) \\ | & & | & | & | \\ (\lambda,\mu) & & Y & T & M_T \end{array} \quad (\text{A}.51)$$

If one would assume SU(3) invariance, all particles in a multiplet would have the same mass, but since the experimental masses of other baryons differ from the nucleon masses by hundreds of MeV, the SU(3) symmetry clearly must be broken.

Dynamical symmetry breaking allows the baryon states to still be classified by (A.51). Following the procedure outlined above and keeping up to quadratic terms, one finds a mass operator of the form

$$\begin{aligned} \hat{M} &= a + b\mathcal{C}_1[\mathrm{U}_Y(1)] + c\mathcal{C}_2[\mathrm{U}_Y(1)] + d\mathcal{C}_2[\mathrm{SU}_T(2)] + e\mathcal{C}_1[\mathrm{SO}_T(2)] \\ &\quad + f\mathcal{C}_2[\mathrm{SO}_T(2)] \end{aligned} \quad (\text{A}.52)$$

with eigenvalues

$$M(Y,T,M_T) = a + bY + cY^2 + dT(T+1) + eM_T + fM_T^2 \quad (\text{A}.53)$$

A further assumption regarding the SU(3) tensor character of the strong interaction [9,11,12] leads to a relation between c and d in (A.53), resulting in the mass formula

$$M'(Y,T,M_T) = a + bY + d\left(T(T+1) - \tfrac{1}{4}Y^2\right) + eM_T + fM_T^2 \quad (\text{A}.54)$$

In Figure A.2 this process of successive dynamical-symmetry breaking is illustrated with the octet representation containing the neutron and the proton and the Λ, Σ, and Ξ baryons. Other hadrons are analogously classified using SU(3) as the dynamical algebra [9]. More general applications of the algebraic approach are illustrated throughout this book, where the steps listed before are implemented for physical systems associated with U(2), U(4), and U(6) models.

A Group Theory and the Algebraic Approach

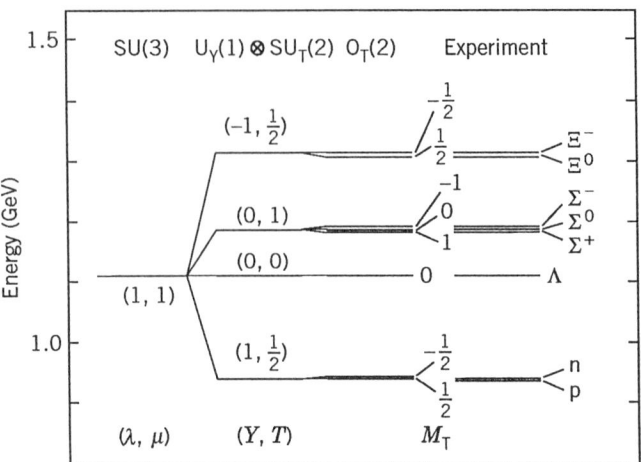

Figure A.2: Mass spectrum of the SU(3) octet $(\lambda, \mu) = (1, 1)$. The column on the left is obtained for an exact SU(3) *symmetry*, which predicts all masses to be the same, while the next two columns represent successive breakings of this symmetry in a dynamical manner. The column under $SO_T(2)$ is obtained with (A.54) with parameters (in MeV) $a = 1111.3$, $b = -189.6$, $d = -39.9$, $e = -3.8$, and $f = 0.9$.

The algebraic approach, both in the sense we have defined here and in its generalizations to other fields of research, has become an important tool in the search for a unified description of physical phenomena. This is illustrated by Figure A.2. The near equality of the neutron and proton masses suggested the existence of isospin multiplets, later confirmed at higher energies for other particles. To find a relationship between these multiplets, the SU(3) dynamical algebra was proposed (and became the basis for the establishment of the quark model). This unification process can be continued: different (λ, μ) multiplets can be unified by means of higher-dimensional algebras such as SU(4) [9].

References

1. R. Gilmore, *Lie Groups, Lie Algebras, and Some of Their Applications*, Wiley-Interscience, New York, 1974.

2. M.E. Rose, *Elementary Theory of Angular Momentum*, Wiley, New York, 1957.
3. B.G. Wybourne, *Classical Groups for Physicists*, Wiley-Interscience, New York, 1974.
4. H.J. Lipkin, *Lie Groups for Pedestrians*, North-Holland, Amsterdam, 1966.
5. V.I. Man'ko, "Invariants and states generating symmetry of nonstationary systems," in *Symmetries in Science V*, edited by B. Gruber, L.C. Biedenharn, and H.D. Doebner, Plenum, New York, 1991, p. 453.
6. A. Messiah, *Quantum Mechanics*, Wiley, New York, 1968.
7. O. Castaños, A. Frank, and R. Lopez–Peña, "Noether's theorem and dynamical groups in quantum mechanics," J. Phys. A **23** (1990) 5141.
8. W. Heisenberg, "Über den Bau der Atomkerne. 1," Z. Phys. **77** (1932) 1.
9. J.P. Elliott and P.G. Dawber, *Symmetry in Physics. I. Principles and Simple Applications*, Oxford University Press, New York, 1979.
10. A. Bohr and B.R. Mottelson, *Nuclear Structure. II. Nuclear Deformations*, Benjamin, New York, 1975.
11. M. Gell-Mann, "Symmetries of baryons and mesons," Phys. Rev. **125** (1962) 1067.
12. S. Okubo, "Note on unitary symmetry in strong interactions," Progr. Theor. Phys. **27** (1962) 949.
13. M. Gell-Mann and Y. Ne'eman, *The Eightfold Way*, Benjamin, New York, 1964.

Appendix B

The Unitary and Orthogonal Algebras

B.1 Commutation relations

The U(n) group is defined [1] by transformations in an n-dimensional field of complex numbers $(z_1, z_2, \ldots, z_n)$

$$z'_i = \sum_{j=1}^{n} U_{ij} z_j, \qquad i = 1, 2, \ldots, n \tag{B.1}$$

where the matrix **U** satisfies

$$\sum_{j=1}^{n} U_{ij} U^*_{kj} = \delta_{ik}, \qquad i, k = 1, 2, \ldots, n \tag{B.2}$$

or in matrix notation,

$$\mathbf{U}\mathbf{U}^\dagger = \mathbf{1} \tag{B.3}$$

We are interested in the infinitesimal unitary transformations, which can be found by expressing **U** as an expansion in the infinitesimal parameter ϵ,

$$\mathbf{U} = \mathbf{1} + i\epsilon \mathbf{S} + \cdots \tag{B.4}$$

It follows that **S** should be a hermitian matrix since

$$\mathbf{U}\mathbf{U}^\dagger \approx (\mathbf{1} + i\epsilon \mathbf{S})(\mathbf{1} - i\epsilon \mathbf{S}^\dagger) \approx \mathbf{1} + i\epsilon(\mathbf{S} - \mathbf{S}^\dagger) = \mathbf{1} \tag{B.5}$$

showing that $\mathbf{S} = \mathbf{S}^\dagger$. The matrix $\mathbf{S}$ can be viewed as an operator $\hat{S}$ which, acting on an arbitrary function $f(z_i)$, gives

$$\hat{S}f(z_i) = f(z_i') = f(z_i + i\epsilon \sum_{j=1}^n S_{ij} z_j) = f(z_i) + i\epsilon \sum_{ij=1}^n S_{ij} z_j \frac{\partial f}{\partial z_i} + \cdots, \tag{B.6}$$

which means that, to first order in ϵ, the infinitesimal unitary transformation is generated by the operator

$$i\epsilon \sum_{ij} S_{ij} \mathcal{G}_j^i \tag{B.7}$$

where

$$\mathcal{G}_i^j \equiv z_i \frac{\partial}{\partial z_j} \tag{B.8}$$

Since $\mathbf{S}$ is an arbitrary hermitian matrix, there are n^2 linearly independent operators (B.8) generating the infinitesimal unitary transformations. Since $[\partial/\partial z_i, z_j] = \delta_{ij}$, they satisfy the commutation relations

$$[\mathcal{G}_i^j, \mathcal{G}_k^l] = \mathcal{G}_i^l \delta_{jk} - \mathcal{G}_k^j \delta_{il}, \qquad i,j,k,l = 1,2,\ldots,n \tag{B.9}$$

which define the U(n) Lie algebra. The U(n) boson realizations used throughout the book are equivalent to (B.8) with the identifications $z_i \to b_i^\dagger$ and $\partial/\partial z_j \to b_j$.

We next consider orthogonal transformations in the n-dimensional space

$$z_i' = \sum_{j=1}^n R_{ij} z_j \tag{B.10}$$

where

$$\sum_{j=1}^n R_{ij} R_{kj} = \delta_{ik} \tag{B.11}$$

or

$$\mathbf{R}\widetilde{\mathbf{R}} = 1 \tag{B.12}$$

with $\widetilde{\mathbf{R}}$ being the transpose of $\mathbf{R}$. Infinitesimal real orthogonal transformations are now expressed as

$$\mathbf{R} = 1 + \epsilon \mathbf{T} + \cdots \tag{B.13}$$

B.2 Casimir invariants

where **T** must be a real antisymmetric matrix since

$$\mathbf{R}\widetilde{\mathbf{R}} \approx (1 + \epsilon\mathbf{T})(1 + \epsilon\widetilde{\mathbf{T}}) \approx 1 + \epsilon(\mathbf{T} + \widetilde{\mathbf{T}}) = 1 \qquad (B.14)$$

or $T_{ji} = -T_{ij}$. A function $g(z_i)$ subject to infinitesimal real orthogonal transformations becomes

$$\begin{aligned}
\hat{T}g(z_i) &= g(z_i') = g(z_i + \epsilon \sum_{j=1}^{n} T_{ij} z_j) \\
&= g(z_i) + \epsilon \sum_{ij=1}^{n} T_{ij} z_j \frac{\partial g}{\partial z_i} + \cdots \\
&= g(z_i) + \epsilon \sum_{i<j=1}^{n} T_{ij} \left(z_j \frac{\partial g}{\partial z_i} - z_i \frac{\partial g}{\partial z_j} \right) + \cdots \qquad (B.15)
\end{aligned}$$

from which we find that the operators

$$\Lambda_{ij} \equiv \mathcal{G}_i^j - \mathcal{G}_j^i \qquad (B.16)$$

are the generators of orthogonal transformations. There are $n(n-1)/2$ independent SO(n) generators (B.16). Using (B.9) we find the commutation relations

$$[\Lambda_{ij}, \Lambda_{kl}] = \Lambda_{il}\delta_{jk} + \Lambda_{jk}\delta_{il} + \Lambda_{lj}\delta_{ik} + \Lambda_{ki}\delta_{jl} \qquad (B.17)$$

which define the SO(n) Lie algebra.

B.2 Casimir invariants

The first-order Casimir invariant of the U(n) algebras is given by

$$\mathcal{C}_1[\mathrm{U}(n)] \equiv \sum_{i=1}^{n} \mathcal{G}_i^i \qquad (B.18)$$

since, from (B.9),

$$[\mathcal{C}_1[\mathrm{U}(n)], \mathcal{G}_k^l] = \sum_{i=1}^{n} [\mathcal{G}_i^i, \mathcal{G}_k^l] = \sum_{i=1}^{n} (\mathcal{G}_i^l \delta_{ik} - \mathcal{G}_k^i \delta_{il}) = 0 \qquad (B.19)$$

No linear invariant exists for the SO(n) algebras, as $\Lambda_{ii} = 0$. The second-order Casimir invariants of U(n) and SO(n) are given by

$$\mathcal{C}_2[\mathrm{U}(n)] \equiv \sum_{ij=1}^{n} \mathcal{G}_i^j \mathcal{G}_j^i \qquad (\mathrm{B.20})$$

and

$$\mathcal{C}_2[\mathrm{SO}(n)] \equiv \tfrac{1}{2} \sum_{ij=1}^{n} \Lambda_{ij} \Lambda_{ji} \qquad (\mathrm{B.21})$$

respectively. Using (B.9) and (B.17), it is straightforward to verify that

$$[\mathcal{C}_2[\mathrm{U}(n)], \mathcal{G}_i^j] = 0 \qquad (\mathrm{B.22})$$

and

$$[\mathcal{C}_2[\mathrm{SO}(n)], \Lambda_{ij}] = 0 \qquad (\mathrm{B.23})$$

Higher-order invariants may be defined by index contraction. For example,

$$\mathcal{C}_3[\mathrm{U}(n)] \equiv \sum_{ijk=1}^{n} \mathcal{G}_i^j \mathcal{G}_j^k \mathcal{G}_k^i \qquad (\mathrm{B.24})$$

is a third-order U(n) invariant. To find the eigenvalues of Casimir invariants it is necessary to define appropriate bases for the irreducible representations of U(n) and SO(n), as is done in the next section.

B.3 Boson bases

It is convenient to define creation and annihilation boson operators $b_{\rho i}^\dagger$ and $b_{\rho i}$ with two indices. The latin indices $i, j, \ldots = 1, 2, \ldots, n$ can be thought of as different vector components, while the greek indices $\rho, \rho', \ldots = 1, 2, \ldots, m$ label different vectors (see also the simplified discussion for U(2) in Chapter 2). These operators satisfy the boson commutation relations

$$[b_{\rho i}, b_{\rho' j}^\dagger] = \delta_{\rho \rho'} \delta_{ij}, \qquad \rho, \rho' = 1, 2, \ldots, m, \quad i, j = 1, 2, \ldots, n \qquad (\mathrm{B.25})$$

with all other commutators being zero. We consider polynomial functions of the $b_{\rho k}^\dagger$, which are denoted by $\mathcal{Q}(b_{\rho k}^\dagger)$. We also define the vacuum state $|0\rangle$ satisfying

$$b_{\rho i} |0\rangle = 0, \qquad \rho = 1, 2, \ldots, m, \quad i = 1, 2, \ldots, n \qquad (\mathrm{B.26})$$

B.3 Boson bases

which allows the definition of a scalar product of $\mathcal{Q}(b^\dagger_{\rho k})$ and $\mathcal{Q}'(b^\dagger_{\rho k})$ as

$$\langle 0| \left(\mathcal{Q}(b^\dagger_{\rho k})\right)^\dagger \mathcal{Q}'(b^\dagger_{\rho k})|0\rangle = \langle 0|\mathcal{Q}(b_{\rho k})\mathcal{Q}'(b^\dagger_{\rho k})|0\rangle \qquad (B.27)$$

which is evaluated using (B.25) repeatedly until all annihilation operators are to the right. At that point we use (B.26) and the normalization $\langle 0|0\rangle = 1$. This procedure shows that the action of $b_{\rho i}$ on $\mathcal{Q}(b^\dagger_{\rho' k})$ is analogous to

$$b_{\rho i}\mathcal{Q}(b^\dagger_{\rho' k})|0\rangle = \frac{\partial}{\partial b^\dagger_{\rho i}}\mathcal{Q}(b^\dagger_{\rho' k})|0\rangle \qquad (B.28)$$

The n^2 operators

$$\mathcal{G}^j_i \equiv \sum_{\rho=1}^m b^\dagger_{\rho i} b_{\rho j}, \qquad i,j = 1,2,\ldots n \qquad (B.29)$$

generate the U(n) algebra and satisfy (B.9). The diagonal generators $\mathcal{G}^i_i$ commute among themselves and can thus be simultaneously diagonalized. We may choose the basis in such a way that the polynomials $\mathcal{Q}(b^\dagger_{\rho k})$ are eigenfunctions of the $\mathcal{G}^i_i$, that is,

$$\mathcal{G}^i_i \mathcal{Q}(b^\dagger_{\rho k})|0\rangle = \omega_i \mathcal{Q}(b^\dagger_{\rho k})|0\rangle \qquad (B.30)$$

and call the set of numbers $(\omega_1, \omega_2, \ldots \omega_n)$ the *weight* of the polynomial $\mathcal{Q}(b^\dagger_{\rho i})$. Note that the sum $\sum_j \omega_j$ gives the degree of the polynomial. If $\mathcal{Q}'(b^\dagger_{\rho k})$ has weight $(\omega'_1, \omega'_2, \ldots, \omega'_n)$, we say that $\mathcal{Q}'(b^\dagger_{\rho k})$ has higher weight than $\mathcal{Q}(b^\dagger_{\rho k})$ if for the set $(\omega'_1 - \omega_1, \omega'_2 - \omega_2, \ldots, \omega'_n - \omega_n)$ the first non-zero entry is positive. It follows that the U(n) generators can be classified in three sets:

$$\begin{aligned}
&\mathcal{G}^i_i && \text{or } \textit{weight generators} \\
&\mathcal{G}^j_i \ (i<j) && \text{or } \textit{raising generators} \\
&\mathcal{G}^j_i \ (i>j) && \text{or } \textit{lowering generators}
\end{aligned} \qquad (B.31)$$

This results from the action of $\mathcal{G}^j_i$ on a polynomial $\mathcal{Q}(b^\dagger_{\rho k})$ with weight $(\omega_1, \omega_2, \ldots, \omega_i, \ldots, \omega_j, \ldots, \omega_n)$, giving rise to a polynomial $\mathcal{Q}'(b^\dagger_{\rho k})$ with weight $(\omega_1, \omega_2, \ldots, \omega_i + 1, \ldots, \omega_j - 1, \ldots, \omega_n)$ for $i < j$, or weight

$(\omega_1, \omega_2, \ldots, \omega_j - 1, \ldots, \omega_i + 1, \ldots, \omega_n)$ for $i > j$. Within a given representation we can now define a highest-weight polynomial $\mathcal{Q}_{\text{hw}}(b^\dagger_{\rho k})$ by the equations

$$\begin{aligned} \mathcal{G}^i_i \mathcal{Q}_{\text{hw}}(b^\dagger_{\rho k})|0\rangle &= h_i \mathcal{Q}_{\text{hw}}(b^\dagger_{\rho k})|0\rangle, & i &= 1,2,\ldots,n \\ \mathcal{G}^j_i \mathcal{Q}_{\text{hw}}(b^\dagger_{\rho k})|0\rangle &= 0, & i &< j, \quad i,j = 1,2,\ldots,n \end{aligned} \quad (\text{B.32})$$

The maximum weight is always associated with a single polynomial of the basis and can thus be used to characterize the irreducible representation.

Computing the scalar product of $\mathcal{G}^j_i \mathcal{Q}_{\text{hw}}(b^\dagger_{\rho k})|0\rangle$ with itself (with $i > j$), we find

$$\begin{aligned} &\langle 0|\mathcal{Q}_{\text{hw}}(b_{\rho k})\mathcal{G}^i_j \mathcal{G}^j_i \mathcal{Q}_{\text{hw}}(b^\dagger_{\rho k})|0\rangle \\ &= \langle 0|\mathcal{Q}_{\text{hw}}(b_{\rho k})[\mathcal{G}^i_j, \mathcal{G}^j_i]\mathcal{Q}_{\text{hw}}(b^\dagger_{\rho k})|0\rangle \\ &= \langle 0|\mathcal{Q}_{\text{hw}}(b_{\rho k})(\mathcal{G}^j_j - \mathcal{G}^i_i)\mathcal{Q}_{\text{hw}}(b^\dagger_{\rho k})|0\rangle \\ &= (h_j - h_i)\langle 0|\mathcal{Q}_{\text{hw}}(b_{\rho k})\mathcal{Q}_{\text{hw}}(b^\dagger_{\rho k})|0\rangle \geq 0 \quad (\text{B.33}) \end{aligned}$$

which proves that

$$h_1 \geq h_2 \geq \cdots \geq h_n \geq 0 \quad (\text{B.34})$$

Defining the degree of the polynomial $\mathcal{Q}_{\text{hw}}(b^\dagger_{\rho k})$ as

$$h_1 + h_2 + \cdots + h_n = N \quad (\text{B.35})$$

establishes that the irreducible representations of the U(n) algebra are characterized by a (Young) partition $[h_1, h_2, \ldots, h_n]$ of the integer N, which satisfies (B.34) and (B.35). A convenient notation is provided by the *Young diagram* or *Young tableau*

$$\begin{array}{c} \overset{h_1}{\overbrace{\square\square\square\cdots\square}} \\ \overset{h_2}{\overbrace{\square\square\cdots\square}} \\ \vdots \\ \overset{h_n}{\overbrace{\square\cdots\square}} \end{array} \quad (\text{B.36})$$

B.3 Boson bases

where the number of blocks in the rows is indicated on top of each row. We now act with (B.20) on the polynomial $\mathcal{Q}_{\text{hw}}(b^\dagger_{\rho k})|0\rangle$ and rewrite the terms

$$\begin{aligned}
&\mathcal{C}_2[\text{U}(n)]\mathcal{Q}_{\text{hw}}(b^\dagger_{\rho k})|0\rangle \\
&= \left(\sum_{i=1}^n \mathcal{G}_i^i \mathcal{G}_i^i + \sum_{i<j=1}^n \mathcal{G}_i^j \mathcal{G}_j^i + \sum_{i>j=1}^n \mathcal{G}_i^j \mathcal{G}_j^i\right) \mathcal{Q}_{\text{hw}}(b^\dagger_{\rho k})|0\rangle \\
&= \left(\sum_{i=1}^n \mathcal{G}_i^i \mathcal{G}_i^i + \sum_{i<j=1}^n [\mathcal{G}_i^j, \mathcal{G}_j^i] + 2\sum_{i>j=1}^n \mathcal{G}_i^j \mathcal{G}_j^i\right) \mathcal{Q}_{\text{hw}}(b^\dagger_{\rho k})|0\rangle \\
&= \left(\sum_{i=1}^n h_i^2 + \sum_{i<j=1}^n (h_i - h_j)\right) \mathcal{Q}_{\text{hw}}(b^\dagger_{\rho k})|0\rangle \quad (B.37)
\end{aligned}$$

where use is made of (B.32) and (B.33). This relation solves the eigenvalue problem for $\mathcal{C}_2[\text{U}(n)]$.

A similar analysis can be carried out for the SO(n) algebra. In order to make the discussion parallel to the one for U(n), it is convenient to define the generators in a spherical basis [2],

$$\Lambda_{mm'} \equiv \mathcal{G}_m^{m'} - \mathcal{G}_{-m'}^{-m} \quad (B.38)$$

with

$$m, m' = \begin{cases} -l, -l+1, \ldots, -1, 0, 1, \ldots, l-1, l & \text{for } n = 2l+1 \\ -l, -l+1, \ldots, -1, 1, \ldots, l-1, l & \text{for } n = 2l \end{cases} \quad (B.39)$$

With this change of basis the commutation relations (B.17) translate into

$$\begin{aligned}
[\Lambda_{mm'}, \Lambda_{m''m'''}] &= \Lambda_{mm'''}\delta_{m''m'} + \Lambda_{-m'-m''}\delta_{mm'''} \\
&\quad + \Lambda_{m''-m}\delta_{-m'm'''} + \Lambda_{-m'''m'}\delta_{m-m''} \quad (B.40)
\end{aligned}$$

Note that now $\Lambda_{mm'} = -\Lambda_{-m'-m}$, so there are still $n(n-1)/2$ independent generators. In particular, it follows from (B.38) that $\Lambda_{m-m} = 0$. Also, from (B.40) we find

$$\begin{aligned}
[\Lambda_{mm}, \Lambda_{m'm'}] &= \Lambda_{mm'}\delta_{m'm} + \Lambda_{-m-m'}\delta_{mm'} \\
&\quad + \Lambda_{m'-m}\delta_{-mm'} + \Lambda_{-m'm}\delta_{m-m'} = 0 \quad (B.41)
\end{aligned}$$

so we may choose a basis where the l operators $\Lambda_{ll}, \Lambda_{l-1\,l-1}, \ldots, \Lambda_{11}$ are diagonal,

$$\Lambda_{mm}\mathcal{Q}(b^\dagger_{\rho k})|0\rangle = \omega_{l-m+1}\mathcal{Q}(b^\dagger_{\rho k})|0\rangle, \qquad m = l, l-1, \ldots, 1 \qquad \text{(B.42)}$$

The subscripts of ω are chosen in such a way that $m = l$ corresponds to ω_1, $m = l - 1$ to ω_2, and so on. The l numbers $(\omega_1, \omega_2, \ldots, \omega_l)$ are called the *weight* of the polynomial $\mathcal{Q}(b^\dagger_{\rho k})$. One may verify that the SO(n) generators can also be classified in the three sets:

$$\begin{array}{ll} \Lambda_{mm} & \text{or } weight\ generators \\ \Lambda_{mm'} \quad (m > m' > -m) & \text{or } raising\ generators \\ \Lambda_{mm'} \quad (m' > m > -m') & \text{or } lowering\ generators \end{array} \qquad \text{(B.43)}$$

Proceeding in analogy with the U(n) case, one can define a highest-weight polynomial for which

$$\begin{aligned} \Lambda_{mm}\mathcal{Q}_{\text{hw}}(b^\dagger_{\rho k})|0\rangle &= \lambda_{l-m+1}\mathcal{Q}_{\text{hw}}(b^\dagger_{\rho k})|0\rangle, \qquad m = l, l-1, \ldots, 1 \\ \Lambda_{mm'}\mathcal{Q}_{\text{hw}}(b^\dagger_{\rho i})|0\rangle &= 0, \qquad\qquad\qquad\qquad\quad m > m' > -m \end{aligned} \qquad \text{(B.44)}$$

with the λ's characterizing the SO(n) irreducible representations. Following the same steps as before, one finds that the results are analogous to the U(n) case except that we need to consider SO($2l+1$) and SO($2l$) separately. We only give the results of the analysis [2]:

1. For SO($2l+1$) the l numbers of the highest weight satisfy

$$\lambda_1 \geq \lambda_2 \geq \cdots \geq \lambda_{l-1} \geq \lambda_l \geq 0 \qquad \text{(B.45)}$$

while the Casimir invariant $\mathcal{C}_2[\text{SO}(2l+1)]$ has the eigenvalue

$$\sum_{k=1}^{l} \lambda_k(\lambda_k + 2l + 1 - 2k) \qquad \text{(B.46)}$$

2. For SO($2l$) the l λ's satisfy

$$\lambda_1 \geq \lambda_2 \geq \cdots \geq \lambda_{l-1} \geq 0, \qquad \lambda_{l-1} \geq |\lambda_l| \qquad \text{(B.47)}$$

and the eigenvalue of the Casimir invariant $\mathcal{C}_2[\text{SO}(2l)]$ is

$$\sum_{k=1}^{l} \lambda_k(\lambda_k + 2l - 2k) \qquad \text{(B.48)}$$

B.4 The Gel'fand bases

Thus the only significant difference between the two cases is that λ_l can be both positive and negative for SO($2l$) [3].

B.4 The Gel'fand bases

The basis states of U(n), classified by the irreducible representation $[h_1, h_2, \ldots, h_n]$, need additional labels in order to distinguish them. These labels are the row indices and may be characterized by the irreducible representations of the subalgebras of U(n). A complete characterization is provided by the chain of algebras [4]

$$\text{U}(n) \supset \text{U}(n-1) \supset \cdots \supset \text{U}(2) \supset \text{U}(1) \tag{B.49}$$

which is known as the *Gel'fand basis*. The classification of the states in terms of the chain (B.49) is complete because no missing labels appear in the reduction. Each of the algebras U(r) has irreducible representations characterized by a Young partition $[h_1^{(r)}, h_2^{(r)}, \ldots, h_r^{(r)}]$, satisfying conditions (B.34) and (B.35),

$$h_1^{(r)} \geq h_2^{(r)} \geq \cdots \geq h_r^{(r)} \geq 0, \qquad h_1^{(r)} + h_2^{(r)} + \cdots + h_r^{(r)} = m \tag{B.50}$$

where m is an integer. To completely characterize the states, we thus need to establish the branching rules for U(r) $\supset$ U($r-1$), that is, the allowed values of $h_i^{(r-1)}, i = 1, 2, \ldots, r-1$, given a partition $h_j^{(r)}, j = 1, 2, \ldots, r$. This problem is analogous to the determination of the allowed m values for a given j in the reduction SO(3) $\supset$ SO(2) [5] and involves the explicit construction of linear combinations of lowering operators of U(r) with the property that, when applied to a highest-weight state in a given irreducible representation of U($r-1$), these combinations give rise to a highest-weight state in a different irreducible representation of U($r-1$) [6]. By means of these operators it is possible to deduce the branching rule [1,2]

$$h_1^{(r)} \geq h_1^{(r-1)} \geq h_2^{(r)} \geq h_2^{(r-1)} \geq \cdots \geq h_{r-1}^{(r)} \geq h_{r-1}^{(r-1)} \geq h_r^{(r)} \tag{B.51}$$

which is valid for all r in (B.49). Every irreducible representation $[h_1^{(r-1)}, h_2^{(r-1)}, \ldots, h_{r-1}^{(r-1)}]$ of U($r-1$) satisfying (B.51) occurs only once.

Note that relation (B.51), applied successively to $U(n) \supset U(n-1)$, $U(n-1) \supset U(n-2)$, and so on, allows the determination of the dimensionality of the irreducible representation $[h_1^{(n)}, h_2^{(n)}, \ldots, h_n^{(n)}]$ of $U(n)$.

The same kind of procedure can be applied to the $SO(n)$ algebra and its Gel'fand chain

$$SO(n) \supset SO(n-1) \supset \cdots \supset SO(3) \supset SO(2) \qquad (B.52)$$

Again, in this case it is necessary to consider the even and odd algebras $SO(2l+1)$ and $SO(2l)$ separately. The results for the branching rules are as follows:

1. For the decomposition $SO(2l+1) \supset SO(2l)$ we consider the irreducible representations $(\lambda_1, \lambda_2, \ldots, \lambda_l)$ and $(\mu_1, \mu_2, \ldots, \mu_l)$, satisfying the conditions (B.45) and (B.47), respectively. The construction of appropriate lowering operators leads to the result [2]

$$\lambda_1 \geq \mu_1 \geq \lambda_2 \geq \mu_2 \geq \cdots \geq \mu_{l-1} \geq \lambda_l \geq |\mu_l| \qquad (B.53)$$

for the irreducible representations $(\mu_1, \mu_2, \ldots, \mu_l)$ of $SO(2l)$ contained in $(\lambda_1, \lambda_2, \ldots, \lambda_l)$ of $SO(2l+1)$.

2. In the $SO(2l) \supset SO(2l-1)$ reduction the corresponding irreducible representations are $(\mu_1, \mu_2, \ldots, \mu_l)$ and $(\lambda_1', \lambda_2', \ldots, \lambda_{l-1}')$, respectively, the branching rule being

$$\mu_1 \geq \lambda_1' \geq \mu_2 \geq \lambda_2' \geq \cdots \geq \mu_{l-1} \geq \lambda_{l-1}' \geq |\mu_l| \qquad (B.54)$$

In both (B.53) and (B.54) the irreducible representations of the subalgebra again occur only once, thus providing a complete characterization of the $SO(n)$ basis states. These rules, applied successively, also permit evaluating the dimensionality of the irreducible representation $(\lambda_1, \lambda_2, \ldots, \lambda_l)$ of $SO(n)$.

In addition to the $U(n) \supset U(n-1)$ and $SO(n) \supset SO(n-1)$ branching rules discussed up to now, the algebraic models often require the knowledge of the $U(n) \supset SO(n)$ reduction, which is associated with the subalgebra generated by the linear combinations (B.16). A closed expression for the corresponding branching rule is not available, due

B.5 Outer products of U(n) representations

to the appearance of missing labels, that is, the possible repetition of an irreducible representation $(\lambda_1, \lambda_2, \ldots, \lambda_l)$ of SO(n) within a given irreducible representation $[h_1, h_2, \ldots, h_n]$ of U(n). The particular case of the fully symmetric U(n) irreducible representation $[N, 0, \ldots, 0]$ is not complicated by this problem, and the corresponding reduction is discussed in Chapter 8. Two-row representations of U(n) have been considered in [2] while general results for small n can be found in [3] and [7].

B.5 Outer products of U(n) representations

To carry out applications of the algebraic models, it is often necessary to couple different representations of the U(n) algebras to arrive at definite irreducible representations of the same algebra U(n). For example, in Chapter 2 it is necessary to compute the $U_1(2) \otimes U_2(2)$ products in their coupling to $U_{12}(2)$, and in Chapter 6 the $U_1(4) \otimes U_2(4) \supset U_{12}(4)$ products are required. These products, corresponding to the coupling of different systems, are usually referred to as the *outer products* or *direct products* of irreducible representations [3]. The well-known case of angular momentum coupling

$$j_1 \otimes j_2 = |j_1 - j_2| \oplus |j_1 - j_2| - 1 \oplus \cdots \oplus j_1 + j_2 \qquad (B.55)$$

is an example (in this case for SU(2)) of such operation.

The general rules for U(n) outer products can be expressed graphically in terms of Young tableaux [3]:

1. The partitions $h_1 \geq h_2 \geq \cdots \geq h_n$ and $k_1 \geq k_2 \geq \cdots \geq k_n$ are first translated into their corresponding tableaux with h_1 boxes in the first row, h_2 boxes in the second, and so on.

2. The second tableau is filled with a's in the first row, b's in the second, and so on.

3. The boxes with a's are added to the first tableau in all possible ways, subject to the rule that the resulting pattern is still a Young partition $h'_1 \geq h'_2 \geq \cdots \geq h'_m$ and that no two a's can appear in the same column.

4. For each new pattern found in 3 repeat the procedure with the b's, then the c's, and so on.

5. When all symbols have been added and one reads from right to left in the first row, then the second row, and so on, at all times the condition $n_a \geq n_b \geq n_c \geq \cdots$ must be satisfied, where n_i indicates the number of times the letter i has appeared at that moment.

We illustrate this procedure with the U(n) product $[2,2] \otimes [2,1]$, which in Young tableaux reads (rules 1 and 2)

$$\begin{array}{c}\square\square \\ \square\square\end{array} \otimes \begin{array}{c}\boxed{a}\boxed{a} \\ \boxed{b}\end{array} \tag{B.56}$$

Application of rule 3 gives three Young tableaux:

$$\begin{array}{c}\square\square\boxed{a}\boxed{a} \\ \square\square\end{array} \,,\, \begin{array}{c}\square\square\boxed{a} \\ \square\square \\ \boxed{a}\end{array} \,,\, \begin{array}{c}\square\square \\ \square\square \\ \boxed{a}\boxed{a}\end{array} \tag{B.57}$$

Note that the pattern

$$\begin{array}{c}\square\square\boxed{a} \\ \square\square\boxed{a}\end{array} \tag{B.58}$$

is not allowed by rule 3. Continuing to allocate the b (rule 4) to each of the Young tableaux in (B.57), we find

$$\begin{array}{c}\square\square\boxed{a}\boxed{a} \\ \square\square\boxed{b}\end{array} \,,\, \begin{array}{c}\square\square\boxed{a}\boxed{a} \\ \square\square \\ \boxed{b}\end{array} \,,\, \begin{array}{c}\square\square\boxed{a} \\ \square\square\boxed{b} \\ \boxed{a}\end{array} \,,\, \begin{array}{c}\square\square\boxed{a} \\ \square\square \\ \boxed{a}\boxed{b}\end{array} \,,\, \begin{array}{c}\square\square\boxed{a} \\ \square\square \\ \boxed{a} \\ \boxed{b}\end{array} \,,\, \begin{array}{c}\square\square \\ \square\square \\ \boxed{a}\boxed{a} \\ \boxed{b}\end{array} \tag{B.59}$$

where it should be noted that the patterns

$$\begin{array}{c}\square\square\boxed{a}\boxed{a}\boxed{b} \\ \square\square\end{array} \,,\, \begin{array}{c}\square\square\boxed{a}\boxed{b} \\ \square\square \\ \boxed{a}\end{array} \,,\, \begin{array}{c}\square\square\boxed{b} \\ \square\square \\ \boxed{a}\boxed{a}\end{array} \tag{B.60}$$

are absent due to rule 5. We thus find

$$[2,2] \otimes [2,1]$$
$$= [4,3] \oplus [4,2,1] \oplus [3,3,1] \oplus [3,2,2] \oplus [3,2,1,1] \oplus [2,2,2,1]$$
(B.61)

The full result is valid for U(n) as long as $n \geq 4$. For $n = 3$ the last two irreducible representations do not exist, while for $n = 2$ only $[4,3]$ is permitted.

References

1. M. Moshinsky, *Group Theory and the Many-Body Problem*, Gordon & Breach, New York, 1968.
2. E. Chacón, *El Grupo Ortogonal en N Dimensiones y la Estructura Nuclear*, doctoral dissertation, Universidad Nacional Autónoma de México, 1968.
3. M. Hamermesh, *Group Theory and Its Application to Physical Problems*, Addison-Wesley, Reading, MA, 1962.
4. I.M. Gel'fand and M.L. Tsetlin, "Matrix elements for the unitary groups," Dokl. Akad. Nauk SSSR **71** (1950) 825.
5. M.E. Rose, *Elementary Theory of Angular Momentum*, Wiley, New York, 1957.
6. J.G. Nagel and M. Moshinsky, "Operators that lower or raise the irreducible vector spaces of U_{n-1} contained in an irreducible vector space of U_n," J. Math. Phys. **6** (1965) 682.
7. B.G. Wybourne, *Symmetry Principles and Atomic Spectroscopy*, Wiley-Interscience, New York, 1970.

Appendix C

Dragt's Theorem

Dragt's theorem [1] is the basic tool for the determination of a coordinate realization for states classified by the chain of algebras

$$\begin{array}{ccc} \mathrm{U}(n) & \supset & \mathrm{SO}(n) \\ | & & | \\ [N] & & \rho \end{array} \quad (C.1)$$

from the corresponding states in a boson representation (or *vice versa*). Consider the polynomials in the creation operators $\mathcal{Q}_\rho(b_k^\dagger)$ which satisfy the equations

$$\sum_{i=1}^{n} b_i b_i \mathcal{Q}_\rho(b_k^\dagger)|0\rangle = 0 \quad (C.2)$$

and

$$\sum_{i=1}^{n} b_i^\dagger b_i \mathcal{Q}_\rho(b_k^\dagger)|0\rangle = \rho \mathcal{Q}_\rho(b_k^\dagger)|0\rangle \quad (C.3)$$

Relation (C.2) establishes that $\mathcal{Q}_\rho(b_k^\dagger)$ is *harmonic*, while (C.3) indicates that the polynomial is homogeneous of degree ρ (see Chapter 8). The term harmonic stems from the equivalent conditions in configuration space, where a harmonic function satisfies the equation $\nabla_n^2 f(x_i) = 0$, where ∇_n^2 is the n-dimensional laplacian. Thus, in three dimensions, the functions $f(r, \theta, \phi) = r^l Y_m^l(\theta, \phi)$ are harmonic, from which the $Y_m^l(\theta, \phi)$ owe their name. The SO(n) Casimir operator

$$\mathcal{C}_2[\mathrm{SO}(n)] \equiv \tfrac{1}{2} \sum_{ij=1}^{n} \Lambda_{ij}\Lambda_{ji} \quad (C.4)$$

with $\Lambda_{ij} \equiv b_i^\dagger b_j - b_j^\dagger b_i$, may be rewritten in the form

$$C_2[\mathrm{SO}(n)] = \hat{N}(\hat{N} + n - 2) - \sum_{i=1}^n b_i^\dagger b_i^\dagger \sum_{j=1}^n b_j b_j \qquad (\mathrm{C.5})$$

where $\hat{N} = \sum_i b_i^\dagger b_i$. Since the $C_2[\mathrm{SO}(n)]$ eigenvalue is $\rho(\rho + n - 2)$ (see Chapter 8), relations (C.2) and (C.3) imply that the polynomials $Q_\rho(b_k^\dagger)|0\rangle$ are $\mathrm{U}(n) \supset \mathrm{SO}(n)$ eigenstates for which $\rho = N$, that is, they are "maximum-weight" states. Dragt's theorem states that a coordinate realization for these $\mathrm{U}(n) \supset \mathrm{SO}(n)$ states is given by

$$Q_\rho(b_k^\dagger)|0\rangle = \pi^{-n/4} 2^{\rho/2} Q_\rho(\alpha_k) e^{-\beta^2/2} \qquad (\mathrm{C.6})$$

where $Q_\rho(\alpha_k)$ is the *same* polynomial as before but in terms of the coordinates

$$\alpha_k = \sqrt{\tfrac{1}{2}}(b_k^\dagger + b_k), \qquad k = 1, 2, \ldots, n \qquad (\mathrm{C.7})$$

and where

$$\beta^2 = \sum_{i=1}^n \alpha_i^2 \qquad (\mathrm{C.8})$$

Note that (C.7) is the n-dimensional analog of the usual definition for the three-dimensional coordinates of a harmonic oscillator. The corresponding momenta, canonically conjugate to the α_k, are then given by

$$\pi_k = \frac{1}{i}\frac{\partial}{\partial \alpha_k} = i\sqrt{\tfrac{1}{2}}(b_k^\dagger - b_k), \qquad k = 1, 2, \ldots, n \qquad (\mathrm{C.9})$$

The proof of (C.6) follows from the one-dimensional oscillator relation (see Chapter 1)

$$(b^\dagger)^r|0\rangle = \pi^{-1/4} 2^{-r/2} H_r(\alpha) e^{-\alpha^2/2} \qquad (\mathrm{C.10})$$

where H_r is a Hermite polynomial whose leading terms in a series expansion are

$$H_r(\alpha) = 2^r \alpha^r - 2^{r-1}\binom{r}{2}\alpha^{r-2} + \cdots \qquad (\mathrm{C.11})$$

C Dragt's Theorem

The n-dimensional oscillator states with $\rho = N$ should then be of the form

$$\begin{aligned}
\mathcal{Q}_\rho(b_k^\dagger)|0\rangle &= \sum_{r_i} A_{r_1 r_2 \ldots r_n} (b_1^\dagger)^{r_1} (b_2^\dagger)^{r_2} \ldots (b_n^\dagger)^{r_n} |0\rangle \\
&= \pi^{-n/4} 2^{-\rho/2} \sum_{r_i} A_{r_1 r_2 \ldots r_n} H_{r_1}(\alpha_1) H_{r_2}(\alpha_2) \ldots H_{r_n}(\alpha_n) e^{-\beta^2/2} \\
&= \pi^{-n/4} 2^{\rho/2} \sum_{r_i} A_{r_1 r_2 \ldots r_n} [(\alpha_1)^{r_1} (\alpha_2)^{r_2} \ldots (\alpha_n)^{r_n} + \cdots] e^{-\beta^2/2}
\end{aligned}$$

(C.12)

where $r_1 + r_2 + \cdots + r_n = \rho$ and the plus sign followed by the dots in the square brackets indicate polynomials in the α_k of degree smaller than ρ. We now express the operators in (C.2) and (C.3) in terms of coordinates, for which we need the inverse of (C.7) and (C.9),

$$b_k^\dagger = \sqrt{\tfrac{1}{2}}(\alpha_k - i\pi_k), \qquad b_k = \sqrt{\tfrac{1}{2}}(\alpha_k + i\pi_k) \qquad \text{(C.13)}$$

We readily find

$$\sum_{i=1}^n b_i^\dagger b_i = \tfrac{1}{2}\beta^2 - \tfrac{1}{2}n - \tfrac{1}{2} \sum_{i=1}^n \frac{\partial^2}{\partial \alpha_i^2} \qquad \text{(C.14)}$$

and

$$\sum_{i=1}^n b_i b_i = \tfrac{1}{2}\beta^2 + \tfrac{1}{2}n + \sum_{i=1}^n \alpha_i \frac{\partial}{\partial \alpha_i} + \tfrac{1}{2} \sum_{i=1}^n \frac{\partial^2}{\partial \alpha_i^2} \qquad \text{(C.15)}$$

Denoting the polynomial in the α_k coordinates by $\mathcal{Q}'(\alpha_k)$, equation (C.12) reads

$$\mathcal{Q}_\rho(b_k^\dagger)|0\rangle = \mathcal{Q}'(\alpha_k) e^{-\beta^2/2} \qquad \text{(C.16)}$$

We can now apply the identities (C.14) and (C.15) on (C.16) acting with the boson operators on the left and with the differential operators on the right. Note that the latter act on both $\mathcal{Q}'(\alpha_k)$ and $e^{-\beta^2/2}$ and that β is the hyperradius defined by (C.8). So we can make use of the relation $\sum_i \alpha_i (\partial f/\partial \alpha_i) = \beta(\partial f/\partial \beta)$ for any function f of α_k. Carrying out the operations on the right-hand side requires some algebra, from which we find that many of the terms cancel out. Finally, using (C.2) and (C.3), we arrive at the relations

$$\sum_{i=1}^{n} b_i^\dagger b_i \mathcal{Q}_\rho(b_k^\dagger)|0\rangle = \left(-\frac{1}{2}\sum_{i=1}^{n} \frac{\partial^2 \mathcal{Q}'}{\partial \alpha_i^2} + \sum_{i=1}^{n} \alpha_i \frac{\partial \mathcal{Q}'}{\partial \alpha_i}\right) e^{-\beta^2/2}$$
$$= \rho \mathcal{Q}_\rho(b_k^\dagger)|0\rangle = \rho \mathcal{Q}'(\alpha_k) e^{-\beta^2/2} \qquad (C.17)$$

and

$$\sum_{i=1}^{n} b_i b_i \mathcal{Q}_\rho(b_k^\dagger)|0\rangle = \frac{1}{2}\sum_{i=1}^{n} \frac{\partial^2 \mathcal{Q}'}{\partial \alpha_i^2} e^{-\beta^2/2} = 0 \qquad (C.18)$$

We see that (C.18) implies that

$$\sum_{i=1}^{n} \frac{\partial^2 \mathcal{Q}'}{\partial \alpha_i^2} = 0 \qquad (C.19)$$

which substituted into (C.17) gives

$$\sum_{i=1}^{n} \alpha_i \frac{\partial \mathcal{Q}'}{\partial \alpha_i} = \rho \mathcal{Q}' \qquad (C.20)$$

These equations indicate that $\mathcal{Q}'(\alpha_k)$ is both harmonic ((C.19)) and homogeneous of degree ρ ((C.20)). The remarkable implication of (C.19) and (C.20) is that $\mathcal{Q}'(\alpha_k)$ satisfies the same two conditions in terms of the α_k coordinates as $\mathcal{Q}(b_k^\dagger)$ does in terms of the bososnic operators, since the annihilation operators can be interpreted as $b_i = \partial/\partial b_i^\dagger$. Hence, returning to (C.12), the coefficients $A_{r_1 r_2 \ldots r_n}$ must be such that only the highest-order term in the expansion is different from zero. In other words,

$$\mathcal{Q}_\rho(b_k^\dagger)|0\rangle = \pi^{-n/4} 2^{\rho/2} \sum_{r_i} A_{r_1 r_2 \ldots r_n} [(\alpha_1)^{r_1}(\alpha_2)^{r_2} \ldots (\alpha_n)^{r_n}] e^{-\beta^2/2}$$
$$= \pi^{-n/4} 2^{\rho/2} \mathcal{Q}_\rho(\alpha_k) e^{-\beta^2/2} \qquad (C.21)$$

which proves (C.6).

References

1. A.J. Dragt, "Classification of three-particle states according to SU_3," J. Math. Phys. **6** (1965) 533.

Index

Algebra, definition of, 446-447
Angular momentum algebra, *see* Rotation algebra
Anharmonic oscillator, 16-18
Anharmonic vibrator, 319
Anticommutation relations:
    in U(2), 75, 102
    in U(4), 265
    in U(6), 393
Associativity of groups, 446

Bending vibrations:
    in C_2H_2, 255
    in X–Y–X, 57
    in X–Y–Z, 222, 224
Born–Oppenheimer approximation, 277
Boson effective charge, 319
Boson gyromagnetic ratio, 374, 385-386
Branching rule, *see also* Multiplication
    in U(2):
        SU(2) $\supset$ U(1), 46
        U(2) $\supset$ SO(2), 18, 23
        U(2) $\supset$ U(1), 18, 20, 46-47, 86
        U(2/2) $\supset$ U(2)$\otimes$U(2), 103-104
        U(4) $\supset$ U(2) $\otimes$ U(2), 110
    in U(4):
        SO(4) $\supset$ SO(3), 150
        U(3) $\supset$ SO(3), 133, 212-213
        U(4) $\supset$ U(3), 133, 214-215
        U(4) $\supset$ SO(4), 150, 226-227
    in U(6):
        SO(5) $\supset$ SO(3), 310-311, 418
        SO(6) $\supset$ SO(5), 327, 418
        Sp(4) $\supset$ SU(2), 406
        SU(4) $\supset$ Sp(4), 406
        U(5) $\supset$ SO(5), 308, 323
        U(6) $\supset$ SO(6), 326, 418
        U(6) $\supset$ SU(3), 345-346
        U(6) $\supset$ U(5), 307
        U(6/4) $\supset$ U(6)$\otimes$U(4), 426
        U(6/12) $\supset$ U(6) $\otimes$ U(12), 431
    for SO(n) $\supset$ SO($n-1$), 472
    for U(n) $\supset$ SO(n), 308-309, 472-473
    for U(n) $\supset$ U($n-1$), 471

Cayley's theorem, 66
Casimir operator:
    definition of, 450
    in U(2):
        of SO(2), 19
        of U(1), 19
        of U(2), 19, 52, 88
        of U(2/2), 103
        of U(4), 110
    in U(4):

of SO(3), 130, 276
of SO(4), 130, 150, 276
of SU(2), 276
of U(3), 127-129, 276
of U(4), 128-129, 237, 276
of U(8), 276
in U(6):
of SO(3), 306
of SO(5), 306
of SO(6), 305-306
of Sp(4), 406
of SU(2), 406
of SU(3), 306, 344-345
of SU(4), 406
of U(3), 342, 344
of U(5), 305
of U(6), 304-305, 369-371
of SO(n), 130, 308-310, 466, 470
of SU(1,1), 144, 323
of U(n), 127-128, 465-466, 469
Chain of algebras, see Limit
Clebsch–Gordan coefficient:
definition of, 32
symmetry properties of, 61
Closure of groups, 446
Commutation relations:
in U(2):
of bosons, 11, 39
of generators, 12, 40, 76, 102
in U(4):
of bosons, 119, 197
of generators, 119, 122, 198
in U(6):
of bosons, 329, 355, 393
of generators, 297, 393-398
of SO(n), 465
of U(n), 464
Complementary algebras, 270
Conjugate representation, 184, 270
Constants of the motion, 449-451

Contraction, 110
Cooper pair, 353
Coordinates:
cartesian, 9
hyperspherical, 153, 329, 331
polar, 10
spherical, 136
Core–particle interaction, 407
Correlation diagram, 234
Coupling coefficient, see also Isoscalar factor
for SO(3) $\supset$ SO(2), see Clebsch–Gordan coefficient
for SU(2) $\supset$ SO(2), see Clebsch–Gordan coefficient
for U(2) $\supset$ U(1), 93-94
notation of, 174

Dimension formula:
of SO(5) representation, 404
of SO(6) representation, 403
of Sp(4) representation, 404
of SU(4) representation, 403
Direct product, definition of, 13
Direct sum, definition of, 13
Discrete group D_3, 65-68
Dragt's theorem, 28, 138, 155-156, 333-335, 477-481
Dunham expansion, 152, 166
Dynamical algebra, definition of, 449
Dynamical symmetry:
breaking, 456-460
definition of, 455

E1 transitions, 168-170
E2 transitions, 319-322, 337-341, 408-409, 415-417, 420-423
Electric dipole transitions, see E1 transitions
Electric quadrupole transitions, see E2 transitions

Index

Elliott SU(3), 1, 302, 344
Euler angles, 25, 31, 173, 297, 329
Exponentiation, 446

F spin, 108-113, 355-366
    algebra, 111, 358-359
    classification, 359-362
    multiplet, 2, 362-366
    symmetry, 364, 373-377

γ-unstable rotor, 327-328
Gegenbauer polynomial, 155, 332
Gel'fand basis, 471-473
Gel'fand labels, 47
Gell-Mann–Okubo mass formula, 458-459
Generators:
    definition of, 447
    in $U(2)$:
        of $SO(2)$, 15
        of $\overline{SO}(2)$, 24
        of $SU(2)$, 12-13, 23, 76
        of $SU(2) \otimes SU(2)$, 40, 81
        of $U(1)$, 15
        of $U(2)$, 12, 44, 75-76, 84
        of $U(2/2)$, 101
        of $U(2) \otimes U(2)$, 40, 81
        of $U(4)$, 109
    in $U(4)$:
        of $SO(4)$, 123, 202
        of $\overline{SO}(4)$, 124
        of $SU(2)$, 266
        of $U(3)$, 123, 202
        of $U(4)$, 119-121, 202, 266
        of $U(4) \otimes U(4)$, 198
        of $U(4) \otimes U(8)$, 268
        of $U(8)$, 265
    in $U(6)$:
        of $SO(3)$, 298
        of $SO(5)$, 299-300
        of $SO(6)$, 300-301
        of $\overline{SO}(6)$, 301-302
        of $SU(3)$, 302-303
        of $U(5)$, 299-300
        of $U(6)$, 297-298, 366
        of $U(6/4)$, 434
        of $U(6/12)$, 434
        of $U(6) \otimes U(6)$, 355
        of $U(6) \otimes U(\Omega)$, 393
        of $U(12)$, 357-358
Gluon, 99
Group, definition of, 446

$\hbar$, 9
Hadron, 3, 458-459
Hamiltonian:
    in $U(2)$:
        boson, 12, 17, 41
        boson–fermion, 82
        fermion, 80
        harmonic oscillator, 9
        of $SO(2)$ limit, 23
        of $SO_{12}(2)$ limit, 53
        of $SO^{BF}(2)$ limit, 88
        of $SO_1(2) \otimes SO_2(2)$ limit, 43
        of $SO^B(2) \otimes SO^F(2)$ limit, 83
        of $U(1)$ limit, 20
        of $U_{12}(1)$ limit, 50
        of $U^{BF}(1)$ limit, 87
        of $U_1(1) \otimes U_2(1)$ limit, 43
        of $U^B(1) \otimes U^F(1)$ limit, 83
    in $U(4)$:
        boson, 125-127, 132, 199
        boson–fermion, 266-268
        for polyatomic molecules, 253
        of $SO(4)$ limit, 148
        of $SO_{12}(4)$ limit, 227
        of $SO_1(4) \otimes SO_2(4)$ limit, 218, 233
        of $SO^{rv-e}(4) \otimes SU^e(2)$ limit, 275

of U(3) limit, 133
of $U_{12}(4)$ limit, 214
of $U_1(4) \otimes U_2(4)$ limit, 210-211
in U(6):
    boson, 303-304, 306
    boson–fermion, 398-399
    of SO(6) limit, 326
    of $SO^{BF}(6) \otimes SU^F(2)$ limit, 419
    of $SO_{\nu\pi}(6)$ limit, 378
    of SU(3) limit, 342
    of $SU^{BF}(4)$ limit, 406
    of U(5) limit, 307, 310
Harmonic oscillator, 9-11
Harmonic function, 477
Hermite polynomial, 10, 478
Highest-weight polynomial:
    of U(2), 48
    of U(n), 468
    of SO(n), 470
Highest-weight state:
    for $U(2) \supset SO(2)$, 27-28
    for $U(3) \supset SO(3)$, 138
    for $U(4) \supset SO(4)$, 155-156
    for $U(5) \supset SO(5)$, 316-317
    for $U(6) \supset SO(6)$, 333-334
Hydride molecules, 280-289
Hypercharge, 459

Identity element, 446
Intrinsic variables β and γ, 329
Invariant operator, see Casimir operator
Inverse element, 446
Irreducible representation, definition of, 453
Isoscalar factor:
    definition of, 181
    for $SO(4) \supset SO(3)$, 179
    for $SO(5) \supset SO(3)$, 373, 421
    for $SO(6) \supset SO(5)$, 373, 421
    for $Sp(4) \supset SU(2)$, 414
    for $SU(4) \supset Sp(4)$, 414-415
    for $U(4) \supset SO(4)$, 184
    for $U(6) \supset SO(6)$, 373, 421
Isomorphism, 451
    definition of, 13
    $SO(4) \simeq SU(2) \otimes SU(2)$, 148-149
    $SO(5) \simeq Sp(4)$, 400
    $SU(2) \simeq SO(3)$, 400
    $SU(4) \simeq SO(6)$, 400
    $U(2) \simeq SU(2) \otimes U(1)$, 13
Isospin, 355-356, 457-458

Jacobi identity, 447, 449

Laguerre polynomial, 11, 137, 154, 330
Lattice of algebras, definition of, 178
Legendre function, 137
Lie, 446
    algebra, see Algebra
    group, see Group
Limits:
    in U(2):
        O(2), 26-27
        SO(2), 23-28
        $SO_{12}(2)$, 45, 53-55
        $SO^{BF}(2)$, 85
        $SO^F(2)$, 79-80
        $SO_1(2) \otimes SO_2(2)$, 42
        $SO^B(2) \otimes SO^F(2)$, 83
        U(1), 20-23
        $U_{12}(1)$, 45, 50-52
        $U^{BF}(1)$, 85
        $U^F(1)$, 79
        $U_1(1) \otimes U_2(1)$, 42
        $U^B(1) \otimes U^F(1)$, 83
    in U(4):
        SO(4), 148-164
        $SO_{12}(4)$, 226-230

Index 485

$SO_1(4) \otimes SO_2(4)$, 218-225, 230-236
$SO^{rv-e}(4) \otimes SU^e(2)$, 273
$U(3)$, 132-148
$U_{12}(3)$, 213-215
$U^{rv-e}(3) \otimes SU^e(2)$, 273
$U_1(3) \otimes U_2(3)$, 210-213
in $U(6)$:
  in IBM-2, 367
  $SO(6)$, 326-337
  $SO_{\nu\pi}(6)$, 372
  $SU(3)$, 342-348
  $SU^{BF}(4)$, 403-404
  $U(5)$, 307-326
  $U^{BF}(6) \otimes SU^F(2)$, 417
  with good F spin, 371-377
Local vibrations, 60-63
  in X–Y–Z molecules, 224
Lowering generators:
  of $SO(n)$, 470
  of $U(2)$, 48
  of $U(n)$, 467
Lowering operator, 22
l-type doubling, 222

M1 transitions, 381, 385-386
Majorana operator, 369-370
Magnetic dipole transitions, see M1 transitions
Matrix elements:
  in $SO(4)$ basis, 161
  in $U(1)$ basis, 22
  in $U(3)$ basis, 142
Missing label, 4, 181, 300, 303, 406
Mixed symmetry, 361
Modified boson triplets, 317
Molecules:
  BeH, 286-289
  BH, 286-289
  CH, 248, 286-289
  C_2D_2, 257
  C_2H_2, 255-258
  C_2HD, 257
  CN, 249
  CO, 249
  CO_2, 56, 247-249
  CS, 249
  DO_2, 247-249
  H_2, 166-168
  HCN, 247-250
  H_2O, 56, 63-64, 247-249
  H_2S, 247-249
  LiH, 286-289
  N_2, 249
  NH, 286-289
  NO, 249
  N_2O, 247-249
  OCS, 247-249
  OH, 286-289,
  SF_6, 70-71
  SO_2, 247-249
  UF_6, 70-71
  WF_6, 70-71
Morse potential, 57-59
Multiplication:
  in $U(2)$:
    of $SU(2)$ representations, 45
    of $U(2)$ representations, 47
  in $U(4)$:
    of $SO(3)$ representations, 210
    of $SO(4)$ representations, 176, 221
    of $U(3)$ representations, 211
    of $U(4)$ representations, 214
  in $U(6)$:
    of $SO(6)$ representations, 405
    of $U(6)$ representations, 361
Multiplicity label, 4, 174, 182
Multipole expansion, 267-268

Non-compact algebra, 144-146
Normal vibrations:
  in C_2H_2, 255

in X_3, 67
in X–Y–X, 57, 60-63
in X–Y–Z, 224
Nuclei:
   Au, 407-408, 416, 423-440
   Cd, 311
   Dy, 363, 365
   Er, 363, 365
   Gd, 386
   Hf, 363, 365
   Hg, 363, 365, 427-428, 435
   Os, 363, 365
   Pt, 327, 337-343, 363, 365, 377-387, 417-440
   W, 363, 365
   Yb, 363, 365

Octahedral group O_h, 69
Octet, 459
Order of group, 446
Orthogonal:
   algebra, 463-475
   transformation, 464-465
Oscillator:
   anharmonic, 16-18
   harmonic, 9-11
Outer product, 473-475
Overtone frequency, 222

Pairing:
   interaction, 2
   operators, 143
Parity:
   in $U(2)$:
      of bosons, 17
      of $SO(2)$ states, 26
      of $SO_{12}(1)$ states, 54
      of $SO_1(2) \otimes SO_2(2)$ states, 43
      of $U(1)$ states, 21
      of $U_{12}(1)$ states, 52
   in $U(4)$:
      of bosons, 119, 198
      of $SO_{12}(4)$ states, 227-228
      of $SO_1(4) \otimes SO_2(4)$ states, 219-221
      of $SO^{rv-e}(4) \otimes SU^e(2)$ states, 273
   in $U(6)$:
      of bosons, 297
Particle–hole conjugation, 204-205, 366-368
Pauli principle, 77-78
Permutation group S_m, 65-70
Pochhammer symbol, 162
Potential approach, 195
Product of representations, see Multiplication
Pseudo-scalar, 231
Pseudo-spin, 396

Quark, 99
Quasi-particle, 427
Quasi-spin, 144, 322-325

Racah's factorization lemma, 181
Raising generators:
   of $SO(n)$, 470
   of $U(2)$, 48
   of $U(n)$, 467
Raising operator, 22
Ramification rule, see Branching rule
Recurrence relations:
   in Gegenbauer polynomials, 157, 159
   in solid spherical harmonics, 158
Reduced matrix element, definition of, 33
Reduction rule, see Branching rule
Representation:
   antisymmetric, 78
   definition of, 453

Index

supersymmetric, 104, 426, 431, 434-435
symmetric, 49
Rigidity parameter, 134, 151
Rotation:
  algebra, 12-13, 31-32, 121
  in 2 dimensions, 447
  in 3 dimensions, 447
  in 4 dimensions, 173
  in n dimensions, 464-465
  of bosons in $U(4)$, 119, 198
  of bosons in $U(6)$, 297
  of equilateral triangle, 65

Scattering, 3
Schmidt orthogonalization, 319
Sector of superalgebra, 101-102
Separated-atoms limit, 264, 284
Skyrmion, 3
Space inversion, see Parity
Spinor, 25
Spinor algebra, 400
Spin–orbit coupling, 267, 396
Stretching vibrations:
  in C_2H_2, 255
  in polyatomic molecules, 63-71
  in X–Y–X, 56-63
  in X–Y–Z, 222, 224
Structure constants:
  of algebra, 447
  of superalgebra, 101
Superalgebra, definition of, 100-101
Supercommutation, 102
Supermultiplet, 104, 427, 431, 435
Superpartners, 104
Supersymmetry, 2, 99-108, 423-440
Symmetry:
  algebra, definition of, 449
  dynamical, see Dynamical symmetry
  internal, 1
  gauge, 1
  geometrical, 1
  group, see Symmetry algebra
  limit, see Limit
  transformation, 448-449
Symplectic algebra, 395

Tensor:
  calculus:
    in $SO(4)$, 172-181, 239-245, 281-282
    in $U(2)$, 30-36, 90-95
    in $U(4)$, 181-190
    in $U(6)$, 338-339, 381-384
  operator:
    in $U(2)$, 31, 91-92, 106-108
    in $U(4)$, 173-174, 182-187
    in $U(6)$, 338, 368, 381-382
Transformation brackets:
  in $U(2)$:
    from $U(1)$ to $SO(2)$, 29-30
    from $U_{12}(1)$ to $SO_{12}(2)$, 55
  in $U(4)$:
    from $U(3)$ to $SO(4)$, 164-165
  in $U(6)$:
    from $U(5)$ to $SO(6)$, 336-337
    from $U(5)$ to $SU(3)$, 347-348
Two-fluid model, 380

Unitary:
  algebra, 463-475
  transformation, 90, 463-464
United-atoms limit, 264, 284

Wave functions:
  in $U(2)$:

O(2) limit, 25
SO(2) limit, bosons, 25
SO(2) limit, coordinates, 11
$\overline{SO(2)}$ limit, 24
$SO_{12}(2)$ limit, 53
$SO^{BF}(2)$ limit, 83
$SO_1(2) \otimes SO_2(2)$ limit, 42
$SO^B(2) \otimes SO^F(2)$ limit, 89
U(1) limit, bosons, 14, 20
U(1) limit, coordinates, 10
$U_{12}(1)$ limit, 51
$U^{BF}(1)$ limit, 83
$U_1(1) \otimes U_2(1)$ limit, 88
$U^B(1) \otimes U^F(1)$ limit, 42
in U(4):
   SO(4) limit, bosons, 157
   SO(4) limit, coordinates, 155
   $SO_{12}(4)$ limit, 227
   $SO_1(4) \otimes SO_2(4)$ limit, 219
   $SO^{rv-e}(4) \otimes SU^e(2)$ limit, 273
   U(3) limit, bosons, 140
   U(3) limit, coordinates, 138
in U(6):
   SO(6) limit, bosons, 334-335
   SO(6) limit, coordinates, 332-333
   $SO_{\nu\pi}(6)$ limit, 372
   $SU^{BF}(4)$ limit, 410
   U(5) limit, bosons, 318
   U(5) limit, coordinates, 333-334
Weak-coupling basis, 399
Weight, 48, 467
Weight generators:
   definition of, 450
   of SO(n), 470
   of U(2), 48
   of U(n), 467
Wigner:
   D function, 25, 31, 55-56, 329
   D matrix, 55-56, 90, 297, 453, 329
   supermultiplet, 1
   U(4), 1
Wigner–Eckart theorem:
   for U(4) ⊃ SO(4) ⊃ SO(3) ⊃ SO(2), 183
   for SO(4) ⊃ SO(3) ⊃ SO(2), 175
   for SU(2) ⊃ SO(2), 32-36
   for U(2) ⊃ U(1), 94-95, 106-108
   in F-spin space, 375

Young diagram, *see* Young tableau
Young partition, 49
Young tableau:
   antisymmetric, 78
   definition of, 49, 468
   multiplication, 473-475
   symmetric, 49